Latest Advances in Robot Kinematics

Jadran Lenarčič • Manfred Husty
Editors

Latest Advances in Robot Kinematics

Springer

Editors
Jadran Lenarčič
J. Stefan Institute
Ljubljana-Vic-Rudnik, Slovenia

Dr. Manfred Husty
Inst. Technische Mathematik, Geometrie und Bauinformatik
Universität Innsbruck
Innsbruck, Austria

ISBN 978-94-007-4619-0 ISBN 978-94-007-4620-6 (eBook)
DOI 10.1007/978-94-007-4620-6
Springer Dordrecht Heidelberg New York London

Library of Congress Control Number: 2012939339

Printed on acid-free paper

Springer is part of Springer Science+Business Media (www.springer.com)

Preface

This book is the eleventh in the series of Advances in Robot Kinematics. All the articles contained within it have been rigorously selected on the basis of a peer-review process. Since the early 1990s these books on Advances in Robot Kinematics have been published every two years, with the publication of each one being followed by a symposium in which the participants exchange their results and opinions. In our eyes, however, all the books represent stand-alone contributions. They are, in principle, independent of the symposia and should not be thought of as simply standard conference proceedings. The articles they contain are a selection that describes the newest and most original achievements in the field and are, from this perspective, identical to a special issue of a scientific journal. In order to guarantee that we have included the latest results, the whole process – from the submission, the reviewing, the selection of the articles, the various revisions, the preparation of the finished articles and the publication of the book – has taken less than six months.

Although one might expect that the research in robot kinematics has lost a little of its "freshness", it is clear that even after 24 years, since the first symposium on Advances in Robot Kinematics took place, the subject still presents an immense number of research challenges. The success of the symposia is due to the fact that since its earliest beginnings it has managed to bring together the best of the world's researchers and scientists. The activity as a whole has been continuously supported by the J. Stefan Institute, and since 1992 it has come under the patronage of the International Federation for the Promotion of Mechanism and Machine Science (IFToMM). The last symposium was organized in collaboration with the University of Innsbruck, which made available its infrastructure and workforce and led most of the scientific activities.

The 56 articles in this book cover the latest topics and methods in the kinematics of robotic systems, including serial, parallel and cable driven, both planar and spatial. The robotic systems range from being less than fully mobile to kinematically redundant and to over-constrained. Emerging areas, such as the design and control of humanoids or their subsystems, man and machine systems, as well as the motion of the human body, are also included.

We are grateful to the authors for their contributions and to the large team of reviewers for their critical and insightful recommendations. In particular we are indebted to Dr. Hans-Peter Schröcker (University of Innsbruck) for his dedication and expertise, Ms. Jolanda Karada (Karada Publishing Services) for her valuable technical contribution, and to the staff of Springer who were responsible for putting the whole book together.

Ljubljana, Slovenia J. Lenarčič
Innsbruck, Austria M.L. Husty

Contents

The Dual Generalized Inverses and Their Applications in Kinematic Synthesis

Jorge Angeles

Abstract The left and right dual Moore–Penrose generalized inverses are the subject of this paper. It is shown that, contrary to the real case, these inverses are not unique, those with *minimum Frobenius norm* being obtained. Their application in kinematic synthesis is discussed. It is shown that, in the case of function-generating RCCC linkages, the left dual generalized inverse leads to a linkage that meets the prescribed input-output relations with both a least-square error and a minimum size. The study concludes with the synthesis of a linkage that approximates a *homokinetic* transmission between shafts with skew, orthogonal axes.

Key words: Dual generalized Moore–Penrose inverses, least-square approximation, minimum-Frobenius-norm, homokinetic joint, skew axes

1 Introduction

Dual numbers are well documented in the literature, an extensive bibliography being available in [1], with 73 entries. The literature is extensive for the scalar case, for vectors and matrices much less so, but some references can be cited, besides the previous one, namely, [2] and [3]. Moreover, dual numbers can be defined over both the real and the complex fields [4]; for the purposes of this paper, real numbers will suffice. The set of dual numbers itself, however, is not a field, but a *ring* [5].

The reason why dual numbers are relevant to kinematics can best be summarized in *The Principle of Transference* [6]:

> *The kinematics and statics relations of spatial linkages and cam mechanisms can be derived upon replacing the real variables occurring in the corresponding relations for spherical linkages by dual numbers.*

Jorge Angeles
McGill University, Montreal, Canada, e-mail: angeles@cim.mcgill.ca

J. Lenarčič, M. Husty (eds.), *Latest Advances in Robot Kinematics*,
DOI 10.1007/978-94-007-4620-6_1,

The theory behind dual numbers is well established, but there are still some applications domains that haven't been fully exploited. This paper is a contribution in this direction. One objective of the paper is to shed light on the handling of overdetermined systems of dual linear equations (DLE), as arising in the *approximate synthesis of linkages*, when the number of prescribed conditions to meet exceeds that of linkage parameters available. In this context, the well-known results of linear least squares are revisited in the realm of dual numbers. It is shown that the least-square approximation of an overdetermined system of DLE admits a solution that can be expressed in the form of the dual-equivalent of the *Moore–Penrose generalized inverse*, often referred to a the *left pseudoinverse*. The author does not subscribe to this terminology because it is misleading: the prefix "pseudo" denotes something "false", which is not the case here. One novel contribution is the result that, contrary to the real case, the left dual generalized inverse is not unique, which allows for minimizing the Frobenius norm of the said inverse, thereby obtaining a unique solution that shows a striking similarity with the dual inverse of a square matrix [2]. The same result is shown to apply to the right counterpart of the left generalized inverse. The concepts discussed in Section 2 are then applied to the approximate synthesis of function-generating RCCC linkages.

As an example, the synthesis of a linkage of this type to approximate a *homokinetic transmission* between two shafts of skew axes and lying at right angles is fully discussed. By virtue of the minimum-norm property of the unique left Moore–Penrose generalized inverse – for conciseness, henceforth the foregoing matrix will be referred to as the "left generalized inverse," with a similar denomination for its right counterpart – the linkage thus obtained is one that not only approximates the prescribed number of conditions with a least-square error, but also does so with a minimum size. Moreover, the slight errors present in the optimum solution can be compensated for by means of computer control, upon resorting to an inverse-kinematics approach that guarantees that the linkage output will follow the prescribed input signal upon modulating the linkage input accordingly.

2 Back to Basics: Algebra of Dual Numbers

While dual algebra is a classic subject, and its bases are well established, there is still room for research contributions in the area of applications. One such area is the approximate synthesis of linkages, which often leads to linear least-square problems, the subject of this paper. Their nonlinear counterparts are manageable once the foundations for linear problems have been established.

An item that has not been duly addressed in the pertinent literature is the definition of the derivative of a dual-valued function of a dual argument, but it was discussed by Kotel'nikov in his original book [7]: given the dual function

$$\hat{f}(\hat{x}) \equiv f(\hat{x}) + \varepsilon f_o(\hat{x}), \quad \hat{x} = x + \varepsilon x_o$$

its derivative with respect to its dual argument can be readily obtained as the limit of a ratio of increments, which yields the relation

$$\frac{d\hat{f}}{d\hat{x}} = \frac{df}{dx} + \varepsilon \frac{df_o}{dx} \tag{1}$$

consistent with Kotel'nikov [7]. This relation will be needed below.

The extension of the foregoing definitions to vectors and matrices follows as a combination of these definitions and the rules for the counterpart operations for vectors and matrices. The inverse of a dual matrix is given in [1, 2]. The former also includes a formula for the *dual left generalized inverse*.[1] As the formulas are displayed in that paper without derivation, the paper misses an important point: the generalized inverse in question is not unique. This issue is made apparent below.

For starters the expression for the dual inverse matrix derived in [2] is recalled: let $\hat{\mathbf{A}} = \mathbf{A} + \varepsilon\mathbf{A}_o$ be a dual matrix, with $\mathbf{A}$, $\mathbf{A}_o \in \mathbf{R}^{n\times n}$, its inverse being defined as long as $\mathbf{A}$ is invertible, although $\mathbf{A}_o$ need not be so. The inverse of $\hat{\mathbf{A}}$ is given by

$$\hat{\mathbf{A}}^{-1} = \mathbf{A}^{-1} - \varepsilon\mathbf{A}^{-1}\mathbf{A}_o\mathbf{A}^{-1} \tag{2}$$

Paraphrasing the derivation of the expression (2) for the dual inverse, not included here for the sake of conciseness, let $\hat{\mathbf{B}} = \mathbf{B} + \varepsilon\mathbf{B}_o$ be the left generalized inverse of a $m \times n$ dual matrix $\hat{\mathbf{A}}$, with $m > n$. As a consequence, $\hat{\mathbf{B}}$ is bound to be of $n \times m$. In the sequel, it will be made apparent that only $\mathbf{A}$ need be of full rank for the desired generalized inverse to exist, but $\mathbf{A}_o$ can be rank-deficient. Then, $\hat{\mathbf{B}}$ verifies $\hat{\mathbf{B}}\hat{\mathbf{A}} = \mathbf{1}_n$, with $\mathbf{1}_n$ denoting the $n \times n$ identity matrix. Upon expansion of the foregoing left-hand side, two real equations are obtained, one for the primal, one for the dual part:

$$\mathbf{B}\mathbf{A} = \mathbf{1}_n, \quad \mathbf{B}_o\mathbf{A} + \mathbf{B}\mathbf{A}_o = \mathbf{O}_n \tag{3}$$

where $\mathbf{O}_n$ denotes the $n \times n$ zero matrix, the first equation leading to the not so unexpected result $\mathbf{B} = (\mathbf{A}^T\mathbf{A})^{-1}\mathbf{A}^T \equiv \mathbf{A}^I$, i.e., the left generalized inverse of $\mathbf{A}$. When the foregoing expression is substituted into the second of the two above equations, a matrix equation for $\mathbf{B}_o$ is derived:

$$\mathbf{B}_o\mathbf{A} = -\mathbf{A}^I\mathbf{A}_o \Rightarrow \mathbf{A}^T\mathbf{B}_o^T = -\mathbf{A}_o^T(\mathbf{A}^I)^T \equiv -\mathbf{A}_o^T\mathbf{A}(\mathbf{A}^T\mathbf{A})^{-1}$$

which is a system of n^2 equations in $mn > n^2$ unknowns, the real components of $\mathbf{B}_o$. The system is, thus, underdetermined, thereby admitting *infinitely many solutions*. The conclusion is, then, that the *dual left generalized inverse is not unique*. Among all that many solutions, one of minimum *Frobenius* norm, $\mathrm{tr}(\mathbf{B}\mathbf{B}^T)$, can be obtained if one resorts to the right generalized inverse of $\mathbf{A}^T$, denoted $(\mathbf{A}^T)^\dagger$ [8]:

$$(\mathbf{A}^T)^\dagger = \mathbf{A}(\mathbf{A}^T\mathbf{A})^{-1} \tag{4}$$

[1] Actually, the authors do not stress the difference between the right and the left generalized inverses; they represent both with the same symbol, $(\,\cdot\,)^+$.

whence, after some obvious manipulations,

$$\mathbf{B}_o = -\mathbf{A}^I\mathbf{A}_o\mathbf{A}^I \tag{5}$$

Therefore, the minimum-Frobenius-norm $\hat{\mathbf{A}}^I$ is

$$\hat{\mathbf{A}}^I = \mathbf{A}^I - \varepsilon\mathbf{A}^I\mathbf{A}_o\mathbf{A}^I \tag{6}$$

which bears a striking similarity with the dual inverse, an expression also displayed in [1].

The right Moore–Penrose generalized inverse of a dual matrix $\hat{\mathbf{C}} = \mathbf{C} + \varepsilon\mathbf{C}_o$, with $\mathbf{C}$, $\mathbf{C}_o \in \mathbf{R}^{m\times n}$ and $m < n$, is defined as the dual matrix $\hat{\mathbf{C}}^\dagger \equiv \hat{\mathbf{D}}$ such that $\hat{\mathbf{C}}\hat{\mathbf{D}} = \mathbf{1}_m$, with $\hat{\mathbf{D}} = \mathbf{D} + \varepsilon\mathbf{D}_o$ and $\mathbf{D}$, $\mathbf{D}_o \in \mathbf{R}^{n\times m}$. The computation of $\mathbf{D}$ and $\mathbf{D}_o$ follows the same pattern as that of $\mathbf{B}$ and $\mathbf{B}_o$ above. The details are not included here for conciseness, but the results are displayed below:

$$\hat{\mathbf{C}}^\dagger = \mathbf{C}^\dagger - \varepsilon\mathbf{C}^\dagger\mathbf{C}_o\mathbf{C}^\dagger \tag{7}$$

a formula that is also displayed in [1], but without a proof. Again, as in the case of $\mathbf{A}^I$, $\hat{\mathbf{C}}^\dagger$ is not unique, the formula displayed above being the one with a minimum Frobenius norm.

Now the left dual generalized inverse is applied to the solution of an overdetermined system of m dual linear equations in $n < m$ dual unknowns, grouped in vector $\hat{\mathbf{x}}$, of the form

$$\hat{\mathbf{A}}\hat{\mathbf{x}} = \hat{\mathbf{b}} \tag{8}$$

where $\hat{\mathbf{A}}$ is assumed as above, to be a dual $m \times n$ matrix, with $m > n$ and with a full-rank primal part, $\hat{\mathbf{x}}$ and $\hat{\mathbf{b}}$ being, respectively, n- and m-dimensional dual vectors. As the system is overdetermined, it is not possible to find a vector $\hat{\mathbf{x}}$ that will verify all m dual equations (8), but it will be shown that it is possible to find the vector $\hat{\mathbf{x}}$ that will render the Euclidean norm of the dual error $\hat{\mathbf{e}}$ a minimum, with $\hat{\mathbf{e}}$ defined as

$$\hat{\mathbf{e}} = \hat{\mathbf{b}} - \hat{\mathbf{A}}\hat{\mathbf{x}} \tag{9}$$

whose Euclidean norm[2] $\|\hat{\mathbf{e}}\|$ is the square root of the scalar product $\hat{\mathbf{e}}^T\hat{\mathbf{e}}$, i.e.,

$$\|\hat{\mathbf{e}}\|^2 = \|\hat{\mathbf{b}}\|^2 - 2\hat{\mathbf{b}}^T\hat{\mathbf{A}}\hat{\mathbf{x}} + \|\hat{\mathbf{A}}\hat{\mathbf{x}}\|^2 \tag{10}$$

The error Euclidean norm is minimized upon zeroing the derivative of $\|\hat{\mathbf{e}}\|^2$ with respect to $\hat{\mathbf{x}}$, which readily leads to the *dual normal equations* (DNE):

$$\hat{\mathbf{A}}^T\hat{\mathbf{A}}\hat{\mathbf{x}} = \hat{\mathbf{A}}^T\hat{\mathbf{b}} \quad \Rightarrow \quad \hat{\mathbf{A}}^T\hat{\mathbf{e}}_0 = \mathbf{0} \tag{11}$$

[2] If $\mathbf{e}$ and $\mathbf{e}_o$ denote the primal and dual parts of $\hat{\mathbf{e}}$, then $\|\hat{\mathbf{e}}\|^2 = \|\mathbf{e}\|^2 + \varepsilon 2\mathbf{e}^T\mathbf{e}_o$.

thereby stating an important theoretical result: the minimum-norm error – i.e., the error $\hat{\mathbf{e}}_0$ of minimum Euclidean norm – lies in the null space of $\hat{\mathbf{A}}^T$, a restatement of the classical *Projection Theorem*, but now in dual space. Another theoretical result is the expression for the least-square solution $\hat{\mathbf{x}}_0$, obtained directly from the normal equations (11):

$$\hat{\mathbf{x}}_0 = \hat{\mathbf{A}}^I \hat{\mathbf{b}} \tag{12}$$

Expression (12) is a representation of the *unique* minimum-norm least-square solution of system (8), but should not be used verbatim to compute $\hat{\mathbf{x}}_0$, because of the frequent ill-conditioning of the product $\hat{\mathbf{A}}\hat{\mathbf{A}}^T$. Instead, the QR decomposition [9] should be applied.

Interestingly, having chosen the dual part $\mathbf{B}_o$ of $\hat{\mathbf{A}}^I$ with minimum norm guarantees that the dual part $\mathbf{x}_{o0}$ of the least-square solution $\hat{\mathbf{x}}_0$ is of minimum Euclidean norm. This property will be exploited in the synthesis of a RCCC linkage intended to approximate a homokinetic transmission between two shafts of skew axes, lying at right angles.

3 Synthesis of a RCCC Linkage

The foregoing results will now be applied to the synthesis of the RCCC linkage shown in Fig. 1, with geometric parameters defined using the original Denavit–Hartenberg notation [10].

The input-output (IO) equation of the RCCC linkage was derived by Yang and Freudenstein [11]. The same equation was more recently cast in a framework that allows its analysis in a unified form applicable to planar, spherical and spatial four-bar linkages [12]. For the sake of brevity, the IO equation is not derived here. It is displayed below, as taken from the foregoing reference:

$$\hat{F}(\hat{\psi},\hat{\phi}) \equiv \hat{k}_1 + \hat{k}_2 \cos\hat{\psi} + \hat{k}_3 \cos\hat{\psi}\cos\hat{\phi} - \hat{k}_4 \cos\hat{\phi} + \sin\hat{\psi}\sin\hat{\phi} = 0 \tag{13}$$

where $\hat{\psi}$, the input angle, has been "hatted", even though this angle is associated with a R joint, which undergoes pure rotations about its axis. In fact, $\hat{\psi} = \psi + \varepsilon b_2$, where b_2 accounts for the location of the common normal between this axis (Z_2) and Z_3. The primal parts of the *dual Freudenstein parameters* (DFP) are given below:

$$k_1 \equiv \frac{\lambda_1\lambda_2\lambda_4 - \lambda_3}{\mu_2\mu_4}, \quad k_2 = \frac{\lambda_4\mu_1}{\mu_4}, \quad k_3 = \lambda_1, \quad k_4 = \frac{\lambda_2\mu_1}{\mu_2} \tag{14}$$

with the definitions $\lambda_i \equiv \cos\alpha_i$ and $\mu_i \equiv \sin\alpha_i \neq 0$, while α_i is displayed in Fig. 1, their dual counterparts being defined as

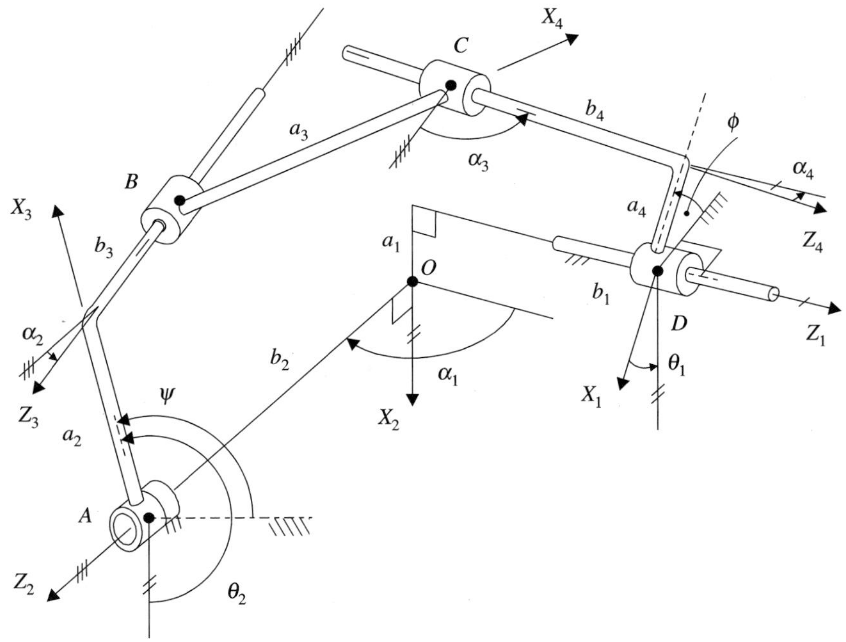

Fig. 1 A generic RCCC linkage.

$$
\begin{aligned}
k_{o1} &= -\frac{a_1\lambda_2\lambda_4\mu_1\mu_2\mu_4 + a_2(\lambda_1\lambda_4-\lambda_2\lambda_3)\mu_4 - a_3\mu_2\mu_3\mu_4 + a_4(\lambda_1\lambda_2-\lambda_3\lambda_4)\mu_2}{\mu_2^2\mu_4^2}\\
k_{o2} &= \frac{a_1\lambda_1\lambda_4\mu_4 - a_4\mu_1}{\mu_4^2}, \quad k_{o3} = -a_1\mu_1, \quad k_{o4} = \frac{a_1\lambda_1\lambda_2\mu_2 - a_2\mu_1}{\mu_2^2}
\end{aligned}
\tag{15}
$$

The synthesis problem can now be formulated as: given a set of input-angle values $\{\psi_i\}_1^m$ and a set of corresponding output values $\{\phi_i, u_i\}_1^m$, where u_i denotes the ith prescribed value of the output variable[3] b_1, *find the linkage parameters* $\{a_i, \alpha_i\}_1^4$ that will produce a RCCC linkage that meets the prescribed IO relations. Since we have m IO conditions to meet, in the form of the dual equations (13), and four dual linkage parameters, when $m = 4$ the prescribed IO values can be met exactly, which corresponds to *exact synthesis*. For $m > 4$, no linkage will possibly meet all m prescribed IO values. However, it is possible to find the linkage that will meet these values with the minimum error, which is known as *approximate synthesis*. Nevertheless, a word of caution is in order: although the error vector defined in eq. (9) has components with two different units, radians and m, its norm is well defined, as per footnote 2. Hence, a linkage can be found that meets the synthesis equations with an error of minimum Euclidean norm, *independent of the units chosen*. The said equations are obtained upon substitution of the input and output variables by

[3] The new variable u_i is introduced with the purpose of avoiding double subscripts.

their m prescribed values in the IO equation:

$$\hat{F}_i(\psi_i,\hat{\phi}_i) \equiv \hat{k}_1 + \hat{k}_2 \cos\hat{\psi}_i + \hat{k}_3 \cos\hat{\psi}_i \cos\hat{\phi}_i - \hat{k}_4 \cos\hat{\phi}_i + \sin\hat{\psi}_i \sin\hat{\phi}_i = 0,\ i = 1,\ldots,m \tag{16}$$

which are linear in the dual Freudenstein parameters $\{\hat{k}_i\}_1^4$. Hence, upon assembling the m foregoing equations, a system of m dual linear equations in the four DFP is obtained:

$$\hat{\mathbf{S}}\hat{\mathbf{k}} = \hat{\mathbf{b}} \tag{17}$$

with

$$\hat{\mathbf{S}} = \underbrace{\begin{bmatrix} 1 & c\psi_1 & c\psi_1 c\phi_1 & -c\phi_1 \\ 1 & c\psi_2 & c\psi_2 c\phi_2 & -c\phi_2 \\ \vdots & \vdots & \vdots & \vdots \\ 1 & c\psi_m & c\psi_m c\phi_m & -c\phi_m \end{bmatrix}}_{\mathbf{S}} + \varepsilon \underbrace{\begin{bmatrix} 0 & -b_2 s\psi_1 & -u_1 c\psi_1 s\phi_1 - b_2 s\psi_1 c\phi_1 & u_1 s\phi_1 \\ 0 & -b_2 s\psi_2 & -u_2 c\psi_2 s\phi_2 - b_2 s\psi_2 c\phi_2 & u_2 s\phi_2 \\ \vdots & \vdots & \vdots & \vdots \\ 0 & -b_2 s\psi_m & -u_m c\psi_m s\phi_m - b_2 s\psi_m c\phi_m & u_m s\phi_m \end{bmatrix}}_{\mathbf{S}_o} \tag{18}$$

with $c(\cdot)$ and $s(\cdot)$ denoting $\cos(\cdot)$ and $\sin(\cdot)$, respectively, while

$$\hat{\mathbf{b}} = -\underbrace{\begin{bmatrix} s\psi_1 s\phi_1 \\ s\psi_2 s\phi_2 \\ \vdots \\ s\psi_m s\phi_m \end{bmatrix}}_{\mathbf{b}} - \varepsilon \underbrace{\begin{bmatrix} u_1 s\psi_1 c\phi_1 + b_2 c\psi_1 s\phi_1 \\ u_2 s\psi_2 c\phi_2 + b_2 c\psi_2 s\phi_2 \\ \vdots \\ u_m s\psi_m c\phi_m + b_2 c\psi_m s\phi_m \end{bmatrix}}_{\mathbf{b}_o} \tag{19}$$

Now, upon equating the primal and the dual parts of eq. (17), two real vector equations are obtained, namely,

$$\mathbf{Sk} = \mathbf{b}, \quad \mathbf{Sk}_o + \mathbf{S}_o\mathbf{k} = \mathbf{b}_o \quad \Rightarrow \quad \mathbf{Sk}_o = \mathbf{b}_o - \mathbf{S}_o\mathbf{k} \tag{20}$$

which amount to two overdetermined linear systems of equations, both with the same matrix coefficient $\mathbf{S}$, one for $\mathbf{k}$, one for $\mathbf{k}_o$. The computation of the least-square solution proceeds in two steps: first the primal equation is solved for $\mathbf{k}$; with the least-square solution thus obtained, $\mathbf{k}_0$, substituted into the dual equation, the least square solution of this equation, $\mathbf{k}_{o0}$, is obtained. Notice that these calculations being done using the QR decomposition, the primal synthesis matrix needs factoring only once. This feature is important if the foregoing procedure is a part of a second, external optimization procedure, that calls for many iterations. It is noteworthy that the DH parameter b_2 is not included in either the primal or the dual part of the DFP, eqs. (14) and (15), respectively, and hence, this parameter has been taken to the right-hand side of the dual synthesis equations (20); b_2 has to be treated not as an unknown, but as a parameter, that can be used to either fine-tune a solution or to optimize an objective function.

Now the RCCC linkage is synthesized so as to approximate a homokinetic transmission for values of the input and the output variables that sweep angles of 120°.

Moreover, the primal synthesis equation leading to a spherical linkage, the associated synthesis procedure is identical to that reported in [13]. In that paper, a search is included on the optimum values of the location of the zeros of the input and output dials, which amount to a translation of the data points $\{\psi_i, \phi_i\}_1^m$ *en masse*, i.e., under a rigid-body translation, in the ϕ-vs.-ψ plane. A continuum of values for the optimum shifts were reported in that paper. The values adopted here are $\xi = 146°$ and $\eta = 34°$, for ψ and ϕ, respectively.

The values ψ_i, for $i = 1, \ldots m$, with $m = 501$ prescribed data *triads*,[4] are uniformly distributed in the interval $86°(= -60° + 146°) \leq \psi \leq 206°$, while their ϕ_i counterparts are distributed likewise in the interval $-26°(= -60° + 34°) \leq \phi \leq 94°$. The shafts to be coupled lying at right angles, $\alpha_1 = 90°$, whence $\lambda_1 = 0$ and $\mu_1 = 1$. Furthermore, given its desired homokinetic behavior, the linkage is assumed symmetric, as the input and output links play the same role, whence $\alpha_4 = \alpha_2$. In this light, the number of unknown primal Freudenstein parameters reduces to only two, for $k_3 = 0$ and $k_4 = k_2$, a consequence of the foregoing assumptions and relations (14). The number of prescribed points led to an overdetermined linear system of 501 equations in two unknowns, whose least-square solution is

$$k_1 = 1.217,\ k_2 = 0.9439 \Rightarrow k_4 = 0.9439,\ \alpha_2 = \alpha_4 = 46.65°,\ \alpha_3 = 132.4° \quad (21)$$

with a rms value of the minimum-norm error equal to 0.01942, or 1.942%.

Next, a_1 is set at 240 mm, as imposed by the design conditions, with $b_2 = a_1$ for symmetry. Further, the values u_i of b_1 at the prescribed values of ϕ_i, which complete the ith triad, were distributed symmetrically around $b_1 = 0$, with $u_1 = -a_1/10$, $u_m = a_1/10$, and following a cycloidal motion program:

$$u_i = -\frac{a_1}{10} + U\left(\frac{i-1}{m-1} - \frac{1}{2\pi}\sin\frac{2\pi(i-1)}{m-1}\right) \quad (22)$$

with amplitude U given as $a_1/5$ in order to limit the output sliding b_1. This program was chosen because it starts smoothly with zero velocity and acceleration, and stops likewise. The second system of eqs. (20), of 501 equations for two unknowns, k_{o1} and k_{o2}, led to the least-square solution $\mathbf{k}_{o0}$, with $k_{o3} = -a_1 = -240$ mm not being part of the unknowns, for its value is fixed from the prescribed values for α_1 and a_1, as per eqs. (15). The optimum values were found to be, for the above-mentioned values of $\mathbf{k}_0$,

$$k_{o1} = 319.0 \text{ mm}, \quad k_{o2} = 154.6 \text{ mm}, \quad k_{o4} = k_{o2} \quad (23)$$

with a rms value of the minimum-norm dual error of 1.119 mm, or 2.33% of the amplitude U. For the record, the *normalized* dual part of the Euclidean norm of the dual error, $\mathbf{e}^T\mathbf{e}_o/\sqrt{m}$, is 0.3260 mm or 0.07% of the amplitude U.

[4] This high number was used with the purpose of bringing the optimum *design error* $\mathbf{e}_0$ as close as possible to the *structural error*, which measures the actual deviation of the synthesized output angle from its prescribed value, as per the results reported in [14].

Fig. 2 A CAD model of the synthesized RCCC linkage.

Computing the DH linkage parameters now is straightforward, as eqs. (15) involve these parameters linearly. The results are displayed below:

$$a_2 = -76.46 \text{ mm}, \quad a_3 = 209.4 \text{ mm}, \quad a_4 = a_2 \tag{24}$$

where, interestingly, a_i being defined as a *length* in the framework of the DH notation, it must be non-negative. However, a negative value for a_2, and hence, for a_4, was obtained above. The interpretation of the negative sign here is well known within the methodology set forth by Freudenstein [15]: should a_2 (a_4) turn out to be negative as a result of the linkage synthesis for function generation, then measure angle ψ (ϕ) not as indicated in Fig. 1, but from its extension, i.e., add 180° to the prescribed input (output) angles. This completes the solution to the synthesis problem, a CAD model thereof being shown in Fig. 2. In this figure, the output motion of the quasi-homokinetic mechanism is the rotation of the splined shaft, which is mounted on the machine frame by means of standard bearings.

4 Conclusions

Some novel results in the realm of the algebra of dual numbers, in connection with dual linear least-square problems were introduced here, then applied to the synthesis of the RCCC function generating linkage. The methodology thus established was then illustrated with the solution of a problem of current interest, the approximate synthesis of a RCCC linkage for homokinetic transmission between shafts with skew axes, lying at right angles.

Acknowledgements The support provided by a James McGill Professorship is dutifully acknowledged. Partial support received from the NSERC (Canada's Natural Sciences and Engineering Research Council) Discovery Grant 4532-2008 is equally acknowledged.

References

1. Pennestrì, E., Stefanelli, R.: Linear algebra and numerical algorithms using dual numbers. Multibody Syst. Dyn. **18**(3), 323–344 (2007)
2. Angeles, J.: The application of dual algebra to kinematic analysis. In: Angeles, J., Zakhariev, E. (eds). Computational Methods in Mechanical Systems, vol. 161, pp. 3–31. Springer-Verlag, Heidelberg (1998)
3. Fischer, I.S.: Dual-Number Methods in Kinematics, Statistics and Dynamics. CRC Press, Boca Raton (1999)
4. Cheng, H.H., Thompson, S.: Dual polynomials and complex dual numbers for analysis of spatial mechanisms. In: Proc. (1996) ASME Design Engineering Technical Conference and Computers in Engineering Conference, Irvine, CA, August 18–22, Vol. 96-DET/MECH-1221 (1996)
5. Simmons, G.F.: Introduction to Topology and Modern Analysis. McGraw-Hill Book Co., New York (1963)
6. Dimentberg, F.M.: The Screw Calculus and Its Applications in Mechanics. Izdat. Nauka, Moscow (1965)
7. Kotel'nikov, A.P.: Screw Calculus and Some of Its Applications to Geometry and Mechanics [in Russian]. Annals of the Imperial University of Kazan (1895). 2006 edition by KomKniga, Moscow
8. Alizadeh, D., Ma, X.Q., Angeles, J.: On the reduction of the normality conditions in equality-constrained optimization problems in mechanics. Meccanica **47**, 755–768 (2012)
9. Golub, G.H., Van Loan, C.F.: Matrix Computations. The Johns Hopkins University Press, Baltimore (1989)
10. Hartenberg, R.S., Denavit, J.: Kinematic Synthesis of Linkages. McGraw-Hill Book Company, New York (1964)
11. Yang, A., Freudenstein, F.: Application of dual-number quaternion algebra to the analysis of spatial mechanisms. J. Appl. Mech. **31**, 300–307 (1964)
12. Bai, S.P., Angeles, J.: A unified input-output analysis of four-bar linkages. Mech. Mach. Theory **43**, 240–251 (2008)
13. Alizadeh, D., Angeles, J., Nokleby, S.: Optimization of a spherical 'homokinetic' linkage with minimum design error and maximum transmission quality. In: Proc. 13th World Congress on the Theory of Machines and Mechanisms, Guanajuato, Gto., Mexico, June 20–22, Paper A-354 (2011)
14. Hayes, M.J.D., Parsa, K., Angeles, J.: The effect of data-set cardinality on the design and structural errors of four-bar function-generators. In: Proc. 10th World Congress on the Theory of Machines and Mechanisms, Oulu, Finland, June 20–24, vol. 1, pp. 437–442 (1999)
15. Freudenstein, F.: Approximate synthesis of four-bar linkages. Trans. ASME **77**, 853–861 (1955)

On the Twist Recovery Methodologies After Failure

Leila Notash

Abstract In this paper, methodologies for investigating the effect of failures on the performance of manipulators are presented, and the correctional input for recovering the lost motion provided by the remaining joints, for minimum Euclidean norm of the correctional and the overall joint velocity vectors, are presented. The procedure is simulated to examine the norm of the overall input before and after a failure, as well as the norm of the correctional input.

Key words: Parallel manipulator, failure, twist recovery

1 Introduction

In parallel manipulators, the mobile platform is connected to the base by a number of legs, e.g., refer to Figure 1. In general, each leg is a kinematic chain of links connected by active and passive joints. For non-redundant actuation, using the one degree of freedom joints such as revolute or prismatic joints, the number of active joints is equal to the degree of freedom (DOF) of the manipulator, e.g., Figure 1(a), with active prismatic joints. To form a kinematically redundant leg, one or more redundant active joints could be added to the leg, e.g., Figure 1(b).

Failure of a link and/or a joint could result in the loss of DOF, actuation, motion constraint, and information in parallel manipulators [6]. If any of these failures affect the performance of manipulator such that the task cannot be completed as desired, then the manipulator is considered failed. From the kinematics point of view, the failure of a joint occurs if the joint is broken, or jammed (its displacement remains constant), or if the displacement/velocity/acceleration of joint is not at the desired level. Redundancy in joint displacement sensing was investigated in [3] to facilitate the joint sensor fault detection, isolation and recovery. The relative manipulativity index was used in [7] to investigate the Jacobian matrices of manipulators fault tolerant to joint failures. In [1], the task space was partitioned to complete the major

Queen's University, Kingston, Canada, e-mail: notash@me.queensu.ca

J. Lenarčič, M. Husty (eds.), *Latest Advances in Robot Kinematics*,
DOI 10.1007/978-94-007-4620-6_2,

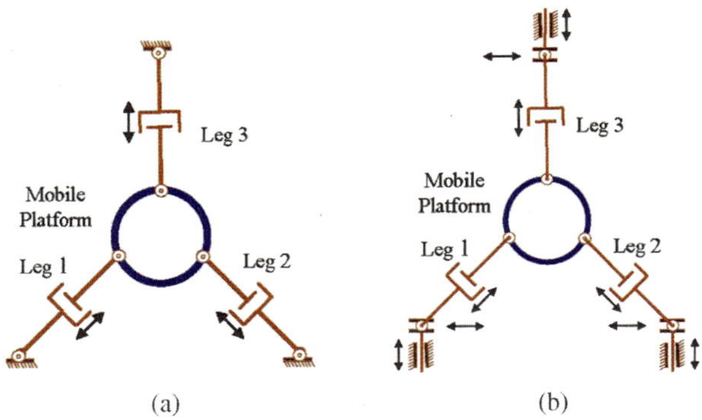

Fig. 1 Planar parallel manipulators: (a) non-redundant; and (b) redundant.

task and optimize a secondary task such as actuator fault tolerance. The effects of joint failures on the force/moment capabilities and motion performance of parallel manipulators were respectively investigated in [4, 5]. In this article, the methodology for recovering the lost motion due to the failure of joints/actuators, presented in [5], is briefly discussed. The properties of the recovered motion are examined in Section 2 by reformulating the correctional and overall velocity vectors for minimum Euclidean norm. The simulation results are reported in Section 3. The article concludes with Section 4.

2 Recovering Lost Velocity

For parallel manipulators, the relation between the $n \times 1$ active joint velocity vector, $\dot{\mathbf{q}}$, and the $m \times 1$ mobile platform twist (velocity vector), $\mathbf{V}$, is given as $\dot{\mathbf{q}} = \mathbf{JV}$, where $m \leq 6$ depending on the dimension of task space, e.g., $m = 3$ for planar motion. The Jacobian matrix of manipulator, $\mathbf{J}$, is an $n \times m$ matrix; with $n = m$ for non-redundant manipulators and $n > m$ for redundant manipulators. For a given $\dot{\mathbf{q}}$, using the generalized inverse (GI) of $\mathbf{J}$, $\mathbf{J}^{\#}$, the platform twist is

$$\mathbf{V} = \mathbf{J}^{\#}\dot{\mathbf{q}} \tag{1}$$

Hence, to provide the required platform twist, $\mathbf{V}$ should belong to the range space of $\mathbf{J}^{\#}$. In addition, each leg of manipulator should allow the platform twist $\mathbf{V}$.

Considering leg i, its $l \times 1$ joint velocity vector, ${}^i\dot{\mathbf{q}} ={}^i [\dot{q}_1\ \dot{q}_2\ \ldots\ \dot{q}_{l-1}\ \dot{q}_l]^T$, and the twist, $\mathbf{V}$, are related by the $m \times l$ Jacobian matrix of the leg, ${}^i\mathbf{J}$, as

$$\mathbf{V} ={}^i \mathbf{J}\,{}^i\dot{\mathbf{q}} = \left[{}^i\mathbf{J}_1\ {}^i\mathbf{J}_2\ \ldots {}^i\mathbf{J}_h\ \ldots {}^i\mathbf{J}_{l-1}\ {}^i\mathbf{J}_l\right]{}^i\dot{\mathbf{q}} = \sum_{k=1}^{l} {}^i\mathbf{J}_k {}^i\dot{q}_k \tag{2}$$

where each column of ${}^i\mathbf{J}$, ${}^i\mathbf{J}_k$, is a screw representing the axis of the corresponding joint of leg i; and $l \geq m$. To provide the platform velocity, $\mathbf{V}$ should be in the range space of all ${}^i\mathbf{J}$, for $i = 1, \ldots, n_l$, where n_l is the number of legs.

After failure, the twist $\mathbf{V}$ should be provided by the active joints of the manipulator, as well as by the joints of each leg. When joint h (active or passive) on leg i is failed its velocity ${}^i\dot{q}_{ch}$ will be different than the desired value ${}^i\dot{q}_h$, and ${}^i\dot{\mathbf{q}}_f =^i [\dot{q}_1 \; \ldots \; \dot{q}_{ch} \; \ldots \; \dot{q}_l]^T$. When ${}^i\dot{q}_{ch} \neq {}^i\dot{q}_h$ the velocity equation for leg i is

$$\mathbf{V}_f =^i \mathbf{J}\,{}^i\dot{\mathbf{q}}_f = \sum_{k=1}^{l} {}^i\mathbf{J}_k\,{}^i\dot{q}_k - {}^i\mathbf{J}_h({}^i\dot{q}_h - {}^i\dot{q}_{ch}) \tag{3}$$

Then, the manipulator would be considered as failed unless the lost motion of platform is in the range space of the Jacobian matrix corresponding to the remaining (healthy) joints of that leg [5]. For full recovery of the lost twist ${}^i\mathbf{J}_h({}^i\dot{q}_h - {}^i\dot{q}_{ch})$, in general the leg with a failed joint should have a redundant joint.

2.1 Correctional Input from Healthy Joints

When joint h has a different velocity the correctional velocity to be provided by the remaining joints of leg i, ${}^i\dot{\mathbf{q}}_{corr} = {}^i[\dot{q}_{corr1} \; \dot{q}_{corr2} \; \ldots \; 0 \; \ldots \; \dot{q}_{corr\,l-1} \; \dot{q}_{corr\,l}]^T$, will compensate for the lost twist partially or completely, where in ${}^i\dot{\mathbf{q}}_{corr}$ entry h is replaced by a zero. Then, the recovered velocity of the platform will be

$$\mathbf{V}_r = {}^i\mathbf{J}\,{}^i\dot{\mathbf{q}}_f +^i \mathbf{J}\,{}^i\dot{\mathbf{q}}_{corr} = {}^i\mathbf{J}\,{}^i\dot{\mathbf{q}}_f +^i \mathbf{J}_f\,{}^i\dot{\mathbf{q}}_{corr} \tag{4}$$

where in ${}^i\mathbf{J}_f = [{}^i\mathbf{J}_1 \; {}^i\mathbf{J}_2 \; \ldots \; \mathbf{0} \; \ldots \; {}^i\mathbf{J}_{l-1} \; {}^i\mathbf{J}_l]$, column h is replaced by zeros.

To fully recover the lost twist, after applying the correctional velocity the change in the twist should be zero, i.e., $\mathbf{V} - \mathbf{V}_r = {}^i\mathbf{J}({}^i\dot{\mathbf{q}} -^i \dot{\mathbf{q}}_f) - {}^i\mathbf{J}_f\,{}^i\dot{\mathbf{q}}_{corr} = \mathbf{0}$. Then, the correctional velocity of the healthy joints will be

$${}^i\dot{\mathbf{q}}_{corr} = {}^i\mathbf{J}_f^{\#}\,{}^i\mathbf{J}_h({}^i\dot{q}_h -^i \dot{q}_{ch}) = {}^i\mathbf{J}_f^{\#}\,{}^i\mathbf{J}({}^i\dot{\mathbf{q}} - {}^i\dot{\mathbf{q}}_f) \tag{5}$$

where ${}^i\dot{\mathbf{q}} -^i \dot{\mathbf{q}}_f = {}^i[0 \; 0 \; \ldots \; (\dot{q}_h - \dot{q}_{ch}) \; \ldots \; 0 \; 0]^T$ is the lost motion due to failure of joint h. Then, the overall joint velocities will be

$${}^i\dot{\mathbf{q}}_{tot} = {}^i\dot{\mathbf{q}}_f + {}^i\dot{\mathbf{q}}_{corr} = {}^i\mathbf{J}_f^{\#}\,{}^i\mathbf{J}\,{}^i\dot{\mathbf{q}} + (\mathbf{I} - {}^i\mathbf{J}_f^{\#}\,{}^i\mathbf{J})^i\,\dot{\mathbf{q}}_f = {}^i\mathbf{J}_f^{\#}\,{}^i\mathbf{J}\,{}^i\dot{\mathbf{q}} \tag{6}$$

If the velocities of g joints of leg i are different than the required values the corresponding g columns of ${}^i\mathbf{J}_f$ will be zero. The lost platform twist will be $\sum {}^i\mathbf{J}_h({}^i\dot{q}_h - {}^i\dot{q}_{ch}) = {}^i\mathbf{J}({}^i\dot{\mathbf{q}} - {}^i\dot{\mathbf{q}}_f)$ and the correctional velocity will be

$${}^i\dot{\mathbf{q}}_{corr} = {}^i\mathbf{J}_f^{\#} \sum {}^i\mathbf{J}_h({}^i\dot{q}_h - {}^i\dot{q}_{ch}) = {}^i\mathbf{J}_f^{\#}\,{}^i\mathbf{J}({}^i\dot{\mathbf{q}} -^i \dot{\mathbf{q}}_f) \tag{7}$$

where the summation is taken over the failed joints.

When ${}^i\mathbf{J}_f$ has full row-rank, i.e., $\mathbf{V}$ belongs to the range space of ${}^i\mathbf{J}_f$, $\mathbf{V} \in \Re({}^i\mathbf{J}_f)$, if ${}^i\dot{\mathbf{q}}$ is not physically consistent, e.g., leg i has a combination of revolute and prismatic joints, the weighted right-generalized inverse [2] of ${}^i\mathbf{J}_f$ is formulated such that ${}^i\dot{\mathbf{q}}^T(\mathbf{W}_{\dot{q}}\,{}^i\dot{\mathbf{q}})$ is physically consistent. Then

$$ {}^i\mathbf{J}^{\#}_{fw} = \mathbf{W}_{\dot{q}}^{-1}\,{}^i\mathbf{J}_f^T\left({}^i\mathbf{J}_f\,\mathbf{W}_{\dot{q}}^{-1}\,{}^i\mathbf{J}_f^T\right)^{-1}. \tag{8} $$

Otherwise, for physically consistent ${}^i\dot{\mathbf{q}}$, ${}^i\mathbf{J}^{\#}_f = {}^i\mathbf{J}_f^T({}^i\mathbf{J}_f\,{}^i\mathbf{J}_f^T)^{-1}$.

The deviation in the platform twist after applying the correctional velocity will be zero when $\mathbf{V} \in \Re({}^i\mathbf{J}_f)$. Hence, the condition for full recovery is

$$ \mathbf{V}_{\Re\perp} = \left(\mathbf{I} - {}^i\mathbf{J}_f\,{}^i\mathbf{J}^{\#}_f\right)\mathbf{V} = \mathbf{0} \tag{9} $$

If some entries of $\mathbf{V}_{\Re\perp} = (\mathbf{I} - {}^i\mathbf{J}_f\,{}^i\mathbf{J}^{\#}_f)\mathbf{V}$ are not zero the corresponding components of the platform twist could not be completely recovered. When some of g failed joints have non-zero velocity $\mathbf{V}^* = \mathbf{V} - \sum \mathbf{J}_h\,{}^i\dot{\mathbf{q}}_{ch}$ could be used in (9). In case ${}^i\mathbf{J}_f$ does not have full row-rank, in general, the lost motion cannot be fully recovered and the platform twist that best approximates the lost motion in the least-square sense is calculated using the weighted left-GI of ${}^i\mathbf{J}_f$.

When the number of failed joints is equal to the number of redundant actuators, $g = l - m$, as long as the leg is not at a singularity, after removing the columns of the $m \times l$ Jacobian matrix corresponding to failed joints, the reduced Jacobian matrix ${}^i\mathbf{J}_r$ will be an $m \times m$ square matrix with rank m and in general there will be a unique solution for the velocity of healthy joints. When $g < l - m$ while the rank of ${}^i\mathbf{J}_f$ is m there will be infinite solutions for ${}^i\dot{\mathbf{q}}_{corr}$ and ${}^i\dot{\mathbf{q}}_{tot}$. In the following subsections, expressions for ${}^i\dot{\mathbf{q}}_{corr}$ and ${}^i\dot{\mathbf{q}}_{tot}$ are derived considering their norms.

2.2 Minimum Norm for Correctional Velocity Vector

When g out of l actuators/joints of leg i are failed, to minimize the jump in the velocity of joints after failure while providing the platform twist, the objective function will be the square of the Euclidean norm of the weighted correctional velocity vector $(\mathbf{W}_{\dot{q}}^{1/2}\,{}^i\dot{\mathbf{q}}_{corr}) \cdot (\mathbf{W}_{\dot{q}}^{1/2}\,{}^i\dot{\mathbf{q}}_{corr}) = {}^i\dot{\mathbf{q}}_{corr} \cdot (\mathbf{W}_{\dot{q}}\,{}^i\dot{\mathbf{q}}_{corr})$. The linear constraint equation in terms of the overall velocity vector ${}^i\dot{\mathbf{q}}_{tot}$ is $\mathbf{V}\,{}^i\mathbf{J}\,{}^i\dot{\mathbf{q}}_{tot} = \mathbf{0}$. The Lagrange function L is formulated by augmenting the constraint equation with the objective function using the Lagrange multiplier vector $\boldsymbol{\lambda}$

$$ L({}^i\dot{\mathbf{q}}_{corr}, \boldsymbol{\lambda}) = \frac{1}{2}\,{}^i\dot{\mathbf{q}}_{corr} \cdot (\mathbf{W}_{\dot{q}}\,{}^i\dot{\mathbf{q}}_{corr}) - \boldsymbol{\lambda}^T(\mathbf{V} - {}^i\mathbf{J}\,{}^i\dot{\mathbf{q}}_{tot}) \tag{10} $$

If ${}^i\dot{\mathbf{q}}_{corr} \cdot (\mathbf{W}_{\dot{q}}\,{}^i\dot{\mathbf{q}}_{corr})$ is a minimum for the original constrained problem at the stationary point $({}^i\dot{\mathbf{q}}_{corr}, \boldsymbol{\lambda})$ the gradient of L vanishes, i.e., $\nabla L({}^i\dot{\mathbf{q}}_{corr}, \boldsymbol{\lambda}) = \mathbf{0}$.

When ${}^i\dot{q}_{ch} = 0$ the constraint equation is $\mathbf{V} - {}^i\mathbf{J}_f\,{}^i\dot{\mathbf{q}}_{tot} = \mathbf{0}$ and

$$\frac{\partial L({}^i\dot{\mathbf{q}}_{corr}, \boldsymbol{\lambda})}{\partial {}^i\dot{\mathbf{q}}_{corr}} = \mathbf{W}_{\dot{q}}\,{}^i\dot{\mathbf{q}}_{corr} - {}^i\mathbf{J}_f{}^T\boldsymbol{\lambda} = \mathbf{0} \quad \Rightarrow \quad {}^i\dot{\mathbf{q}}_{corr} = \mathbf{W}_{\dot{q}}^{-1}\,{}^i\mathbf{J}_f{}^T\boldsymbol{\lambda} \tag{11}$$

$$\frac{\partial L({}^i\dot{\mathbf{q}}_{corr}, \boldsymbol{\lambda})}{\partial \boldsymbol{\lambda}} = {}^i\mathbf{J}_f\,{}^i\dot{\mathbf{q}}_{tot} - \mathbf{V} = \mathbf{0} \quad \Rightarrow \quad {}^i\mathbf{J}_f\,{}^i\dot{\mathbf{q}}_f + {}^i\mathbf{J}_f\,{}^i\dot{\mathbf{q}}_{corr} = \mathbf{V} \tag{12}$$

then $\boldsymbol{\lambda} = ({}^i\mathbf{J}_f\,\mathbf{W}_{\dot{q}}^{-1}\,{}^i\mathbf{J}_f{}^T)^{-1}(\mathbf{V} - {}^i\mathbf{J}_f\,{}^i\dot{\mathbf{q}}_f)$, ${}^i\mathbf{J}\,{}^i\dot{\mathbf{q}}_f = {}^i\mathbf{J}_f\,{}^i\dot{\mathbf{q}}_f$ for ${}^i\dot{q}_{ch} = 0$

$${}^i\dot{\mathbf{q}}_{corr} = {}^i\mathbf{J}^{\#}_{fw}\,{}^i\mathbf{J}({}^i\dot{\mathbf{q}} - {}^i\dot{\mathbf{q}}_f) \tag{13}$$

and ${}^i\mathbf{J}^{\#}_{fw} = \mathbf{W}_{\dot{q}}^{-1}\,{}^i\mathbf{J}_f{}^T({}^i\mathbf{J}_f\,\mathbf{W}_{\dot{q}}^{-1}\,{}^i\mathbf{J}_f{}^T)^{-1}$ is the weighted right-GI of ${}^i\mathbf{J}_f$.

When ${}^i\dot{q}_{ch} \neq 0$ the constraint equation is $\mathbf{V} - {}^i\mathbf{J}\,{}^i\dot{\mathbf{q}}_{tot} = \mathbf{0}$. To ensure zero correctional velocity for the failed joints, ${}^i\mathbf{J}^T\,\boldsymbol{\lambda}$ is replaced with ${}^i\mathbf{J}_f{}^T\,\boldsymbol{\lambda}$ for ${}^i\dot{\mathbf{q}}_{corr}$

$$\frac{\partial L({}^i\dot{\mathbf{q}}_{corr}, \boldsymbol{\lambda})}{\partial {}^i\dot{\mathbf{q}}_{corr}} = \mathbf{W}_{\dot{q}}\,{}^i\dot{\mathbf{q}}_{corr} - {}^i\mathbf{J}^T\boldsymbol{\lambda} = \mathbf{0} \quad \Rightarrow \quad {}^i\dot{\mathbf{q}}_{corr} = \mathbf{W}_{\dot{q}}^{-1}\,{}^i\mathbf{J}_f{}^T\boldsymbol{\lambda} \tag{14}$$

$$\frac{\partial L({}^i\dot{\mathbf{q}}_{corr}, \boldsymbol{\lambda})}{\partial \boldsymbol{\lambda}} = {}^i\mathbf{J}\,({}^i\dot{\mathbf{q}}_f + {}^i\dot{\mathbf{q}}_{corr}) - \mathbf{V} = \mathbf{0} \quad \Rightarrow \quad {}^i\mathbf{J}\,{}^i\dot{\mathbf{q}}_f + {}^i\mathbf{J}\,{}^i\dot{\mathbf{q}}_{corr} = \mathbf{V}. \tag{15}$$

Then $\boldsymbol{\lambda} = ({}^i\mathbf{J}\,\mathbf{W}_{\dot{q}}^{-1}\,{}^i\mathbf{J}_f{}^T)^{-1}(\mathbf{V} - {}^i\mathbf{J}\,{}^i\dot{\mathbf{q}}_f)$, and as ${}^i\mathbf{J}_f{}^T({}^i\mathbf{J}\,{}^i\mathbf{J}_f{}^T)^{-1} = {}^i\mathbf{J}_f{}^T({}^i\mathbf{J}_f\,{}^i\mathbf{J}_f{}^T)^{-1}$

$${}^i\dot{\mathbf{q}}_{corr} = {}^i\mathbf{J}^{\#}_{fw}(\mathbf{V} - {}^i\mathbf{J}\,{}^i\dot{\mathbf{q}}_f) = {}^i\mathbf{J}^{\#}_{fw}\,{}^i\mathbf{J}({}^i\dot{\mathbf{q}} - {}^i\dot{\mathbf{q}}_f) \tag{16}$$

As indicated by equations (13) and (16), the failure recovery methodology of Section 2.1 results in minimum 2-norm solution for the correctional velocity vector. To have physically consistent twist $(\mathbf{W}_V^{1/2}\mathbf{V}) \cdot (\mathbf{W}_V^{1/2}\mathbf{V}) = \mathbf{V} \cdot (\mathbf{W}_V\mathbf{V})$, using the reformulated constraint equation $\mathbf{W}_V^{1/2}\mathbf{V} = \mathbf{W}_V^{1/2}\;{}^i\mathbf{J}\;\mathbf{W}_{\dot{q}}^{1/2}\;\mathbf{W}_{\dot{q}}^{-1/2}\dot{\mathbf{q}}_{tot}$, the weighted GI will be ${}^i\mathbf{J}^{\#}_{fw} = \mathbf{W}_{\dot{q}}^{-1/2}(\mathbf{W}_V^{1/2}\;{}^i\mathbf{J}_f{}^T\;\mathbf{W}_{\dot{q}}^{-1/2})^{\#}\,\mathbf{W}_V^{1/2}$, which will result in ${}^i\dot{\mathbf{q}}_{corr}$ of equations (13) and (16) when ${}^i\mathbf{J}_f$ has full row-rank, and ${}^i\mathbf{J}^{\#}_{fw} = ({}^i\mathbf{J}_f{}^T\;\mathbf{W}_V\;{}^i\mathbf{J}_f)^{-1}\,{}^i\mathbf{J}_f{}^T\;\mathbf{W}_V$ when ${}^i\mathbf{J}_f$ has full column-rank.

2.3 *Minimum Norm for Overall Velocity Vector*

To minimize the actuation energy after failure while providing the required platform twist, the objective function will be the square of the Euclidean norm of the weighted overall velocity vector, ${}^i\dot{\mathbf{q}}_{tot} \cdot (\mathbf{W}_{\dot{q}}\,{}^i\dot{\mathbf{q}}_{tot})$. The linear constraint equation in terms of ${}^i\dot{\mathbf{q}}_{tot}$ is $\mathbf{V} - {}^i\mathbf{J}\,{}^i\dot{\mathbf{q}}_{tot} = \mathbf{0}$. The Lagrange function L is

$$L({}^i\dot{\mathbf{q}}_{tot}, \boldsymbol{\lambda}) = \frac{1}{2}{}^i\dot{\mathbf{q}}_{tot} \cdot (\mathbf{W}_{\dot{q}}\,{}^i\dot{\mathbf{q}}_{tot}) - \boldsymbol{\lambda}^T(\mathbf{V} - {}^i\mathbf{J}\,{}^i\dot{\mathbf{q}}_{tot}) \tag{17}$$

and when ${}^i\dot{\mathbf{q}}_{tot} \cdot (\mathbf{W}_{\dot{q}}\,{}^i\dot{\mathbf{q}}_{tot})$ is a minimum $\nabla L({}^i\dot{\mathbf{q}}_{tot}, \boldsymbol{\lambda}) = \mathbf{0}$.

When ${}^i\dot{q}_{ch} = 0$, to have zero overall (and correctional) velocity for the failed (jammed) joints, the constraint equation is $\mathbf{V} - {}^i\mathbf{J}_r\, {}^i\dot{\mathbf{q}}_{tot\,r} = \mathbf{0}$, where ${}^i\mathbf{J}_r$ and ${}^i\dot{\mathbf{q}}_{tot\,r}$ are respectively obtained by removing the columns and entries of ${}^i\mathbf{J}$ and ${}^i\dot{\mathbf{q}}_{tot}$ corresponding to the failed joints.

$$\frac{\partial L({}^i\dot{\mathbf{q}}_{tot\,r}, \boldsymbol{\lambda})}{\partial\, {}^i\dot{\mathbf{q}}_{tot\,r}} = \mathbf{W}_{\dot{q}}\, {}^i\dot{\mathbf{q}}_{tot\,r} - {}^i\mathbf{J}_r{}^T \boldsymbol{\lambda} = \mathbf{0} \quad \Rightarrow \quad {}^i\dot{\mathbf{q}}_{tot\,r} = \mathbf{W}_{\dot{q}}^{-1}\, {}^i\mathbf{J}_r{}^T \boldsymbol{\lambda} \tag{18}$$

$$\frac{\partial L({}^i\dot{\mathbf{q}}_{tot\,r}, \boldsymbol{\lambda})}{\partial \boldsymbol{\lambda}} = {}^i\mathbf{J}_r\, {}^i\dot{\mathbf{q}}_{tot\,r} - \mathbf{V} = \mathbf{0} \quad \Rightarrow \quad {}^i\mathbf{J}_r\, {}^i\dot{q}_{tot\,r} = \mathbf{V} \tag{19}$$

then $\boldsymbol{\lambda} = ({}^i\mathbf{J}_r\, \mathbf{W}_{\dot{q}}^{-1}\, {}^i\mathbf{J}_r{}^T)^{-1}\mathbf{V}$ and

$$^i\dot{\mathbf{q}}_{tot\,r} = \mathbf{W}_{\dot{q}}^{-1}\, {}^i\mathbf{J}_r{}^T\, ({}^i\mathbf{J}_r\, \mathbf{W}_{\dot{q}}^{-1}\, {}^i\mathbf{J}_r{}^T)^{-1}\mathbf{V} = {}^i\mathbf{J}_{rw}^{\#}\mathbf{V} = {}^i\mathbf{J}_{rw}^{\#}\, {}^i\mathbf{J}\, {}^i\dot{\mathbf{q}} \tag{20}$$

and the correctional velocity from the healthy joints will be ${}^i\dot{\mathbf{q}}_{corr\,r} = {}^i\dot{\mathbf{q}}_{tot\,r} - {}^i\dot{\mathbf{q}}_{f\,r}$, where the reduced joint velocity vector after failure ${}^i\dot{\mathbf{q}}_{f\,r}$ is obtained by removing the zero entries of ${}^i\dot{\mathbf{q}}_f$ corresponding to failed joints.

When ${}^i\dot{q}_{ch} \neq 0$ the constraint equation is $\mathbf{V} - {}^i\mathbf{J}\, {}^i\dot{\mathbf{q}}_{tot} = \mathbf{0}$. To calculate the minimum norm overall joint velocity vector, first the portion of the platform twist provided by the failed joints with ${}^i\dot{q}_{ch} \neq 0$ should be removed from the required platform twist. When gc out of g failed joints have non-zero velocity

$$\mathbf{V}^* = \mathbf{V} - \sum_{gc} \mathbf{J}_k\, {}^i\dot{q}_{ck} \tag{21}$$

The Lagrange function and its partial derivatives in terms of the overall velocity of healthy joints are

$$L({}^i\dot{\mathbf{q}}_{tot\,r}, \boldsymbol{\lambda}) = \frac{1}{2}{}^i\dot{\mathbf{q}}_{tot\,r} \cdot (\mathbf{W}_{\dot{q}}\, {}^i\dot{\mathbf{q}}_{tot\,r}) - \boldsymbol{\lambda}^T(\mathbf{V}^* - {}^i\mathbf{J}_r\, {}^i\dot{\mathbf{q}}_{tot\,r}) \tag{22}$$

$$\frac{\partial L({}^i\dot{\mathbf{q}}_{tot\,r}, \boldsymbol{\lambda})}{\partial\, {}^i\dot{\mathbf{q}}_{tot\,r}} = \mathbf{W}_{\dot{q}}\, {}^i\dot{\mathbf{q}}_{tot\,r} - {}^i\mathbf{J}_r{}^T \boldsymbol{\lambda} = \mathbf{0} \quad \Rightarrow \quad {}^i\dot{\mathbf{q}}_{tot\,r} = \mathbf{W}_{\dot{q}}^{-1}\, {}^i\mathbf{J}_r{}^T \boldsymbol{\lambda} \tag{23}$$

$$\frac{\partial L({}^i\dot{\mathbf{q}}_{tot\,r}, \boldsymbol{\lambda})}{\partial \boldsymbol{\lambda}} = {}^i\mathbf{J}_r\, {}^i\dot{\mathbf{q}}_{tot\,r} - \mathbf{V}^* = \mathbf{0} \quad \Rightarrow \quad {}^i\mathbf{J}_r\, {}^i\dot{\mathbf{q}}_{tot\,r} = \mathbf{V}^*. \tag{24}$$

Then $\boldsymbol{\lambda} = ({}^i\mathbf{J}_r\, \mathbf{W}_{\dot{q}}^{-1}\, {}^i\mathbf{J}_r{}^T)^{-1}\mathbf{V}^*$ and the overall velocity of healthy joints is

$$^i\dot{\mathbf{q}}_{tot\,r} = {}^i\mathbf{J}_{rw}^{\#}\mathbf{V}^* = {}^i\mathbf{J}_{rw}^{\#}\left({}^i\mathbf{J}\, {}^i\dot{\mathbf{q}} - \sum_{gc} \mathbf{J}_k\, {}^i\dot{\mathbf{q}}_{ck}\right) \tag{25}$$

${}^i\dot{\mathbf{q}}_{tot}$ is obtained by incorporating the velocities of failed joints (zero and non-zero velocities). As indicated by equations (20) and (25), the recovery methodology of Section 2.1 results in the minimum 2-norm solution for the overall velocity vector.

3 Case Study

Considering the manipulator of Figure 1(a), to form kinematically redundant legs, two active prismatic joints, with axes in the Y and X directions, are added to each leg between the base and the first revolute joint; the $\underset{\rightarrow}{\mathrm{P}}\,\underset{\rightarrow}{\mathrm{P}}\,\mathrm{R}\,\underset{\rightarrow}{\mathrm{P}}\,\mathrm{R}$ layout of Figure 1(b). The platform twist $\mathbf{V}$ is related to the joint velocities of leg i, ${}^i\dot{\mathbf{q}}$, as

$$\begin{bmatrix} v_x \\ v_y \\ \dot{\varphi} \end{bmatrix} = \begin{bmatrix} 0 & 1 & l_i s\alpha_i + r_{Bi/P}\, s(\alpha_i+\beta_i) & -c\alpha_i & r_{Bi/P}\, s(\alpha_i+\beta_i) \\ 1 & 0 & -l_i c\alpha_i - r_{Bi/P}\, c(\alpha_i+\beta_i) & -s\alpha_i & -r_{Bi/P}\, c(\alpha_i+\beta_i) \\ 0 & 0 & 1 & 0 & 1 \end{bmatrix} \begin{bmatrix} \dot{d}y_i \\ \dot{d}x_i \\ \dot{\alpha}_i \\ \dot{l}_i \\ \dot{\beta}_i \end{bmatrix}$$

$$= {}^i\mathbf{J}\, {}^i\dot{\mathbf{q}} \tag{26}$$

where $c\alpha_i = \cos\alpha_i$, $s\alpha_i = \sin\alpha_i$, $c(\alpha_i+\beta_i) = \cos(\alpha_i+\beta_i)$ and so on.

The coordinates of the base attachment points of leg i, $A_i, i = 1, \ldots, 3$, are $(-2, -1.5)$, $(2, -1.5)$ and $(0, 1.5)$, respectively. The position of connection points of leg i, B_i, on the platform is set at a constant radius of $r_{Bi/P} = 0.25$ meters with angular coordinates, θ_i, of $-150°, -30°$ and $90°$. When the platform pose is $\mathbf{p} = [0\ \ 0]^T$ meter and $\varphi = -30°$ leg 3 is in the Y direction with the joint displacements of ${}^3\mathbf{q} = [0\ \ 0.125\ \ 90\ \ 1.283\ \ -30]^T$. Then, the leg Jacobian matrix is

$${}^3\mathbf{J} = \begin{bmatrix} 0 & 1.0 & 1.5 & 0 & 0.217 \\ 1.0 & 0 & -0.125 & -1.0 & -0.250 \\ 0 & 0 & 1.0 & 0 & 1.0 \end{bmatrix} \tag{27}$$

For the twist of $\mathbf{V} = [1\ \ 0.5\ \ 0]^T$, the minimum norm vector of joint velocity is

$${}^3\dot{\mathbf{q}} = {}^3\mathbf{J}^{\#}\mathbf{V} = [0.250\ \ 0.548\ \ 0.352\ \ -0.250\ \ -0.352]^T \tag{28}$$

with a magnitude of $\|{}^3\dot{\mathbf{q}}\|_2 = 0.821$. When the second active joint ($h = 2$) of leg 3 is jammed there remain four joints (two active prismatic and two passive revolute joints) for a 3 DOF task. Then ${}^3\dot{\mathbf{q}}_f = [0.250\ \ 0\ \ 0.352\ \ -0.250\ \ -0.352]^T$ with $\|{}^3\dot{\mathbf{q}}_f\|_2 = 0.611$, and the platform twist is $\mathbf{V}_f = {}^3\mathbf{J}\, {}^3\dot{\mathbf{q}}_f = [0.452\ \ 0.5\ \ 0.0]^T$.

The failure of this active joint could be fully recovered as $(\mathbf{I} - {}^3\mathbf{J}_f\, {}^3\mathbf{J}_f^{\#})\,\mathbf{V} = \mathbf{0}$. Using an identity weighting matrix, the correctional velocity is

$${}^3\dot{\mathbf{q}}_{corr} = {}^3\mathbf{J}_f^{\#}\, {}^3\mathbf{J}_2\, \dot{d}x_3 = [0\ \ 0\ \ 0.427\ \ 0\ \ -0.427]^T \tag{29}$$

with $\|{}^3\dot{\mathbf{q}}_{corr}\|_2 = 0.604$. In this configuration of leg 3, because the first and third prismatic joints axes are collinear (and the first and second prismatic joints axes are always perpendicular), the motion of the failed second prismatic joint is fully recovered by the two passive revolute joints. Then, the overall joint velocities are

$${}^3\dot{\mathbf{q}}_{tot} = {}^3\dot{\mathbf{q}}_f + {}^3\dot{\mathbf{q}}_{corr} = [0.250\ \ 0\ \ 0.779\ \ -0.250\ \ -0.779]^T \tag{30}$$

which is identical to the overall joint velocity vector calculated with the Lagrange multiplier method ${}^3\dot{\mathbf{q}}_{tot\,r} = {}^3\mathbf{J}_r^{\#}\mathbf{V}$ with $\|{}^3\dot{\mathbf{q}}_{tot}\|_2 = 1.157$.

At this pose and required twist, if the first two joints were jammed the leg would reduce to the one in Figure 1(a) and $\mathbf{V}_f = [0.452\ \ 0.250\ \ 0.0]^T$. The unique solution for recovery would be ${}^3\dot{\mathbf{q}}_{corr} = [0\ \ 0\ \ 0.427\ \ -0.250\ \ -0.427]^T$ with $\|{}^3\dot{\mathbf{q}}_{corr}\|_2 = 0.654$. Then, ${}^3\dot{\mathbf{q}}_{tot} = [0\ \ 0\ \ 0.779\ \ -0.500\ \ -0.779]^T$ with $\|{}^3\dot{\mathbf{q}}_{tot}\|_2 = 1.210$, i.e., the third prismatic joint and the two revolute joints would respectively recover the motion of the failed first and second prismatic joints.

4 Conclusion

In this article, methodologies for recovering the lost motion of manipulators due to joint/actuator failures were presented. When a joint (active or passive) is failed the required platform twist should be provided by adjusting the motion of the remaining active joints of the manipulator, as well as the active and passive joints of the leg with failed joint(s). The method discussed here examined the motion of a leg with failed joint(s) utilizing the Jacobian matrix of the leg. A similar process could be adapted to investigate the motion of active joints using the Jacobian matrix of manipulator. It was shown that the procedure based on the projection of the lost joint motion onto the orthogonal complement of the null space of the reduced Jacobian matrix results in minimum Euclidean norm for the correctional velocity vector and the overall velocity vector.

References

1. Chen, Y., McInroy, J., Yi, Y.: Optimal, fault-tolerant mappings to achieve secondary goals without compromising primary performance. IEEE Trans. Robot. Autom. **19**(4), 680–691 (2003)
2. Doty, K., Melchiorri, C., Bonivento, C.: A theory of generalized inverses applied to robotics. Int. J. Robot. Res. **12**(1), 1–19 (1995)
3. Notash, L.: Joint sensor fault detection for fault tolerant parallel manipulators. J. Robot. Syst. **17**(3), 149–157 (2000)
4. Notash, L.: A methodology for actuator failure recovery in parallel manipulators. Mech. Mach. Theory **46**(4), 454–465 (2011)
5. Notash, L.: Motion recovery after joint failure in parallel manipulators. Trans. Can. Soc. Mech. Eng. **35**(4), 559–571 (2011)
6. Notash, L., Huang, L.: On the design of fault tolerant parallel manipulators. Mech. Mach. Theory **38**(1), 85–101 (2003)
7. Roberts, R., Yu, H., Maciejewski, A.: Fundamental limitations on designing optimally fault tolerant redundant manipulators. IEEE Trans. Robot. **24**(5), 1224–1237 (2008)

A Loop-Based Approach for Rigid Subchain Identification in General Mechanisms

Shuxian Xia, Huafeng Ding and Andres Kecskemethy

Abstract The determination of rigid or overconstrained subsystems is an important task in the creative design of robotic mechanisms and in the processing of CAD-generated models. While for planar mechanisms with planar graphs a number of methods have been proposed, the case of general spatial mechanisms is still an open topic. In this paper, a novel method for identifying rigid subsystems is presented. The method uses the independent loops as building blocks of a graph, called kinematical network, which describes the overall transmission behavior. The detection of rigid subsystems can then be realized by finding the minimal cutsets in the solution flow of the kinematical network. The method is independent of the subspace in which the bodies are moving, i.e., it is possible to mix planar, spherical and spatial systems. Moreover, it is fast, as only the implicitly coupled loops need to be processed, which comprise much less elements than the number of bodies.

Key words: Rigid subsystems, degrees of freedom, degenerate kinematic chains

1 Introduction

The topic of rigid subsystem detection has attracted scientists in the robotics and mechanisms community for many decades. The problem is to detect whether the overall Grübler sum of degrees of freedom (DoF) for the mechanism is composed by substructures with positive DoF (i.e. movable subsystems) and negative DoF (i.e. overconstrained subsystems). Such cases arise for example in the automatic generation of candidate mechanisms in creative design of robotic devices [3], or when the constraint graphs for the mechanisms are automatically generated by CAD systems [10]. An example is shown in Fig. 1. Here, a loop with an internal DoF $= 1$ is attached to a subsystem of internal DoF $= -1$. Thus, the overall Grübler count suggests that the mechanism has DoF $= 0$, which is not correct, as the system

Shuxian Xia · Huafeng Ding · Andres Kecskemethy
Chair for Mechanics and Robotics, University of Duisburg-Essen, Duisburg, Germany,
e-mail: shuxian.xia@uni-due.de, dhf@ysu.edu.cn, andres.kecskemethy@uni-due.de

J. Lenarčič, M. Husty (eds.), *Latest Advances in Robot Kinematics*,
DOI 10.1007/978-94-007-4620-6_3,

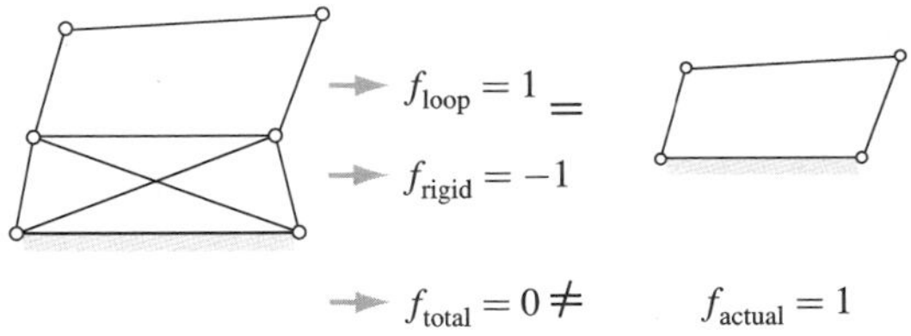

Fig. 1 Example of a rigid subchain.

actually has a DoF of 1 and the lower subsystem can be exchanged by a single rigid body.

Starting with the rigidity test by Laman's theorem [9], a number of methods for planar mechanisms [2, 5, 6, 12, 15], as well as simple spatial systems [10, 11] have been proposed. However, as shown in [14], these methods are restricted to mechanisms with planar graphs, and they also do not cover the case of general spatial mechanisms with arbitrary joints. Ding et al. [4] presented an algorithm for detecting rigid subchains for planar mechanisms with also non-planar graphs. They use the "smallest" independent loop as a starting point of a set of stepwise extended clusters for which rigidity is tested. As a rigid subsystem does not necessarily contain the smallest loop, the algorithm only works when using each independent loop as a starting point, making the method very involving. Also recently, Sunkari [13] presented a polynomial-time algorithm for degeneracy testing using cut sets which may be applicable both to planar and spatial mechanisms. However, in spite of a high complexity of $O(|V|^4)$, where $|V|$ is the number of bodies in the graph, the algorithm is only able to detect the rigid subchain with the highest number of overconstraint conditions, and cannot compute the mobility type.

The idea of the present method is to base the rigidity testing on the coupling of the independent loops of the mechanism. By looking at the global transmission properties, search space can be restricted to subsystems where rigidity can occur, and general mechanisms containing any mixture of spatial, planar, or other special loops can be processed. In the following, first, the basic concept of representing general mechanisms as networks of connected loops is recollected, and subsequently the method for finding rigid subsystems based on the kinematical network is discussed.

2 Description of Mechanisms as Networks of Connected Loops

The concept of mechanism description using a network of linearly connected loops was introduced in [7], and a fully automatic implementation in the symbolic formula manipulation software Mathematica for general planar, spherical, translational and spatial cases is described in [8]. The following planar examples are used for illustration of the method, but the concepts apply one-to-one also to spatial cases. In this approach, as a first step, a minimal cycle basis comprising a set of smallest independent loops L_i of the mechanism is determined, where 'length' is measured in terms of the number of involved joint variables. Each loop has a local DoF f_{Li}, which is the number of loop joint coordinates minus the dimension of the subgroup

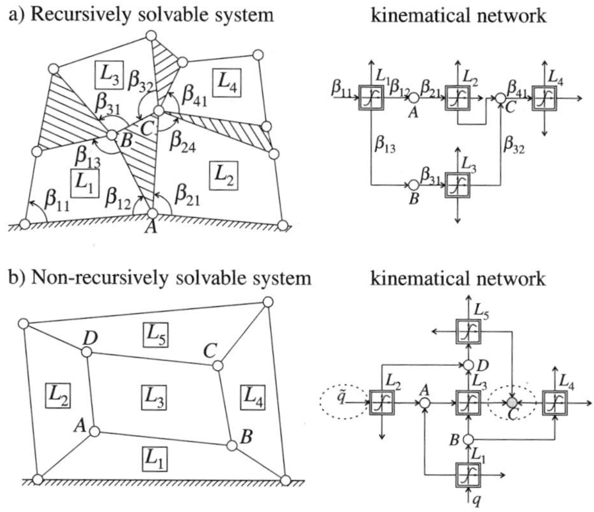

Fig. 2 Mapping of mechanisms to kinematical networks.

of Euclidean motion in which the bodies of the loop locally move (spatial, planar, spherical, translational, etc.). Between the loops there will be couplings at exactly those joints in which the number of incident loops is equal to or larger than the number of incident bodies. They correspond to a balance of inner joint coordinates of the incident loops and some constants. For example, in joint A of Fig. 2a, the sum of joint coordinates β_{12} and β_{21} plus a constant is equal to 360°, yielding a linear coupling between the loops L_1 and L_2. The thus connected loops form the so-called kinematical network in which the loops represent local nonlinear transmission elements and the couplings between the loops represent the global interrelationships. From the kinematical network, the DoF can be obtained as the sum of local DoF of all loops minus the sum of linear couplings. In Fig. 2a, there are four four-bar loops with local DoF $= 1$ each, and three linear couplings. Thus the overall DoF is 1. In Fig. 2b, there are five four-bar loops with local DoF $= 1$ each, and four linear couplings, thus the global DoF is again 1. This is identical to the usual Grübler count. Note that here, however, one can easily combine planar, spatial, etc., loops.

An advantage of the kinematical network is that one can easily recognize recursively solvable substructures by the so-called *sink method* [7]: here, one searches iteratively for elements in the kinematical network for which the number of edges is less than or equal to the local degree of freedom of the element. After finding such an element, all edges are oriented into the element, the element together with all ingoing edges is removed, and the procedure is re-applied to the rest of the system. If the procedure covers all loops, one obtains a recursive solution flow (Fig. 2a). If no such elements can be found, the network must be solved iteratively (Fig. 2b). In this case, one chooses additional inputs and iterates the equations resulting at linear couplings in which all edges are oriented towards the linear coupling (in Fig. 2b, $\tilde{q}$ is the additional input and C is the iteration coupling).

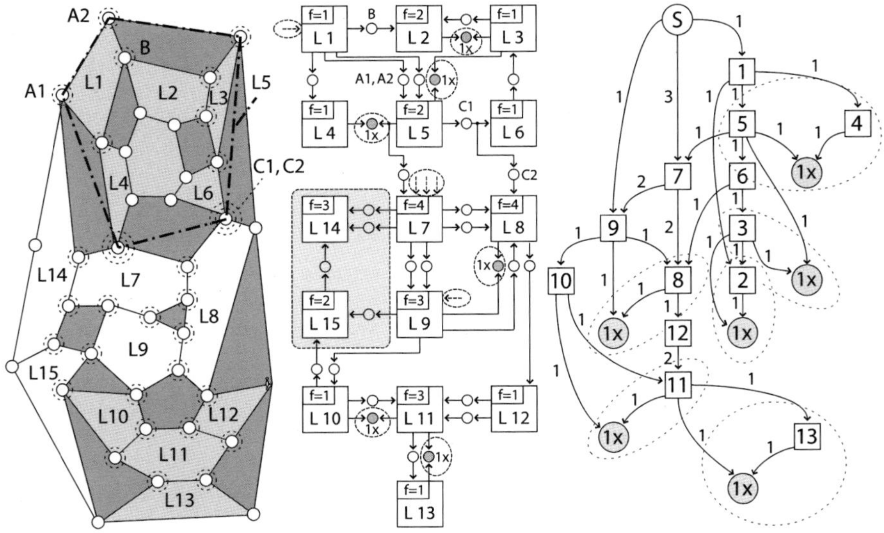

a) Linkage b) Kinematical network c) Loop Connection Graph

Fig. 3 Different levels of abstraction for a given linkage.

3 Loop-Based Procedure for Rigid Subsystem Detection

In the following, the procedure for the detection of rigid subsystems is illustrated with the example of Fig. 3a. The example mechanism is planar but all steps are equally valid for general spatial or mixed planar/spatial mechanisms.

Initialization step

This first step consists in "initializing" the algorithm by mapping the concrete linkage to a corresponding abstract 'loop connection graph' from which the 'flow' of degrees of freedom is more clearly visible.

As a first substep, the minimal cycle basis and the corresponding kinematical network is established (Fig. 3b). For better visibility, all coupling joints are marked in Fig. 3a by a dashed circle, and four of the 31 coupling conditions (A1, A2, B, C1, C2) have been tagged in the corresponding kinematical network Fig. 3b. Moreover, the local degree of freedom of each loop is displayed in the upper left box of the loop. Counting all local loop DoF together and subtracting the number of loop coupling conditions, one obtains the global DoF $= 30 - 31 = -1$, which is identical to the classical Grübler count (29 planar bodies, 44 revolute joints).

As a second substep, all recursively solvable subsystems are removed from the kinematical network, as they are by definition movable and thus cannot contribute to the rigidity of any other subsystem. In the present example, these are loops L_{14} and L_{15}, which can be selected as sinks in this order. In general, it can be expected that this step will remove a substantial number of loops from the system to be further analyzed.

The third substep consists in orienting the edges of the kinematical network. Typically, as in the example Fig. 2b, the edge orientation will require additional inputs (not true independent inputs) leading to implicit constraints at some summing junctions. In Fig. 3b, the additional inputs have been encircled by dashed ellipses, while the implicit summing junctions are encircled by dashed ellipses and displayed as gray-filled circles with the number of implicit constraints marked as λx (here everywhere with $\lambda = 1$) within the circle. In the example, the start was chosen at loop L_1 and then the network was traversed such that each loop has sufficient inputs either from predecessor loops or from further additional inputs. Note that a different input selection will lead to a different edge orientation, but the global DoF and that of any substructure will remain the same, as these do not depend on the order of equation solving. One verifies that for this example one has six sink summing junctions and five additional inputs, leading again to the global DoF $= -1$.

As a final substep, the corresponding oriented acyclic '*Loop Connection Graph*' describing the level of dependency of the individual loops and the sink summing-junctions is established. In this graph, only loops and sink summing junctions are retained as nodes. The corresponding dependencies are described by weighted edges, the weight w_i of each edge describing the number of joint variables involved in the connection. Additionally, a global 'source' node 'S' is introduced, from which the external inputs for the loops are extracted. Starting from the source node, the 'depth' of each branch leading to a node describes the level of recursive dependency of the node variables with respect to the inputs. Fig. 3c) shows the loop connection graph for the example at hand. For better visibility, loop nodes are displayed as boxes, while sink summing nodes are displayed as gray-filled circles.

The rest of the algorithm consists in evaluating directed 'cuts' through this acyclic graph such that the source is on one side, the sink is on the other, and all edges through which the cut passes are directed from the 'upstream' to the 'downstream' side. Such cuts can be easily determined using state-of-art graph-theoretic methods [1], which are not further discussed here due to lack of space. The sum of weights of the cut edges is termed the *weight* of the cut, while the sum of implicit equations in the downstream side of the cut is termed the *absorbing degree* of the cut. The DoF of the cut is the equal to its weight minus its absorbing degree. This DoF is equal to the DoF of the nodes on the downstream side of the cut. Whenever the DoF of the cut is less or equal to zero, the downstream subsystem will represent a rigid or an overconstrained subsystem. The cut with the minimal DoF will determine the most overconstrained substructure.

Contraction step

In order to simplify the minimal cut search, the loop connection graph can be further simplified by the repeated application of the following two rules (see Fig. 4):

I) *Sink node contraction.* For a sink node for which the sum of input weights is larger than its absorbing degree, any minimal cut will not be altered if the sink node 'absorbs' all its direct predecessor nodes at the layer of greatest depth, i.e. all loops and absorbing degrees of one direct predecessor, or those of several predecessor nodes at the same level. In such a case, the absorbed loops are

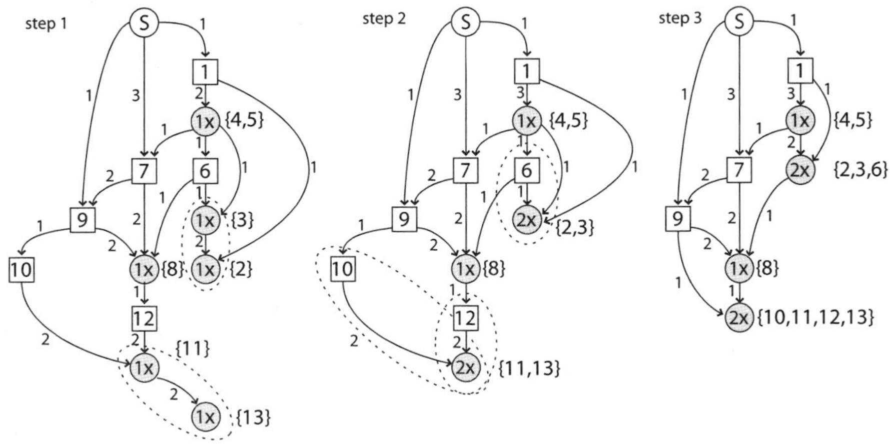

Fig. 4 Loop connection graph after three contraction steps.

tagged as '$\{i,k,\ldots\}$' to the right of the node. The resulting weight of the contracted node is equal to the union of input weights of the absorbed nodes and those of the sink node, minus the common edges between the absorbed nodes and the sink node, where parallel edges can be summed together. Note that after application of this contraction step, implicit nodes need not to be sinks anymore. Application of this rule is shown by the several dashed ellipses in Fig. 3 leading to 'step 1' in Fig. 4.

II) *Binary branch contraction.* For a binary branch connecting a non-binary start node V_S with a non-binary end node V_E, one can determine the 'weakest link' as that node closest to the start node for which a cut at its input produces the minimum value of weight at the cut minus the sum of implicit equations in the upstream part of the branch, including the absorbing degree of the end node. The branch is then replaced by a single edge with the weight of the cut, where all loops and absorbing degrees in the upstream part of the cut are included in the end node, while all loops and absorbing degrees of the downstream part are included in the start node. Application of this rule is seen in the transition from 'step 2' to 'step 3' in Fig. 4 for the loops L_{10}, L_{12} and $\{L_{11}, L_{13}\}$.

Pruning and cutting step

In order to find the minimal cuts, one 'pruned' graph is established for each sink. The pruned graphs are obtained by removing all vertices that do not lie on the path from the source to the selected sink (Fig. 5). For each of the pruned graphs, all cuts X_i having non-positive DoF are selected. In Fig. 5, these are the cuts X_1, X_2, X_3, X_4 with DoFs $0, -1, -2, 0$, respectively. For each cut X_i, there is an associated set D_i with the loops (i.e. bodies and joints) contained in the downstream side of the cut. Such sets may include each other, and may have a varying degree of non-positive DoF. By sorting the sets according to their DoF, the rigid subchains can be determined as follows:

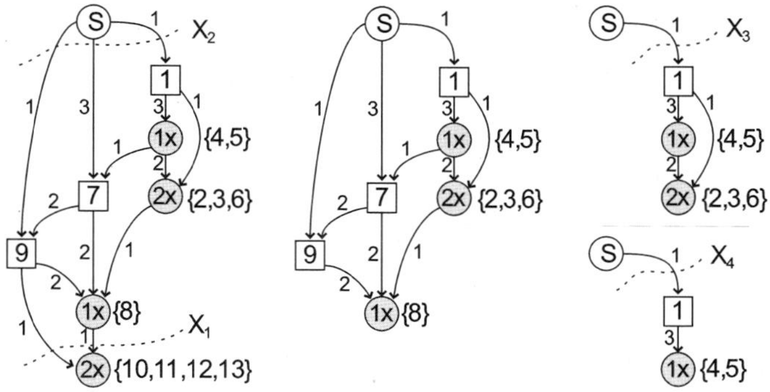

Fig. 5 Pruned graphs and minimal cuts X_i for each of the four sinks from Fig. 4.

1) Select the cut X_i^* with the most negative DoF. The corresponding downstream set D_i^* is an overconstrained or rigid subassembly. In the example of Fig. 4, the minimal cut is X_3 with corresponding loop set $D_3^* = \{L_1, \ldots, L_6\}$ and DoF $= -2$.
2) For all cuts X_k with $D_k \subset D_i^*$, the sets D_k inherit the DoF of D_i^*. Thus these sets can be ignored. In the present example, this is true for cut X_4 which inherits the DoF $= -2$ from X_3.
3) For all sets D_m which are supersets of D_i^*, the actual degree of freedom cannot be determined from the cut, as there may be further internal DoF count compensations. Thus these cuts must be removed from the list. In the present this applies to the cut X_2.
4) If there are any cuts left, continue with step (1). In the present example, this corresponds to cut X_1 with loop set $D_1 = \{L_{10}, \ldots, L_{13}\}$ and DoF $= 0$.

Eventually, no further cuts remain in the list. Then, it may be necessary to run again the algorithm with the rigid and overconstrained subsystems replaced by single rigid bodies. However, this is seldom necessary (as in the present example), so that usually the algorithm terminates here.

Note that while a negative DoF of a cut means that the subsystem D_i at the downstream side of the cut will be overconstrained, for the surrounding system, this subsystem will behave like a regular rigid body. This means that negative DoFs are not propagated through the linkage and that the remaining DoFs will be decreased only as if the overconstrained subgraph has DoF $= 0$. The replacement of an overconstrained subgraph by a rigid body in the loop connection graph can thus be easily realized by removing the subgraph altogether with all its outgoing edges, and increasing the number of implicit equations at the start node of each edge through the cut by the weight of this edge. In the graph of Fig. 5, the complete subgraph of the cut X_3 can be removed, and removal of the subgraph of cut X_1 increases the implicit equation count of loops L_9 and L_8 by one, respectively. Hence, the remaining subgraph consists of loops L_7, L_8 and L_9 with 4 global inputs and 3 implicit constraints. This yields the correct DoF of 1 for loops L_7, L_8 and L_9, and also of the initially removed recursively solvable loops L_{14} and L_{15}. Thus it can be seen that the rigidity detection algorithm also correctly determines the mobility of the remaining system.

4 Conclusions and Acknowledgments

By the proposed loop-based approach, the detection of rigid subsystems in general linkages is reduced to the determination of minimal cut sets in the corresponding loop connection graphs. This (1) reduces search space considerably, and (2) allows one to mix planar and spatial subsystems without any restrictions. Moreover, the method automatically provides the mobility of the remaining system. This may be used for automatic generation of the kinematical transmission equations at position, velocity and acceleration level [8]. The method has been fully implemented in Mathematica, and all examples for rigidity testing from the literature have been correctly processed. Further steps will be to implement efficient restart procedures after replacement of overconstrained subsystems by rigid bodies and to include known single loop special cases such as Bennett or Bricard configurations by pattern matching.

The funding of a research stay at the University of Duisburg-Essen of the second author as a Humboldt research fellow is greatly acknowledged. Moreover, special thanks are due to Andreas Scholz for his valuable help in the drawing of the graphs.

References

1. Abel, U., Bicker, R.: Determination of all minimal cut-sets between a vertex pair in an undirected graph. IEEE Trans. Reliab. **R-31**(2), 167–171 (1982)
2. Agrawal, V., Rao, J.: Fractionated freedom kinematic chains and mechanisms. Mech. Mach. Theory **22**(2), 125–130 (1987)
3. Ding, H., Kecskemethy, A., Huang, Z.: Synthesis of the whole family of planar 1-DOF kinematic chains and creation of their atlas database. Mech. Mach. Theory **47**(1), 1–15 (2012)
4. Ding, H., Zhao, J., Huang, Z.: Unified structural synthesis of planar simple and multiple joint kinematic chains. Mech. Mach. Theory **45**(4), 555–568 (2008)
5. Hwang, W.M., Hwang, Y.W.: An algorithm for the detection of degenerate kinematic chains. Math. Comput. Modelling **15**(11), 9–15 (1991)
6. Jacobs, D., Hendrickson, B.: An algorithm for two dimensional rigidity percolation: The pebble game. J. Comput. Phys. **137**, 346–365 (1997)
7. Kecskeméthy, A.: On closed form solutions of multiple-loop mechanisms. In: Angeles, J., Hommel, G., Kovacs, P. (eds.) Computational Kinematics, pp. 263–274 (1993)
8. Kecskeméthy, A., Krupp, T., Hiller, M.: Symbolic processing of multiloop mechanism dynamics using closed-form kinematics solutions. Multibody Syst. Dyn. **1**, 23–45 (1997)
9. Laman, G.: On graphs and rigidity of plane skeletal structures. J. Engrg. Math. **4**(4), 331–340 (1970)
10. Lee, A.: Geometric constraint systems with applications in CAD and biology. PhD thesis (2008)
11. Lee, A., Streinu, I., Theran, L.: Analyzing rigidity with pebble games. In: Proc. 24th Ann. Symp. on Computational Geometry, SCG'08, pp. 226–227. ACM, New York (2008)
12. Moukarzel, C.: An efficient algorithm for testing the generic rigidity of graphs in the plane. J. Phys., A, Math. Gen. **29**(24), 8079–8098 (1996)
13. Sunkari, R.P.: Structural synthesis and analysis of planar and spatial mechanisms satisfying Gruebler's degrees of freedom equation. PhD thesis (2006)
14. Sunkari, R.P., Schmidt, L.C.: Critical review of existing degeneracy testing and mobility type identification algorithms for kinematic chains. ASME Conf. Proc. **2005**(47446), 255–263 (2005)
15. Tay, T.S.: Rigidity of multi-graphs. I. Linking rigid bodies in n-space. J. Comb. Theory, Ser. B **36**(1), 95–112 (1984)

Self-Motions of Planar Projective Stewart Gough Platforms

G. Nawratil

Abstract In this paper, we study self-motions of non-architecturally singular parallel manipulators of Stewart Gough type, where the planar platform and the planar base are related by a projectivity. By using mainly geometric arguments, we show that these manipulators have either so-called elliptic self-motions or pure translational self-motions. In the latter case, the projectivity has to be an affinity $\mathbf{a} + \mathbf{A}\mathbf{x}$, where the singular values s_1 and s_2 of the 2×2 transformation matrix $\mathbf{A}$ with $0 < s_1 \leq s_2$ fulfill the condition $s_1 \leq 1 \leq s_2$.

Key words: Self-motion, Stewart Gough platform, Borel Bricard problem

1 Introduction

The geometry of a planar Stewart Gough (SG) platform is given by the six base anchor points M_i with coordinates $\mathbf{M}_i := (A_i, B_i)^T$ with respect to the xy-plane π_{M} of the fixed system Σ_0 and by the six platform anchor points m_i with coordinates $\mathbf{m}_i := (a_i, b_i)^T$ with respect to the xy-plane π_{m} of the moving system Σ. If the geometry of the manipulator is given as well as the six leg lengths R_i, then the SG platform is in general rigid, but under particular conditions the manipulator can perform an n-parametric motion ($n > 0$), which is called self-motion. Note that such motions are also solutions to the famous Borel Bricard problem (cf. [1–3]).

It is well known that planar SG platforms which are singular in every possible configuration, possess self-motions in each pose. These so-called architecturally singular planar SG platforms were extensively studied in [4–7]. Therefore, we are only interested in self-motions of planar SG platforms, which are not architecturally singular.

G. Nawratil
Institute of Discrete Mathematics and Geometry, Vienna University of Technology, Vienna, Austria, e-mail: nawratil@geometrie.tuwien.ac.at

J. Lenarčič, M. Husty (eds.), *Latest Advances in Robot Kinematics*,
DOI 10.1007/978-94-007-4620-6_4,

In this paper, we discuss the case where the base anchor points M_i and the platform anchor points m_i are related by a non-singular projectivity κ. [1] For the remainder of this article we call such manipulators planar projective SG platforms. Note that a projectivity is the most general linear mapping between two projective extended planes, and that κ is uniquely determined by corresponding quadrangles.

It is well known (cf. Chasles [8]), that a planar projective SG platform is architecturally singular if and only if one set of anchor points is located on a conic section. Under consideration of this result the theorem given by Karger in Sec. 3 of [9] can be rewritten as follows:

Theorem 1. *A singular configuration of a planar projective SG platform, which is not architecturally singular, does not depend on the distribution of the anchor points in the platform and the base, but only on the mutual position of the planes π_{M} and π_{m} and on the correspondence between them. The configuration is singular iff either one of the legs can be replaced by a leg of zero length or two legs can be replaced by aligned legs.*

A non-singular projectivity which maps ideal points onto ideal points is a non-singular affinity. The subcase of planar parallel manipulators of SG type with affinely equivalent platform and base (= planar affine SG platforms) was studied by Karger in [9–11]. It should also be noted that according to Mielczarek et al. [12], one can attach a two-parametric set of additional legs to planar affine SG platforms without restricting the direct kinematics, whereas the correspondence between the anchor points is given by the affinity itself.

As we want to study planar projective SG platforms we have to consider the projective extension of the carrier planes of the platform and base anchor points, i.e.

$$(a_i, b_i) \mapsto (w_i : x_i : y_i) \quad \text{and} \quad (A_i, B_i) \mapsto (W_i : X_i : Y_i). \tag{1}$$

Note that ideal points are characterized by $w_i = 0$ and $W_i = 0$, respectively.

2 Basic Results

Lemma 1. *One can attach a two-parametric set of additional legs to planar projective SG platforms without changing the forward kinematics and singularity surface.*

Proof. For the proof we can use the homogenized version of the criterion given in Eq. (12) of [13] which corresponds with the criterion for the solvability of the inhomogeneous system of equations given in Eq. (30) of [12].

Assume a non-architecturally singular planar manipulator $\mathsf{m}_1, \ldots, \mathsf{M}_6$ is given. Then one can add a further leg (with anchor points m_7 and M_7) to the originally legs (without changing the direct kinematics and the singularity surface) if the following rank condition holds (see also Remark 1 of Röschel and Mick [6]):

[1] If κ is singular, one set of anchor points would collapse into a line or a point, which yields trivial cases of architecturally singular manipulators.

$$rk(\mathbf{Q}) = 6 \quad \text{with} \quad \mathbf{Q} = (\mathbf{q}_1, \ldots, \mathbf{q}_7)^T \quad \text{and}$$
$$\mathbf{q}_i = (w_iW_i, w_iX_i, w_iY_i, x_iW_i, x_iX_i, x_iY_i, y_iW_i, y_iX_i, y_iY_i)^T. \tag{2}$$

Now it can easily be checked by the use of MAPLE that $rk(\mathbf{Q}) = 6$ holds true for

$$(W_i, X_i, Y_i)^T := \mathbf{P}(w_i, x_i, y_i)^T \quad \text{for} \quad i = 1, \ldots, 7 \tag{3}$$

where **P** is the matrix of the projectivity (**P** is a regular 3×3 matrix). □

Remark 1. Due to Lemma 1 it is clear why a singular configuration of a planar projective SG platform does not depend on the distribution of the anchor points in π_{M} and π_{m} (cf. Thm. 1). ◇

Theorem 2. *Self-motions of planar projective SG platforms, which are not architecturally singular, can only be of the following type:*

1. *Spherical self-motion with rotation center* $\mathsf{m}\kappa = \mathsf{m}$,
2. *Schönflies self-motion, where the direction of the rotation axis is parallel to the planes* π_{M} *and* π_{m},
3. *Elliptic self-motion.*

Proof. We start by denoting the line of intersection of π_{M} and π_{m} by s in the projective extension of the Euclidean 3-space. As in any pose of a self-motion of a planar projective SG platform, the manipulator has to be in a singular configuration, we can apply Thm. 1. Therefore the manipulator is singular if and only if one of the following cases hold:

a. π_{M} and π_{m} coincide,
b. $\mathsf{S} = \mathsf{S}\kappa$ holds, where S is the intersection point of s and $\mathsf{s}\kappa$,
c. $\mathsf{s} = \mathsf{s}\kappa$.

It is well known that every projectivity of the projective extension of the Euclidean plane onto itself has at least one real fixed point $\mathsf{F} = \mathsf{F}\kappa$. Therefore, if one pose of the self-motion is singular due to item (a), this already implies item (1) if F is a finite point or item (2) if F is an ideal point. Clearly, this also holds for item (b) with respect to the fixed point $\mathsf{S} = \mathsf{S}\kappa$. For the study of item (c) we consider again only one singular configuration of the self-motion. As $\mathsf{s} = \mathsf{s}\kappa$ holds the projectivity from s onto itself can be (i) hyperbolic, (ii) parabolic or (iii) elliptic.

Item (i) immediately implies that the self-motion can only be a pure rotation about the finite axis s which is a special case of item (1) and (2), respectively. If s is the ideal line ($\Rightarrow \pi_{\mathsf{M}} \parallel \pi_{\mathsf{m}}$) then the self-motion is a pure translation, which is a special case of item (2).

For item (ii) we have one fixed point and we end up with item (2) and (1), respectively, depending on the circumstance if the fixed point is an ideal point or not.

Item (iii) corresponds to the case of Thm. 1, where two legs can be replaced by collinear legs, as we cannot attach a leg with zero length (over $\mathbb{R}$) without changing the direct kinematics and singularity surface. Therefore the following definition finishes the proof of Thm. 2.

Definition 1. A self-motion of a planar projective SG platform is called *elliptic*, if each pose of this motion is singular due to item (c,iii). □

Due to Thm. 2 we only have to investigate spherical self-motions with rotation center $\mathsf{m}\kappa = \mathsf{m}$ (cf. Sec. 3), Schönflies self-motions with the rotation axis parallel to π_{M} and π_{m} (cf. Sec. 4) and elliptic self-motions (cf. Sec. 5).

3 Spherical Self-Motions

If a planar projective SG platform has a spherical self-motion about $\mathsf{m}\kappa = \mathsf{m}$, then the spherical image of this manipulator with respect to the unit sphere S^2 centered in $\mathsf{m}\kappa = \mathsf{m}$ also has to have a self-motion. Therefore the problem reduces to the determination of non-degenerated[2] spherical 3-dof RPR manipulators with self-motions, where the three base and platform anchor points are located on great circles. The following result is proven in Appendix A of the corresponding technical report [14]:

Lemma 2. *A non-degenerated spherical 3-dof RPR manipulator, where the base anchor points* M_1°, M_2°, M_3° *and the platform anchor points* m_1°, m_2°, m_3° *are located on great circles, can only have a self-motion if two platform points* $\mathsf{m}_1^\circ = \mathsf{m}_3^\circ$ *coincide (after relabeling of anchor points and interchange of platform and base) and if the spherical lengths* R_i° *of the legs equal* $R_1^\circ = \overline{\mathsf{M}_1^\circ \mathsf{M}_2^\circ}$, $R_2^\circ = \overline{\mathsf{m}_1^\circ \mathsf{m}_2^\circ}$, $R_3^\circ = \overline{\mathsf{M}_3^\circ \mathsf{M}_2^\circ}$.

The self-motion of the manipulator given in Lemma 2 is a pure rotation about the axis $\mathsf{a} := [\mathsf{m}\kappa = \mathsf{m}, \mathsf{m}_1^\circ = \mathsf{m}_3^\circ = \mathsf{M}_2^\circ]$ (cf. Fig. 1a). Trivially, we can only add an additional leg (with anchor points m_4° and M_4°) to this manipulator without restricting the self-motion if $\mathsf{m}_4^\circ = \mathsf{m}_1^\circ$ or $\mathsf{M}_4^\circ = \mathsf{M}_2^\circ$ holds. This has the following consequence for the corresponding planar projective SG platform: κ has to map all platform anchor points $\notin \mathsf{a}$ on points of a. Therefore κ cannot be a bijection and we get the contradiction. This proves the following theorem:

Theorem 3. *Planar projective SG platforms, which are not architecturally singular, do not have spherical self-motions with rotation center* $\mathsf{m}\kappa = \mathsf{m}$.

4 Schönflies Self-Motions

The Schönflies motion group is a four-dimensional subgroup of the Euclidean motion group and consists of all translations combined with all rotations about a fixed direction d, which in our case is parallel to π_{M} and π_{m}. Moreover, it is well known (e.g. [15]) that platform points being on lines parallel to d have congruent trajectories in a Schönflies motion. Therefore we can translate every leg of the manipulator in direction d during a Schönflies self-motion without changing this motion. This property is important for the following argumentation.

[2] Neither all platform anchor points nor all base anchor points collapse into one point.

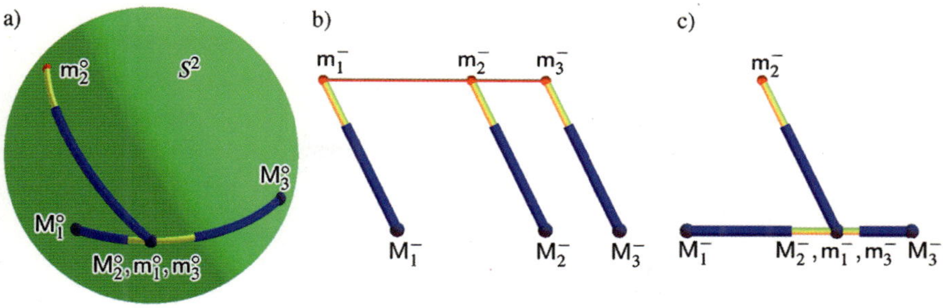

Fig. 1 (a) Non-degenerated spherical 3-dof RPR manipulator with a self-motion. (b,c) Non-degenerated planar 3-dof RPR manipulators with self-motions: Circular translation (b) and a pure rotation about the point $\mathsf{M}_2^- = \mathsf{m}_1^- = \mathsf{m}_3^-$ (c), which is the planar analogue of (a).

We choose the y-axis of the moving and the fixed frame parallel to the direction d. Moreover, we choose a line $\mathsf{g} \in \pi_{\mathsf{m}}$ which is orthogonal to d (cf. Fig. 2a). Under the projectivity κ the platform anchor points $\mathsf{m}_g \in \mathsf{g}$ are mapped to the corresponding base anchor points $\mathsf{M}_g := \mathsf{m}_g\kappa$ on the line $\mathsf{g}\kappa \in \pi_{\mathsf{M}}$, which cannot be parallel to d (cf. Fig. 2b). Note that the lines $[\mathsf{m}_g, \mathsf{M}_g]$ belong to a regulus $\mathscr{R}$.

Now we choose a platform point $\mathsf{m} \notin \mathsf{g}$ and denote the footpoint on g with respect to m by m_f. Then τ denotes the signed distance of m_f and m with respect to the direction d. Due to the above considerations we can also add the leg $[\mathsf{m}, \mathsf{m}\tau]$ (beside the leg $[\mathsf{m}, \mathsf{m}\kappa]$) without restricting the self-motion, where $\mathsf{m}\tau$ denotes the point which we get by translating $\mathsf{m}_f\kappa$ about τ in direction d. If this construction is done for all points of a line $\mathsf{h} \parallel \mathsf{g}$ through m we get the line $\mathsf{h}\tau$. We distinguish two cases:

- $\mathsf{h}\kappa \neq \mathsf{h}\tau$: Now every point $\mathsf{m} \in \mathsf{h}$ (with exception of $\mathsf{m}_e := \{\mathsf{h}\kappa \cap \mathsf{h}\tau\}\kappa^{-1}$) can only rotate about the line $[\mathsf{m}\tau, \mathsf{m}\kappa] \parallel \mathsf{d}$ (cf. Fig. 2a,b). Therefore the platform cannot move in direction d during the self-motion and the problem reduces to the following planar one: Determine all non-degenerated 3-dof RPR manipulators with collinear platform anchor points m_1^-, m_2^-, m_3^- and collinear base anchor points M_1^-, M_2^-, M_3^- possessing a self-motion.
  It is well known, that there only exists the so-called *circular translation* (cf. Fig. 1b) beside the planar analogue (cf. Fig. 1c) of the spherical self-motion given in Lemma 2, which yields for the same arguments as in the spherical case no solution to our problem. The *circular translation* implies that the projectivity κ with matrix $\mathbf{P} = (p_{ij})$ has to be an affinity of the following form:

$$\mathbf{M}_i = \begin{pmatrix} p_{21} \\ p_{31} \end{pmatrix} + \begin{pmatrix} 1 & 0 \\ p_{32} & p_{33} \end{pmatrix} \mathbf{m}_i \quad \text{with} \quad p_{33} \in \mathbb{R} \setminus \{0, 1\} \quad \text{and} \quad p_{21}, p_{31}, p_{32} \in \mathbb{R}. \tag{4}$$

  This can be seen as follows: As the pencil of lines through the ideal point of d has to be mapped onto an identical pencil of lines through the ideal point of d, the ideal line has to be mapped onto itself ($\Rightarrow p_{12} = p_{13} = 0$). As the entries of $\mathbf{P}$ are still homogeneous, we can set $p_{11} = 1$ without loss of generality (w.l.o.g.), as $p_{11} = 0$ implies that κ is singular. Moreover, the fact that the above mentioned

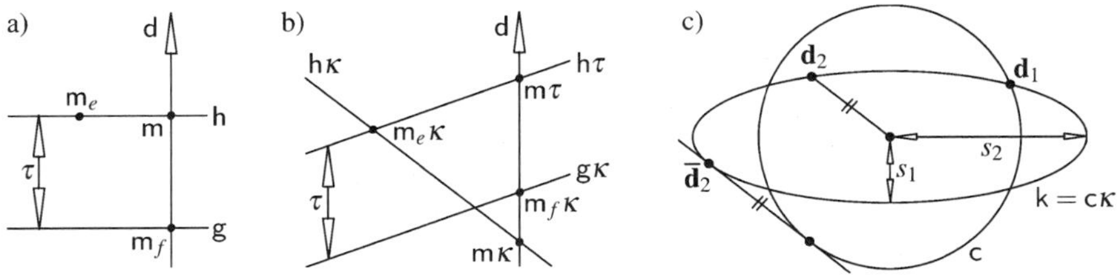

Fig. 2 Sketch of the mappings κ and τ between the platform (a) and the base (b). In (c) the proof of Thm. 5 is illustrated: common point and tangent of an ellipse k and the unit circle c.

pencils are identical yields $p_{22} = 1$. Finally, we get $p_{23} = 0$ from the condition that the ideal point in direction d is fixed under κ. Moreover, $p_{33} \neq \{0, 1\}$ has to hold, because otherwise the affinity is singular resp. $h\kappa = h\tau$ holds, a contradiction.

- $h\kappa = h\tau$: Now this has to hold for all distances τ because otherwise we get the above case. As a consequence the projectivity κ with matrix $\mathbf{P} = (p_{ij})$ has to be an affinity of the following form:

$$\mathbf{M}_i = \begin{pmatrix} p_{21} \\ p_{31} \end{pmatrix} + \begin{pmatrix} p_{22} & 0 \\ p_{32} & 1 \end{pmatrix} \mathbf{m}_i \quad \text{with} \quad p_{22} \in \mathbb{R} \setminus \{0\} \quad \text{and} \quad p_{21}, p_{31}, p_{32} \in \mathbb{R}. \tag{5}$$

This can be seen as follows: The condition that the ideal point of the y-axis of the moving frame is mapped onto the ideal point of the y-axis of the fixed frame yields $p_{13} = p_{23} = 0$. Now we invest the property that the anchor points of a leg, during its translation in direction d, always have to correspond one another within the projectivity. This can be expressed as follows:

$$\mathbf{P} \begin{pmatrix} 1 \\ u \\ v \end{pmatrix} + \begin{pmatrix} 0 \\ 0 \\ \tau \end{pmatrix} = \mathbf{P} \begin{pmatrix} 1 \\ u \\ v + \tau \end{pmatrix}. \tag{6}$$

As the first two rows are fulfilled identically, only the third row yields a condition, which reads as follows: $\tau(p_{11} - p_{33} + p_{12}u) = 0$. This equation can only be fulfilled for all $u \in \mathbb{R}$ if $p_{12} = 0$ ($\Rightarrow \kappa$ is an affinity) and $p_{11} = p_{33}$ hold. As the entries of $\mathbf{P}$ are still homogeneous, we can set $p_{11} = 1$ w.l.o.g., as $p_{11} = 0$ implies that κ is singular. Moreover, $p_{22} \neq 0$ has to hold, because otherwise the affinity is also singular.

These considerations prove the first two sentences of the following theorem:

Theorem 4. *A planar projective SG platform, which is not architecturally singular, can only have a Schönflies self-motion with the direction* d *of the rotation axis parallel to* π_M *and* π_m*, if it belongs to the subset of planar affine SG platforms. Moreover, if we choose the y-axis of the moving and the fixed frame in direction of* d*, the affinity* κ *has to be of the form given in Eqs. (4) or (5). In addition, all self-motions of these manipulators are pure translations and the self-motion is two-dimensional only if the platform and the base are congruent and all legs have equal length.*

The last sentence of Thm. 4, which was already known to Karger [9], can easily be proved by direct computations, which are given in Appendix B of the corresponding technical report [14]. Moreover, we can give a geometric characterization of all non-architecturally singular planar affine SG platforms with self-motions:

Theorem 5. *Assume a planar affine SG platform, which is not architecturally singular, is determined by* $\mathbf{M}_i = \mathbf{a} + \mathbf{A}\mathbf{m}_i$*. Then this manipulator has a self-motion if and only if the singular values* s_1 *and* s_2 *of* $\mathbf{A}$ *with* $0 < s_1 \leq s_2$ *fulfill* $s_1 \leq 1 \leq s_2$*.*

Proof. First of all, we prove that a planar affine SG platform cannot have an elliptic self-motion. If $\mathsf{s} = \mathsf{s}\kappa$ is not the ideal line, then the projectivity from s onto itself has at least one fixed point, namely the ideal point of $\mathsf{s} = \mathsf{s}\kappa$. Therefore $\mathsf{s} = \mathsf{s}\kappa$ has to be the ideal line during the whole self-motion. This implies that the elliptic self-motion is a Schönflies motion, where the direction of the rotation axis is orthogonal to $\pi_{\mathsf{M}} \parallel \pi_{\mathsf{m}}$. As all points of the platform have to run on spherical paths, this Schönflies motion can only be the Borel Bricard motion (cf. [1, 2]) due to [15]. Therefore the corresponding points of the platform and base have to be related by an inversion. As an inversion is no projectivity, we get a contradiction.

Under consideration of this result and Thms. 2 and 3, planar affine SG platforms can only have self-motions given in Thm. 4. We consider the image of the unit vectors $\mathbf{c} = (\cos\varphi, \sin\varphi) \in \pi_{\mathsf{m}}$ for $\varphi \in [0, 2\pi]$ under κ. Clearly, the tie points of the vectors $\mathbf{Ac}$ are located on an ellipse k (including the special case of a circle).

- ⋆ The necessary and sufficient condition for an affinity of the form Eq. (5) is that a vector $\mathbf{d}_1$ of $\mathbf{Ac}$ has length 1. This corresponds geometrically to the common points of k and the unit circle c (cf. Fig. 2c).
- ⋆ The necessary and sufficient condition for an affinity of the form Eq. (4) is that a vector $\mathbf{d}_2$ of $\mathbf{Ac}$ exists, which has distance 1 from the ellipse tangent in its conjugate point $\overline{\mathbf{d}}_2$ on k. This corresponds geometrically to the determination of common tangents of k and c (cf. Fig. 2c).

If we choose a new coordinate system in the base and platform such that the y-axis is parallel to $\mathbf{d}_i$ and $\mathbf{A}^{-1}\mathbf{d}_i$, respectively, we end up with an affinity of the form given in Eq. (5) for $i = 1$ resp. Eq. (4) for $i = 2$. Clearly, we only get real common points and tangents of k and c if the singular values s_1 and s_2 of $\mathbf{A}$ fulfill $s_1 \leq 1 \leq s_2$. □

Remark 2. Note, that Thm. 5 also implies the result of [10] that planar equiform SG platforms cannot have a self-motion if they are not architecturally singular, as $s_1 = s_2 \neq 1$ holds. Finally, it should also be mentioned that all planar affine SG platforms given in Eq. (4) and Eq. (5) are Schönflies-singular manipulators due to item (3) and item (2), respectively, of Thm. 3 given by the author in [16]. ⋄

Example 1. We verify Thm. 5 at hand of the planar affine SG platform with a self-motion given by Karger on page 162 of [9]. The first three pairs of anchor points are determined by $a_1 = b_1 = b_2 = A_1 = B_1 = B_2 = 0$, $a_2 = 1$, $a_3 = 5$, $b_3 = -4$ and $A_2 = A_3 = B_3 = 2$. For this example $\mathbf{a}$, $\mathbf{A}$, s_1 and s_2 are given by:

$$\mathbf{a} = \begin{pmatrix} 0 \\ 0 \end{pmatrix}, \quad \mathbf{A} = \begin{pmatrix} 2 & 2 \\ 0 & -\frac{1}{2} \end{pmatrix}, \quad s_1 = \frac{\sqrt{41} - 5}{4} \approx 0.35, \quad s_2 = \frac{\sqrt{41} + 5}{4} \approx 2.85.$$

5 Conclusion and Future Research

We proved that non-architecturally singular planar projective SG platforms have either elliptic self-motions (Def. 1) or pure translational self-motions (Thms. 2–4). The latter are the only self-motions of planar affine SG platforms (Thm. 5).

The study of elliptic self-motions is dedicated to future research. It remains open whether these self-motions even exist, as no example is known to the author so far.

Acknowledgements This research is supported by Grant No. I 408-N13 of the Austrian Science Fund FWF within the project "Flexible polyhedra and frameworks in different spaces", an international cooperation between FWF and RFBR, the Russian Foundation for Basic Research.

References

1. Borel, E.: Mémoire sur les déplacements à trajectoires sphériques, Mém. présenteés par divers savants. Paris **33**(2), 1–128 (1908)
2. Bricard, R.: Mémoire sur les déplacements à trajectoires sphériques. J. École Polytech. **11**(2), 1–96 (1906)
3. Husty, M.: E. Borel's and R. Bricard's papers on displacements with spherical paths and their relevance to self-motions of parallel manipulators. In: Ceccarelli, M. (ed.) Int. Symp. on History of Machines and Mechanisms, pp. 163–172. Kluwer (2000)
4. Karger, A.: Architecture singular planar parallel manipulators. Mech. Mach. Theory **38**(11), 1149–1164 (2003)
5. Nawratil, G.: On the degenerated cases of architecturally singular planar parallel manipulators. J. Geom. Graph. **12**(2), 141–149 (2008)
6. Röschel, O., Mick, S.: Characterisation of architecturally shaky platforms. In: Lenarcic, J., Husty, M.L. (eds.) Advances in Robot Kinematics: Analysis and Control, pp. 465–474. Kluwer (1998)
7. Wohlhart, K.: From higher degrees of shakiness to mobility. Mech. Mach. Theory **45**(3), 467–476 (2010)
8. Chasles, M.: Sur les six droites qui peuvent étre les directions de six forces en équilibre. C. R. Séances Acad. Sci. **52**, 1094–1104 (1861)
9. Karger, A.: Singularities and self-motions of a special type of platforms. In: Lenarcic, J., Thomas, F. (eds.) Advances in Robot Kinematics: Theory and Applications, pp. 155–164. Springer (2002)
10. Karger, A.: Singularities and self-motions of equiform platforms. Mech. Mach. Theory **36**(7), 801–815 (2001)
11. Karger, A.: Parallel manipulators with simple geometrical structure. In: Ceccarelli, M. (ed.) Proc. of the 2nd European Conference on Mechanism Science, pp. 463–470. Springer (2008)
12. Mielczarek, S., Husty, M.L., Hiller, M.: Designing a redundant Stewart-Gough platform with a maximal forward kinematics solution set. In: Proc. of the International Symposion of Multibody Simulation and Mechatronics (MUSME), Mexico City, Mexico (2002)
13. Borras, J., Thomas, F., Torras, C.: Singularity-invariant leg rearrangements in doubly-planar Stewart-Gough platforms. In: Proc. of Robotics Science and Systems, Zaragoza, Spain (2010)
14. Nawratil, G.: Self-motions of planar projective Stewart Gough platforms, Technical Report No. 221, Geometry Preprint Series, TU Vienna (2011) www.geometrie.tuwien.ac.at/nawratil
15. Husty, M.L., Karger, A.: Self motions of Stewart-Gough platforms: an overview. In: Gosselin, C.M., Ebert-Uphoff, I. (eds.) Proc. of the Workshop on Fundamental Issues and Future Research Directions for Parallel Mechanisms and Manipulators, pp. 131–141 (2002)
16. Nawratil, G.: Special cases of Schönflies-singular planar Stewart Gough platforms. In: Pisla, D., et al. (eds.) New Trends in Mechanisms Science, pp. 47–54. Springer (2010)

Asymptotic Singularities of Planar Parallel 3-RPR Manipulators

Michel Coste

Abstract We study the limits of singularities of planar parallel 3-RPR manipulators as the lengths of their legs tend to infinity, paying special attention to the presence of cusps. These asymptotic singularities govern the kinematic behaviour of the manipulator in a rather large portion of its workspace.

Key words: Parallel robots, singularities

1 Introduction

Planar 3-RPR manipulators have been extensively studied, see for instance [1–10]. Much attention has been paid to the analysis of the singularities of these manipulators. These singularities govern the kinematic properties of the manipulators, for instance their ability to perform singularity-free assembly mode changing.

It has been reported in several papers (see for instance [8]) that the picture of the singularities in the actuated joint space stabilizes when the lengths of the legs are sufficiently large. Actually, in several examples, this stabilization appears already when the lengths of legs are two or three times the lengths of the sides of the base and platform triangles. Figure 1 exemplifies this phenomenon.

We study in this paper this stable configuration, introducing the limits of singularities of planar parallel 3-RPR manipulators as the lengths of their legs tends to infinity. These asymptotic singularities govern the kinematic behaviour of the manipulator in a rather large portion of its workspace. We classify the 3-RPR manipulators with respect to the features of this stable configuration of singularities. We pay special attention to the presence of cusps in the singular locus, since their presence is crucial for the possibility of singularity-free assembly mode changing.

Michel Coste
IRMAR, Université de Rennes 1, CNRS, Rennes, France, e-mail: michel.coste@univ-rennes1.fr

J. Lenarčič, M. Husty (eds.), *Latest Advances in Robot Kinematics*,
DOI 10.1007/978-94-007-4620-6_5,

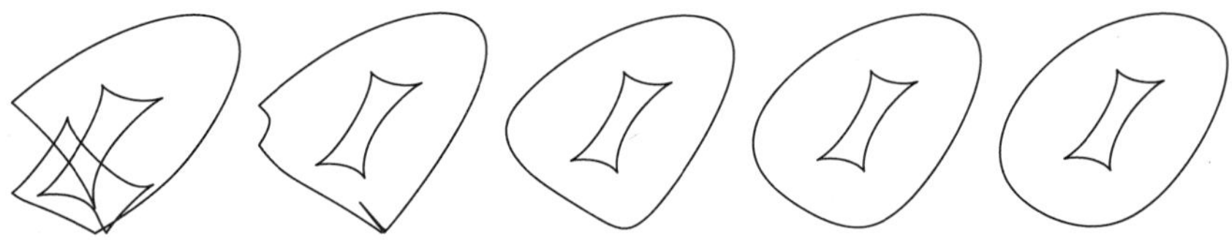

Fig. 1 Asymptotic stability of the section of the singularity surface of a 3-RPR manipulator: sections at $r_1 = 20, 30, 40, 60, 100$.

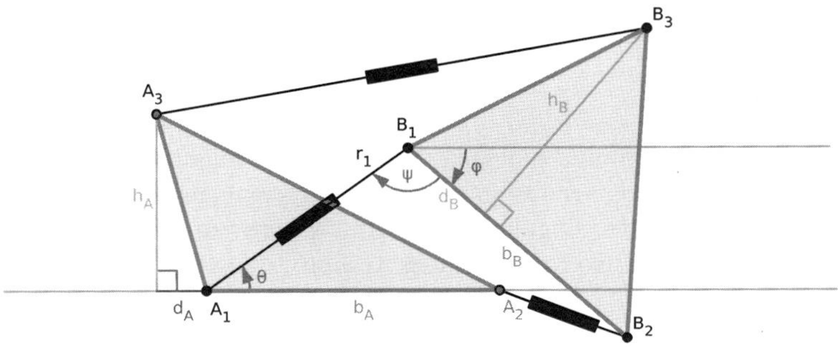

Fig. 2 3-RPR, parameters and variables used.

2 Mathematical Modelling

The notations, the parameters and the variables we use are explained in Figure 2.

The geometry of the base (resp. platform) is described by the parameters b_A, h_A, d_A (resp. b_B, h_B, d_B and a sign $\varepsilon = \pm 1$). In the direct orthonormal frame attached to the base with origin A_1, A_2 has coordinates $(b_A, 0)$ $(b_A > 0)$ and A_3 has coordinates (d_A, h_A) $(h_A > 0)$. In the direct orthonormal frame attached to the platform with origin B_1, B_2 has coordinates $(b_B, 0)$ $(b_B > 0)$ and B_3 has coordinates $(d_B, \varepsilon h_B)$ $(h_B > 0)$; the sign ε is -1 when the triangle $B_1B_2B_3$ is oriented clockwise. We always assume that neither triangle is flat, i.e. $b_A h_A b_B h_B \neq 0$. The position of the vertex B_1 of the platform is given in polar coordinates by the angle θ and the length r_1 of the leg A_1B_1. The orientation of the platform is given by the angle φ. We shall also use the angle $\psi = \theta - \varphi + \pi$ instead of φ in order to simplify some equations.

We denote by r_2 and r_3 the lengths of the legs A_2B_2 and A_3B_3. These lengths are given by

$$r_2^2 = r_1^2 + 2r_1\left((b_B\cos\varphi - b_A)\cos\theta + b_B\sin\varphi\sin\theta\right) - 2b_Ab_B\cos\varphi + b_A^2 + b_B^2$$
$$r_3^2 = r_1^2 + 2r_1\left((d_B\cos\varphi - \varepsilon h_B\sin\varphi - d_A)\cos\theta + (d_B\sin\varphi + \varepsilon h_B\cos\varphi - h_A)\sin\theta\right)$$
$$- 2(d_Ad_B + \varepsilon h_Ah_B)\cos\varphi + 2(\varepsilon d_Ah_B - h_Ad_B)\sin\varphi + d_A^2 + d_B^2 + h_A^2 + h_B^2 \quad (1)$$

3 Asymptotic Direct Kinematic Problem and Singularities

We are interested in the limits of $r_1 - r_2$ and $r_1 - r_3$ as r_1 tends to infinity. We denote these limits by ℓ_2 and ℓ_3, respectively. They are easily obtained from Equations (1) as

$$\begin{aligned} \ell_2 &= (b_A - b_B\cos\varphi)\cos\theta - b_B\sin\varphi\sin\theta \\ \ell_3 &= (d_A + \varepsilon h_B\sin\varphi - d_B\cos\varphi)\cos\theta + (h_A - d_B\sin\varphi - \varepsilon h_B\cos\varphi)\sin\theta \end{aligned} \quad (2)$$

Equations (2) are to be seen as describing the asymptotic inverse kinematic mapping (IKM). The asymptotic direct kinematic problem (DKP) is solving Equations (2) for θ and φ. This can be done by writing Equations (2) in terms of $u = \tan(\varphi/2)$, eliminating u and writing the resultant obtained in $t = \tan(\theta/2)$; this process gives the following quartic characteristic equation in t:

$$\begin{aligned} 0 = &\left((d_Bl_2 - b_Bl_3)(t^2+1) + 2h_Ab_Bt + (b_Ad_B - d_Ab_B)(t^2-1)\right)^2 \\ &+ h_B^2\left((l_2(t^2+1) + b_A(t^2-1))^2 - b_B^2(t^2+1)^2\right) \end{aligned} \quad (3)$$

If we have a value of θ such that $t = \tan(\theta/2)$ is a solution of Equation (3), carrying this value in Equations (2) yields a linear system in $\cos\varphi$ and $\sin\varphi$, whose determinant $-\varepsilon b_Bh_B$ does not vanish. So we obtain a unique solution (θ, φ) to the asymptotic DKP.

The asymptotic DKP has degree 4 and has no more than 4 solutions.

The Jacobian determinant of the asymptotic IKM (2) is

$$\begin{aligned} \mathrm{Jac} = &\left(h_Ab_B\cos\theta^2 + (b_Ad_B - d_Ab_B)\cos\theta\sin\theta + \varepsilon b_Ah_B\sin\theta^2\right)\sin\varphi \\ &+ \left((\varepsilon b_Ah_B - h_Ab_B)\cos\theta\sin\theta + (d_Ab_B - b_Ad_B)\sin\theta^2\right)\cos\varphi \end{aligned} \quad (4)$$

The equation $\mathrm{Jac} = 0$ can easily be solved in $\tan(\varphi)$:

$$\tan(\varphi) = \frac{(h_Ab_B - \varepsilon b_Ah_B)\tan(\theta) + (b_Ad_B - d_Ab_B)\tan(\theta)^2}{h_Ab_B + (b_Ad_B - d_Ab_B)\tan(\theta) + \varepsilon b_Ah_B\tan(\theta)^2} \quad (5)$$

There is no surprise that we get in this way the curve of poles of the rational parametrization $r_1^{\mathrm{Sing}}(\varphi, \theta)$ of the singularity surface in the workspace obtained in [10]. This singularity curve in the torus of variables (φ, θ) was already studied in the quoted paper, where it is proved that it has two branches (connected components),

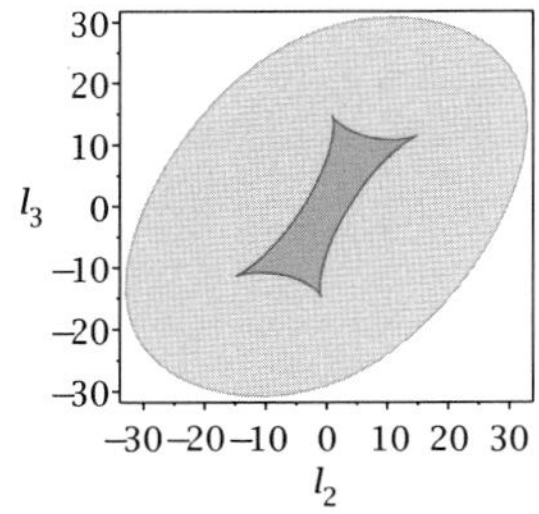

Fig. 3 The asymptotic image singularity curve for the Innocenti-Merlet manipulator (diamond).

one passing through the point $\varphi = 0, \theta = 0$ and the other passing through $\varphi = \pi, \theta = 0$ obtained from the first by a translation of π on the angular variable φ.

The image by the asymptotic IKM of the singularity curve in the plane of variables (ℓ_2, ℓ_3) consists also of two branches. This curve is generically an irreducible curve of degree 8, not rational. However, it is possible to obtain parametrizations by analytic functions of θ of each of the branches by substituting in Equations (2) the expressions for the trigonometric function of φ extracted from Equation (5). One has to be careful in the determination of the angle φ from its tangent, in order to ensure the continuity. One obtains for the first branch (the inner branch)

$$\begin{aligned} \ell_2^{\text{in}} &= \frac{(b_A\sqrt{\delta} - h_A b_B^2)\cos\theta - b_B(b_A d_B - d_A b_B)\sin\theta}{\sqrt{\delta}} \\ \ell_3^{\text{in}} &= \frac{(d_A\sqrt{\delta} - h_A b_B d_B)\cos\theta + (h_A\sqrt{\delta} - d_B(b_A d_B - d_A b_B) - b_A h_B^2)\sin\theta}{\sqrt{\delta}} \end{aligned} \quad (6)$$

where $\delta = (h_A b_B \cos\theta + (b_A d_B - d_A b_B)\sin\theta)^2 + b_A^2 h_B^2 \sin\theta^2$. Note that δ is always > 0, which ensures the analyticity in θ of the formulas (6). For the second branch (the outer branch), one replaces all occurrences of $\sqrt{\delta}$ by $-\sqrt{\delta}$ in formulas (6).

4 Examples

We show here a few examples of asymptotic image singularity curves in the plane (ℓ_2, ℓ_3). The curves are drawn using formulas (6) and their variants for the outer branch.

The first example (Figure 3) is the Innocenti–Merlet manipulator [7], with $b_A = 15.9$, $h_A = 10$, $d_A = 0$, $b_B = 17$, $\varepsilon h_B = 16.1$, $d_B = 13.2$; this is also the example of Figure 1, and one recovers the stable section of the singularity surface for large lengths of legs. The light gray indicates two solutions for the DKP, the dark gray four solutions. The inner branch has four cusps and no double point; we call this configuration a *diamond*.

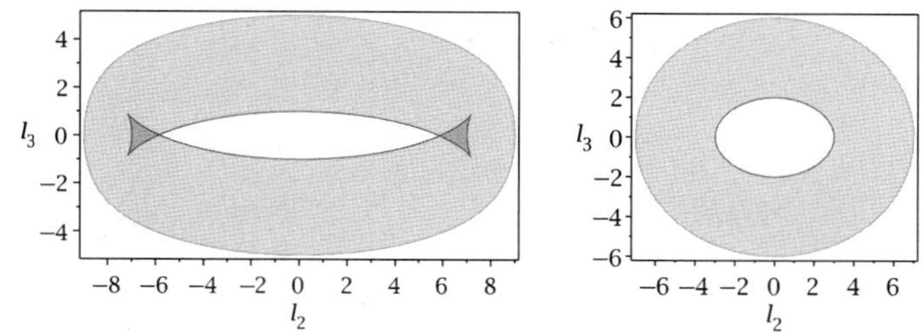

Fig. 4 Other configurations for the asymptotic image singularity curve (swallowtail and annulus).

The next examples (Figure 4) show two other configurations for the image singularity curves, with the same gray scale code. In both examples we have $d_A = d_B = 0$. The example on the left is for $b_A = 8$, $h_A = 3$, $b_B = 1$, $\varepsilon h_B = 2$. The inner branch in this case has four cusps and two double points; we call this configuration a *swallowtail*. The example on the right is for $b_A = 5$, $h_A = 4$, $b_B = 2$, $\varepsilon h_B = 2$. Here the inner branch is just an oval; we call this configuration an *annulus*.

We shall explain in the next section why these three examples show all possible stable configurations for the asymptotic image singularity curve: diamond, swallowtail or annulus.

5 Asymptotic Kinematics and Wave Fronts of an Ellipse

Equations (2) for the asymptotic IKM take a more symmetric form when they are written, after substituting φ with $\theta - \psi + \pi$, as:

$$\begin{pmatrix} \ell_2 \\ \ell_3 \end{pmatrix} = \begin{pmatrix} b_A & 0 \\ d_A & h_A \end{pmatrix} \begin{pmatrix} \cos\theta \\ \sin\theta \end{pmatrix} + \begin{pmatrix} b_B & 0 \\ d_B & \varepsilon h_B \end{pmatrix} \begin{pmatrix} \cos\psi \\ \sin\psi \end{pmatrix} . \tag{7}$$

The two summands of the right-hand side of Equation (7) are parametrizations of ellipses $\mathcal{E}_A(\theta)$ and $\mathcal{E}_B(\psi)$. By a linear change of coordinates, we can assume that one of the ellipses (say $\mathcal{E}_B$) is a circle of radius ρ. We are then in the situation of the paradigmatic example for wave fronts [11]: the wave front of an ellipse, which is the envelope of the family of circles with radius ρ centred in a point of the ellipse (the image of $\mathcal{E}_A$ under the coordinate change). The outer branch of the wave front is always a smooth convex curve, whereas the shape of the inner branch of the wave front bifurcates at values $\rho = \beta^2/\alpha,\ \beta,\ \alpha,\ \alpha^2/\beta$ (where α and β are respectively the semi-major and semi-minor axes lengths of the ellipse). The stable configurations of the wave front outside of the bifurcation values are the ones we encountered in the preceding section: diamond, swallowtail and annulus.

There is a cusp of the inner branch of the wave front, where it is tangent to the circle of radius ρ centred at P on the ellipse, when ρ is equal to the curvature radius of the ellipse at P. Hence, there is no cusp on the inner branch of the wave front

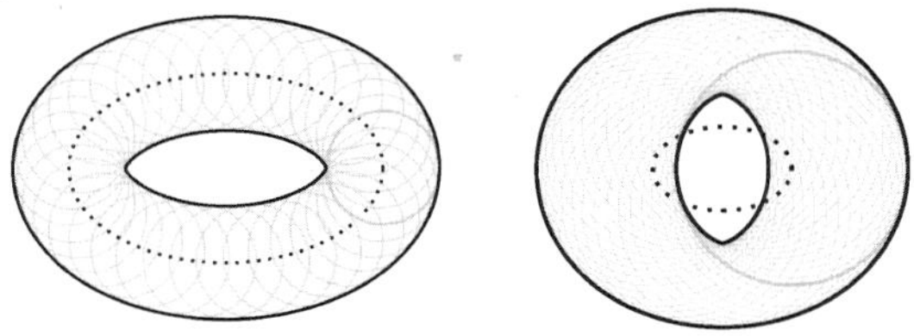

Fig. 5 Bifurcation between annulus and swallowtail for $\rho = \beta^2/\alpha$ and $\rho = \alpha^2/\beta$.

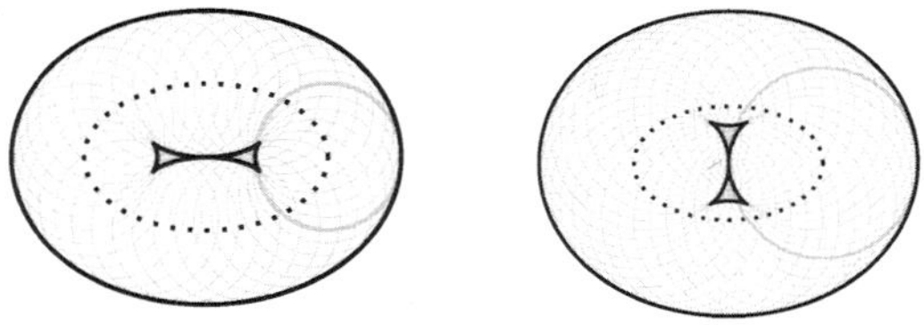

Fig. 6 Bifurcation between diamond and swallowtail for $\rho = \beta$ and $\rho = \alpha$.

when ρ is below the minimum or above the maximum of the curvature radius along the ellipse, which are respectively β^2/α and α^2/β. These values correspond to the bifurcation between annulus and swallowtail, as shown in Figure 5 (ellipse in dotted line, wave front in solid line, circles in light gray).

The inner branch of the wave front has four cusps and no double points for values of ρ between β and α. These values correspond to the bifurcation between diamond and swallowtail, as shown in Figure 6.

In conclusion, we know what are the stable configurations (diamond, swallowtail, annulus) for the asymptotic DKP and we can characterize the bifurcations between these stable configurations. We shall make precise the classification of 3-R$\underline{\text{P}}$R manipulators according to these stable configurations in the following section.

6 Classification of 3-R$\underline{\text{P}}$R Manipulators

We use the analysis of the bifurcation of the wave front of an ellipse recalled in the preceding section. In order to do that, we change coordinates in the (ℓ_2, ℓ_3)-plane so that the ellipse $\mathscr{E}_B$ of Equation 7 is a circle of radius 1 in the new coordinates. We set $M_A = \begin{pmatrix} b_A & 0 \\ d_A & h_A \end{pmatrix}$, $M_B = \begin{pmatrix} b_B & 0 \\ d_B & \varepsilon h_B \end{pmatrix}$. The new coordinates (u,v) are given by $\begin{pmatrix} u \\ v \end{pmatrix} = M_B^{-1} \begin{pmatrix} \ell_1 \\ \ell_2 \end{pmatrix}$ and the equation of the ellipse $\mathscr{E}_A$ in variables (u,v) is

$$(u\ v)\, {}^t(M_A^{-1} M_B) M_A^{-1} M_B \begin{pmatrix} u \\ v \end{pmatrix} = 1 \qquad (8)$$

We denote by α and β the semi-major and semi-minor axes lengths of the ellipse described by Equation (8). The eigenvalues of the positive symmetric matrix ${}^t(M_A^{-1}M_B)M_A^{-1}M_B$ are $1/\alpha^2$ and $1/\beta^2$, whence we derive an equation whose roots are α^2 and β^2:

$$S_B^2\lambda^2 - U_{A,B}\lambda + S_A^2 = 0\,, \tag{9}$$

with $S_A = b_A h_A/2$ and $S_B = b_B h_B/2$ the areas of the triangles and

$$\begin{aligned} U_{A,B} &= \frac{1}{4}\left(b_A^2 h_B^2 + h_A^2 b_B^2 + (b_A d_B - d_A b_B)^2\right) \\ &= \frac{1}{8}\left(a_A^2(b_B^2 + c_B^2 - a_B^2) + b_A^2(c_B^2 + a_B^2 - b_B^2) + c_A^2(a_B^2 + b_B^2 - c_B^2)\right)\,, \end{aligned} \tag{10}$$

where a_A, c_A, a_B, c_B denote the lengths of the sides A_3A_1, A_2A_3, B_3B_1, B_2B_3 respectively,. The second expression for $U_{A,B}$ makes clear that it is symmetric w.r.t. the three sides of the triangles. The inequality $U_{A,B} \geq 2S_AS_B$, expressing the fact that the discriminant of Equation (9) is always non-negative, is known in geometry as the Neuberg-Pedoe inequality [12]; the equality case is precisely when the two triangles are similar.

The diamond configuration occurs when $\beta < 1 < \alpha$, i.e. when $(1-\alpha^2)(1-\beta^2) < 0$. The annulus configuration occurs when $1 < \beta^2/\alpha$ or $1 > \alpha^2/\beta$, i.e. when $(1-\alpha^4/\beta^2)(1-\beta^4/\alpha^2) > 0$. We transform these inequalities using Equation (9) and obtain the following classification result.

The configuration of the asymptotic singularities of a 3-RPR manipulator is

- *a diamond when* $U_{A,B} > S_A^2 + S_B^2$,
- *an annulus when* $U_{A,B}^3 < S_A^2 S_B^2(3U_{A,B} + S_A^2 + S_B^2)$,
- *a swallowtail when* $U_{A,B} < S_A^2 + S_B^2$ *and* $U_{A,B}^3 > S_A^2 S_B^2(3U_{A,B} + S_A^2 + S_B^2)$.

The inequality $U_{A,B} \geq S_A^2 + S_B^2$ has a geometric interpretation [13]: it is satisfied if and only if there is a pose of the manipulator such that the three legs A_1B_1, A_2B_2 and A_3B_3 are parallel. No such geometric interpretation seems to be known for the inequality $U_{A,B}^3 < S_A^2 S_B^2(3U_{A,B} + S_A^2 + S_B^2)$.

The three stable configurations we have found (diamond, annulus and swallowtail) only have the stable singularities [11] of a projection of a surface to a plane: folds, cusps and transversal intersection of folds. Hence these configurations of asymptotic singularities remain the same for the sections at large values of r_1 (or r_2, or r_3) of the singularity surface in the actuated joint space; the precise meaning of "large" depends, of course, on the geometry of the manipulator.

There are non-generic manipulators for which the asymptotic singularities do not belong to a stable configuration. This is so for "symmetric" manipulators [14]: in this case the inner branch of the asymptotic singularity curve is reduced to a point. We plan to study the stable perturbations of such manipulators in a future work.

7 Conclusion

We studied the asymptotic DKP for 3-R$\underline{P}$R manipulators. We have shown that this is a fourth-degree problem. We have established that there are only three possible stable configurations for the asymptotic singularities (diamond, swallowtail and annulus), which remain unchanged for sufficiently large values of the lengths of the legs. We also gave an explicit classification of 3-R$\underline{P}$R manipulators with respect to their asymptotic singularities.

References

1. Hunt, K.H.: Structural kinematics of in-parallel actuated robot arms. J. Mech. Transm. Autom. Des. **105**(4), 705–712 (1983)
2. Innocenti, C., Parenti-Castelli, V.: Singularity-free evolution from one configuration to another in serial and fully-parallel manipulators. In: Proc. ASME Design Technical Conferences, Spatial Mechanisms and Mechanical Systems, DE-vol. 45, pp. 553–560 (1992)
3. Gosselin, C.M., Merlet, J.-P.: On the direct kinematics of planar parallel manipulators: Special architectures and number of solutions. Mech. Mach. Theory **29**(8), 1083–1097 (1994)
4. Wenger, P., Chablat, D.: Workspace and assembly-modes in fully-parallel manipulators: A descriptive study. In: Advances in Robot Kinematics and Computational Geometry, pp. 117–126. Kluwer Academic Publishers (1998)
5. McAree, P.R., Daniel, R.W.: An explanation of never-special assembly changing motions for 3-3 parallel manipulators. Int. J. Robot. Res. **18**(6), 556–574 (1999)
6. Bonev, I., Zlatanov, D., Gosselin, C.: Singularity analysis of 3-DOF planar parallel mechanisms via screw theory. ASME J. Mech. Des. **125**(3), 573–581 (2003)
7. Merlet, J.-P.: Parallel Robots. Springer (2006)
8. Zein, M., Wenger, P., Chablat, D.: Singular curves in the joint space and cusp points of 3-RPR parallel manipulators. Robotica **25**(6), 717–724 (2007)
9. Husty, M.L.: Non-singular assembly mode change in 3-RPR-parallel manipulators. In: Computational Kinematics: Proceedings of the 5th International Workshop on Computational Kinematics, pp. 51–60. Springer Verlag (2009)
10. Coste, M.: A simple proof that generic 3-RPR manipulators have two aspects. J. Mech. Robot. **4**(1), 011008 (2012)
11. Arnol'd, V.I., Guseĭn-Zade, S.M., Varchenko, A.N.: Singularities of Differentiable Maps, Vol. I. Birkhäuser (1985)
12. Pedoe, D.: Thinking geometrically. Amer. Math. Monthly **77**, 711–721 (1970)
13. Mitrinović, D.S., Pečarić, J.E.: About the Neuberg–Pedoe and the Oppenheim inequalities. J. Math. Anal. Appl. **129**(1), 196–210 (1988)
14. Coste, M., Wenger, P., Chablat, D.: Singular surfaces and cusps in symmetric planar 3-RPR manipulators. In: Proc. of IEEE/RSJ International Conference on Intelligent Robots and Systems, San Francisco, pp. 1453–1458 (2011)

Classification of the Singularity Loci of m-n Fully-Parallel Manipulators

Raffaele Di Gregorio

Abstract Singularity analysis of fully-parallel manipulators (FPMs) produced a wide literature that tried to overcome the difficulty of algebraically calculating the determinant of general FPM's Jacobian. An early work of this author addressed this problem by using Laplace expansion, and proposed an analytic expression of general FPM's singularity locus which contains ten terms easy to compute and geometrically interpret. Such an expression is exploited here to classify the singularity loci of all the m-n FPM architectures.

Key words: Fully-parallel manipulator, instantaneous kinematics, singular configuration, singularity locus

1 Introduction

Singularity analysis of parallel manipulators (PMs) is a challenging subject that faced and explained many unforeseen behaviors of these machines (see [1] for references), and it is central in their design for avoiding the machine breakdown and the use of machine components that are unnecessarily over-sized.

m-n Fully-parallel manipulators (FPMs) [2] are an important class of PMs consisting of a platform connected to a base by means of six kinematic chains (legs) of type S<u>P</u>S[1]. In each leg, the distance (leg length) between the two S-pair centers, henceforth referred to as leg endings, is the actuated-joint variable. Endings of different legs may coalesce into a multiple attachment point in the base (platform), thus generating different FPM architectures that can be collected into families, named m-n, where m and n are the numbers of separated attachment points in the base and in the platform, respectively.

Raffaele Di Gregorio
Department of Engineering, University of Ferrara, 44122 Ferrara, Italy,
e-mail: raffaele.digregorio@unife.it

[1] S and P stand for spherical pair and prismatic pair, respectively, and the underscore indicates the actuated pair.

J. Lenarčič, M. Husty (eds.), *Latest Advances in Robot Kinematics*,
DOI 10.1007/978-94-007-4620-6_6,

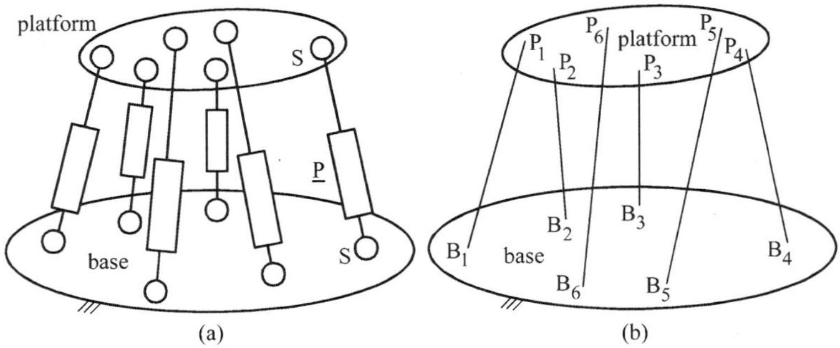

Fig. 1 6-6 FPM architecture: (a) kinematic scheme, (b) notation.

The 6-6 FPM architecture (Fig. 1), often named general FPM, is the only one where the attachment points are single both in the base and in the platform. Since all the other m-n FPM architectures are obtained from the 6-6 one by suitably setting some geometric constants, all the formulas, properties, etc. applicable to the general FPM can be easily adapted for each particular m-n FPM architecture.

The singularity analysis of the general FPM was addressed from many points of view (see, for instance [3–6]) that tried to overcome the difficulty of algebraically calculating the determinant of its left-Jacobian[2] which is a 6×6 matrix with entries that are explicit functions of manipulator's configuration. An early work [4] of this author addressed this problem by using Laplace expansion [7], and proposed an analytic expression of general FPM's singularity locus which contains ten terms easy to compute and geometrically interpret. Such an expression is exploited here to classify the singularity loci of all the m-n FPM architectures.

The next section will determine all the m-n FPM architectures by using combinatorial calculus. Section 3 will show that all the singularity locus expressions can be collected into three particular forms which allow the geometric classification of FPM singularities. Eventually, Section 4 will draw the conclusions.

2 m-n FPM Architectures

In the literature, m-n FPM architectures have been enumerated according to the characteristics that are relevant for the solution of the particular problem to address. For instance, in [2], the focus was on the solution of FPMs' direct position analysis, and the adopted criterion was the presence of connected sets of leg sequences in the FPM architecture to study. That approach brought to determine 21 FPM architectures collected into 10 m-n FPM families. Here, the problem to address is FPMs'

[2] The instantaneous input-output relationship of the general FPM relates platform's twist (output) to leg-lengths' rates (input). It is a linear and homogeneous mapping which contains two 6×6 matrices, one, here referred to as left-Jacobian, multiplies platform's twist and the other multiplies the 6-tuple collecting leg-lengths' rates [4].

singularity analysis; thus, the particular architectural features, that simplify the general analytic expression reported in [4] for the singularity locus of the 6-6 FPM, has to be considered. Since the most important simplifications of that expression occur when two or more leg endings coalesce into a multiple attachment point either in the base or in the platform, the possible presence of multiple attachment points together with their multiplicity is the feature to be considered here for classifying all the m-n FPM architectures. As it will be stressed later, the resulting enumeration is different from the one presented in [2], and it includes that one.

The determination of all the m-n FPM architectures with different types of attachment points must take into account the following lemmas:

(i) *There are always six leg endings in the base (platform), a number of which may coalesce to form multiple attachment points. Moreover, due to static reasons, less than three attachment points either in the base or in the platform yield a singular structure.* As a consequence, the number, k, of attachment points, either in the base or in the platform, must belong to the set $\{3, 4, 5, 6\}$.

(ii) *The four elements of the set* $\{3, 4, 5, 6\}$ *can be combined into couples, which include the combinations of each element with itself, to form the following* $5 \times 4/2$ $(= 10)$ *couples: 6-6, 6-5, 6-4, 6-3, 5-5, 5-4, 5-3, 4-4, 4-3, 3-3.* Thus, each of these 10 couples can identify a subset of FPM architectures, hereafter named m-n subset, which collects all the FPM architectures with m attachment points in the base and n attachment points in the platform. And the union of these 10 disjoint subsets yields the set of all the m-n FPM architectures.

(iii) *More than three leg endings which coalesce into one multiple spherical pair yield an hyperstatic substructure*[3]. Thus, the leg-ending coalescences can involve either two or three leg endings (i.e., only double or triple spherical pairs are allowed), which implies that the multiplicity, μ_i, of the ith attachment point ($i = 1,\ldots,k$ with $k \in \{3,4,5,6\}$) must belong to the set $\{1, 2, 3\}$ and satisfy the condition: $\mu_1 + \mu_2 + \cdots + \mu_k = 6$.

(iv) *If two legs share both the endings, they will constitute an hyperstatic substructure.* Thus, two legs can share only one ending either in the base or in the platform.

In order to find all the m-n FPM architectures, one just needs to identify how many elements, which satisfy (iii) and (iv), belong to the above mentioned 10 m-n subsets. Hereafter, a type of platform (base) with k attachment points will be identified by the non-ordered k-tuple $(\mu_1, \mu_2, \ldots, \mu_k)$ where μ_i, for $i = 1,\ldots,k$, is the multiplicity of the ith attachment point, and all the k-tuples obtained by permuting the entries denote the same type of platform (base).

The determination of all the types of platform (base) that have the same number, k, of attachment points is the first step of this combinatorial analysis. According to the above considerations, the possible types of platform (base) are:

- if $k = 3$, the types of platform (base) will be identified by the 3-tuples

[3] In the platform (base), a multiple spherical pair with multiplicity greater than three allows the redistribution of a single transmitted force along more than three directions, what brings to an indeterminate static problem with an infinite number of solutions.

$$(\mu_1,\mu_2,\mu_3) \in \{(\mu_1,\mu_2,\mu_3) \mid \mu i \in \{1,2,3\} \& \ \mu_1+\mu_2+\mu_3=6\} \quad (1)$$

and it is easy to demonstrate that $\{(\mu_1,\mu_2,\mu_3) \mid \mu_i \in \{1,2,3\} \& \ \mu_1+\mu_2+\mu_3=6\}$ coincides with $\{(3,2,1),(2,2,2)\}$.

- If $k=4$, the types of platform (base) will be identified by the 4-tuples

$$(\mu_1,\mu_2,\mu_3,\mu_4) \in \{(\mu_1,\mu_2,\mu_3,\mu_4) \mid \mu_i \in \{1,2,3\} \& \ \mu_1+\mu_2+\mu_3+\mu_4=6\} \quad (2)$$

and it is easy to demonstrate that $\{(\mu_1,\mu_2,\mu_3,\mu_4) \mid \mu_i \in \{1,2,3\} \& \ \mu_1+\mu_2+\mu_3+\mu_4=6\}$ coincides with $\{(3,1,1,1),\ (2,2,1,1)\}$.

- If $k=5$, the types of platform (base) will be identified by the 5-tuples

$$(\mu_1,\mu_2,\mu_3,\mu_4,\mu_5) \in \{(\mu_1,\mu_2,\mu_3,\mu_4,\mu_5) \mid \mu_i \in \{1,2,3\} \\ \& \ \mu_1+\mu_2+\mu_3+\mu_4+\mu_5=6\} \quad (3)$$

and proving that $\{(\mu_1,\mu_2,\mu_3,\mu_4,\mu_5) \mid \mu_i \in \{1,2,3\} \& \ \mu_1+\mu_2+\mu_3+\mu_4+\mu_5=6\}$ coincides with $\{(2,1,1,1,1)\}$ is easy.

- If $k=6$, the types of platform (base) will be identified by the 6-tuple

$$(\mu_1,\mu_2,\mu_3,\mu_4,\mu_5,\mu_6) \in \{(\mu_1,\mu_2,\mu_3,\mu_4,\mu_5,\mu_6) \mid \mu_i \in \{1,2,3\} \\ \& \ \mu_1+\mu_2+\mu_3+\mu_4+\mu_5+\mu_6=6\} \quad (4)$$

and proving that $\{(\mu_1,\mu_2,\mu_3,\mu_4,\mu_5,\mu_6) \mid \mu_i \in \{1,2,3\} \& \ \mu_1+\mu_2+\mu_3+\mu_4+\mu_5+\mu_6=6\}$ coincides with $\{(1,1,1,1,1,1)\}$ is easy.

Summarizing, there are only 6 types of platform (base): two with $k=3$, two with $k=4$, one with $k=5$, and one with $k=6$. By combining these 6 types of platform (base) into couples, that include the combinations of each element with itself, the following $7\times 6/2$ $(=21)$ couples, which one-to-one correspond to as many different FPM architectures, are obtained:

3-3 FPMs: **3-3/I**) (3,2,1)-(2,2,2), **3-3/II**) (2,2,2)-(2,2,2), **3-3/III**) (3,2,1)-(3,2,1);
4-4 FPMs: **4-4/I**) (3,1,1,1)-(2,2,1,1), **4-4/II**) (3,1,1,1)-(3,1,1,1),
4-4/III) (2,2,1,1)-(2,2,1,1);
4-3 FPMs: **4-3/I**) (3,1,1,1)-(2,2,2), **4-3/II**) (3,1,1,1)-(3,2,1),
4-3/III) (2,2,1,1)-(2,2,2), **4-3/IV**) (2,2,1,1)-(3,2,1);
5-5 FPMs: (2,1,1,1,1)-(2,1,1,1,1);
5-4 FPMs: **5-4/I**) (2,1,1,1,1)-(3,1,1,1), **5-4/II**) (2,1,1,1,1)-(2,2,1,1);
5-3 FPMs: **5-3/I**) (2,1,1,1,1)-(3,2,1), **5-3/II**) (2,1,1,1,1)-(2,2,2);
6-6 FPMs: (1,1,1,1,1,1)-(1,1,1,1,1,1);
6-5 FPMs: (1,1,1,1,1,1)-(2,1,1,1,1);
6-4 FPMs: **6-4/I**) (1,1,1,1,1,1)-(3,1,1,1), **6-4/II**) (1,1,1,1,1,1)-(2,2,1,1);
6-3 FPMs: **6-3/I**) (1,1,1,1,1,1)-(3,2,1), **6-3/II**) (1,1,1,1,1,1)-(2,2,2).

The comparison of the found FPM architectures with those reported in [2] reveals that every architecture reported in [2] is referable to one of the architectures identified here, but the architectures identified here are not all present in the enumeration [2]. In particular, the subsets 6-6, 6-5, 6-4, 6-3 and 5-3 are the same in both the lists, whereas the remaining m-n subsets are different. Moreover, it is worth noting that the three 3-3 architectures found here are the ones used in [8] to generate the family of "flagged parallel manipulators". Eventually, one can easily check that all the 35 architectures listed in [9] are referable to those identified here.

3 Singularity Loci Classification

The left-Jacobian transpose, **J**, of the general FPM can be written (see [4] for details and Figure 1(b) for the notation) as follows:

$$\mathbf{J} = \begin{bmatrix} \mathbf{U} \\ \mathbf{V} \end{bmatrix} \tag{5}$$

where $\mathbf{U} = [\mathbf{u}_1, \mathbf{u}_2, \mathbf{u}_3, \mathbf{u}_4, \mathbf{u}_5, \mathbf{u}_6]$, and $\mathbf{V} = [\mathbf{v}_1, \mathbf{v}_2, \mathbf{v}_3, \mathbf{v}_4, \mathbf{v}_5, \mathbf{v}_6]$ with $\mathbf{u}_i = (\mathbf{P}_i - \mathbf{B}_i)$, and $\mathbf{v}_i = (\mathbf{P}_i - \mathbf{P}_1) \times \mathbf{u}_i$, for $i = 1, \ldots, 6$.

With reference to (5), the singularity condition of the general FPM is $det(\mathbf{J}) = 0$, where $det(\mathbf{J})$ can be computed through the Laplace expansion [4, 7] as follows:

$$\begin{aligned} det(\mathbf{J}) = {} & u_{123}v_{456} - u_{124}v_{356} + u_{125}v_{346} - u_{126}v_{345} + \\ & + u_{134}v_{256} - u_{135}v_{246} + u_{136}v_{245} + u_{145}v_{236} - u_{146}v_{235} + u_{156}v_{234} \end{aligned} \tag{6}$$

where $u_{ijk} = det([\mathbf{u}_i, \mathbf{u}_j, \mathbf{u}_k]) \equiv \mathbf{u}_i \cdot \mathbf{u}_j \times \mathbf{u}_k$, and $v_{ijk} = det([\mathbf{v}_i, \mathbf{v}_j, \mathbf{v}_k]) \equiv \mathbf{v}_i \cdot \mathbf{v}_j \times \mathbf{v}_k$ with $i, j, k = 1, \ldots, 6$. Even though expression (6) was deduced [4] for the 6-6 FPM, it holds for all the 21 FPM architectures identified in the previous section where it can be further simplified by suitably choosing the attachment point, P_1, to use as reference point in the formulas. In fact, a multiple attachment point, when chosen as reference point, can make more than one $\mathbf{v}_i$ vector null; and, since the leg numbering is arbitrary, the only presence of a multiple attachment point allows this choice.

Since attachment points' multiplicity can be at most three, the following three cases can be distinguished: (a) FPMs with only single S pairs, (b) FPMs with at least one double S pair and no triple S pair, and (c) FPMs with at least one triple S pair. Among the FPM architectures identified in the previous section, *one* has only single S pairs (i.e., the 6-6); *nine* have at least one double S pair and no triple S pair (i.e., 3-3/II, 4-4/III, 4-3/III, 5-5, 5-4/II, 5-3/II, 6-5, 6-4/II, 6-3/II); and *eleven* have at least one triple S pair (i.e., 3-3/I, 3-3/III, 4-4/I, 4-4/II, 4-3/I, 4-3/II, 4-3/IV, 5-4/I, 5-3/I, 6-4/I, 6-3/I).

3.1 FPMs with Only Single S Pairs

This FPM family contains only *one* architecture, the 6-6, and needs the use of expression (6) as it is. This expression, referred to 6-6 FPMs, was discussed in [4] mainly to highlight its consistency with previous geometrically-obtained results and to determine which type of algebraic form assumes the singularity locus equation. For the sake of conciseness, that discussion will not be summarized here.

Nevertheless, in addition to the comments reported in [4], it is worth stressing that expression (6) can be factorized as follows

$$det(\mathbf{J}) = d_1 d_2 d_3 d_4 d_5 d_6 (a_{123}b_{456} - a_{124}b_{356} + a_{125}b_{346} - a_{126}b_{345} + \\ + a_{134}b_{256} - a_{135}b_{246} + a_{136}b_{245} + a_{145}b_{236} - a_{146}b_{235} + a_{156}b_{234}) \quad (7)$$

with $a_{ijk} = \mathbf{a}_i \cdot \mathbf{a}_j \times \mathbf{a}_k$, and $b_{ijk} = \mathbf{b}_i \cdot \mathbf{b}_j \times \mathbf{b}_k$, for $i, j, k = 1, \ldots, 6$, where $\mathbf{a}_i$ is the unit vector of the ith leg axis and $\mathbf{b}_i = (\mathbf{P}_i - \mathbf{P}_1) \times \mathbf{a}_i$ is the moment of the same line[4]; whereas $d_i = |\mathbf{u}_i|$ is the length of the ith leg. Expression (7) separates the leg lengths from the factor in round brackets, which is a geometric invariant referable to the relative positions of the six leg axes. This invariant expresses a particular property of a six-line set that can be enunciated as follows "*if this invariant is equal to zero, the six lines cannot be the lines of action of as many reaction forces that have to equilibrate any force system applied to a rigid body; moreover, at parity of force system that must be equilibrated, the greater its value is, the lower the product of the magnitudes of the six reaction forces is.*" In short, if six lines make the factor in round brackets equal to zero, they can be called "statically" dependent, otherwise they can be called "statically" independent.

3.2 FPMs with at least One Double S Pair and No Triple S Pair

This FPM family contains *nine* architectures (i.e., 3-3/II, 4-4/III, 4-3/III, 5-5, 5-4/II, 5-3/II, 6-5, 6-4/II, 6-3/II). In all these architectures, the presence of a double S pair allows the legs to be so numbered that $P_1 \equiv P_2$ what makes $\mathbf{v}_2 = \mathbf{0}$. The zeroing of $\mathbf{v}_2$ implies that, in expression (6), all the v_{2jk} are equal to zero. As a consequence, for all these architectures, expression (6) can be reduced as follows

$$det(\mathbf{J}) = u_{123}v_{456} - u_{124}v_{356} + u_{125}v_{346} - u_{126}v_{345} \equiv \mathbf{n}_{12} \cdot \mathbf{s}_{3456} \quad (8)$$

where

$$\mathbf{n}_{12} = \mathbf{u}_1 \times \mathbf{u}_2 \equiv d_1 d_2 (\mathbf{a}_1 \times \mathbf{a}_2) \quad (9a)$$

$$\mathbf{s}_{3456} = \mathbf{u}_3 v_{456} - \mathbf{u}_4 v_{356} + \mathbf{u}_5 v_{346} - \mathbf{u}_6 v_{345} \equiv d_3 d_4 d_5 d_6 \\ (\mathbf{a}_3 b_{456} - \mathbf{a}_4 b_{356} + \mathbf{a}_5 b_{346} - \mathbf{a}_6 b_{345}) \quad (9b)$$

Vector $\mathbf{n}_{12}$ is perpendicular to the triangle, $B_1 P_1 B_2$, formed by the two legs that share the double spherical pair; whereas, vector $\mathbf{s}_{3456}$ can be further elaborated to reach analytic expressions that exploit the particular geometry of the architecture to

[4] The 6-tuple $\$_i = (\mathbf{a}_i^T, \mathbf{b}_i^T)^T$ is the screw of the ith leg axis, and identifies the location of this axis in the space.

be analyzed. For instance, the two expressions reported in [10] and [11], for $det(\mathbf{J})$, straightforwardly provide the two particular expressions of $\mathbf{s}_{3456}$ that hold for the architectures 6-3/II and 6-4/II, respectively. The second expression of $det(\mathbf{J})$ given by (8) states that all the singular configurations of these architectures can be referred to three geometric conditions: i) $\mathbf{n}_{12}$ is perpendicular to $\mathbf{s}_{3456}$; ii) $\mathbf{n}_{12}$ is a null vector (i.e., the triangle $B_1P_1B_2$ degenerates into a segment); and iii) $\mathbf{s}_{3456}$ is a null vector, that is, the following vector relationship is satisfied

$$d_3d_4d_5d_6(\mathbf{a}_3b_{456} - \mathbf{a}_4b_{356} + \mathbf{a}_5b_{346} - \mathbf{a}_6b_{345}) = 0 \tag{10}$$

Condition (ii) occurs when the substructure constituted by the two legs that share the double S pair assumes a singular configuration. Condition (iii) occurs when the substructure constituted by the legs 3, 4, 5, and 6 plus platform and base, with the platform connected to a fixed frame by an S pair having P_1 as center, and with the base connected to the same frame trough a revolute pair whose axis passes through the points B_1 and B_2, assumes a singular configuration. Eventually, condition (i) occurs when the two above-mentioned substructures are not singular, but they are so assembled that form a singular structure.

3.3 FPMs with at least One Triple S Pair

This FPM family contains *eleven* architectures (i.e., 3-3/I, 3-3/III, 4-4/I, 4-4/II, 4-3/I, 4-3/II, 4-3/IV, 5-4/I, 5-3/I, 6-4/I, 6-3/I). In all these architectures, the presence of a triple S pair allows the legs to be so numbered that $P_1 \equiv P_2 \equiv P_3$ what makes $\mathbf{v}_2 = \mathbf{v}_3 = \mathbf{0}$. The zeroing of $\mathbf{v}_2$ and $\mathbf{v}_3$ implies that, in expression (6), all the v_{ijk} where at least one of the indices i, j, and k is equal to 2 or 3 are equal to zero. As a consequence, for all these architectures, expression (6) can be reduced to a single term, as follows

$$det(\mathbf{J}) = u_{123}v_{456} \equiv d_1d_2d_3d_4d_5d_6(a_{123}b_{456}) \tag{11}$$

Expression (11) makes it possible to state that all the singularities of these architectures can be referred to two geometric conditions: i) $u_{123} = 0 \Rightarrow$ the tetrahedron, $B_1B_2B_3P_1$, formed by the three legs that share the triple S pair, degenerates into a triangle, and ii) $v_{456} = 0 \Rightarrow$ the intersection of the three planes, which the center, P_1, of the triple S pair and the axes of the legs 4, 5, and 6 lie on, is a line. The occurrence of condition (i) makes point P_1 able to perform infinitesimal displacements; whereas, the occurrence of condition (ii) makes the platform able to perform infinitesimal rotations around an axis passing through P_1 (i.e., around the line which is the common intersection of the above-mentioned three planes).

4 Conclusion

All the m-n FPM architectures have been classified according to a criterion that makes it possible to collect them into families which share similar singularity locus equations. Then, the particular forms of these equations have been deduced.

The results of this analysis is that 21 FPM architectures can be distinguished, but only three forms of singularity locus equation exist: the first contains 10 terms and must be used only for one architecture; the second contains 4 terms and must be used for other nine architectures; and the third contains only one term and must be used for the remaining eleven architectures. Moreover, the second and the third forms of singularity locus equation provide an easy to use geometric criterion for classifying the singularities of all the architectures they hold for.

Acknowledgements This work has been developed at the Laboratory of Advanced Mechanics (MECH-LAV) of Ferrara Technopole, supported by UNIFE funds and by Regione Emilia Romagna (District Councillorship for Productive Assets, Economic Development, Telematic Plan) POR-FESR 2007–2013, Attività I.1.1.

References

1. Conconi, M., Carricato, M.: A new assessment of singularities of parallel kinematic chains. IEEE Trans. Robot. **25**(4), 757–770 (2009)
2. Innocenti, C., Parenti-Castelli, V.: Exhaustive enumeration of fully-parallel kinematic chains. In: Proc. of the 1994 ASME Int. Winter Annual Meeting, Chicago, USA. DSC-Vol. 55-2, pp. 1135–1141 (1994)
3. St-Onge, B.M., Gosselin, C.M.: Singularity analysis and representation of the general Gough–Stewart platform. Int. J. Robot. Res. **19**(3), 271–288 (2000)
4. Di Gregorio, R.: Singularity-locus expression of a class of parallel mechanisms. Robotica **20**, 323–328 (2002)
5. Li, H., Gosselin, C.M., Richard, M.J., St-Onge, B.M.: Analytic form of the six-dimensional singularity locus of the general Gough–Stewart platform. ASME J. Mech. Des. **128**(1), 279–287 (2006)
6. Ben-Horin, P., Shoham, M.: Singularity analysis of a class of parallel robots based on Grassmann–Cayley algebra. Mech. Mach. Theory **41**(8), 958–970 (2006)
7. Eberly, D.: The Laplace expansion theorem: computing the determinants and inverses of matrices. Geometric Tools, LLC (2008). http://www.geometrictools.com/Documentation/LaplaceExpansionTheorem.pdf
8. Alberich-Carramiñana, M., Thomas, F., Torras, C.: Flagged parallel manipulators. IEEE Trans. Robot. **23**(5), 1013–1023 (2007)
9. Faugère, J.C., Lazard, D.: Combinatorial classes of parallel manipulators. Mech. Mach. Theory **30**(6), 765–776 (1995)
10. Di Gregorio, R.: Analytic formulation of the 6-3 fully-parallel manipulator's singularity determination. Robotica **19**, 663–667 (2001)
11. Di Gregorio, R.: Singularity locus of 6-4 fully-parallel manipulators. In: J. Lenarčič, M.M. Stanišić (eds.) Advances in Robot Kinematics: Motion in Man and Machine, pp. 437–445. Springer (2010)

Compensation of Compliance Errors in Parallel Manipulators Composed of Non-perfect Kinematic Chains

Alexandr Klimchik, Anatol Pashkevich, Damien Chablat and Geir Hovland

Abstract The paper is devoted to the compliance errors compensation for parallel manipulators under external loading. Proposed approach is based on the non-linear stiffness modeling and reduces to a proper adjusting of a target trajectory. In contrast to previous works, in addition to compliance errors caused by machining forces, the problem of assembling errors caused by inaccuracy in the kinematic chains is considered. The advantages and practical significance of the proposed approach are illustrated by examples that deal with groove milling with Orthoglide manipulator.

Key words: Parallel robots, nonlinear stiffness modeling, compliance error compensation, non-perfect manipulators

1 Introduction

In many robotic applications such as machining, grinding, trimming etc., the interaction between the workpiece and technological tool causes essential deflections that significantly decrease the processing accuracy and quality of the final product. To overcome this difficulty, it is possible to modify either control algorithm or the prescribed trajectory, which is used as the reference input for a control system [1]. This paper focuses on the second approach that is considered to be more realistic in the practice. In contrast to the previous works, the proposed compliance error com-

Alexandr Klimchik · Anatol Pashkevich
Institut de Recherche en Communications et Cybernetique de Nantes, Nantes, France and Ecole des Mines de Nantes, Nantes, France, e-mail: {alexandr.klimchik,anatol.pashkevich}@mines-nantes.fr

Damien Chablat
Institut de Recherche en Communications et Cybernetique de Nantes, Nantes, France,
e-mail: damienl.chablat@irccyn.ec-nantes.fr

Geir Hovland
University of Agder, Grimstad, Norway, e-mail: geir.hovland@uia.no

J. Lenarčič, M. Husty (eds.), *Latest Advances in Robot Kinematics*,
DOI 10.1007/978-94-007-4620-6_7,

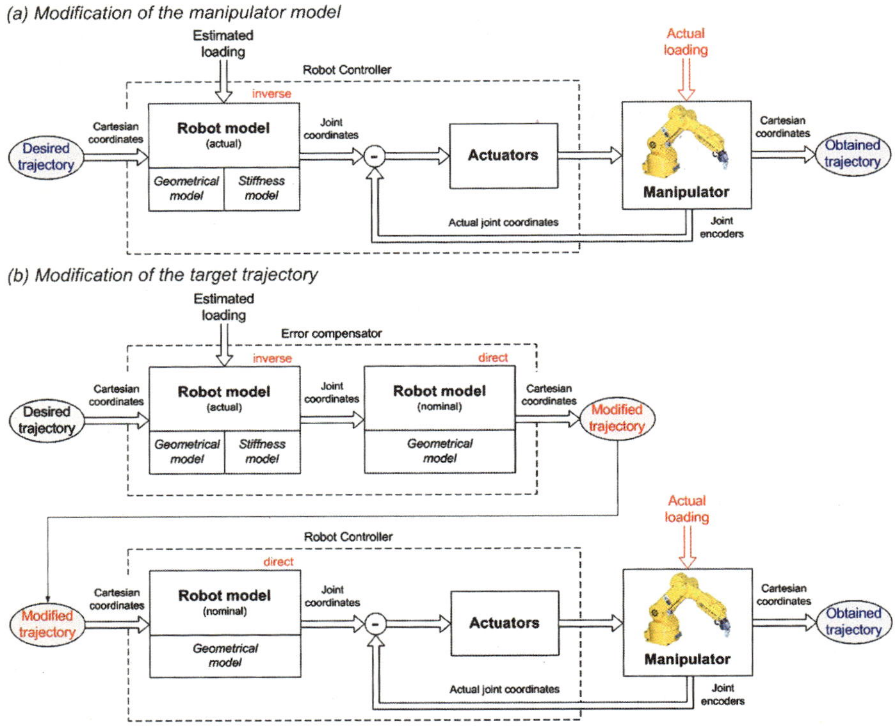

Fig. 1 Robot error compensation methods.

pensation technique is based on the non-linear stiffness model of the manipulator that is able to take into account significant external loading [2].

Usually, the problem of the robot error compensation can be solved in two ways that differ in degree of modification of the robot control software:

(a) by modification of the manipulator model (Fig. 1a) which better suits to the real manipulator and is used by the robot controller (in simple case, it can be limited by tuning of the nominal manipulator model, but may also involve essential model enhancement by introducing additional parameters, if it is allowed by the robot manufacturer);
(b) by modification of the robot control program (Fig. 1b) that defines the prescribed trajectory in Cartesian space (here, using relevant error model, the input trajectory is generated in a such way that under the loading the output trajectory coincides with the desired one, while input trajectory differs from the target one).

It is clear that the first approach can be implemented in on-line mode, while the second one requires preliminary off-line computations. But in practice it is rather unrealistic to include the stiffness model in a commercial robot controller where all transformations between the joint and Cartesian coordinates are based on the manipulator geometrical model. In contrast, the off-line error compensation, based on the second approach, is attractive for industrial applications.

For the geometrical errors, relevant compensation techniques are already well developed. Comprehensive review of related works is given in [3]. In the frame of this work, it is assumed that the geometrical errors are less essential compared to the non-geometrical ones caused by the interaction between the machining tool and workpiece. So, the main attention will be paid to the compliance errors and their compensation techniques.

2 Problem of Compliance Error Compensation

For the compliance errors, the compensation technique must rely on two components. The first of them describes distribution of the stiffness properties throughout the workspace and is defined by the stiffness matrix as a function of the joint coordinates or the end-effector location [2]. The second component describes the forces/torques acting on the end-effector while the manipulator is performing its manufacturing task (manipulator loading). In this work, it is assumed that the second component is given and can be obtained either from the dedicated technological process model (that take into account the tool wear, type of machining process, cutting speed, rake angle, cutting fluid, workpiece shape etc.) or by direct measurements using the force/torque sensor integrated into the end-effector.

The stiffness matrix required for the compliance errors compensation highly depends on the robot configuration and essentially varies throughout the workspace. From general point of view, full-scale compensation of the compliance errors requires essential revision of the manipulator model embedded in the robot controller. In fact, instead of conventional geometrical model that provides inverse/direct coordinate transformations from the joint to Cartesian spaces and vice versa, here it is necessary to employ the so-called kinetostatic model [4]. It is essentially more complicated than the geometrical model and requires intensive computations.

If the compliance errors are relatively small, composition of conventional geometrical model and the stiffness matrix give rather accurate approximation of the modified mapping from the joint to Cartesian space. In this case, for the first compensation scheme (see Fig. 1a), the kinetostatic model can be easily implemented on-line if there is an access to the control software modification. Otherwise, the second scheme (see Fig. 1b) can be easily applied. Moreover, with regard to the robot-based machining, there is a solution that does not require force/torque measurements or computations [1] where the target trajectory for the robot controller is modified by applying the "mirror" technique. However, this approach is only suitable for the large-scale production where the manufacturing task and the workpiece location remains the same. Hence, to be applied to the robotic-based machining, the existing compliance errors compensation techniques should be essentially revised to take into account essential forces and torques as well as some other important error sources (inaccuracy in serial chains, for instance).

3 Nonlinear Technique for Compliance Error Compensation

In industrial robotic controllers, the manipulator motions are usually generated using the inverse kinematic model that allows us to compute the input signals for actuators $\boldsymbol{\rho}_0$ corresponding to the desired end-effector location $\mathbf{t}_0$, which is assigned assuming that the compliance errors are negligible. However, if the external loading $\mathbf{F}'$ is essential, the kinematic control becomes non-applicable because of changes in the end-effector location. It can be computed from the nonlinear compliance model as

$$\mathbf{t}_{\mathrm{F}} = f^{-1}(\mathbf{F}|\mathbf{t}_0) \tag{1}$$

where the subscripts 'F' and '0' refer to the loaded and unloaded modes respectively, and '|' separates arguments and parameters of the function $f()$. Some details concerning this function are given in our previous publication [2]. It should be mentioned that function (1) takes into account loop-closure constraints and validates both for serial and parallel manipulators.

To compensate this undeterred end-effector displacement from $\mathbf{t}_0$ to $\mathbf{t}_{\mathrm{F}}$, the target point should be modified in a such way that, under the loading $\mathbf{F}$, the end-effector is located in the desired point $\mathbf{t}_0$. This requirement can be expressed using the stiffness model in the following way

$$\mathbf{F} = f\left(\mathbf{t}_0|\mathbf{t}_0^{(\mathrm{F})}\right) \tag{2}$$

where $\mathbf{t}_0^{(\mathrm{F})}$ denotes the modified target location. Hence, the problem is reduced to the solution of the nonlinear equation (2) for $\mathbf{t}_0^{(\mathrm{F})}$, while $\mathbf{F}$ and $\mathbf{t}_0$ are assumed to be given. It is worth mentioning that this equation completely differs from the equation $\mathbf{F} = f(\mathbf{t}|\mathbf{t}_0)$, where the unknown variable is $\mathbf{t}$. It means that here the compliance model does not allow us to compute the modified target point $\mathbf{t}_0^{(\mathrm{F})}$ straightforwardly, while the linear compensation technique directly operates with Cartesian compliance matrix [5].

Since $\mathbf{t}_0$ and $\mathbf{t}_0^{(\mathrm{F})}$ are close enough, to solve equation (2) for $\mathbf{t}_0^{(\mathrm{F})}$, the Newton-Raphson technique can be applied. It yields the following iterative scheme

$$\mathbf{t}_0^{(\mathrm{F})'} = \mathbf{t}_0^{(\mathrm{F})} + \mathbf{K}_{\mathrm{t.p.}}^{-1}(\mathbf{t}_0|\mathbf{t}_0^{(\mathrm{F})})\left(\mathbf{F} - f(\mathbf{t}_0|\mathbf{t}_0^{(\mathrm{F})})\right) \tag{3}$$

where the prime corresponds to the next iteration and $\mathbf{K}_{\mathrm{t.p.}}(\mathbf{t}_0|\mathbf{t}_0^{(\mathrm{F})})$ is the stiffness matrix computed with respect to the second argument of the function $\mathbf{F} = f(\mathbf{t}|\mathbf{t}_0)$ at the original target point (i.e. for $\mathbf{t} = \mathbf{t}_0$) assuming that unloaded configuration is modified and corresponds to the end-effector location $\mathbf{t}_0^{(\mathrm{F})}$. Here $\mathbf{F}$ stands for the solution of equation (2), while the function $f(\mathbf{t}_0|\mathbf{t}_0^{(\mathrm{F})})$ defines the loading for the current end-effector location under the loading $\mathbf{t}_0^{(\mathrm{F})}$.

To overcome computational difficulties related to the evaluation of the matrix $\mathbf{K}_{\mathrm{t.p.}}(\mathbf{t}_0|\mathbf{t}_0^{(\mathrm{F})})$, it is possible to use its simple approximation that does not change from iteration to iteration. In particular, assuming that $\mathbf{t}$ and $\mathbf{t}_0$ are close enough and the stiffness properties do not vary substantially in their neighborhood, the stiffness

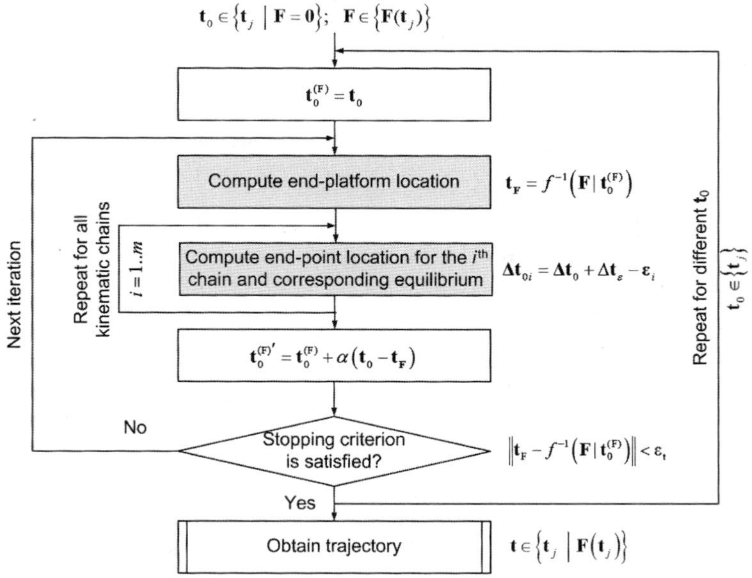

Fig. 2 Procedure for compensation of compliance errors in parallel manipulator.

model (2) can be approximated by a linear expression $\mathbf{F} = \mathbf{K}_{\mathrm{C}}(\mathbf{t} - \mathbf{t}_0)$, which includes the conventional Cartesian stiffness matrix $\mathbf{K}_{\mathrm{C}}$. This allows us to replace the above derivative matrix $\mathbf{K}_{\mathrm{t.p.}}$ by $-\mathbf{K}_{\mathrm{C}}$ and to present the iterative scheme (3) as

$$\mathbf{t}_0^{(\mathrm{F})'} = \mathbf{t}_0^{(\mathrm{F})} - \alpha \mathbf{K}_{\mathrm{C}}^{-1}(\mathbf{t}_0|\mathbf{t}_0^{(\mathrm{F})}) \left(\mathbf{F} - f(\mathbf{t}_0|\mathbf{t}_0^{(\mathrm{F})})\right) \tag{4}$$

where $\alpha \in (0, 1)$ is the scalar parameter ensuring the convergence. Using the non-linear compliance model (1), this idea can also be implemented in an iterative algorithm

$$\mathbf{t}_0^{(\mathrm{F})'} = \mathbf{t}_0^{(\mathrm{F})} + \alpha \left(\mathbf{t}_0 - f^{-1}(\mathbf{F}|\mathbf{t}_0^{(\mathrm{F})})\right) \tag{5}$$

which does not include stiffness matrices $\mathbf{K}_{\mathrm{C}}$ or $\mathbf{K}_{\mathrm{t.p.}}$. Obviously, this is the most computationally convenient solution and it will be used in the next section.

It should be mentioned that the considered case deals with a perfect parallel manipulator where end-points of all kinematic chains are aligned and matched. However, in practice, kinematic chains may include some errors that do not allow us to assemble them in a parallel manipulator with the same end-effector location. In this case it is required to compensate two types of errors (caused by the external loading $\mathbf{F}$ and inaccuracy in the serial chains). The second source of errors can be taken into account by changing of target location $\Delta\mathbf{t}_{0i}$ for each kinematic chain

$$\Delta\mathbf{t}_{0i} = \Delta\mathbf{t}_0 + \Delta\mathbf{t}_\varepsilon - \boldsymbol{\varepsilon}_i \tag{6}$$

where $\Delta\mathbf{t}_\varepsilon$ is the end-effector deflections due to assembling of non-perfect kinematic chains and $\boldsymbol{\varepsilon}_i$ is shifting of the end-point location of ith kinematic chain because of geometrical errors. Using the principle of virtual work it can be proved that $\Delta\mathbf{t}_\varepsilon$ can be computed as

$$\Delta\mathbf{t}_\varepsilon = \left(\sum_{i=1}^{m}\mathbf{K}_\mathrm{C}^{(i)}\right)^{-1}\sum_{i=1}^{m}\left(\mathbf{K}_\mathrm{C}^{(i)}\boldsymbol{\varepsilon}_i\right) \tag{7}$$

where $\mathbf{K}_\mathrm{C}^{(i)}$ defines the Cartesian stiffness matrix of i-th kinematic chain that can be computed using techniques proposed in [2] and m is the number of kinematic chains in the parallel manipulator. More detailed presentation of the developed iterative routines is given in Fig. 2.

Hence, using the proposed computational techniques, it is possible to compensate the essential compliance errors by proper adjusting the reference trajectory that is used as an input for robotic controller. In this case, the control is based on the inverse kinetostatic model (instead of kinematic one) that takes into account both the manipulator geometry and elastic properties of its links and joints. Efficiency of this technique is confirmed by an example presented in the next section.

4 Illustrative Example: Compliance Error Compensation for Milling

Let us illustrate the compliance errors compensation technique by an example of the circle groove milling with Orthoglide manipulator (Fig. 3). Detailed specification of this manipulator can be founded in [6]. According to [7], such technological process causes the loading $F_r = 215$ N; $F_t = -10$ N; $F_z = -25$ N that together with angular parameter $\varphi = [0, 360°]$ define the forces F_x and F_y (Figs. 3b,c). Here, the tool length h is equal to 100 mm. It is assumed that the manipulator has two sources of inaccuracy:

(a) Orthoglide manipulator

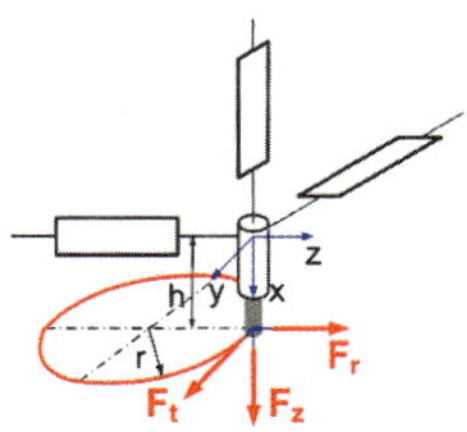

(b) groove milling with Orthoglide

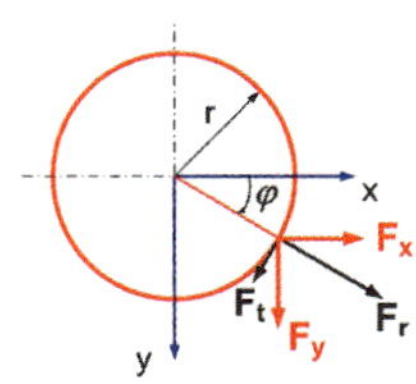

(c) milling forces

Fig. 3 Milling forces and trajectory location for groove milling using Orthoglide manipulator.

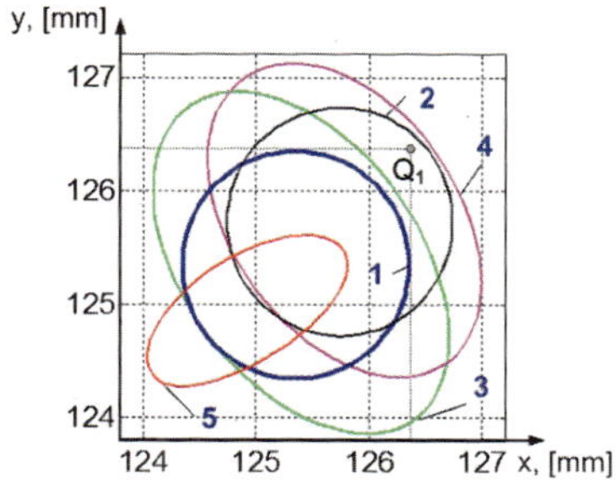

(1) Target trajectory;

(2) Shifting of target trajectory caused by errors in serial chains (assembling errors);

(3) Shifting of target trajectory caused by cutting force (compliance errors);

(4) Shifting of target trajectory caused by cutting force and errors in serial chains;

(5) Adjusted trajectory, that insure following the target trajectory while machining.

Fig. 4 Influence of different error sources on the machining trajectory.

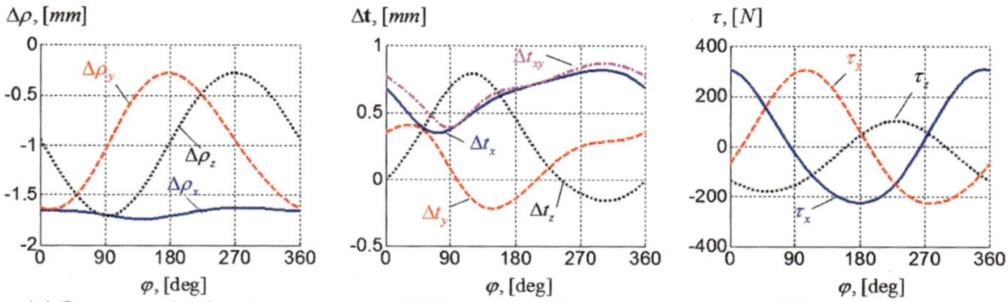

(a) Compensation in actuators (b) Compensated error (c) Forces in actuators

Fig. 5 Compliance error compensation for Orthoglide milling application.

1. the assembling errors in the kinematic chains (assembling errors in actuator angular locations of about 1° around the corresponding actuated axis) causing internal forces and relevant deflections in joints and links;
2. the external loading $\|\mathbf{F}\| = 217$ N which generates essential compliance deflections causing non-desirable end-platform displacement.

In order to illustrate influence of different error sources on the machining trajectory, let us focus on the 1 mm radius of the circle that should be machined. In this case, the stiffness matrix is almost the same along the trajectory. Modeling results for the neighborhood of point Q_1 (see [2] for details) are presented in Fig. 4. They show the influence of different error sources on the machining trajectory without compensation and the revised machining trajectory that should be implemented in robot controller in order to follow the target trajectory while machining. Here, path 5 compensates the effects seen in path 4 such that circle 1 is achieved. It can be seen that the center of path 5 is on the opposite side of circle 1 compared to path 4. It can also be seen that the main elliptic direction in path 4 becomes the smallest elliptic direction in path 5. It should be mentioned that because of the torque induced by the cutting forces (tool length 100 mm), the target trajectory and shifted trajectory under the cutting forces are intersecting.

Figure 5 presents results for the milling of the 50 mm circle. In this case, without compensation, the compliance errors can exceed 0.8 mm. After compensation, the above mentioned errors are reduced to zero (it is obvious that in practice, the compensation level is limited by the accuracy of the stiffness model). This compensation is achieved due to the modification of the actuator coordinates $\boldsymbol{\rho}$ along the machining trajectory. Compared to the relevant values computed via the inverse kinematics, the actuator coordinates differ up to 1.7 mm. Corresponding forces in actuators can reach 300 N. Some more results on the compliance errors compensation are presented in Fig. 5, which includes plots showing modifications of the actuator coordinates $\Delta\boldsymbol{\rho}$, values of compensated end-effector displacement $\Delta\mathbf{t}$ and the torques in actuators τ. It should be mentioned that while implementing target trajectory in the robot controller additional control errors may arise.

Hence, the developed algorithm demonstrates good convergence. It is able to compensate the compliance errors and can be efficient both for off-line trajectory planning and for on-line errors compensation.

5 Conclusions

The paper presents a new technique for on-line and off-line compensation of the compliance errors caused by external loadings in parallel manipulators (including over-constrained ones) composed of both perfect and non-perfect serial chains. In contrast to previous works this technique is based on nonlinear stiffness model (inverse kinetostatic model) that gives essential benefits for robotic-based machining, where the elastic deflections can be essential. The advantages and practical significance are illustrated by groove milling with Orthoglide manipulator.

Acknowledgements The work presented in this paper was partially funded by the Region "Pays de la Loire", France and by the project ANR COROUSSO, France.

References

1. Dépincé, P., Hascoët, J.-Y.: Active integration of tool deflection effects in end milling. Part 2. Compensation of tool deflection. Int. J. Mach. Tools Manuf. **46**(9), 945–956 (2006)
2. Pashkevich, A., Klimchik, A., Chablat, D.: Enhanced stiffness modeling of manipulators with passive joints. Mech. Mach. Theory **46**(5), 662–679 (2011)
3. Hollerbach, J., Khalil, W., Gautier, M.: Model identification. Chapter 14. In: Siciliano B., Khatib O. (eds.) Springer Handbook of Robotics, pp. 321–344. (2008)
4. Su, H.-J., McCarthy, J.M.: A polynomial homotopy formulation of the inverse static analyses of planar compliant mechanisms. J. Mech. Des. **128**(4), 776–786 (2006)
5. Gong, Ch., Yuan, J., Ni, J.: Nongeometric error identification and compensation for robotic system by inverse calibration. Int. J. Mach. Tools Manuf. **40**(14), 2119–2137 (2000)
6. Chablat, D., Wenger, Ph.: Architecture optimization of a 3-DOF parallel mechanism for machining applications, the Orthoglide. IEEE Trans. Robot. Autom. **19**(3), 403–410 (2003)
7. Majou, F., Gosselin, C., Wenger, P., Chablat, D.: Parametric stiffness analysis of the Orthoglide. Mech. Mach. Theory **42**(3), 296–311 (2007)

Motion Planning of the Multi-Bar System: The Imbalanced Jacobian Algorithm

Janusz Jakubiak, Krzysztof Tchoń and Mariusz Janiak

Abstract We study the motion planning problem for the multi-bar system composed of a series of rigid bars connected by spherical joints, moving in $\mathbb{R}^3$ and subject to non-holonomic constraints. The control system representation of the system takes the form of a driftless control system with 3 inputs. In order to avoid representation singularities, the constrained motion planning problem is addressed and solved using the imbalanced Jacobian algorithm. Performance of the algorithm is illustrated with the example of the 2-bar system.

Key words: Non-holonomic system, Jacobian motion planning, constraints

1 Introduction

The multi-bar system is defined as a chain of rigid bars connected by spherical joints that moves in the m-dimensional Euclidean space, subject to non-holonomic constraints resulting from the assumption that the instantaneous velocity of the source point of each bar is aligned with that bar. This kind of system that has appeared for the first time in [1], recently is playing a significant role in the study of geometry and flatness of non-holonomic control systems [5–7]. Specifically, in [6] it has been proved that, locally, the multi-bar system is feedback equivalent to the chained form, therefore differentially flat. Furthermore, it turns out that its only minimal flat output is the position of the source point of the initial bar. To our best knowledge, the multi-bar system does not have any physical realization, except for $m = 2$ when it specializes to the multi-trailer system in the plane [4]. Potentially, the multi-bar system moving in $\mathbb{R}^3$ may represent the motion of a chain of rigid bars in a very viscous liquid or the flight of a formation of aircraft. However, in this paper the multi-bar

Janusz Jakubiak · Krzysztof Tchoń · Janusz Jakubiak
Institute of Computer Engineering, Control and Robotics, Wrocław University of Technology, Wrocław, Poland, e-mail: {janusz.jakubiak, krzysztof.tchon, mariusz.janiak}@pwr.wroc.pl

J. Lenarčič, M. Husty (eds.), *Latest Advances in Robot Kinematics*,
DOI 10.1007/978-94-007-4620-6_8,

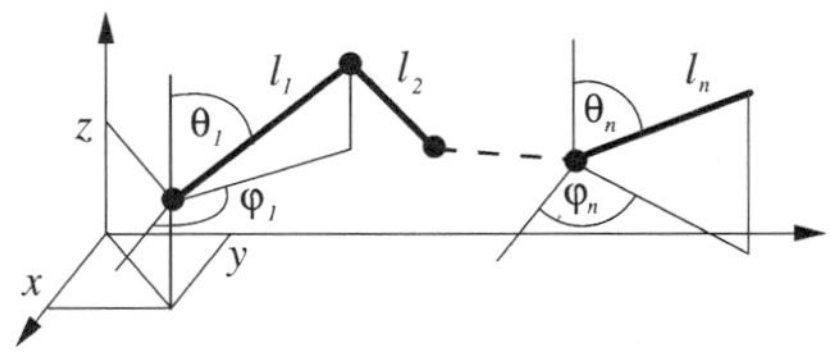

Fig. 1 Multi-bar system.

system should be merely regarded as a 3-input, non-holonomic testbed for motion planning algorithms.

In this paper we shall study the motion planning problem of the multi-bar system moving in 3-dimensional Euclidean space. Using the spherical coordinates we derive the control system representations of the kinematics for 1-, 2- and 3-bar systems. To account for the representation singularities, we formulate a constrained motion planning problem, and solve it using the imbalanced Jacobian motion planning algorithm [3]. Performance of the algorithm will be illustrated with computer simulations of the 2-bar system. The main contribution of this paper lies in providing a constrained motion planning algorithm for a new class of 3-input non-holonomic systems, beyond the trident snake robot [2].

The remaining part of this paper is organized as follows. Section 2 defines the kinematics of the multi-bar systems. The imbalanced Jacobian motion planning algorithm is described in Section 3. Section 4 presents results of computations. Conclusions are given in Section 5.

2 Kinematics of Multi-Bar System

We shall study the multi-bar system composed of n rigid bars connected to each other by spherical joints, moving in the 3-dimensional Euclidean space, schematically shown in Fig. 1. It is assumed that the bar number i has length l_i, and connects the joints number $i-1$ and i, for $i = 1,\ldots,n$, therefore the end point of the ith bar moves on the 2-dimensional sphere of radius l_i. The source point of the initial bar (the bar number 1), treated as the 0-joint, can move freely in $\mathbb{R}^3$. The configuration of the multi-bar system will be defined by means of the Cartesian position of the source point of the initial bar and the spherical coordinates (φ_i, θ_i) of the end point of the bar i (note that the angles $\theta_i = 0, \pi$ correspond to representation singularities). For the notational convenience, the absolute angles have been chosen, measured with respect to the axes of the inertial coordinate frame. A configuration of the system is described by

$$\mathbf{q} = (x, y, z, \varphi_1, \theta_1, \ldots, \varphi_n, \theta_n)^T \in \mathbb{R}^N, \tag{1}$$

where $N = 2n+3$. The velocity constraints are imposed on the motion of the multi-bar system by requiring that the instantaneous velocity of the ith joint is aligned with the $(i+1)$st bar, leading to a set of constraint equations for $i = 1,\ldots,n$ (for brevity of notation we shall use s and c for sine and cosine functions, respectively)

$$-\begin{pmatrix} s_{\theta_i}c_{\varphi_i} \\ s_{\theta_i}s_{\varphi_i} \\ c_{\theta_i} \end{pmatrix} \times \begin{pmatrix} \dot{x} \\ \dot{y} \\ \dot{z} \end{pmatrix} + \sum_{j=1}^{i-1} l_j \begin{bmatrix} s_{\theta_j}c_{\varphi_j}c_{\theta_i} & c_{\theta_j}s_{\varphi_j}c_{\theta_i} + s_{\theta_j}s_{\varphi_i}s_{\theta_i} \\ s_{\theta_j}s_{\varphi_j}c_{\theta_i} & -c_{\theta_j}c_{\varphi_j}c_{\theta_i} - s_{\theta_j}c_{\varphi_i}s_{\theta_i} \\ -s_{\theta_j}c_{\varphi_i - \varphi_j}s_{\theta_i} & c_{\theta_j}s_{\varphi_i - \varphi_j}s_{\theta_i} \end{bmatrix} \begin{pmatrix} \dot{\varphi}_j \\ \dot{\theta}_j \end{pmatrix} = \mathbf{0}. \quad (2)$$

Note that in (2) there are only 2 independent equations.

2.1 Kinematics of 1-Bar System

It is instructive to look in some details at the case of $n = 1$. In this case the configuration variable $\mathbf{q} = (x, y, z, \varphi_1, \theta_1)^T \in \mathbb{R}^5$. Constraints (2) take the following form

$$(\sin\theta_1 \cos\varphi_1, \sin\theta_1 \sin\varphi_1, \cos\theta_1) \times (\dot{x}, \dot{y}, \dot{z}) = \mathbf{0}$$

that is tantamount to

$$\begin{bmatrix} 0 & \cos\theta_1 & -\sin\theta_1 \sin\varphi_1 & 0 & 0 \\ -\cos\theta_1 & 0 & \sin\theta_1 \cos\varphi_1 & 0 & 0 \\ \sin\theta_1 \cos\varphi_1 & -\sin\theta_1 \cos\varphi_1 & 0 & 0 & 0 \end{bmatrix} \dot{\mathbf{q}} = \mathbf{0}. \quad (3)$$

It is easily observed that since the constraint matrix contains a skew-symmetric 3×3 submatrix, its rank is equal to 2. This means that the kinematics of the 1-bar system can be represented by the driftless control system

$$\dot{\mathbf{q}} = \mathbf{G}(\mathbf{q})\mathbf{u} = \mathbf{g_1}(\mathbf{q})u_1 + \mathbf{g_2}(\mathbf{q})u_2 + \mathbf{g_3}(\mathbf{q})u_3. \quad (4)$$

The control vector fields $\mathbf{g_1}(\mathbf{q}) = (\sin\theta_1 \cos\varphi_1, \sin\theta_1 \sin\varphi_1, \cos\theta_1, 0, 0)^T$, $\mathbf{g_2}(\mathbf{q}) = \mathbf{e_4}$, $\mathbf{g_3}(\mathbf{q}) = \mathbf{e_5}$, where $\mathbf{e_i}$ stands for the ith unit vector in $\mathbb{R}^5$. The control u_1 equals the magnitude of the bar's linear velocity, while control variables (u_2, u_3) denote the rates of change of the spherical coordinates. A computation shows that the growth vector of the distribution associated with the control system (4) is equal to $(3,5)$ whenever $\sin\theta_1 \neq 0$, and to $(3,4,5)$ in the case of $\sin\theta_1 = 0$. This means that the constraints (3) are non-holonomic, i.e. the system (4) is controllable, however, in this case the representation singularities give rise to the posture singularities of the system [8]. Observe that when the 1-bar system becomes planar (i.e. $z = 0$, $\theta_1 = \frac{\pi}{2}$), the system (4) describes the kinematics of the unicycle.

2.2 Kinematics of 2- and 3-Bar Systems

For the 2-bar system the configuration variable $\mathbf{q} = (x, y, z, \varphi_1, \theta_1, \varphi_2, \theta_2)^T \in \mathbb{R}^7$, and the constraints (2) yield the following kinematics representation

$$\dot{\mathbf{q}} = \mathbf{G}(\mathbf{q})\mathbf{u} = \mathbf{g_1}(\mathbf{q})u_1 + \mathbf{g_2}(\mathbf{q})u_2 + \mathbf{g_3}(\mathbf{q})u_3, \tag{5}$$

where the control vector fields are defined as

$$\mathbf{g_1}(\mathbf{q}) = \left(l_1 d_2(\mathbf{q}) s_{\theta_1} c_{\varphi_1}, l_1 d_2(\mathbf{q}) s_{\theta_1} s_{\varphi_1}, l_1 d_2(\mathbf{q}) c_{\theta_1}, \frac{s_{\theta_2} s_{\varphi_2 - \varphi_1}}{s_{\theta_1}}, f_2(\mathbf{q}), 0, 0\right)^T,$$

$\mathbf{g_2} = \mathbf{e_6}$, $\mathbf{g_3} = \mathbf{e_7}$, $\mathbf{e_i}$ being the ith unit vector in $\mathbb{R}^7$. Above, for brevity, we have set $d_2(\mathbf{q}) = c_{\theta_1} c_{\theta_2} + s_{\theta_1} s_{\theta_2} c_{\varphi_2 - \varphi_1}$ and $f_2(\mathbf{q}) = c_{\theta_1} s_{\theta_2} c_{\varphi_2 - \varphi_1} - s_{\theta_1} c_{\theta_2}$. The system (5) is well defined on condition that $\sin\theta_1 \neq 0$. It can be shown that the identity $d_2(\mathbf{q}) = 0$ means that the bars are mutually perpendicular (a posture singularity), and that $f_2(\mathbf{q}) = d_2(\mathbf{q} + \frac{\pi}{2}\mathbf{e_5})$. Apparently, for the planar 2-bar system ($z = 0$, $\theta_1 = \theta_2 = \frac{\pi}{2}$) the control system (5) represents the kinematic car.

For the 3-bar system we get $\mathbf{q} = (x, y, z, \varphi_1, \theta_1, \varphi_2, \theta_2, \varphi_3, \theta_3)^T \in \mathbb{R}^9$, and derive from (2) the kinematics representation

$$\dot{\mathbf{q}} = \mathbf{G}(\mathbf{q})\mathbf{u} = \mathbf{g_1}(\mathbf{q})u_1 + \mathbf{g_2}(\mathbf{q})u_2 + \mathbf{g_3}(\mathbf{q})u_3, \tag{6}$$

whose control vector fields are the following

$$\mathbf{g_1}(\mathbf{q}) = \Big(l_1 l_2 d_2(\mathbf{q}) d_3(\mathbf{q}) s_{\theta_1} c_{\varphi_1}, l_1 l_2 d_2(\mathbf{q}) d_3(\mathbf{q}) s_{\theta_1} s_{\varphi_1}, l_1 l_2 d_2(\mathbf{q}) d_3(\mathbf{q}) c_{\theta_1},$$
$$l_2 d_3(\mathbf{q}) \frac{s_{\theta_2} s_{\varphi_2 - \varphi_1}}{s_{\theta_1}}, l_2 d_3(\mathbf{q}) f_2(\mathbf{q}), l_1 \frac{s_{\theta_3} s_{\varphi_3 - \varphi_2}}{s_{\theta_2}}, l_1 f_3(\mathbf{q}), 0, 0\Big)^T,$$

$\mathbf{g_2}(\mathbf{q}) = \mathbf{e_8}$, $\mathbf{g_3}(\mathbf{q}) = \mathbf{e_9}$. The terms $d_2(\mathbf{q})$ and $f_2(\mathbf{q})$ are defined, whereas $d_3(\mathbf{q}) = c_{\theta_2} c_{\theta_3} + s_{\theta_2} s_{\theta_3} c_{\varphi_3 - \varphi_2}$ and $f_3(\mathbf{q}) = c_{\theta_2} s_{\theta_3} c_{\varphi_3 - \varphi_2} - s_{\theta_2} c_{\theta_3}$. The system (6) is well defined provided that both $\sin\theta_1$ and $\sin\theta_2$ are non-zero. Vanishing of $d_2(\mathbf{q})$ means that the bars 1 and 2 are perpendicular, while $d_3(\mathbf{q}) = 0$ results in the perpendicularity of the bars 2 and 3 (posture singularities). By analogy to $f_2(\mathbf{q})$ we have $f_3(\mathbf{q}) = d_3(\mathbf{q} + \frac{\pi}{2}\mathbf{e_7})$. Again, when the 3-bar system gets planar ($z = 0$, $\theta_1 = \theta_2 = \theta_3 = \frac{\pi}{2}$), the system (6) converts to the kinematic car towing a trailer.

3 Motion Planning

Given the control system representation of kinematics of the n-bar system

$$\dot{\mathbf{q}} = \mathbf{G}(\mathbf{q})\mathbf{u} = \sum_{i=1}^{m} \mathbf{g_i}(\mathbf{q})u_i, \tag{7}$$

$\mathbf{q} \in \mathbb{R}^N$, let $\mathbf{q}(t) = \varphi_{\mathbf{q_0},t}(\mathbf{u}(\cdot))$ denote its trajectory initialized at $\mathbf{q_0}$ and steered by $\mathbf{u}(t)$. The motion planning problem for the system (7) consists in defining a control function $\mathbf{u}(t)$ such that at a prescribed time instant T the system trajectory reaches a desired point $\mathbf{q_d} \in \mathbb{R}^N$, i.e. $\mathbf{q}(T) = \mathbf{q_d}$. Additionally, to prevent the system trajectory from entering into the representation singularities, it will be required that at any time instant $t \in [0,T]$ the angles $\theta_i(t)$, $i = 1,\ldots,n$ are suitably bounded.

This constrained motion planning problem will be solved by means of the imbalanced Jacobian algorithm described in [3]. For the reader's convenience we shall concisely recollect the main idea of this algorithm. Given the trajectory $\mathbf{q}(t)$, we introduce the end point map of the system (7)

$$\mathbf{K}_{\mathbf{q_0},T}(\mathbf{u}(\cdot)) = \mathbf{q}(T) = \varphi_{\mathbf{q_0},T}(\mathbf{u}(\cdot)). \tag{8}$$

Now, the motion planning problem is equivalent to determining a control $\mathbf{u}(t)$ such that $\mathbf{K}_{\mathbf{q_0},T}(\mathbf{u}(\cdot)) = \mathbf{q_d}$, while the instantaneous values of the θ_i angles for $i = 1,\ldots,n$ are lower and upper bounded

$$\theta_i^{lb} \leq \theta_i(t) \leq \theta_i^{ub}. \tag{9}$$

In order to include the constraints (9) into the system (7), we shall describe them using the plus function $(x)_+ = \max\{x,0\}$, so that the constraints will be satisfied, whenever the functions $(\theta_i - \theta_i^{ub})_+$ and $(-\theta_i + \theta_i^{lb})_+$ vanish for every $t \in [0,T]$. Because the plus function is nonnegative, this is guaranteed when the sum of integrals over $[0,T]$ of these functions is zero. Furthermore, for the sake of smoothness the plus function will be approximated by a function

$$(x)_+ \cong p(x,\alpha) = x + \frac{1}{\alpha}\ln(1+\exp(-\alpha x)),$$

parametrized by $\alpha > 0$, that approaches $(x)_+$ along with α growing up to $+\infty$. The constraints (9) will be added to the system (7) by extending the system by an extra state variable q_{N+1}, resulting in

$$\left.\begin{array}{l} \dot{\mathbf{q}} = \mathbf{G}(\mathbf{q})\mathbf{u}, \\ \dot{q}_{N+1} = \sum_{i=1}^{n} p(\theta_i - \theta_i^{ub},\alpha) + p(-\theta_i + \theta_i^{lb},\alpha) \end{array}\right\} \cong \dot{\mathbf{q}}_{ext} = \mathbf{G}_{ext}(\mathbf{q}_{ext})\mathbf{u}. \tag{10}$$

The motion planning problem in the extended system (10) is unconstrained, and amounts to defining a control $\mathbf{u}(t)$ such that $\mathbf{q}_{ext}(T) = (\mathbf{q_d},0)$. In principle, this problem can be solved by means of a Jacobian motion planning algorithm [3]. However, in to ascertain regularity of the Jacobian, we need to regularize the extended system. This will be achieved by perturbing the $\dot{q}_{N+1}$ equation with a function $r(\mathbf{q})$, e.g. a quadratic function. The resulting regularized system takes the following form

$$\left.\begin{array}{l} \dot{\mathbf{q}} = \mathbf{G}(\mathbf{q})\mathbf{u}, \\ \dot{q}_{N+1} = \sum_{i=1}^{n} p(\theta_i - \theta_i^{ub},\alpha) + p(-\theta_i + \theta_i^{lb},\alpha) + r(\mathbf{q}) \end{array}\right\} \cong \dot{\mathbf{q}}_{reg} = \mathbf{G}_{reg}(\mathbf{q}_{reg})\mathbf{u}. \tag{11}$$

A feature of the imbalanced Jacobian motion planning algorithm is that the inverse Jacobian operator associated with the regularized system operates on the error computed for the extended system. Having chosen the Jacobian pseudo inverse algorithm this means that the control function solves the functional differential equation

$$\frac{d\mathbf{u}_{\vartheta}(t)}{d\vartheta} = -\gamma \mathbf{B}^T_{reg\vartheta}(t)\boldsymbol{\Phi}^T_{reg\vartheta}(T,t)\mathscr{G}^{-1}_{reg}(\mathbf{u}_{\vartheta}(\cdot))\mathbf{e}_{ext}(\vartheta). \tag{12}$$

The number $\gamma > 0$ defines the speed of convergence of the algorithm. The data appearing in (12) are computed in the following way. Given a control $\mathbf{u}_{\vartheta}(t)$ parametrized by $\vartheta \in \mathbb{R}$, one first computes trajectories $\mathbf{q}_{ext\vartheta}(t)$ of the extended system and $\mathbf{q}_{reg\vartheta}(t)$ of the regularized system. Then, the linearization of the regularized system along $(\mathbf{u}_{\vartheta}(t), \mathbf{q}_{reg\vartheta}(t))$ is found in the form

$$\dot{\xi}(t) = \mathbf{A}_{reg\vartheta}(t)\xi + \mathbf{B}_{reg\vartheta}(t)v,$$

where

$$\mathbf{A}_{reg\vartheta}(t) = \frac{\partial \mathbf{G}_{reg}(\mathbf{q}_{reg\vartheta}(t))\mathbf{u}_{\vartheta}(t)}{\partial \mathbf{q}_{reg}} \quad \text{and} \quad \mathbf{B}_{reg\vartheta}(t) = \mathbf{G}_{reg}(\mathbf{q}_{reg\vartheta}(t)).$$

The fundamental matrix $\boldsymbol{\Phi}_{reg\vartheta}(t,s)$ of the linearization obeys the evolution equation

$$\frac{\partial \boldsymbol{\Phi}_{reg\vartheta}(t,s)}{\partial t} = \mathbf{A}_{reg\vartheta}(t)\boldsymbol{\Phi}_{reg\vartheta}(t,s)$$

with the initial condition $\boldsymbol{\Phi}_{reg\vartheta}(s,s) = \mathbf{I}_{N+1}$. The system (12) contains the inverse Gram matrix

$$\mathscr{G}_{reg}(\mathbf{u}_{\vartheta}(\cdot)) = \int_0^T \boldsymbol{\Phi}_{reg\vartheta}(T,s)\mathbf{B}_{reg\vartheta}(s)\mathbf{B}^T_{reg\vartheta}(s)\boldsymbol{\Phi}^T_{reg\vartheta}(T,s)ds$$

of the linearization. The error $\mathbf{e}_{ext}(\vartheta) = \mathbf{q}_{ext}(T) - \mathbf{q}_{ext\,d}$ is computed in the extended system. If $\mathbf{u}_{\vartheta}(t)$ denotes the solution of (12) initialized at a certain $\mathbf{u}_0(t)$, then $\mathbf{u}(t) = \lim_{\vartheta \to +\infty} \mathbf{u}_{\vartheta}(t)$ provides a solution to the constrained motion planning problem for the n-bar system.

In computations a discrete form of the system (12) has been exploited, thus starting from the initial control $\mathbf{u}_0(t)$, the control is updated in accordance with

$$\mathbf{u}_{\vartheta+1}(t) = \mathbf{u}_{\vartheta}(t) - \gamma \mathbf{B}^T_{reg\vartheta}(t)\boldsymbol{\Phi}^T_{reg\vartheta}(T,t)\mathscr{G}^{-1}_{reg}(\mathbf{u}_{\vartheta}(\cdot))\mathbf{e}_{ext}(\vartheta), \quad \vartheta = 0,1,2,\ldots.$$

Furthermore, to improve efficiency of the computations, a finite-dimensional parametrization of control functions will be employed.

4 Computations

For illustration we shall solve the constrained motion planning problem for the 2-bar system whose kinematics are represented by the system (5). The unit bar lengths $l_1 = l_2 = 1$ are assumed. The control functions are selected as truncated Fourier series containing a constant term and 3 harmonics, so for every $i = 1,2,3$

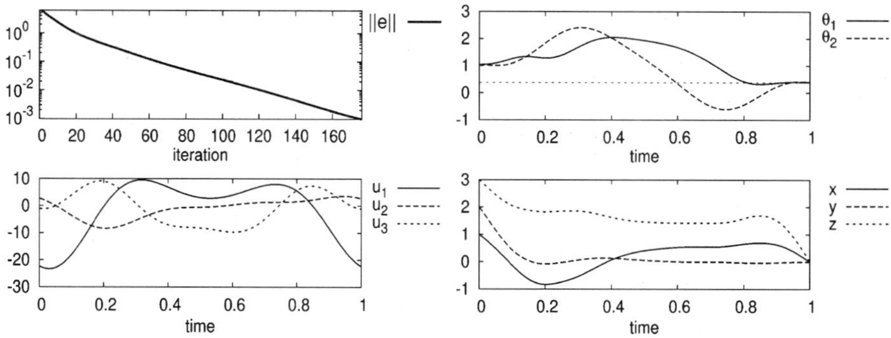

Fig. 2 Unconstrained algorithm: error convergence, θ_1, θ_2 angles, controls, xyz-trajectories.

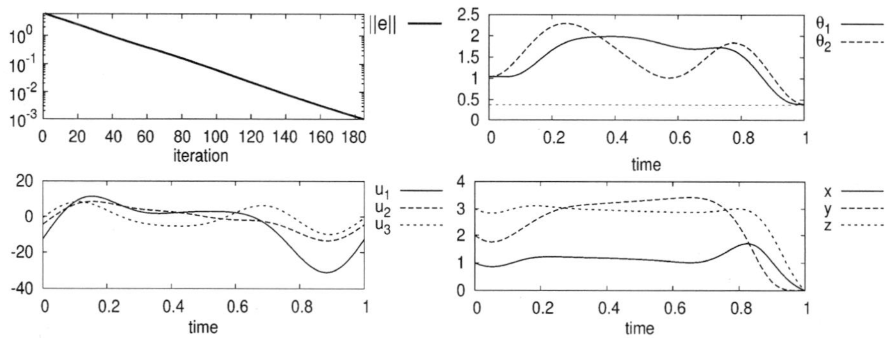

Fig. 3 Imbalanced Jacobian algorithm: error convergence, θ_1, θ_2 angles, controls, xyz-trajectories.

$$u_i(t) = \lambda_{i0} + \sum_{j=1}^{3} \lambda_{i2j-1} \sin(j\omega t) + \lambda_{i2j} \cos(j\omega t).$$

In this way the control space becomes finite dimensional, of dimension 21. In computations the desired configuration has been set to $\mathbf{q}_d = (0,0,0,0,\frac{\pi}{8},0,\frac{\pi}{8})^T$, the initial configuration $\mathbf{q}_0 = (1,2,3,\frac{\pi}{3},\frac{\pi}{3},\frac{\pi}{3},\frac{\pi}{3})$. The bounds imposed on the θ_1, θ_2 angles are $\theta_1^{lb} = \theta_2^{lb} = 0.12\pi$ and $\theta_1^{ub} = \theta_2^{ub} = 0.88\pi$. The initial value of the control parameters $\lambda_0 = (1.5,0,0,0,0,0,0,1.5,0,0,0,0,0,0,1.5,0,0,0,0,0,0)^T$. The perturbation function regularizing Jacobian $r(t) = \mathbf{q}^T(t)\mathbf{q}(t)$, and the plus function approximation parameter $\alpha = 90$. The convergence ratio γ equals 0.05, and the computations stop when the error norm drops below 10^{-3}. For comparison, firstly the Jacobian pseudo inverse algorithm without constraints is applied, and then, after adding the constraints, the imbalanced Jacobian algorithm is used. The results are displayed in Figs. 2 and 3. One can observe that in the unconstrained example in Fig. 2 both θ_1 (slightly) and θ_2 (substantially) violate the assumed lower bound, while after the application of the imbalanced Jacobian, as in Fig. 3, both these angles remain within bounds for the whole motion time. This has been achieved at the cost of slightly increasing the number of iterations of the motion planning algorithm.

5 Conclusions

For specific multi-bar systems we have addressed and solved a constrained motion planning problem using the imbalanced Jacobian motion planning algorithm. Our result extends applicability of this kind of algorithms to a new class of 3 input nonholonomic systems.

Acknowledgements This research was supported by a statutory grant from Wrocław University of Technology.

References

1. Jakubczyk, B.: Invariants of dynamic feedback and free systems. In: Proc. ECC'93, pp. 1510–1513. Groningen, The Netherlands (1993)
2. Jakubiak, J., Tchoń, K., Janiak, M.: Motion planning of the trident snake robot: An endogenous configuration space approach. In: Robot Design, Dynamics, and Control, pp. 159–166. Springer-Verlag, Wien (2010)
3. Janiak, M., Tchoń, K.: Constrained motion planning of nonholonomic systems. Syst. Control Lett. **60**, 625–631 (2011)
4. Laumond, J.P.: Controllability of a multibody mobile robot. IEEE Trans. Robot. Autom. **9**, 755–763 (1993)
5. Li, S.J., Respondek, W.: A kinematic model of the nonholonomic n-bar system: Geometry and flatness. In: Proc. 8th IFAC Symposium NOLCOS, pp. 689–694. Bologna, Italy (2010)
6. Li, S.J., Respondek, W.: The geometry, controllability and flatness property of the n-bar system. Int. J. Control **84**(5), 834–850 (2011)
7. Slayman, M., Pelletier, F.: Articulated arm and special multi-flags. JMSAA **8**(1), 9–41 (2011)
8. Tchoń, K.: On kinematic singularities of nonholonomic robotic systems. In: Theory and Practice of Robots and Manipulators, pp. 75–84. Springer-Verlag, Wien (2000)

Robotic Fish Kinectics Design Based on a Fuzzy Control

Pei-Jun Lee and Wen-June Wang

Abstract This paper shows an application of fuzzy logic in designing and fabricating an intelligent robotic fish with multiple actuators which can swim freely and autonomously avoid obstacles in water. The multiple actuators on the robotic fish consist of two pectoral fins, one on each side of the fish, two tail sections, a center of gravity adjuster for the head of the fish, and a pump used to draw in or expel water. The fish can then avoid the obstacle autonomously when it approaches the obstacle. Obstacle avoidance is achieved using fuzzy control technique. As shown in the simulation and practical experiment, the path of obstacle avoidance using fuzzy control is much smoother than that using intuitive control. All of the above motion controls are implemented by FPGA with the aids of several sensors.

Key words: Robotic fish, multiple actuators, fuzzy logic, obstacle avoidance, FPGA

1 Introduction

There have been many studies to investigate robotic fish. In [1, 2] the authors studied the swimming of manta rays and cownose rays and found that forward motion is achieved through the use of large-scale, flexible, triangular pectoral fins. Fish of the Carangidae and Cyprinidae families which produce forward motion by swinging their tails were studied in [3, 4]. Experimental analysis shows that, there are many types of fish tail including those that can swing with one, two, or multiple degrees of freedom. Multiple actuators swinging multiple fins leads to multiple degrees of freedom, therefore, if the robotic fish has multiple actuators to swing multiple fins, the robotic fish must have high agility in water. In our previous work [7], the robotic fish with a multimedia processing system had only two actuators. In order to in-

Pei-Jun Lee
Department of E.E., Chi Nan University, Puli, Taiwan, e-mail: pjlee@ncnu.edu.tw

Wen-June Wang
Department of E.E., National Central University, Jhongli City, Taiwan, e-mail: wjwang@cc.ncu.edu.tw

J. Lenarčič, M. Husty (eds.), *Latest Advances in Robot Kinematics*,
DOI 10.1007/978-94-007-4620-6_9,

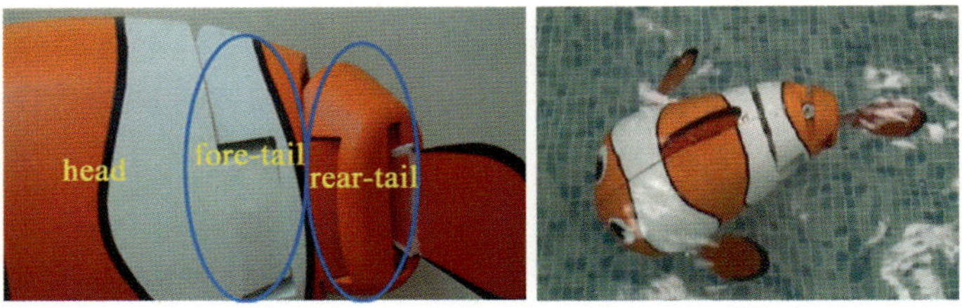

(a) (b)

Fig. 1 (a) The three parts of the robotic fish. (b) The robotic fish.

crease swimming speed and agility, the implemented robotic fish in this study has multiple actuators. The multiple actuators of the robotic fish include two pectoral fins, one on each side of the fish, two tail sections, a center of gravity adjuster for the head of the fish, and a pump used to draw in or expel water. In this study, the robotic fish also uses swinging pectoral fins and center of gravity alteration to increase agility. All electrical elements are housed within a box, the outside of which has been given a fish-like appearance in order to reduce water resistance. The most important issue for the robotic fish is that of the movement controllers. In [5] a low frequency wireless remote controller is used which was originally an aircraft remote control. In [8] the authors used a microcontroller to control the swinging of the fish tail in such a way that the fish could swim freely.

Based on the above, this study attempts to implement a robotic fish. The robotic fish has three sections; the head, the fore-tail and the rear-tail. There are multiple actuators on the robotic fish consisting of two pectoral fins, one on each side of the fish, two tail sections, a center of gravity adjuster for the head of fish, and a pump used to draw in or expel water. The robotic fish has the following functions, basic forward swimming, turning right and left, sinking, and rising. Obstacle avoidance uses infrared sensors. Fuzzy control is the main control method used to achieve obstacle avoidance. As in [6], this study uses VHDL to implement the fuzzy control system in the FPGA embedded system.

2 Swimming Functions of the Robot Fish

In order to investigate the swimming motions and increase the agility of the robotic fish, we must first realize the functions of the various parts of biological fish. Turning and forward swimming functions are controlled by the tail; the dorsal fin is used to maintain balance whilst swimming; the pectoral fins play an important role when the fish either swims forward, turns, or stops in the water; the ventral fin controls upward and downward motion of the fish; and the anal fin has a similar functions to the dorsal fin. In order to imitate the swimming of a biological fish, the robotic fish is fabricated in three sections; the head, the fore-tail, and the rear-tail as shown in Figures 1(a) and (b).

Two actuators (servo motors) drive the fore-tail forepart and rear-tail sections, one actuator per section. While the robotic fish swims at on a fixed horizontal depth,

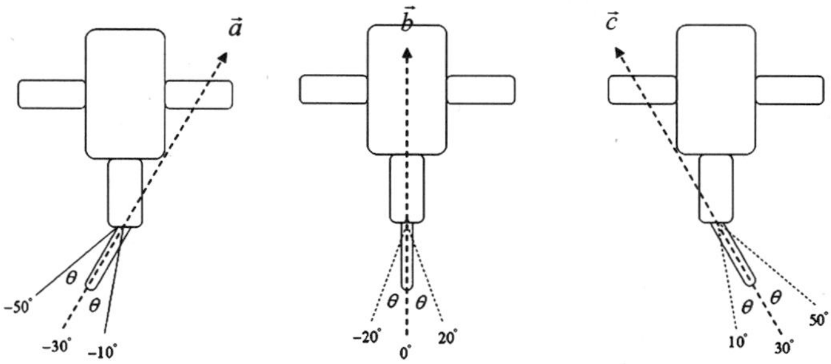

(a) Turning left (b) Swimming forward (c) Turning right

Fig. 2 Synchronous swinging of the fore-tail and rear-tail sections.

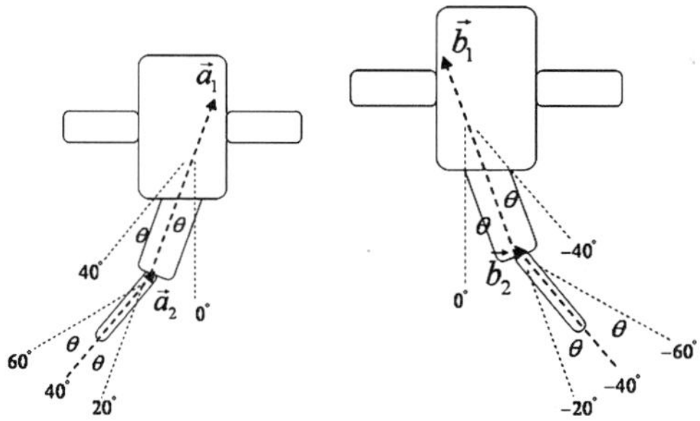

(a) Turning left (b) Turning right

Fig. 3 The fore-tail and rear-part tails swinging with different degrees.

the motor drives the rear-tail swing within the degree interval $[\phi - \theta, \phi + \theta]$ where ϕ is the central shaft angle and θ is the swing amplitude angle. For instance when $\phi = -30$ (or $+30$) and $\theta = 20$, the fish will turn left. Certainly, if $\phi = 0$ the fish will swim forward. If the servo motor drives the fore-tail to swing synchronously with the rear-tail, the swimming direction in Figures 2(a), (b) and (c), respectively; the swimming speed will be much faster than the central shaft angle $\phi = 0°$.

It should be noted that the fore-tail and the rear-tail swing synchronously but they may have different angles of swing, for example, if the fore-tail swings within the degree interval $[\hat{\phi} - \hat{\theta}, \hat{\phi} + \hat{\theta},]$, but the rear-tail swings within the degree interval $[\phi - \theta, \phi + \theta]$ (see Figures 3(a) or (b)), then the fish will turn left or right much faster than those in Figure 2.

At the same time, the pectoral fins assist the robotic fish in swimming stably and smoothly. When the rear-tail swings as in Figure 2(b), and both pectoral fins swing up and down within the degree interval (70, 100) (see Figure 4), the robotic fish swims straight forward. Furthermore, when the rear-tail swings as in Figure 2(a) (or (c)), but only the right (or left) pectoral fin swings up and down within the degree interval (70, 100) (see Figures 4(a) or (c)), the fish turns left (or right).

In addition to swimming at a fixed horizontal depth, the robotic fish should be able to sink and rise in water. There are two ways to achieve the sinking and rising

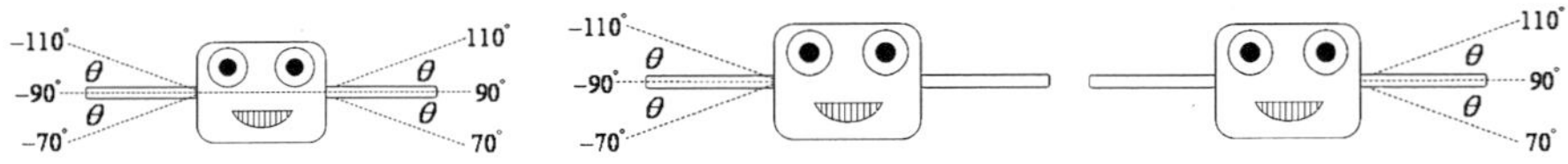

(a) Swimming forward (b) Turning left (c) Turning right

Fig. 4 Pectoral fin swing.

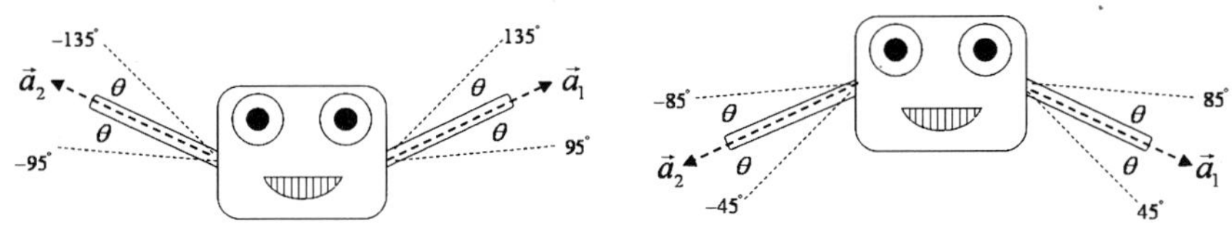

(a) Pectoral fins angled for sinking (b) Pectoral fins swinging for rising

Fig. 5 Sinking and rising.

motions; drawing in or expelling water [7] (i.e., changing the density of the fish), or swinging the pectoral fins [2]. A pump can be used to draw water into, or expel water from a bladder within the fish. The advantage of this method is that the sinking and rising motions are performed well even when the fish is swimming very slowly or has stopped. However, this method requires a large space inside the fish's body in which to put the pump and bladder, and also involves a long response time. In this paper, we swing both pectoral fins to achieve the upward or downward motion. Each pectoral fin is driven by an actuator (servo motor). When the robotic fish swims downward, the swinging of the pectoral fins is as shown in Figure 5(a), in which the pectoral fins swing up and down within the degree interval (95, 135). On the other hand, when the robotic fish swims upward, the swinging of the pectoral fins is as shown in Figure 5(b), in which the pectoral fins swing up and down within the degree interval (45, 85). Furthermore, there is a gravity center adjusting structure inside the fish body [7], in which a mechanism controls the position of rolling balls thus changing the fish's center of gravity so that the pitch of the fish body causes the fish to sink or rise. During upward swimming the gravity center is relocated toward the rear of the body causing the fish head to pitch up. Conversely, during downward swimming, the center of gravity is relocated toward the front of the body causing the fish head to pitch down.

The control core of the robot fish is an embedded system (FPGA) which handles the motion control on the actuators of the fore-tail, rear-tail, pectoral fins, pump, and gravity center adjuster. Furthermore, the communication between the infrared sensors and wireless transmission signals is also processed by FPGA.

3 Fuzzy Control

In this paper we apply fuzzy control to the robotic fish in order to achieve two tasks, obstacle avoidance. In our experiment, the robotic fish swims autonomously. If there is an obstacle detected, the robot fish is able to autonomously avoid the

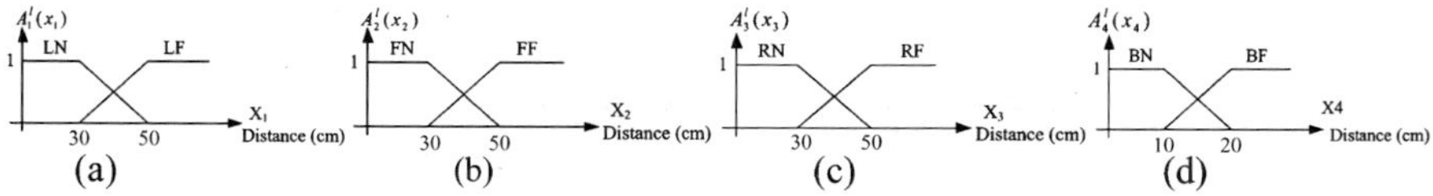

Fig. 6 Fuzzy sets of antecedent variables. (a) Membership function of the fuzzy sets on X_1. (b) Membership function of the fuzzy sets on X_2. (c) Membership functions of the fuzzy sets on X_3. (d) Membership functions of the fuzzy sets on X_4.

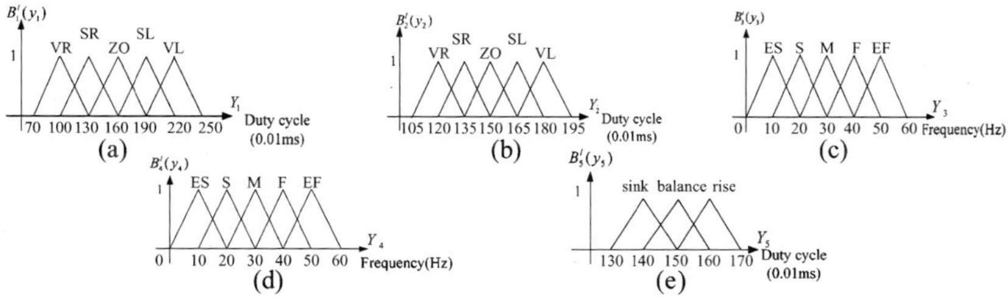

Fig. 7 Consequent fuzzy sets (a) Rear-tail fuzzy sets. (b) Fore-tail fuzzy sets. (c) Right pectoral fin fuzzy sets. (d) Left pectoral fin fuzzy sets. (e) Gravity center adjustor fuzzy sets.

obstacle. To detect obstacles, infrared sensors (Gp2D120) are utilized, which can measure distances precisely within the 10∼80 cm range (the effective range in water is 10∼60 cm). The analog signals from the infrared sensors are then converted by an ADC0831 analog/digital converter into 8-bit digital signals that can be used by the FPGA chips.

There are three infrared sensors positioned on the fish's head, one on the left; another on the right, and the other at the front. There is an additional infrared sensor set under the fish's body. According to the actual test in water, if the obstacle is to the front, left, or right of the robotic fish at a distance of around 40 cm, it will be reliably detected by the infrared sensors. However, if the obstacle is below the fish then the range of reliable detection drops to 20 cm. In the following, we introduce the fuzzy control for obstacle avoidance.

Step 1. Determining the antecedent part and the consequent part

The distance between the obstacle and the fish is detected and measured by four infrared sensors on the fish. Let the distance measured by the left sensor be $x_1 \in X_1 = [0,\infty]$, by the front sensor be $x_2 \in X_2 = [0,\infty]$, by the right sensor be $x_3 \in X_3 = [0,\infty]$, and by the lower sensor be $x_4 \in X_4 = [0,\infty]$. The above four distances are the antecedent variables of the fuzzy rule. Through frequent experiments, the following fuzzy sets with membership functions $A_i^1(x_i)$, $i = 1,2,3,4$, are shown in Figure 6.

There are five controlled plants in the robotic fish with the two tail sections, the left and right pectoral fins and the gravity center adjustment mechanism each having one motor All consequent parts of the swing are shown in Figure 7. There are five fuzzy sets representing five different ranges of fish tail swing. For instance, the fuzzy set VL (very left) denotes the fish tail swing within the central shaft angle $\phi \in [190°,\ 250°]$; meaning that the duty cycle range of the motor is $[190,250]$ 0.01 ms.

Similarly, the other four fuzzy sets for the fore-tail swing within the central shaft angle are SL (slightly left), ZO (zero), SR (slightly right) and VR (very right) as shown in Figure 7(b). In the other two figures, Figures 7(c) and 7(d), the fuzzy sets for the swing frequency of the pectoral fins are EF (extremely fast), F (fast), M (medium), S (slow), and ES (extremely slow). The motor that alters the center of gravity enabling the fish body to balance, rise and sink is shown in Figure 7(e).

Step 2. Establishing a fuzzy rule base
The establishment of a fuzzy rule base simply follows a logical process as below.

Rule	x_1	x_2	x_3	x_4	y_1	y_2	y_3	y_4	y_5
$R^{(1)}$	LF	FF	RF	BF	ZO	ZO	M	M	balance
$R^{(2)}$	LN	FF	RF	BF	SR	SR	S	F	balance
$R^{(3)}$	LF	FN	RF	BF	VL	VL	EF	ES	balance
$R^{(4)}$	LN	FN	RF	BF	VR	VR	ES	EF	balance
$R^{(5)}$	LF	FF	RN	BF	SL	SL	F	S	balance
$R^{(6)}$	LN	FF	RN	BF	ZO	ZO	M	M	balance
$R^{(7)}$	LF	FN	RN	BF	VL	VL	EF	ES	balance
$R^{(8)}$	LN	FN	RN	BF	ZO	ZO	M	M	sink
$R^{(9)}$	LF	FF	RF	BN	ZO	ZO	M	M	sink
$R^{(10)}$	LN	FF	RF	BN	SR	SR	S	F	rise
$R^{(11)}$	LF	FN	RF	BN	VL	VL	EF	ES	rise
$R^{(12)}$	LN	FN	RF	BN	VR	VR	VL	EF	rise
$R^{(13)}$	LF	FF	RN	BN	SL	SL	F	S	rise
$R^{(14)}$	LN	FF	RN	BN	ZO	ZO	M	M	rise
$R^{(15)}$	LF	FN	RN	BN	VL	VL	EF	ES	rise
$R^{(16)}$	LN	FN	RN	BN	VR	VR	ES	EF	rise

Step 3. Using a fuzzy inference mechanism and a defuzzification
By using a fuzzy minimum inference engine and the center of gravity [9], we can obtain the final output y_i^*, $i = 1,2,3,4,5$.

4 Simulation and Experiments

Let the characters L, S, N, and F denote four different distances between the fish and the obstacle. L and S represent distances which are respectively larger and smaller than 40 cm. Similarly, F and S represent distances which are respectively larger and smaller than 20 cm. The other way to achieve obstacle avoidance is the fuzzy control mentioned in Section 3, which compares to the intuitive control of the robotic fish. Using MatLab software, Figure 8(a) shows the swimming path of the fish during obstacle avoidance using intuitive control. Figure 8(b) shows the swimming path of the fish during obstacle avoidance using fuzzy control. It can be seen that when turning the corner, the path of the robotic fish using fuzzy control is much smoother than that using intuitive control.

The following figures are real experiments for the comparison of intuitive control and fuzzy control during obstacle avoidance. Figure 9(a) is with fuzzy control and Figure 9(b) shows swimming with intuitive control. It should be noted that the robotic fish with fuzzy control can turn right smoothly, whereas the fish with intuitive control turns right very close to the obstacle as highlighted in frames 7 and 8 of both Figures 9(a) and (b).

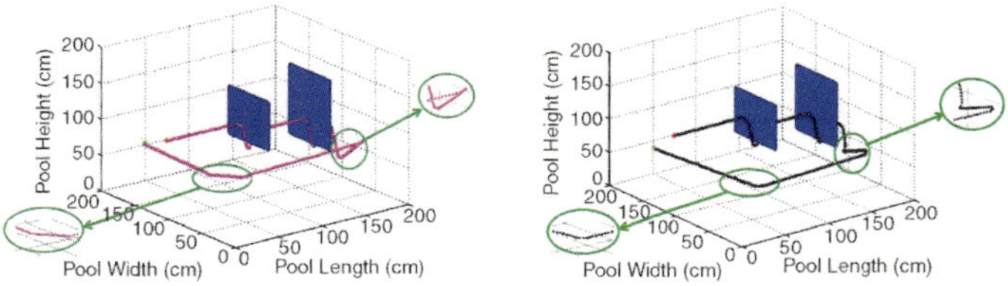

(a) Using intuitive control (b) Using fuzzy control

Fig. 8 Swimming paths during obstacle avoidance.

(a)Using fuzzy control (b)Using intuitive control

Fig. 9 Comparison during obstacle avoidance.

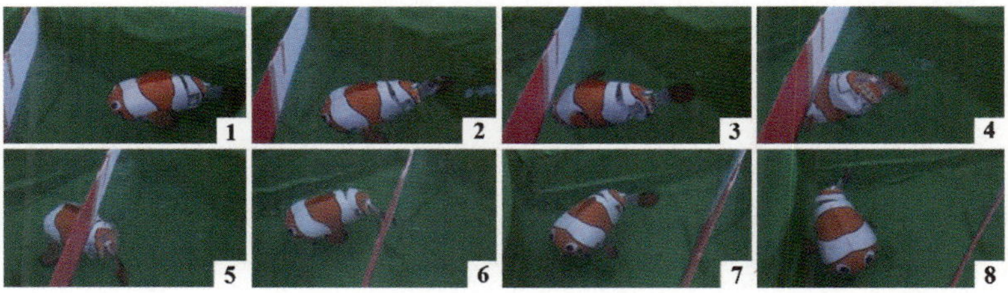

Fig. 10 Robotic fish surrounded by obstacles.

The following is an experiment where there are obstacles surrounding the robotic fish as shown in Figure 10. The surrounding obstacles are detected by the right, left and front infrared sensors of the fish. Thus, the robotic fish will use rising and sinking to swim away from the closed environment. However, if the bottom sensor detects the bottom of the pool while the fish is sinking, the fish will start to rise.

Table 1 shows the comparison data between the proposed robot fish and that in the paper [7], where the test environment is an indoor pool measuring $120 \times 120 \times 120$ cm.

The swimming tests are not limited to indoor pools. The experiments can be also performed in three outdoor environments; an outdoor fish pool, an adult swimming pool, and a pool under a waterfall. All the above experiments are shown in the

Table 1 Data comparisons of robot fish.

	The proposed robot fish	The robot fish in [7]
Speed	60 cm/s	10 cm/s
Turning radius	10 cm/s	30 cm/s
Maximum operating depth	110 cm	50 cm
Vertical speed	20 cm/s	5 cm/s

video [10]. The video is divided into four sections. The first section covers controlling the fish with the remote control and performing basic swimming movements. The second and the third sections cover obstacle avoidance using fuzzy control.

5 Conclusions

This study has completed a robotic fish which can perform several basic swimming motions, obstacle avoidance. The above motions are achieved by multiple actuators with fuzzy control methods on the fish. The technologies used to implement this robotic fish include mechanism design, circuit design, FPGA signal integration, wireless communication, distance sensing, and actuator controlling. When fuzzy control is applied to the robotic fish, the fish has a much smoother swimming motion for obstacle avoidance. Finally, we have shown many experiments performing all functions proposed in this study.

References

1. Wang, Z.L., Wang, Y.W., Li, J., Hang, G.R.: A micro biomimetic manta ray robot fish actuated by SMA. In: Proc. of IEEE International Conference on Robotics and Biomimetics, pp. 1809–1813 (2009)
2. Jun, G., Bi, S.S., Li, J., Liu, C.: Design and experiments of robot fish propelled by pectoral fins. In: Proc. of IEEE International Conference on Robotics and Biomimetics, pp. 445–450 (2009)
3. Ichiklzaki, T., Yamamoto, I.: Development of robotic fish with various swimming functions. In: Proc. of Symposium on Underwater Technology and Workshop on Scientific Use of Submarine Cables and Related Technologies, pp. 378–383 (2007)
4. Xia, D., Chen, W., Liu, J., Liu, Y.: System and experimental research on biomimetic robot fish. In: Proc. of International Conference on Mechatronics and Automation, pp. 111–116 (2007)
5. Low, K.H.: Parametric study of modular and reconfigurable robotic fish with oscillating caudal fin mechanisms. In: Proc. of IEEE International Conference on Mechatronics and Automation, pp. 123–128 (2007)
6. Montiel, O., Maldonado, Y., Sepulveda, R., Castillo, O.: Simple tuned fuzzy controller embedded into an FPGA. In: Proc. of NAFIPS on Fuzzy Information Processing Society, pp. 1–6 (2009)
7. Lee, P.J., Yen, C.H., Chan, C.L., Lee, M.S., Wang, R.C.: Implementation of a fuzzy control based intelligent robot fish. Int. J. Fuzzy Syst. **11**(4), 287–296 (2009)
8. Chan, W.L., Kang, T., Lee, Y.J., Sung, S.K., Yoon, K.J.: Swimming study on an ostraciiform fish robot. In: Proc. of IEEE International Conference on Control, Automation and Systems, pp. 700–705 (2007)
9. Wang, L.X.: A course in fuzzy systems and control. Prentice Hall PTR (1977)
10. http://video.ee.ncnu.edu.tw/mv/video/vd02.htm

Self-Calibration of Redundantly Actuated PKM Based on Motion Reversal Points

Andreas Müller and Maurizio Ruggiu

Abstract In this paper a novel self-calibration method for redundantly actuated parallel manipulators (RA-PKM) is proposed. It does not require external measurements or encoders in passive joints but only uses actuator joint measurements. The method uses motion reversal points (MRP) as calibration landmarks. MRP are characterized by vanishing velocity of one actuator while the RA-PKM performs a continuous motion. In other word the MRP are the input-singularities of an associated non-redundantly actuated PKM. The general calibration concept is introduced in the paper and applied to a planar 4R$\underline{\text{P}}$R RA-PKM. Numerical simulation results are presented that confirm the feasibility of this method. The selection of appropriate MRP is briefly discussed.

Key words: Kinematic self-calibration, parallel manipulators, actuation redundancy

1 Introduction

Actuation redundancy improves the kinematic and dynamic properties of PKM as reported in several publications owed, however, to a more complex control that is sensitive to geometric uncertainties. But on the other hand *actuation redundancy implies sensor redundancy*. This feature gives rise to a novel calibration concept described in this paper.

Essential for any kinematic calibration method is the acquisition of redundant measurements. The redundancy is due to inherent constraints within the system to be

Andreas Müller
Institute of Mechatronics, Chemnitz, Germany, e-mail: andreas.mueller@ifm-chemnitz.de

Maurizio Ruggiu
Department of mechanical engineering, University Cagliari, Cagliari, Italy, e-mail: ruggiu@dimeca.unica.it

J. Lenarčič, M. Husty (eds.), *Latest Advances in Robot Kinematics*,
DOI 10.1007/978-94-007-4620-6_10,

identified. Representing such constraints by a mathematical model allows for adapting the model parameters so to fit to the measurements. Commonly redundant measurements are achieved by combining actuator sensor information with additional external metrology measurement devices temporarily attached to the end-effector (EE) [6, 12, 15, 16], which makes the calibration expensive and prohibits simple repetition. In order to avoid the use of expensive external measurement devices it was proposed to equip passive joints with additional sensors. Such semiautonomous calibration schemes are reported in [22] for a Gough-Stewart platform where additional sensors are located at the universal joints, and in [18] where sensors were attached to each one of the five passive joints of one strut. A fully autonomous calibration shall not need any additional sensors and so provide a cost-efficient and non-intrusive solution. The redundant sensor information is usually acquired by restraining the mobility of the EE or by locking some passive joints [1, 8, 14, 17, 19] so that the manipulator can be actuated by a subset of the actuators while measuring the motion of those actuators that become redundant due to the locking.

Actuation redundancy implies sensor redundancy since the number of measured joint coordinates exceeds the DOF of the PKM. This redundancy was exploited for kinematic calibration of RA-PKM in [2, 3, 5, 20, 21]. Yiu, Meng and Li [20] calibrated a planar 2DOF manipulator by application of standard calibration methods. Cong et al. [2] and Zhang et al. [21] developed a calibration method based on the tracking error projected to the null-space of the forward kinematics Jacobian. Since, for a perfect match of model and plant, the tracking error should belong to the null-space, the strategy is to minimize the projected tracking error by adjusting the kinematic model parameters.

In this paper an alternative approach to kinematic calibration of RA-PKM is presented that does not make straightforward use of redundant measurements but uses specific calibration landmarks. It originates from the singularity-based calibration method proposed for a planar 3-DOF non-redundantly actuated PKM in [9–11] that was later applied in [13] to the calibration of a spatial 3$\underline{P}$RS PKM. The basic idea of that method was to identify passive input singularities (type II singularities/force singularities [4]) in the kinematic model (analytically) and in the plant (measurement). Passive singularities are characterized by a reversal motion, i.e. a zero velocity, of some actuator coordinates for a continuous EE-motion, which allows to detect them without additional sensors. Due to uncertainties in the geometric parameters the singularities of the model and plant occur at different locations. This mismatch was minimized by an ad hoc parameter adaptation method so to match the values inferred from the model with the encoder readings. Apparently the need to enter input-singularities is a problematic aspect of this method when applied to non-redundant PKM. In the following a calibration method for RA-PKM is introduced making use of the basic idea behind the singularity-based calibration method.

Throughout the paper configurations where the velocity of one actuator becomes zero, while the EE performs continuous motion, will be called *motion reversal points* (MRP) and δ denotes the manipulator's DOF.

2 Calibration Method Based on Motion Reversal Points

2.1 *Basic Concept*

The problematic point of the calibration scheme reported in [9–11, 13], originally proposed for non-redundantly actuated PKM, is that input singularities are critical in the sense that the PKM cannot be fully restrained by means of the actuators. Another crucial point is to ensure that the PKM actually passes through the anticipated singularities forming a closed low-dimensional subset in joint space. To this end $\delta - 1$ actuators were locked so that only motions of one free actuator, for which a MRP (i.e. a zero crossing of its velocity) is expected, are possible. Thus the problem of detecting the input singularities reduces to a one-dimensional search. The most problematic aspect of this method is that, in the input singularity, the one-dimensional motion cannot be controlled by the only one actuator left free. For this reason the PKM was given an initial motion so to passively swing through the input singularity and so that the motion reversal of the one free joint can be observed.

Such MRP can also be observed in RA-PKM but without meeting input singularities thus without passing through critical configurations. This is best understood by recalling that the input-singularities of a non-redundantly actuated PKM with DOF δ can be eliminated by introducing $m > \delta$ redundant actuators that can always control the PKM motion even when certain δ actuators would encounter a singularity. In other words, actuation redundancy allows controlling the RA-PKM through points that are input-singularities, i.e. MRP, of the PKM when controlled by some δ actuators. Consequently a RA-PKM can be safely controlled through MRP that provide kinematic *calibration landmarks*.

As an example consider the 4R$\underline{\text{P}}$R RA-PKM with DOF $\delta = 3$ in Fig. 1. The manipulator is redundantly actuated by the $m = \delta + 1 = 4$ prismatic joints whose translation variables are denoted with $q^1, \ldots, q^4$. Now $\delta - 1 = 2$ of the actuators could be locked, and one of the two remaining actuators be actuated so to observe a MRP of the free fourth actuator. This is schematically shown in Fig. 1a) where actuators 1 and 2 are locked resembling a 1-DOF mechanism. This mechanism can be controlled either by actuator 3 or 4. The platform is confined to rotate about the attachment point of actuator 1 and 2 at the platform. A continuous motion of actuator 3 leads to reversing motion of actuator 4, i.e. $\dot{q}^4 = 0$. Such MRP is attained when the line of action of the fourth leg passes through the attachment point of actuator 1 and 2 at the platform, as indicated in Fig. 1a). Analogously if prismatic joint 4 is actuated, a corresponding MRP of actuator 3 can be observed (Fig. 1b)). In these cases the platform can only rotate about a fixed point, which gives rise to a simple analytic expression of the MRP in terms of q^1, q^2 and the geometric parameters. Locking actuators 1 and 4 the attachment point of the third actuator moves on the coupler curve of the 4-bar mechanism consisting of actuators 1 and 4 and the platform (Fig. 2). Moving actuator 2 the MRP of actuator 3 is attained at the point of the coupler curve that is closest to the attachment point of actuator 3 on the ground, which can also be determined analytically.

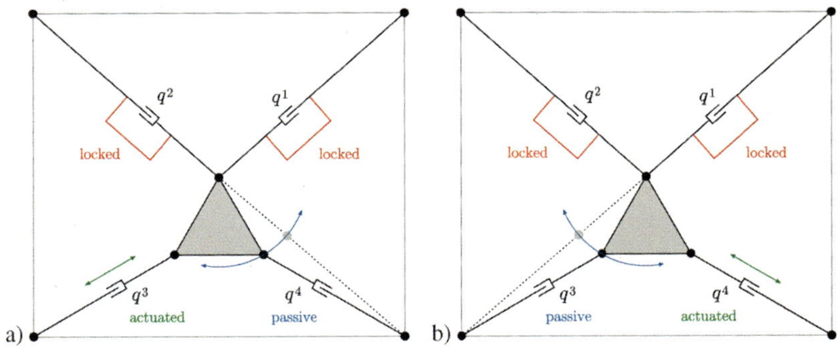

Fig. 1 MRP of actuator 4 and 3 of the 4R$\underline{\text{P}}$R RA-PKM when actuators 1 and 2 are locked.

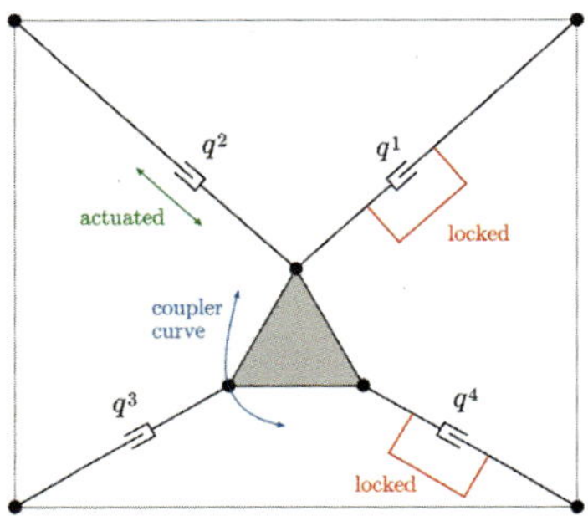

Fig. 2 MRP of actuator 3 when the actuators 1 and 4 are locked.

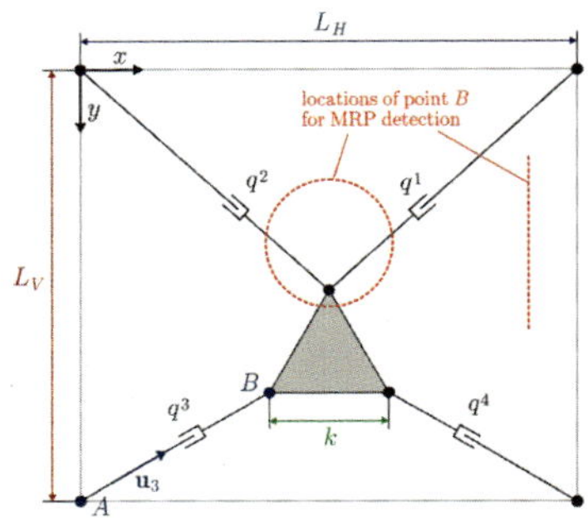

Fig. 3 Definition of geometric model parameters for the 4R$\underline{\text{P}}$R RA-PKM.

The sensor readings for which a MRP of actuator i is detected will differ from those inferred from the kinematic model. Minimizing this calibration error by adapting geometric model parameters gives rise to a MRP-based calibration method. Clearly the drives for which MRP are to be detected must be backdrivable.

The preceding discussion can easily be generalized for RA-PKM with actuation redundancy $\rho = m - \delta$. The MRP detection requires to restrain the RA-PKM by locking $\delta - 1$ actuators leaving $m - \delta + 1$ movable actuators. One of the latter is used to drive the restrained RA-PKM thus moving the remaining $\rho = m - \delta$ actuators. In general there are $C^m_{\delta-1} = \binom{m}{\delta-1}$ different possibilities to lock $\delta - 1$ of the m actuators. For the sake of simplicity $m = \delta + 1$, i.e. $\rho = 1$, is assumed in the following, as it was for the above example, so that there are $C^m_{m-2} = \frac{(m-1)m}{2}$ different possibilities. Then, fixing $\delta - 1$ actuators leaves two free, of which either one can be controlled so to detect the MRP of the other actuator. Thus, in total there are $2C^m_{m-2}$ (in general $(\rho + 1)C^m_{\delta-1}$) different actuation schemes to detect MRP. For each of the C^m_{m-2} possibilities denote with $\mathbf{q}_{\mathrm{a},l}, l = 1, \ldots, C^m_{m-2}$ the vector of locked actuator coordinates and, for a particular choice, with q^j_{a} and q^i_{a} the joint coordinate of the controlled and the free passive actuator, respectively. The vector of actuator coordinates, $\mathbf{q}_{\mathrm{a}}$, is thus split to $\mathbf{q}_{\mathrm{a},l}$, q^j_{a} and q^i_{a}.

In Fig. 1a) for instance $\mathbf{q}_{\mathrm{a},l} = (q^1, q^2)$, $q_{\mathrm{a}}^j = q^3$, and $q_{\mathrm{a}}^i = q^4$, while in Fig. 1b) $\mathbf{q}_{\mathrm{a},l} = (q^1, q^2)$, $q_{\mathrm{a}}^j = q^4$, and $q_{\mathrm{a}}^i = q^3$. In order to detect MRP of actuator i by moving actuator j introduce an indicator function $F_{ij}^l(q_{\mathrm{a}}^j, \mathbf{q}_{\mathrm{a},l}; \boldsymbol{\pi})$ such that $F_{ij}^l = 0$ if and only if $\dot{q}_{\mathrm{a}}^i = 0$ for any $\dot{q}_{\mathrm{a}}^j$. The vector $\boldsymbol{\pi}$ summarizes the geometric model parameters. The condition $F_{ij}^l = 0$ allows to infer the value $q_{\mathrm{a},0}^j$ of the controlled actuator at which the MRP of the free passive actuator i occurs in the model for given $\mathbf{q}_{\mathrm{a},l}$ and geometry $\boldsymbol{\pi}$.

The proposed calibration method for a RA-PKM with DOF δ and actuation redundancy $\rho = 1$ can be summarized as follows:

1. Lock $\delta - 1$ actuator coordinates, $\mathbf{q}_{\mathrm{a},l}$, at selected points in the workspace. This leaves a 1-DOF system with $\rho + 1 = 2$ movable actuator coordinates.
2. Select one actuator coordinate, q_{a}^j, from the two movable actuator coordinates to control the 1-DOF system.
3. Detect MRP in the plant: perform a continuous motion of q_{a}^j so to pass through the expected MRP of the second (passive) actuator coordinate q_{a}^i.
   The MRP will in general not be detected exactly due to the time discrete sampling of sensor readings and due to the finite sensor resolution. To approach the MRP, at which $\dot{q}_{\mathrm{a}}^i = 0$, perform a one-dimensional search by controlling q_{a}^j and observing the sign reversal of $\dot{q}_{\mathrm{a}}^i$. This yields an estimate for the controlled actuator joint coordinate at which the MRP of joint i occurs, denoted $\widehat{q}_{\mathrm{a},0}^j$.
4. The calibration error for this MRP is $e^j := \widehat{q}_{\mathrm{a},0}^j - q_{\mathrm{a},0}^j$. This error is a function of the geometric model parameters ($q_{\mathrm{a},0}^j$ depends on $\boldsymbol{\pi}$). Find π minimizing e^j.

2.2 Adaptation Method

The steps 1–4 are repeated for a representative set of samples corresponding to different poses of the RA-PKM. For each one of the $2C_{m-2}^m$ possible combinations of locked and controlled actuators the MRP of the free actuators are detected in the plant as well as determined from the model. Denote with $\widehat{q}_{\mathrm{a},0}^{j,n}$, $n = 1, \ldots, N$ the overall set of MRP of actuator j detected in the plant, and with $q_{\mathrm{a},0}^{j,n}$ the set of MRP deduced from the model (with current parameter value $\boldsymbol{\pi}$) obtained for preselected MRP. The overall 'calibration error' can be expressed as

$$J(\boldsymbol{\pi}) := \frac{1}{2} \sum_{j=1}^{C_{m-2}^m} \sum_{n=1}^{N} \left(\widehat{q}_{\mathrm{a},0}^{j,n} - q_{\mathrm{a},0}^{j,n} \right)^2. \tag{1}$$

The geometry of the model is encoded in the parameter vector $\boldsymbol{\pi} \in \Pi$, where Π is a ν-dimensional parameter space of feasible geometries. Thus the calibration problem is transformed into the non-linear minimization problem

$$J(\boldsymbol{\pi}) \rightarrow \min \quad \boldsymbol{\pi} \in \Pi \tag{2}$$

in the ν geometric model parameters. This is in general a highly non-linear problem. However, the numerical minimization of (2) should rapidly converge since the actual geometry of the plant can be assumed to be closed to the nominal geometry π_0, which serves as initial value. Clearly the convergence speed depends on the sensitivity of the MRP to variations of the geometric parameters. It is thus crucial for the proposed calibration method that the MRP are selected so that a variation of the geometric parameters has a significant effect on the MRP.

3 Example: 4RPR Mechanism

MRP Indicator Function: Due to space limitations only the MRP of actuator 3 of the 4RPR RA-PKM for locked actuators 1 and 2 are discussed here. Then the PKM is controlled by q^4 as in Fig. 1b). The indicator function F_{34} is constructed from the manipulator Jacobian

$$\mathbf{J}^3 = (\mathbf{J}_1^3, \mathbf{J}_2^3, \mathbf{J}_4^3) \tag{3}$$

of a 3RPR PKM actuated by q^1, q^2, and q^4 with EE located at point B (Fig. 3). The column $\mathbf{J}_i^3$ corresponds to actuator i. The twist of a frame attached at B is

$$\begin{pmatrix} \mathbf{v}_3 \\ \omega_3 \end{pmatrix} = \mathbf{J}^3 \begin{pmatrix} \dot{q}^1 \\ \dot{q}^2 \\ \dot{q}^4 \end{pmatrix}. \tag{4}$$

Denote with $\mathbf{J}_v^3 = (\mathbf{J}_{v1}^3, \mathbf{J}_{v2}^3, \mathbf{J}_{v4}^3)$ the linear velocity Jacobian corresponding to $\mathbf{v}_3$ extracted from $\mathbf{J}^3$. Now, if actuator 1 and 2 are fixed, i.e. $\dot{q}^1 = \dot{q}^2 = 0$, then the linear velocity of point B is $\mathbf{v}_3 = \mathbf{J}_{v4}^3 \dot{q}^4$. The actuator velocity $\dot{q}^3$ is the linear velocity $\mathbf{v}_3$ of B projected onto the line AB. Hence the MRP indicator function for q_3 is

$$F_{34} := \mathbf{u}_3 \cdot \mathbf{J}_{v4}^3 \tag{5}$$

where $\mathbf{u}_3$ is a unit vector along the line AB. Zero crossing of F_{34} indicates the MRP of q^3 when the PKM is driven by q^4 while q^1, q^2 are fixed. This approach can be applied analogously to all four actuator coordinates.

Numerical Simulation Results: A numerical simulation of the calibration method was performed to estimate the deviation of the dimensions L_H, L_V, and k (Fig. 3) in the plant from the nominal lengths $L_H^0 = L_V^0 = 0.5$ m, and $k^0 = 0.15$ m. The moving platform is an equilateral triangle with nominal side length k^0. Only the MRP of actuator 3 when driven by 4 are considered, using the above indicator function F_{34}. The three plant parameters are subject to an imperfection with magnitude ΔL and set to $L_H = L_H^0 - \Delta L$, $L_V = L_V^0 - \Delta L$, $k = k^0 + \Delta L$. Simulation is carried out for the three values $\Delta L = 10^{-4}, 10^{-3}, 10^{-2}$ m, and the calibration performance is studied for encoder resolutions $\Delta x = 10^{-5}, 10^{-4}, 10^{-3}$ m. Fig. 4a) shows the average absolute calibration error $\varepsilon := (|L_H^0 - \widehat{L}_H| + |L_V^0 - \widehat{L}_V| + |k^0 - \widehat{k}|)/3$ when 20 MRP are selected with EE locations on the circle in Fig. 3. The calibration converges, and remarkably the average error is always smaller than the encoder

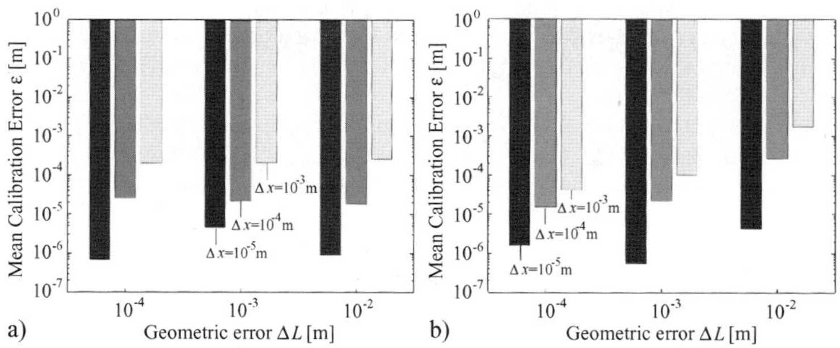

Fig. 4 Calibration error for MRP (a) on the circle and (b) on the vertical line at $x = 0.485$ m.

resolution. The latter may be explained as average effect similar to the interpolation between discrete samples. To show the effect of different MRP Fig. 4b) shows the results when MRP are selected on the vertical line at $x = 0.485$ m indicated in Fig. 3. Apparently the achieved calibration errors are in average higher then for MRP on the circle, which can be explained by the lower sensitivity of MRP closed to the workspace boundary. The minimization (2) was performed ad hoc using Matlab's `fminunc` for non-smooth objectives. The minimization converged after less than 500 steps. A faster convergence is to expected when the MRP of different actuators are included in the calibration process.

4 Summary

In this paper a novel calibration method is proposed that differs from the standard calibration methods in that it does not directly use redundant measurements but rather employes actuation redundancy to acquire motion reversal points (MRP) that serve as calibration landmarks. It is discussed for a 4R$\underline{P}$R but applicable to general RA-PKM. This calibration method should allow for an automatic non-intrusive in situ online self-calibration that could be combined with standard calibration methods that use the m redundant measurements. It must be noted that the drives of which MRP are detect must be backdrivable. Here the feasibility of the proposed method is demonstrated by means of numerical simulations. The statistical error analysis will be presented in an upcoming publication after the method has been implemented in a prototype. Future work will focus on strategies for selecting optimal MRP and on tailored methods for solving (2). MRP should be selected so to maximize the sensitivity of their location to variations in the geometry. This requires an analytic sensitivity analysis. In general any geometric parameter can be identified that has an effect on the MRP locus.

References

1. D.J. Bennett, J.M. Hollerbach: Autonomous Calibration of a single loop closed kinematic chain formed by manipulators with passive endpoint constraints. IEEE Trans. Robot. Autom. **7**(5), 597–605 (1989)

2. S. Cong, C. Feng, Y. Zhang, Z. Li, S. Jiang: Kinematic Parameters Auto-Calibration of Redundant Planar 2-Dof Parallel Manipulator. In: Parallel Manipulators, New Developments, pp. 241–268. I-Tech Education and Publishing, Vienna, Austria, (2008)
3. G. Ecorchard, R. Neugebauer, P. Maurine: Elasto-geometrical modeling and calibration of redundantly actuated PKMs. Mech. Mach. Theory **45**, 795–810 (2010)
4. C. Gosselin, J. Angeles: Singularity analysis of closed loop kinematic chains. IEEE Trans. Robot. Autom. **6**(3), 281–290 (1990)
5. J. Hesselbach et al.: Passive-joint sensors for parallel robots. Mechatronics **15**, 43–65 (2005)
6. J.M. Hollerbach, C.W. Wampler: The calibration index and taxonomy for robot kinematic calibration methods. Int. J. Robot. Res. **15**(6), 573–591 (1996)
7. D. Jeon et al.: A calibration method of redundantly actuated parallel mechanism machines based on projection technique. CIRP Ann. - Manuf. Technol. **59**, 413–416 (2010)
8. W. Khalil, S. Besnard: Self-calibration of Stewart–Gough parallel robots without extra sensors. IEEE Trans. Robot. Autom. **15**(6), 1116–1121 (1999)
9. P. Last, C. Budde, J. Hesselbach: Self-calibration of the HEXA-parallel-structure. In: Proc. IEEE Int. Conf. Aut. Science and Eng. (CASE), Edmonton, Canada, August 1–2, pp. 393–398 (2005)
10. P. Last, D. Schütz, A. Raatz, J. Hesselbach: Singularity based calibration of 3-DOF fully parallel planar manipulators. In: Proceedings of the 12th IFToMM World Congress, Besancon, France, June, pp. 18–21 (2007)
11. P. Last, A. Raatz, J. Hesselbach: Singularity-based calibration – A novel approach for absolute-accuracy enhancement of parallel robots. In: A. Lazinica, H. Kawai (eds.) Robot Manipulators New Achievements, InTech Publishing (2010)
12. A. Nahvi, J.M. Hollerbach, V. Hayward: Calibration of parallel robot using multiple kinematic closed loops. In: Proc. IEEE Int. Conf. Robot. Autom. (ICRA), pp. 407–412 (1994)
13. S.M. O'Brien, J.A. Carretero, P. Last: Self calibration of 3-PRS manipulator without redundant sensors. Trans. Can. Soc. Mech. Eng. **31**(4), 483–494 (2007)
14. H. Ota, T. Shibukawa, T. Tooyama, M. Uchiyama: Forward kinematic calibration method for parallel mechanism using pose data measured by a double ball bar system. In: Proc. Year 2000 Parallel Kinematic Mach. Int. Conf., pp. 57–62 (2000)
15. A. Rauf, S.G. Kim, J. Ryu: Complete parameter identification of parallel manipulators with partial pose information using a new measurement device. Robotica **22**(6), 689–695 (2004)
16. A. Rauf, A. Pervez, J. Ryu: Experimental results on kinematic calibration of parallel manipulators using a partial pose measurement device. IEEE Trans. Rob. Control **22**(2), 379–384 (2006)
17. J. Ryu, A. Rauf: A new method for fully autonomous calibration of parallel manipulators using a constraint link. In: Proc. IEEE/ASME Int. Conf. Advanced Intelligent Mechatronics (AIM), Como, Italy, July 8–12, pp. 141–146 (2001)
18. C.W. Wampler, J.M. Hollerbach, T. Arai: An implicit loop method for kinematic calibration and its application to closed-chain mechanisms. IEEE Trans. Robot. Autom. **11**(5), 710–724 (1995)
19. M. Weck, D. Staimer: Accuracy issues of parallel kinematic machine tools: Compensation and calibration. In: Int. Conf. Parallel Kinematic Machines, pp. 35–41 (2000)
20. Y.K. Yiu, J. Meng, Z.X. Li: Auto-calibration for a parallel manipulator with sensor redundancy. In: Proc. IEEE Int. Conf. Rob. Automat. (ICRA), Taipei, Sept. 14–19, pp. 3660–3665 (2003)
21. Y.X. Zhang, S. Cong, W.W. Shang, Z.X. Li, S.L. Jiang: Modeling, identification and control of a redundant planar 2-DOF parallel manipulator. Int. J. Control. Autom. Syst. **5**(5), 559–569 (2007)
22. H. Zhuang: Self-calibration of parallel mechanisms with a case study on Stewart platforms. IEEE Trans. Robot. Autom. **13**(3), 387–397 (1995)

Trajectory Planning for Systems with Homotopy Class Constraints

Soonkyum Kim, Koushil Sreenath, Subhrajit Bhattacharya and Vijay Kumar

Abstract There are various applications where homotopy constraints are useful in trajectory generation for mobile robots. In this paper, we present a method to generate an optimal trajectory restricted to a particular homotopy class, which is specified by a given representative trajectory. The optimality is achieved by formulating the trajectory generation problem as a Mixed-Integer Quadratic Program (MIQP). We partition the configuration space into nonoverlapping cells and model each cell in the partition with integer variables and inequality constraints. We associate with any sequence of integer variables a *word*, so that each trajectory can be mapped to a word. We then construct a set of all words that are homotopically equivalent to a given word. For each word, we fix the integer variables of the MIQP to find the optimal time distribution in each cell, by solving a QP for each iteration, to obtain the locally optimal trajectory in the specified homotopy class. We illustrate an example of minimum acceleration trajectory generation on a plane with different homotopy class constraints.

Key words: Trajectory planning, homotopy constraint, optimal trajectory

1 Introduction

Early attempts at classifying homotopy classes in two dimensions include geometric methods [5, 6], homotopy preserving probabilistic road-map constructions [10], and triangulation-based path planning [3]. Two trajectories are said to be homotopic if one can be continuously deformed to another without any intersection with obstacles. Each set of trajectories that are homotopic forms an equivalence class, called a homotopy class (see Figure 1(a)). A particular homotopy class can be specified by a representative trajectory in the homotopy class. Thus, trajectory generation with a

Soonkyum Kim · Koushil Sreenath · Subhrajit Bhattacharya · Vijay Kumar
Mechanical Engineering and Applied Mechanics, University of Pennsylvania, Philadelphia, USA,
e-mail: {soonkyum,koushils,subhrabh,kumar}@seas.upenn.edu

J. Lenarčič, M. Husty (eds.), *Latest Advances in Robot Kinematics*,
DOI 10.1007/978-94-007-4620-6_11,

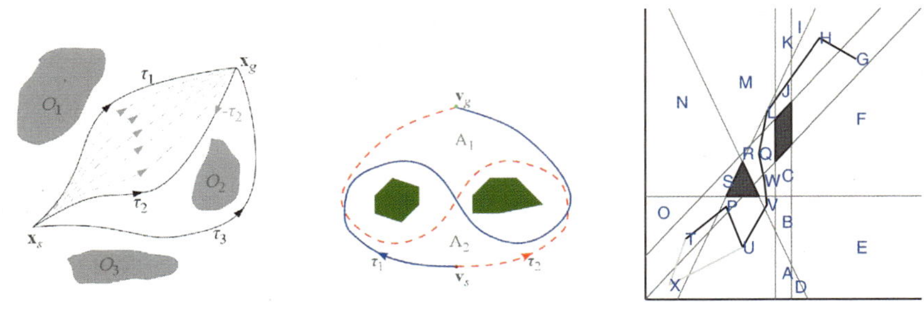

(a) (b) (c)

Fig. 1 (a) τ_1 is homotopic to τ_2 since there is a continuous sequence of trajectories representing deformation of one into the other. τ_3 belongs to a different homotopy class since it cannot be continuously deformed into any of the other two. (b) Example where the trajectories (τ_1 and τ_2) are homologous, but not homotopic. (c) An example of a trajectory corresponding to the word TPUVWQLJHG.

homotopy class constraint consists of finding an optimal trajectory in the desired homotopy class, specified by the given representative trajectory, that also respects the kinematic constraints. One can think of applications ranging from multi-robot exploration, where it may be beneficial to deploy each robot in a different homotopy class to ensure maximal coverage and minimal congestion, to single arm motion planning where one may seek paths that go around obstacles one way or the other way based on the specific task.

In this paper, we use mixed-integer quadratic programming [8] to partition the configuration space into non overlapping cells and represent each cell by a label or a *letter*. We use the notion of *words*, constructed out of the letters, to coarsely represent trajectories and relate them with their homotopy classes. First, all words corresponding to a particular homotopy class are constructed. Then, for each word, a quadratic program (QP) is solved to find a trajectory that spends equal amounts of time in each cell specified by the letters of the word. Finally, this trajectory is iteratively refined to obtain the locally optimal trajectory in the specified homotopy class.

2 Preliminaries

Our objective in this paper is to design an optimal trajectory for a robot that minimizes an integral cost functional (which depends on the trajectory), while also respecting kinematic constraints of the system, avoiding obstacles, and constraining the trajectory to a particular homotopy class. Although several of these subproblems have been solved separately, (see [1, 2, 4, 7, 11, 12]), there is no literature, to our knowledge, that addresses the combined problem described above.

We will start by assuming that the required homotopy class is specified by an initial representative trajectory in the homotopy class. Specifically, we will derive a locally optimal trajectory that is *homotopic* to the representative trajectory while also satisfying the kinematic and geometric constraints. We represent a planar trajectory, $q(t)$, by the points $[q_x(t), q_y(t)]$ parametrized by t.

2.1 Homotopy and Homology Classes for Trajectories

We begin by defining *homologous* trajectories and illustrate the difference between homology and homotopy. Two trajectories, q_1 and q_2, connecting the same start and end points are homologous if and only if the closed loop formed by them, $q_1 \bigsqcup -q_2$ (*i.e.*, q_1 together with q_2 with opposite orientation), forms the boundary of a 2-dimensional region on the plane not containing/intersecting any obstacles. It is well known that homology is a "coarser" representation of homotopy [2], with trajectories that are homotopic being also homologous. Figure 1(b) shows a good example of two trajectories, which are homologous but not homotopic.

A compact formulation for computing the homology class as *h-signatures* using Cauchy integral theorems from complex analysis is carried out by Bhattacharya et al. [1, 2]. The h-signature of a trajectory, q, with respect to obstacle o_j is defined as

$$H_j(q) = \int \frac{1}{z - z_j} dz = \int_{t_0}^{t_f} \frac{1}{q_x(t) + iq_y(t) - z_j} (\dot{q}_x(t) + i\dot{q}_y(t))\, dt \qquad (1)$$

where $z(t) = q_x(t) + iq_y(t)$ is the complex representation of the trajectory, and z_j is the complex representation of an arbitrary point inside obstacle o_j. Then the h-signature about all obstacles is given by $H = [H_1, \ldots, H_{n_o}]^T$ with n_o obstacles.

2.2 Optimal Trajectory Generation

We consider trajectory planning in a compact subset $Q \subset \mathbb{R}^2$ on a plane. Let $O = \{o_1, o_2, \cdots, o_{n_o}\}$ be a set of convex pair-wise disjoint *obstacles* in Q (The requirement of convexity of obstacles can be relaxed by considering an arbitrarily-shaped obstacle as a union of convex obstacles.) Each obstacle $o_i \in O$ can be represented by a n_i-sided convex polygon, whose faces define hyperplanes that partition Q into two half-spaces. A binary variable can be used to represent either side of the hyperplane as in [8]. So a point $q \in Q$ will be feasible and will avoid collision with an obstacle o_i if there is at least one face $f \in [1, \ldots, n_i]$ satisfying $n_{i,f} \cdot q \leq s_{i,f}$. Where $n_{i,f}$ is a normal vector to the fth face of obstacle o_i pointing inward and $s_{i,f} = n_{i,f} \cdot p$, for any point p on the fth face. For the obstacle o_i, all points $q \in Q$ outside the obstacle must satisfy [9]:

$$n_{i,f} \cdot q \leq s_{i,f} + Mb_{i,f} \quad \text{for} \quad b_{i,f} \in \{0,1\}, \quad f = 1, \ldots, n_i \qquad (2)$$
$$\sum_{f=1}^{n_i} b_{i,f} \leq n_i - 1,$$

where M is a large positive number. The second inequality in (2) implies that the point should be feasible about at least one face; at least one f such that $b_{i,f} = 0$. This formulation breaks up Q into overlapping regions.

The problem of finding a trajectory, $q(t)$, that avoids obstacles, respects kinematic constraints and is restricted to a specified homotopy class, is formulated as follows. The trajectory is obtained by splicing N_s sub-trajectories, each parametrized by linear combination of $N+1$ basis functions,

$$q(t) = \sum_{k=0}^{N} c_{j,k}\, e_k(t) \quad \text{for} \quad t_j \leq t < t_{j+1}, \qquad (3)$$

for $j \in [0, \ldots, N_s - 1]$, $0 = t_0 \leq t_1 \leq \ldots \leq t_{N_s} = t_f$. Where $e_k(t)$ is a suitably chosen basis function and $c_{j,k}$ are coefficients. Throughout this paper, we use $e_k(t) = (t - t_j)^k$. The trajectory is restricted to be k_r-times continuously differentiable at each of the intermediate points, $q(t_j)$, for $j \in [0, \ldots, N_s - 1]$. Further, obstacle avoidance is achieved by enforcing (2) at some equally distributed intermediate points on each subtrajectories. The cost function is the integration of norm of rth-derivative of trajectory:

$$J(c) = \int_{t_0}^{t_f} \left\| \frac{d^r q(t)}{dt^r} \right\|^2 dt \;=\; c^T H c. \qquad (4)$$

where $c = [c_0^T, \ldots c_{N_s-1}^T]^T$. The optimal trajectory generation problem can then be simplified as the following mixed-integer quadratic program,

$$\min_{c,\, b} \quad c^T H c \qquad (5)$$
$$\text{s.t.} \quad A_f c + D_f b \leq g_f, \quad A_b b \leq g_b, \quad A_{eq} c = 0$$

where b is the vector of binary variables. The first inequality captures the feasibility constraints of (2) for the intermediate points, the second inequality captures the constraint on sum of binary variables in (2), and $A_{eq}c = 0$ imposes rth order differentiability at the intermediate points as well as the boundary condition of initial configuration, $q(0) = q_0$ and final configuration $q(t_f) = q_f$.

To find an optimal trajectory in a specific homotopy or homology class, we now incorporate topological constraints. If we add a constraint on h-signature, which we defined earlier, such that the h-signature of the trajectory, $H(q)$, should be some desired H_d, the quadratic program (5) becomes a nonlinear problem. Furthermore, the gradient of the new constraint, $H = H_d$ will be zero almost everywhere, because the value of h-signature does not change within a particular homology class (*i.e.* the range of the h-signature is a set of discrete values). So, the resulting problem is NP and it is numerically hard to find a solution based on gradients of cost and constraints. So, we need another way to enforce topological constraints, and this is carried out as follows.

Although (2) is a sufficient condition for feasibility, we introduce an additional inequality so as to obtain a partition of Q.

$$-n_{i,f} \cdot q \leq -s_{i,f} + M(1 - b_{i,f}) \quad \text{for} \quad b_{i,f} \in \{0,1\}, \ f = 1, \ldots, n_i. \tag{6}$$

The first inequality of (2) only guarantees that the point q is on one side of face when $b_{i,f} = 0$. With the additional constraint (6), the binary variable, $b_{i,f}$, uniquely determines in which half space a point q is on. So we can divide the work space with hyperplanes of obstacles by value of binary variables.

As a result, a set of connected *cells* is built, whose union is the feasible space, Q, and the intersection is only the extended lines of faces of the obstacles (see Figure 1(c)). Each cell can be identified by a unique *letter*, representing the vector of binary variables with one binary variable for each face of each obstacle. Every point in a particular cell will have the same letter representation. It must be noted that not all binary vectors define valid cells, and hence letters. The collection of all possible valid letters is defined as an *alphabet*.

Determining homotopy class of a trajectory is non-trivial. However, we use location information of intermediate intermediate point, each represented by a letter in the alphabet. Assembling the sequence of letters corresponding to each intermediate point of the trajectory and removing trivial repetitions will results in a *word*, which is a coarse representation of the trajectory. For example, the path shown in Figure 1(c) can be represented by the word $TPUVWQLJHG$. This can then be used to restrict trajectories to a homotopy class as will be seen in Section 3.

3 Algorithm Description

We have broken the problem of optimal trajectory generation into two parts. First we find a *word* that is a coarse representation of the trajectory and use this to restrict the homotopy class of the trajectory, and next find an optimal trajectory with this restriction. The following sections present the algorithm in more detail.

3.1 Finding Words in the Same Homotopy Class

To find an optimal trajectory satisfying a given homotopy class constraint, we first construct W_h, the set of words of the same homotopy class with the required one. We construct W_h by starting with the word for the given initial trajectory; $W_h = \{w_0\}$. Then we choose a word $w_c \in W_h$ and expand the chosen word as follows. For example let $w_c = TPUVWQLJHG$ as in Figure 1(c). We choose two letters, say T and U. If there is an alternative path, like TXU(the gray plot in Figure 1(c)), for the path TPU, we construct the closed loop by reversing the new path, and obtain

$TPUXT$ after removing duplicating letters. If the length of the closed loop is less than six, no obstacle lies in the closed loop (since we need to visit at least six cells to encircle a triangle). So we replace the path between the two chosen letters with the new path, and an expanded word representing the same homotopy class is achieved, $w_1 = TXUVWQLJHG$. The new word is added into W_h. We repeat this expansion until there are no more new words.

3.2 Finding the Optimal Trajectory

For a given word, $w_c \in W_h$, we parameterize the trajectory with N_s subtrajectories, where N_s is same as the length of w_c. Each subtrajectory is restricted to be in a particular cell specified by the corresponding letter in the word. Thus, all the binary variables, b_c, of the trajectory generation problem of (5) are fixed by the given word w_c, to reduce the optimization problem to

$$\begin{aligned} \min_{c} \quad & c^T H c \\ \text{s.t.} \quad & A_f c \leq \tilde{g}_f, \quad A_{eq} c = 0, \end{aligned} \tag{7}$$

which is obtained by substituting b_c in (5) and $\tilde{g}_f = g_f - D_f b_c$. As the resulting problem (7) is a quadratic program, we can find the global optimal trajectory for all words in W_h, which are in the given homotopy class.

However, it is not trivial to find the spending time in each cell to minimize the cost of the whole trajectory. To refine the trajectory further, we can adjust the time spent in each cell. With the final time, t_f, fixed, we can find an optimal time distribution by solving

$$\begin{aligned} \min_{t_j} \quad & \min_{c} \quad c^T H c \quad \left(\text{s.t. } A_f c \leq \tilde{g}_f, \; A_{eq} c = 0\right) \\ \text{s.t.} \quad & t_j \leq t_{j+1} \quad \text{for} \quad j = [0, \ldots, N_s - 1], \quad t_0 = 0, \quad t_{N_s} = t_f. \end{aligned} \tag{8}$$

As this problem is a nonlinear program, we cannot guarantee the global minimum. However, the trajectory is iteratively refined by starting with $\Delta t_j = t_{j+1} - t_j = \frac{t_f}{n_w}$ for $j \in [0, \ldots, N_s - 1]$ and solving (8) by an interior-point method. Although we can find an initial solution without iteration, a better trajectory can be obtained by iterating the time distribution. Moreover, since the optimization cost reduces with more iterations, this method can be considered as an anytime algorithm that produces better solutions with more time.

4 Simulation Results

To illustrate how the suggested algorithm works, we performed some simulations to generate optimal trajectories with various homotopy classes. In this simulation, we

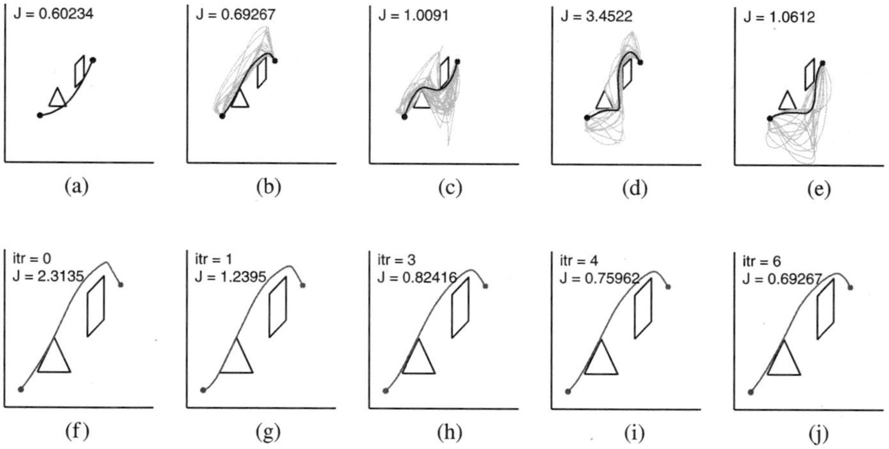

Fig. 2 (a) Optimal trajectory without homotopy constraints. (b)–(e) Trajectories with four different homotopy class constraints. The thick black curve is the optimal trajectory in each homotopy class and thin gray curves are the suboptimal trajectories for each word. The cost (J) for each case is specified on the upper left corners of plots. (f)–(j) Effect of varying the time distribution in each cell through iterations of the optimization (8). The number of iterations (itr) and cost are also specified on the upper left corner of each plot. Note that the cost converges to the local optimal cost of the case (b) in 6 iterations.

fix the final time $t_f = 10$ and find optimal trajectories for four homotopy classes. To reduce the computation time, we limit the maximum length of word to twelve.

The plot of Figure 2(a) shows the result of solving (5) without homotopy class constraints, resulting in an optimal cost of 0.60234. The plots of Figure 2(b)–2(e) show the result of solving (8) with four different homotopy class constraints, resulting in optimal costs that are greater than the global optimal one. When we search for trajectories with the same homotopy as the optimal trajectory achieved without homotopy class constraints (Figure 2(a)), the obtained optimal trajectory (Figure 2(e)) is a local optimal one with a larger cost. This disparity occurs due to restricting the trajectory to pass through certain cells and the fact that it is hard to find global optimal time distribution in each cell. The most optimal trajectory with homotopy class constraints lies in a different homotopy class from the global optimal one (Figure 2(b)). However, this is due to the symmetric arrangement of initial/final location of the trajectory and arrangement of obstacles.

With a fixed time distribution for each cell, the optimization reduces to a quadratic program for each word, which can be solved efficiently. To see the effect of optimizing the time distribution, we begin with a trajectory in the particular homotopy class of Figure 2(b) with equal time distribution over all the cells and iteratively optimize time distribution. The plots of Figure 2(f)–2(j) illustrate the changes in the trajectory and the corresponding cost with each iteration. Although this nested optimization is computationally expensive, with each iteration we get closer to the local optimal solution, resulting in an algorithm with *anytime* properties.

5 Conclusion

In this paper, we have presented a method to find an optimal trajectory subject to kinematic constraints, obstacle avoidance, and restricted to a specific homotopy class. This has been achieved by suitably modifying a MIQP to partition the configuration space and by constructing a coarser representation of the trajectory in the form of a word to represent the homotopy class. The set of all words representing the same homotopy class is constructed, and a nested optimization is carried out to find a locally optimal trajectory restricted to a homotopy class. Although we have only illustrated examples in the plane, work in [2] suggests obvious extensions to 3-D.

References

1. Bhattacharya, S., Kumar, V., Likhachev, M.: Search-based path planning with homotopy class constraints. In: AAAI Conf. on Artificial Intelligence (2010)
2. Bhattacharya, S., Likhachev, M., Kumar, V.: Identification and representation of homotopy classes of trajectories for search-based path planning in 3D. In: Robotics: Science and Systems (2011)
3. Demyen, D., Buro, M.: Efficient triangulation-based pathfinding. In: National Conf. on Artificial Intelligence, pp. 942–947 (2006)
4. Donald, B., Xavier, P., Canny, J., Reif, J.: Kinodynamic motion planning. J. ACM **40**(5), 1048–1066 (1993)
5. Grigoriev, D., Slissenko, A.: Polytime algorithm for the shortest path in a homotopy class amidst semi-algebraic obstacles in the plane. In: Int. Symposium on Symbolic and Algebraic Computation, pp. 17–24 (1998)
6. Hershberger, J., Snoeyink, J.: Computing minimum length paths of a given homotopy class. Comput. Geom. Theory Appl. **4**, 331–342 (1991)
7. Lavalle, S.M., Kuffner, J.J.: Rapidly-exploring random trees: Progress and prospects. In: Donald B.R., Lynch K.M., Rus D. (eds.) Algorithmic and Computational Robotics: New Directions, pp. 293–308. A K Peters (2001)
8. Richards, A., How, J.P.: Aircraft trajectory planning with collision avoidance using mixed integer linear programming. In: American Control Conf., vol. 3, pp. 1936–1941 (2002)
9. Richards, A., How, J.P., Schouwenaars, T., Feron, E.: Plume avoidance maneuver planning using mixed integer linear programming. In: AIAA Guidance Navigation and Control Conference (2001)
10. Schmitzberger, E., Bouchet, J.L., Dufaut, M., Wolf, D., Husson, R.: Capture of homotopy classes with probabilistic road map. In: IEEE/RSJ Int. Conf. on Intelligent Robots and Systems, pp. 2317–2322 (2002)
11. Tedrake, R., Manchester, I.R., Tobenkin, M.M., Roberts, J.W.: LQR-trees: Feedback motion planning via sums of square verification. Int. J. Robot. Res. **29**(8), 1038–1052 (2010)
12. Tovar, B., Cohen, F., LaValle, S.: Sensor beams, obstacles, and possible paths. In: Chirikjian G., Choset H., Morales M., Murphey T. (eds.) Algorithmic Foundation of Robotics VIII. Springer Tracts in Advanced Robotics, vol. 57, pp. 317–332. Springer, Berlin/Heidelberg (2009)

Validation of a Power Grasping Algorithm for an Anthropomorphic Robotic Hand on the Basis of Human Grasping Action

F. Cordella, L. Zollo, A. Salerno, E. Guglielmelli and B. Siciliano

Abstract It is generally acknowledged that performing a natural and reliable grasp with a robotic hand is a challenging task not yet completely solved. Further, the robotic hands designed until now do not have the same kinematic characteristics of a human hand, especially regarding the thumb. The main purpose of this paper is to use information obtained from the analysis of the human grasping action for validating and improving a reach-and-grasp algorithm we proposed for determining the optimal hand position for grasping a cylindrical object. Algorithm effectiveness has been tested on a real arm-hand robotic system.

Key words: Robotic grasping, hand kinematics

1 Introduction

The human hand represents the most dexterous part of the human body, both for its complex mechanical structure and for the versatility of its possible activities. In particular, grasping is one of the human skills that robotic researchers mostly attempt at imitating. One possible approach to reduce the complexity of the control that ensures stability of grasping consists of optimizing grasping configuration. During the pre-shaping phase, on the basis of the physical characteristics of the object to be grasped, such as shape and weight, the hand, while approaching the object, attains the most suitable configuration for seizing. Thus, pre-shaping plays a fundamental role in order to guarantee a stable grasp.

F. Cordella · B. Siciliano
PRISMA Lab, Dipartimento di Informatica e Sistemistica, Università di Napoli Federico II, Napoli, Italy, e-mail: {francesca.cordella, bruno.siciliano}@unina.it

L. Zollo · A. Salerno · E. Guglielmelli
Laboratory of Biomedical Robotics and Biomicrosystems, Università Campus Bio-Medico, Roma, Italy, e-mail: {l.zollo, antonino.salerno, e.guglielmelli}@unicampus.it

J. Lenarčič, M. Husty (eds.), *Latest Advances in Robot Kinematics*,
DOI 10.1007/978-94-007-4620-6_12,

Many approaches can be found in the literature to the purpose of searching an optimization criterion for selecting the optimal grasp configuration for a given object. In [1] a review of various techniques has been proposed for identifying the optimal grasp within the space of feasible grasps for a given set of task constraints, like object shape, type of task, hand degrees of freedom (DOFs), maximum force to be applied, etc. The main difficulty of this approach is the choice of the parameters to be used as constraints and to be included in the objective function [2]. A different approach consists in determining the hand-object contact points that guarantee a firm grip. Joint positions and forces that a robotic hand should apply for a stable grasp could be provided by a neural system [3] implying a high computational cost. Contact points can also be determined by means of grasp quality measures, which are used to evaluate grasp performance. In [4] two groups of grasp quality measurements have been considered and the authors propose a method for satisfying them. In [5], after heuristically generating a set of feasible grasp candidates, grasp quality measures, computed with an ad hoc method, have been used to choose the best grasp. Finally, methods for finding an optimal grasp configuration based on grasping force determination can be mentioned. They rely on the concepts of wrench matrix and friction cone [6].

By analyzing the robotic literature, it is evident a sizeable lack of information about the thumb behavior, despite its fundamental role during the grasping action. In fact, the robotic hands designed so far [7–9] are not able to replicate the motor capabilities of the human hand.

From the analysis of the grasping action performed by human beings and from the study of the anatomy of the human hand and of its behavior during grasping, it is possible to obtain useful information for developing human-like grasping algorithms so as to improve knowledge about hand kinematics on which this work is based. Objective of this work is to validate a recently proposed algorithm [10] that guarantees a stable power grasp and to pave the way for future works devoted to improving knowledge about the thumb behavior, giving some hints for enhancing the existing robotic hands. An optoelectronic motion analysis system has been used for collecting information about finger behavior during grasping. The obtained results have been used for improving and validating the optimization algorithm that has been also tested on a real robotic arm-hand system during reach-and-grasp tasks of cylindrical objects.

2 Experimental Analysis of Human Grasping Action

The Vicon optoelectronic motion analysis system has been used for collecting information about the fingers during grasping. The system is composed of 7 InfraRed (IR) cameras, with a frame rate of 100 Hz.

Seven human subjects, 31.7 years old in the average (8.75 Standard Deviation), five men and two women, all right handed, volunteered to participate in this study. Subjects were asked to grasp a cylindrical object of 6 cm diameter with a diagonal

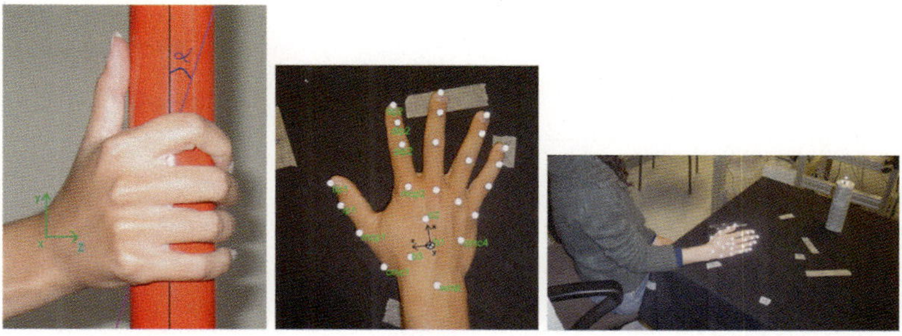

(a) (b) (c)

Fig. 1 (a) Human grasp of a cylindrical object with a diagonal volar grasp. (b) Marker configuration with reference frame in black; (c) Subject starting position. The markers on the object allows identifying its position.

volar grasp (Fig. 1(a)) for 10 times each. 25 reflective markers of 6 mm diameter were placed on their right hand as shown in Fig. 1(b). All subjects gave informed consent to participate in the study.

According to results in the literature [11, 12], the protocol for positioning markers on the hand has been chosen in order to minimize artifacts, due to skin movements or marker occlusion, and for obtaining information about the wrist position (a reference point for the grasping algorithm).

The participants were seated in front of a table on which the cylindrical object was located in a-priori known position. Hand starting position and initial posture were the same for all the participants (Fig. 1(c)). The hand starting configuration was the one in which the four fingers were fully extended and the thumb was adducted. The marker positions were recorded in this starting position and during all the trial until the hand grasps the object. The object position was identified by 4 markers placed on the top of the cylindrical object, as shown in Fig. 1(c). Every subject was asked to grasp the object, without lifting it, ten times. Before starting the data acquisition, each participant was asked to grasp the object five times for training.

2.1 Data Analysis and Results

The Vicon Nexus 1.6.1 Software package has been used to reconstruct marker Cartesian positions with the Vicon system and a link model of the hand has been constructed. In order to understand if a common behavior among subjects can be observed during the grasping action, two performance parameters have been extracted from the collected data: (i) the radius of curvature of every finger (i.e. the radius of the obsculator circles tangent, at each joint, the spline passing throughout the finger joints); (ii) the adduction/abduction angle between the fingers (i.e. the angle between two adjacent fingers, subtended by the distance between the PIP joints of the two fingers). For calculating the above parameters, the last frame of each trial

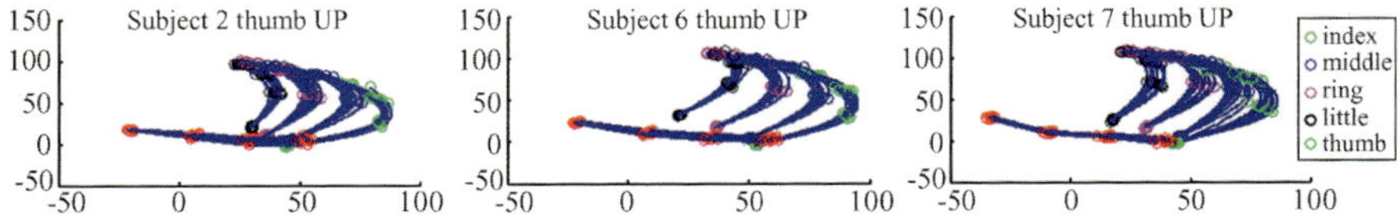

Fig. 2 Curvature of the long fingers during diagonal volar grasp. For space reasons only 3 samples are shown.

Table 1 Mean and standard deviation of radius of curvature in diagonal volar grasp for the index finger.

	PIP joint		DIP joint	
Subject #	Mean (mm)	SDV	Mean (mm)	SDV
1	28.96	3.01	32.83	5.34
2	22.81	0.93	30.97	8.30
3	25.32	1.97	26.36	1.48
4	33.54	2.32	27.19	2.44
5	29.10	1.27	36.35	4.23
6	29.80	1.66	34.06	4.22
7	31.98	3.51	30.95	2.64

Table 2 Mean and standard deviation of the adduction/abduction angle (in degrees) of the MCP joints in diagonal volar grasp.

	MCP_{2-3}		MCP_{3-4}		MCP_{4-5}	
Subject #	Mean	SDV	Mean	SDV	Mean	SDV
1	31	2.7	29.11	2.18	25.64	1.49
2	37.8	1.49	24.42	1.28	20.88	1.70
3	36.98	1.38	20.71	0.62	23.34	3.25
4	36.44	4.19	21.31	1.34	21.05	0.85
5	36.88	1.36	20.74	0.18	24.67	0.45
6	32.09	1.52	26.40	0.61	29.76	2.4
7	35.48	3.37	22.98	1.41	28.24	3

has been considered for each subject. They are expected to provide useful information about the configuration of all the fingers during grasping. Figure 2 shows the curvature of all the fingers when grasping the cylindrical object with a diagonal volar grasp. The blue lines are the splines passing throughout the hand joints outlined with different colors, as explained in the figure legend. Furthermore, in Table 1 mean value and standard deviation of finger curvature radii during a diagonal volar grasp are listed. The MetaCarpoPhalangeal (MCP) and fingertip (TIP) radii values are not reported since they are obviously very high. The mean is calculated for each subject during the 10 trials. A quite similar behavior among different subjects can be observed. In Table 1 only the behavior of the index finger is reported for brevity, but the results are similar also for the other fingers. In Table 2 mean and standard deviation of the adduction/abduction angle (in degrees) of the MCP joints are listed. In particular, MCP_{2-3}, MCP_{3-4}, MCP_{4-5} are the angles between the index finger and the middle finger, between the middle finger and the ring finger and between the ring finger and the little finger, respectively.

These findings allow us to say that subjects with different hand size grasp the cylindrical object with the same long finger configuration independently of the hand dimension. Therefore, it is possible to use the hand joint position for validating the grasping algorithm and for introducing the MCP adduction angles.

3 Bio-Inspired Grasping Algorithm for a Robotic Hand

The wrist plays a fundamental role in the grasping action, i.e. it guides the hand to the grasp position and adjusts the orientation on the basis of the grasping type and task. This is the reason why one marker has been positioned on the hand of the examined human subjects during the motion analysis and why a point on the wrist, called CarpoMetaCarpal (CMC) joint, is considered in the following as reference point for the bio-inspired grasping algorithm for determining the position of the other hand joints.

The algorithm was proposed in our previous work [10] and had the purpose of predicting the optimal hand configuration for stably grasping a cylindrical object, given the size of the object and its location in the space. The starting hypothesis is that the long finger optimal configurations for grasping a cylindrical object with a diagonal volar grasp is the one that minimizes the sum of the distances between the hand joints and the object surface. In particular, the position of the CMC joint that guarantees a stable grasp configuration can be obtained by minimizing the objective function given by the sum of the distances of all finger joints from the object surface. Providing the CMC coordinates and the distances of the joints from the object surface, in addition to some geometrical considerations (such as the adduction/abduction angles obtained from the motion analysis system), all the joint coordinates are computed for the four long fingers with a human-like optimal grasp configuration.

4 Experimental Validation of the Grasping Algorithm

The experimental validation of the algorithm mentioned in Section 3 on a real arm-hand robotic system has been carried out. The experimental platform (Fig. 3(a)) is composed of the MIT-Manus planar robot, acting as the arm to realize the reaching task, and the DLR-HIT-Hand II mounted at the MIT-Manus end-effector, which is responsible for preshaping and grasping.

The five-fingered dexterous robotic hand DLR-HIT-Hand II has five identical modular fingers with four DOFs each. The last two joints (proximal and distal) are $1:1$ coupled, meaning that the corresponding flexion/extension angles are equal. The thumb is mechanically constrained to assume a fixed opposition of 35.51^o in the xy–plane with an inclination, with respect to z-axis, of 44.13^o; this only enables transverse volar grasps with a fixed thumb inclination. Therefore, during the experiments, the thumb has been disabled in order to perform a power grasp with only the four long fingers.

The object to be grasped is a cylinder with given shape, weight and position. The initial configuration of the hand joints, as well as the optimal configuration from the algorithm have been provided in the MIT-Manus reference frame. Being the MIT-Manus planar, the arm and hand height from the table could not be varied. Consequently, the object was properly located in order to allow closing the middle

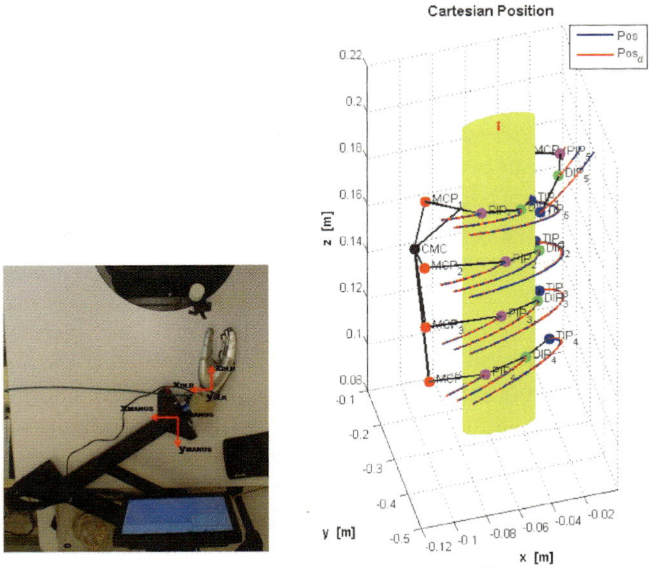

(a) (b)

Fig. 3 (a) Experimental setup. The DLR-HIT-Hand II and MIT Manus reference frames are shown. (b) Hand joints trajectory in the Cartesian space for the DLR-HIT-Hand II grasping the cylindrical object with radius 0.020 m.

finger at half of the object height. This assumption was coherent with findings of our experiment.

Given the object position, the optimal CMC Cartesian position and the final hand configuration for grasping the object were obtained through the MATLAB function $fminsearch(f, [initialcondition])$. During the reaching movement, the hand has been moved by the arm towards the optimal CMC position. Thus, the hand has been controlled in order to reach the final MCP, PIP, DIP joint angles, also provided by the optimization algorithm. During reaching, the hand cannot change orientation, being arm motion planar: z_{DLR}-axis is always parallel to y_{MANUS}.

A fifth order polynomial function has been used to plan the MIT-Manus linear motion from initial position up to final position. Then, a proportional-derivative (PD) torque control in the Cartesian space has been used to control arm position (and consequently CMC position) in the plane.

As regards preshaping, final MCP, PIP and DIP joint positions, provided by the optimization algorithm, have been taken as reference for the DLR-HIT-Hand II motion controller. A third-degree polynomial function has been used to plan joint motion up to the final reference value and a PD torque control in the joint space enabled reaching the desired final angles. It is worth noticing that, in the DLR-HIT-Hand II, the DIP and PIP joints are coupled with $\theta_{DIP} = \theta_{PIP}$, thus providing a constraint to the final position of PIP and TIP, that was often slightly different with respect to the desired one. The obtained results are shown in Fig. 3, where the joint trajectories are reported in the Cartesian space. For the arm, a point-to-point movement has been performed in 3.0 s for each trial, starting from the initial position $P_i = [-0.1 \quad 0.1]^T$ m to final position $P_f = [-0.0975 \quad -0.1245]^T$ m. The final position, reported only for grasping an object with radius of 0.020 m for sake of brevity, takes into account the CMC position supplied by the algorithm $CMC = [-0.0975 \quad -0.179 \quad 0.15]^T$ m as well as the offset between the arm end-effector and the hand CMC due to the flange that connects the DLR-HIT-Hand II to the MIT-Manus robotic arm. The cylindrical objects to grasp have radii 0.0225 m and 0.020 m and have been located in $(-0.051, -0.257)$ m in the MIT-Manus reference frame.

Table 3 reports the actual Cartesian coordinates for each joint of the robotic hand performing the task of grasping an object with radius of 0.0225 m and compares them with the Cartesian coordinates produced by the optimization algorithm and with those measured by the Vicon system. The values in the last column regard an object with radius of 0.03 m. Moreover, the reference frame is not coincident with the hand wrist, but is positioned as shown in Fig. 1(b). This is the reason why the z-coordinate of the values taken with the Vicon system are different from the values obtained with the other two systems. In particular, the z-coordinate of the hand wrist is -0.056 m in this reference frame. By taking into account this offset, it is possible to note a quite similar behavior among the algorithm output, the robot position and the Vicon results. The small differences in the Cartesian coordinates for the robotic hand are due to its mechanical structure that constrains the DIP joint motion in a different way with respect to the human hand. The obtained results allow us to conclude that the hand configuration obtained with the optimization algorithm is similar to the stable one assumed by the human beings.

Table 3 Joint Cartesian coordinates resulting from the optimization algorithm, measured on the robotic hand and obtained by the Vicon system.

		"Cartesian position"		
Finger	Joint	Algorithm	Robotic hand	Vicon system
index	MCP	$[0 \quad 0.045 \quad 0.088]^T$ m	$[-0.0025 \quad 0.0368 \quad 0.1078]^T$ m	$[-0.0143 \quad 0.011 \quad 0.043]^T$ m
	PIP	$[0.0485 \quad 0.0468 \quad 0.1040]^T$ m	$[0.0264 \quad 0.0343 \quad 0.1545]^T$ m	$[0.029 \quad 0.014 \quad 0.073]^T$ m
	DIP	$[0.0672 \quad 0.0454 \quad 0.0877]^T$ m	$[0.0513 \quad 0.0321 \quad 0.1526]^T$ m	$[0.057 \quad 0.018 \quad 0.073]^T$ m
middle	MCP	$[0.0012 \quad 0.0100 \quad 0.0977]^T$ m	$[-0.0037 \quad 0.0100 \quad 0.1178]^T$ m	$[0.0037 \quad -0.009 \quad 0.042]^T$ m
	PIP	$[0.0541 \quad 0.0100 \quad 0.1126]^T$ m	$[0.0437 \quad 0.0100 \quad 0.1457]^T$ m	$[0.044 \quad -0.008 \quad 0.063]^T$ m
	DIP	$[0.0716 \quad 0.0100 \quad 0.0947]^T$ m	$[0.0655 \quad 0.0100 \quad 0.1335]^T$ m	$[0.074 \quad 0.008 \quad 0.057]^T$ m
ring	MCP	$[0 \quad -0.0250 \quad 0.0925]^T$ m	$[-0.0025 \quad -0.0168 \quad 0.1126]^T$ m	$[0.011 \quad -0.026 \quad 0.031]^T$ m
	PIP	$[0.0515 \quad -0.0268 \quad 0.1093]^T$ m	$[0.0409 \quad -0.0130 \quad 0.1462]^T$ m	$[0.054 \quad -0.019 \quad 0.046]^T$ m
	DIP	$[0.0695 \quad -0.0253 \quad 0.0922]^T$ m	$[0.0641 \quad -0.0110 \quad 0.1371]^T$ m	$[0.080 \quad -0.061 \quad 0.042]^T$ m
little	MCP	$[-0.0035 \quad -0.0600 \quad 0.0735]^T$ m	$[0.0010 \quad -0.0434 \quad 0.0936]^T$ m	$[0.021 \quad -0.041 \quad 0.020]^T$ m
	PIP	$[0.0365 \quad -0.0635 \quad 0.0905]^T$ m	$[0.0313 \quad -0.0380 \quad 0.1391]^T$ m	$[0.056 \quad -0.031 \quad 0.030]^T$ m
	DIP	$[0.0584 \quad -0.0615 \quad 0.0794]^T$ m	$[0.0559 \quad -0.0337 \quad 0.1409]^T$ m	$[0.073 \quad -0.026 \quad 0.032]^T$ m

5 Conclusion

Experiments for validating a biologically inspired approach for finding the optimal grasp configuration have been performed. The algorithm has been improved with some geometric considerations obtained from the analysis of the human grasping action. The realized experiments have been described and the obtained data have been analyzed. They allowed us to maintain that the hand configuration obtained by the optimization algorithm is similar to the stable one recorded with the Vicon motion analysis system. The algorithm has been validated through experimental trials on a real arm-hand robotic system, composed of the MIT-Manus robot arm and the DLR-HIT-Hand II. Experimental results on the described robotic platform have proven its feasibility and reliability but have also shown limitations in the grasping tasks due to the constraints imposed by the mechanical structure, not adequately similar to the human structure.

References

1. Cutkosky, M.R.: On grasp choice, grasp models, and the design of hands for manufacturing tasks. IEEE Trans. Robot. Autom. **5**(3), 269–279 (1989)
2. Kerr, J., Roth, B.: Analysis of multifingered hands. Int. J. Robot. Res. **4**(4), 3–17 (1986)
3. Folgheraiter, M., Baragiola, I., Gini, G.: Teaching grasping to a humanoid hand as a generalization of human grasping data. In: Knowledge Exploration in Life Science Informatics, pp. 139–150. Springer (2004)
4. Xue, Z., Zoellner, J.M., Dillmann, R.: Grasp planning: Find the contact points. In: Proceedings of the 2007 IEEE International Conference on Robotics and Biomimetics, pp. 835–840 (2007)
5. Borst, Ch., Fischer, M., Hirzinger, G.: A fast and robust grasp planner for arbitrary 3D objects. In: Proceedings of the 1999 IEEE International Conference on Robotics & Automation, vol. 3, pp. 1890–1896 (1999)
6. Bicchi, A., Kumar, V.: Robotic grasping and contact: A review. In: Proceedings of the 2000 IEEE International Conference on Robotics and Automation, pp. 348–353 (2000)
7. Jacobsen, S.C., Wood, J.E., Knutti, D.F., Biggers, K.B.: The Utah-MIT dextrous hand: Work in progress. Int. J. Robot. Res. **3**(4), 21–50 (1984)
8. Lovchik, C.S., Diftler, M.A.: The robonaut hand. A dextrous robot hand for space. In: Proceedings of the 1999 IEEE International Conference on Robotics and Automation, vol. 2, pp. 907–912 (1999)
9. Butterfass, M., Grebenstein, H., Liu, H., Hirzinger, G.: DLR-Hand II. Next generation of a dextrous robot hand. In: Proceedings of the 2001 IEEE International Conference on Robotics and Automation, vol. 1, pp. 109–114 (2001)
10. Cordella, F., Zollo, L., Guglielmelli, E., Siciliano, B.: A bio-inspired grasp optimization algorithm for an anthropomorphic robotic hand. International Journal of Interactive Design and Manufacturing, in press (2012)
11. Carpinella, I., Mazzoleni, P., Rabuffetti, M., Thorsen, R., Ferrarin, M.: Experimental protocol for the kinematic analysis of the hand: Definition and repeatability. Gait Posture **23**, 445–454 (2006)
12. Metcalf, C.D., Notley, S.V., Chappell, P.H., Burridge, J.H., Yule, V.T.: Validation and application of a computational model for wrist and hand movements using surface markers. IEEE Trans. Biomed. Eng. **55**, 1199–1210 (2008)

On the Vertical Darboux Motion

Chung-Ching Lee and Jacques M. Hervé

Abstract A vertical Darboux motion (*VDM*) is a 1-DoF motion. It is a special case of a general Darboux motion in which all the trajectories in the moving body are planar ellipses. An axis is globally invariant in a *VDM*, which is a special cylindrical motion. Beyond the transformation of point coordinates, which is available in the literature, some mechanical generators of a *VDM* are revealed. These *VDM* generators are systematically synthesized by the parallel arrangement of one cylindrical (C) pair and one generator of a 5-dimensional (5D) manifold of rigid-body displacements, in which one point is compelled to move in a fixed plane. For instance, the center of a spherical S pair in a serial PPS chain moves in a plane that is parallel to the prismatic P pairs. Replacing the PPS chain with other generators of planar-spherical motion yields isoconstrained generators of *VDM*. In addition, overconstrained realizations of *VDM* generators with exceptional mobility are also unveiled.

Key words: Vertical Darboux motion, mechanical generator, isoconstrained, overconstrained, exceptional mobility

1 Introduction

In 1881 Darboux [1–3] introduced a non-planar motion in which all trajectories are planar curves. Darboux' explanations are based on the transformation of Cartesian coordinates in a privileged frame of reference and he proved that the planar trajectories are ellipses. A Darboux motion is a special kind of Schoenflies motion

Chung-Ching Lee
National Kaohsiung University of Applied Sciences, Kaohsiung, Taiwan ROC,
e-mail: cclee@cc.kuas.edu.tw

Jacques M. Hervé
Ecole Centrale des Arts et Manufacturess, Paris, France, e-mail: jacques.herve07@orange.fr

J. Lenarčič, M. Husty (eds.), *Latest Advances in Robot Kinematics*,
DOI 10.1007/978-94-007-4620-6_13,

with one DoF. In other words, a set of Darboux displacements is a one-dimensional (1D) subset of a 4D Schoenflies-motion group. A derivation of the general Darboux motion by using spatial instantaneous invariants has been given by Veldkamp [4]. When one axis is globally invariant, the special Darboux motion was called in [5], a "vertical" Darboux motion (abbreviated as *VDM*), whose mechanical generation is the main topic of our article. It is worthwhile noticing that, recently, this kind of special motion was taken as an example of a line symmetric motion in [6].

A *VDM* is mathematically modeled by a 1D submanifold of a 2D Lie group of cylindrical motions. Hence, it could be called "cylindrical" Darboux motion. The parallel arrangement of a 2-DoF cylindrical pair (C) and a 5-DoF generator of planar-spherical motion (*Pl-Sph*) such as a PPS open chain, forms a single-loop chain, which is a mechanical generator of a *VDM*. Two main families of single-loop isoconstrained *VDM* generators are then obtained by replacing a PPS chain with other planar-spherical motion generators [7, 8]. Four subfamilies of the first type and two subfamilies of the second one are further proposed. Moreover, overconstrained realizations of *VDM* generator, which belong to the exceptional category of chains [9, 10] are revealed too.

2 Vertical Darboux Motion

The *VDM* axis is globally invariant and is assumed to be vertical. That geometric constraint is obviously realized by the cylindrical pair denoted C. The set of the relative displacements between the pair of bodies is a 2D Lie group of vertical cylindrical motions. Therefore, a *VDM* is mathematically modeled by a 1D submanifold of a 2D group of cylindrical motions. In all types of Darboux motion, any point moves in a fixed plane. As for the special case of a *VDM*, any point moves also on a fixed revolute cylinder having the axis of the cylindrical motion. It is straightforward to verify that the trajectories are ellipses.

In Fig. 1a, a rigid body is connected in parallel by a 2-DoF C pair and a 5-DoF PPS open chain to a fixed base. This forms a new initiatory model of a C-SPP chain. The C pair axis is parallel to the unit vector **k** and the spherical S pair is centered at A. Point Q is the foot of the perpendicular from A to the C axis. The center A of the S pair is compelled to move in a fixed plane parallel to both prismatic P pairs. This plane is orthogonal to the unit vector **u** and intersects the C axis at the point O. Hence, the motion of the foregoing body is a cylindrical motion and one of its point moves in a plane. Clearly, the motion has one DoF. In any two mechanisms that are congruent at their home configurations, the relative displacements between any couple of bodies are conjugate by the displacement which transforms one mechanism into the other one [11]. The group of cylindrical motions is commutative (or Abelian) and, therefore, any of its subsets is self-conjugate (or normal) in the group. Using the invariance of any subset of cylindrical motions through the conjugation by any element of the group of cylindrical motions, we establish that the C-SPP mechanism, which is obtained from the first one (Fig. 1a), through any cylindrical

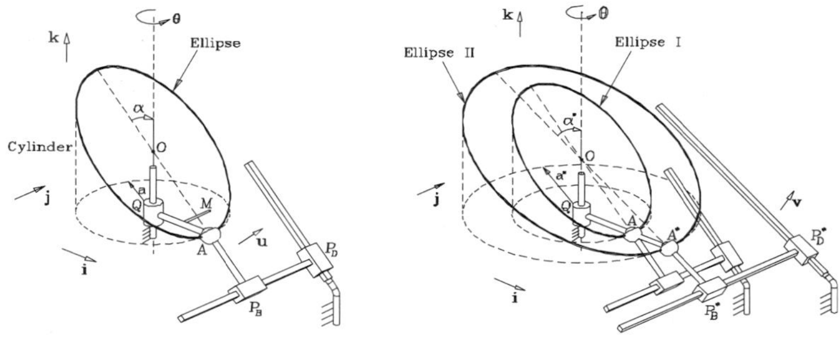

(a) a C-SPP chain. (b) another C-SPP chain

Fig. 1 Initiatory C-SPP models.

displacement of the whole C-SPP mechanism, generates the same cylindrical motion. In the cylindrical displacement of the whole C-SPP mechanism, the axis $(Q, \mathbf{k})$ of the C pair does not change and two C pairs with the same axis are equivalent to one C. The S center of the first C-SPP mechanism becomes another point, which is located on the revolute cylinder with the C axis and passes through the first S center. The angle between the C axis and the PP plane does not change. That way, we prove that all the points belonging to a revolute cylinder move in planes.

Further considerations are needed to establish that the points that do not belong to the initiatory cylinder move also on ellipses. Frame-free vector calculations lead to the geometric characterization of all elliptical trajectories. Figure 1a shows a vertical plane determined by the axis $(Q, \mathbf{k})$ and A. Thus, we write: vector $(\mathbf{QA}) = \mathbf{ai}$ and $(\mathbf{i}, \mathbf{j}, \mathbf{k})$ is an orthornormal vector base. Using the exponential formulation of a rotation, we have $\mathbf{i} = \exp(\alpha\mathbf{j}\times)\mathbf{u} \Leftrightarrow \mathbf{u} = \exp(-\alpha\mathbf{j}\times)\mathbf{i} = \cos\alpha\mathbf{i} + \sin\alpha\mathbf{k}$. When the point A is not constrained to move in a plane, a cylindrical motion provided by the C pair is expressed by the point transform:

$$\forall \text{ point } M \in \text{moving body}, M \to M' = Q + t\mathbf{k} + \exp(\theta\mathbf{k}\times)(\mathbf{QM})$$

in which t and θ are two canonical variable parameters of the 2-DoF motion. The t and θ are not independent when A is constrained to move in a plane. One way to obtain the equation tying t and θ, is to write: $A \to A' = Q + t\mathbf{k} + \exp(\theta\mathbf{k}\times)(\mathbf{QA})$ together with $(\mathbf{AA'}) \cdot \mathbf{u} = \mathbf{0}$. Thus, $A' = Q + t\mathbf{k} + \mathbf{a}\exp(\theta\mathbf{k}\times)\mathbf{i} = \mathbf{Q} + \mathbf{tk} + \mathbf{a}\cos\theta\mathbf{i} + \mathbf{a}\sin\theta\mathbf{j}$

$$\Rightarrow (\mathbf{AA'}) = Q + t\mathbf{k} + a\cos\theta\mathbf{i} + a\sin\theta\mathbf{j} - Q - a\mathbf{i} = a(\cos\theta - \mathbf{1})\mathbf{i} + a\sin\theta\mathbf{j} + t\mathbf{k} \quad (1)$$

and $(\mathbf{AA'}) \cdot \mathbf{u} = [a(\cos\theta - \mathbf{1})\mathbf{i} + a\sin\theta\mathbf{j} + t\mathbf{k}] \cdot (\cos\alpha\mathbf{i} + \sin\alpha\mathbf{k}) = a\cos\alpha(\cos\theta - \mathbf{1}) + t\sin\alpha$. The geometric condition $(\mathbf{AA'}) \cdot \mathbf{u} = \mathbf{0}$ produces the relation

$$a\cos\alpha(\cos\theta - 1) + t\sin\alpha = 0 \Leftrightarrow t = a(\cos\alpha)(1 - \cos\theta)/\sin\alpha \quad (2)$$

It is straightforward to verify that $(\mathbf{QO}) = (a/\tan\alpha)\mathbf{k} = a(\cos\alpha/\sin\alpha)\mathbf{k}$ from elementary geometry. Hence, the vertical Darboux motion can be expressed by

$$M \to M' = Q + (1-\cos\theta)(\mathbf{QO}) + \exp(\theta\mathbf{k}\times)(\mathbf{QM}) \tag{3}$$

where the angle θ is the variable parameter of the special 1-DoF motion. Furthermore, we consider another point A^* such as $(\mathbf{QA}^*) = a^*\mathbf{i}$; A^* belongs to the line QA. In a C-SPP chain indicated by $QA^*PB^*PD^*$, A^* can be constrained to move in the plane perpendicular to the vertical plane of (Q, **k**) and A, which passes through A^* and O, as shown in Fig. 1b. As mentioned above, all the points that lie on the revolute cylinder of axis (Q, **k**), which contains the point A^* moves also on fixed planes. The radius of the cylinder is a^* and can be chosen arbitrarily. Hence, we obtain the following theorem: if, in a subset of cylindrical motions, one point moves in a fixed plane, then all points move in fixed planes. A particular *VDM* can then be specified by its axis (Q, **k**) together with its vertical amplitude that is the signed length 2(QO).

3 Isoconstrained Mechanical Generators of *VDM*

The serial arrangement of a planar pair (G) and a spherical pair (S) is called a generator of a 5D planar-spherical (*Pl-Sph*) motion. An enumeration of the serial arrays of pairs without redundant mobility, which are equivalent to generate a (*Pl-Sph*) motion, was introduced in [7, 8]. In the previous section, a PPS open chain is the simplest realization of *Pl-Sph* motion generator. There are two main families of *Pl-Sph* motion generators. One is the *Pl-(RR)* family and the other is the family of *RR-Sph*. The underline indicates a planar subsystem and a parenthesis denotes a spherical subsystem. The mechanical generators of a *Pl-Sph* motion including helical H pairs [7] are applied hereinafter. Replacing the PPS subchain in the closed loop C-SPP chain by any generator of *Pl-Sph* motion leads to a proper mechanical generator of a *VDM*, which is an isoconstrained chain [12]. Two main categories of *VDM* generators: C-[(RR)-*Pl*] and C-[*Sph*-RR] are presented in Table 1. The corresponding architectures are graphically displayed in Figs. 2 and 3. It is worth noting that depending on the architectural type of *VDM* generator and the size of its links, the range of the mechanically generated *VDM* may be limited.

In three generators of the first main category, which are C-(RR)PRR (Fig. 2c), C-(RR)PPR (Fig. 2e), and C-(RR)PRP (Fig. 2f), contiguous R) and P pairs may be parallel and, then, the R)P subchain can be replaced by a C. We obtain three subcategories: C-(RC)RR, C-(RC)PR, and C-(RC)RP. Furthermore, the C pair can be replaced by one of its kinematic equivalences, namely by either a *HH*, *RH*, *HR*, *PH*, *HP* or *PR* subchain, which differs from the original RP subchain that is equivalent to a C pair (Table 2). In the obtained combinations C-(R*H*)*P*PR, C-(R*H*)*P*RR and C-(R*H*)*P*RP, the *H* and the *P* are not necessarily parallel, which leads to the addition of the C-(RH)*Pl* family (fourth row in Table 2). One representative generator in each subcategory is graphically displayed in Fig. 4. In two generators of the second main

Table 1 Isoconstrained *VDM* generators of the two main categories.

Category	Isoconstrained *VDM* generators	Amount
C-[(RR)-*Pl*]	C-(RR)RRR C-(RR)RPR C-(RR)PRR C-(RR)RRP C-(RR)PPR C-(RR)PRP C-(RR)RPP	7
C-[*Sph*-RR]	C-(RRR)RR C-(RRR)PR C-(RRR)PP C-(RRR)RP	4

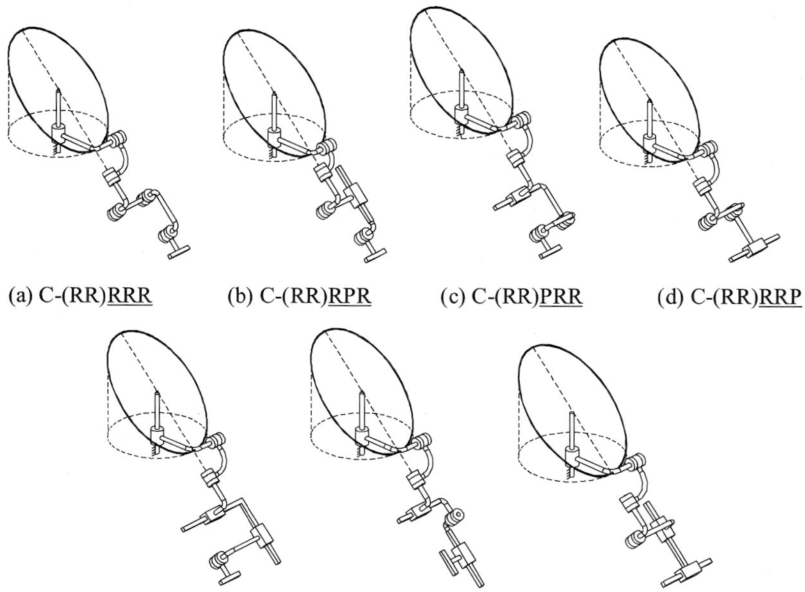

(a) C-(RR)RRR (b) C-(RR)RPR (c) C-(RR)PRR (d) C-(RR)RRP

(e) C-(RR)PPR (f) C-(RR)PRP (g) C-(RR)RPP

Fig. 2 A *VDM* generators of the first main category *Pl*-(*RR*).

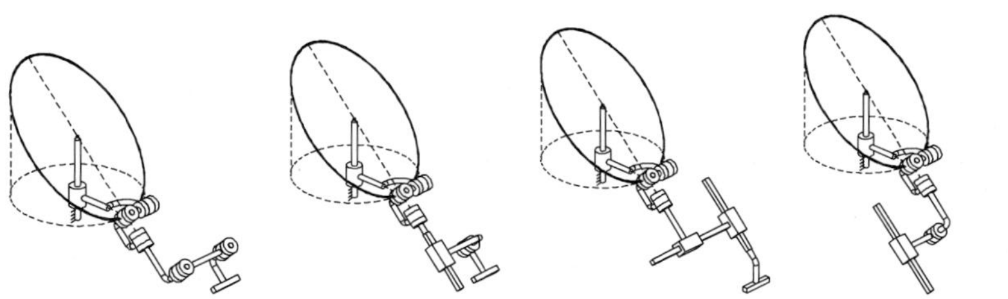

(a) C-(RRR)RR (b) C-(RRR)PR (c) C-(RRR)PP (d) C-(RRR)RP

Fig. 3 *VDM* generators of the second main category *RR*-*Sph*.

category, which are C-(RRR)PP (Fig. 3c), C-(RRR)PR (Fig. 3b), contiguous R) and P pairs may be parallel and, then, the R)P subchain can be replaced by either a C or an equivalence of C that differs from the original RP. We obtain two subcategories: C-(RRC)P and C-(RRC)R (Table 3). Typical architectures are shown in Fig. 5.

Table 2 Isoconstrained *VDM* generators of subcategories of the first main category.

Subcategory	Isoconstrained *VDM* generators	Amount
C-(RC)*RR*	C-(R*P*)*R*RR C-(R*H*)*P*RR C-(R*H*)*H*RR C-(R*P*)*H*RR C-(R*H*)*R*RR C-(R*R*)*H*RR	(1+6)
C-(RC)*PR*	C-(R*P*)*R*PR C-(R*H*)*P*PR C-(R*H*)*H*PR C-(R*P*)*H*PR C-(R*H*)*R*PR C-(R*R*)*H*PR	(1+6)
C-(RC)*RP*	C-(R*P*)*R*RP C-(R*H*)*P*RP C-(R*H*)*H*RP C-(R*P*)*H*RP C-(R*H*)*R*RP C-(R*R*)*H*RP	(1+6)
C-(RH)-*Pl*	C-(RH)RRR C-(RH)RPR C-(RH)PRR C-(RH)RRP C-(RH)PPR C-(RH)PRP C-(RH)RPP	7

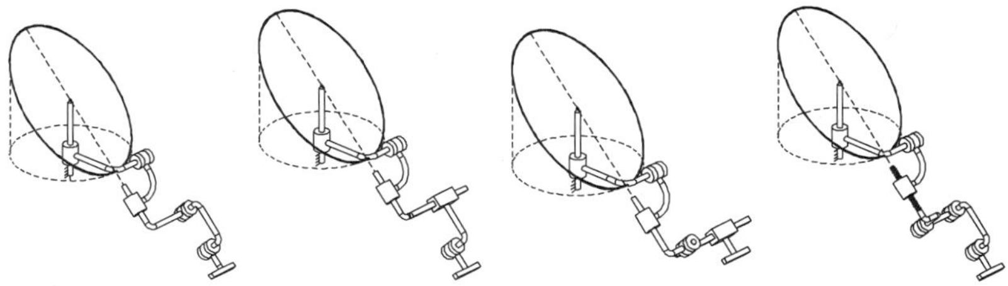

(a) C-(RC)RR (b) C-(RC)PR (c) C-(RC)RP (d) C-(RH)-*Pl*

Fig. 4 Representative *VDM* generators for each subcategory of C-[(RR)-*Pl*].

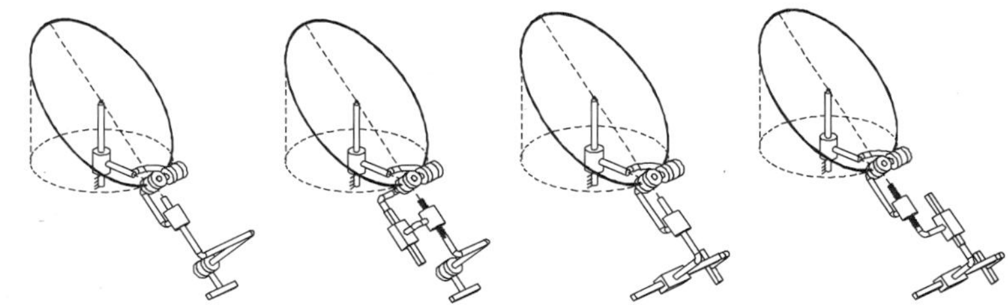

(a) C-(RRC)R (b) C-(RR*P*)*H*R (c) C-(RRC)P (d) C-(RR*H*)*PP*

Fig. 5 Representative *VDM* generators for each subcategory of C-[*Sph*-RR] type.

Table 3 Isoconstrained *VDM* generators of two subcategories of C-[*Sph*-RR] category.

Subcategory	Possible isoconstrained *VDM* generators	Amount
C-(RRC)R	C-(RR*H*)*P*R C-(RR*P*)*H*R C-(RR*H*)*H*R C-(RR*H*)*R*P C-(RR*R*)*H*R	(1+5)
C-(RRC)P	C-(RR*H*)*P*P C-(RR*P*)*H*P C-(RR*H*)*H*P C-(RR*H*)*R*P C-(RR*R*)*H*P	(1+5)

4 Overconstrained *VDM* Generators

In Fig. 6, two open chains connect in parallel an end-effector to the fixed frame. One CR chain embodies the product $\mathscr{C}(Q,\mathbf{k})\mathscr{R}(A,\mathbf{k})$, which is a 3D submanifold $\mathscr{X}_{-1}(\mathbf{k})$ of the 4D group $\mathscr{X}(\mathbf{k})$ of Schoenflies motions. The other HHH chain is a generator of the 3D group $\mathscr{Y}(\mathbf{u},p)$ of pseudo-planar motions [13]; $\mathbf{u} \neq \mathbf{k}$.

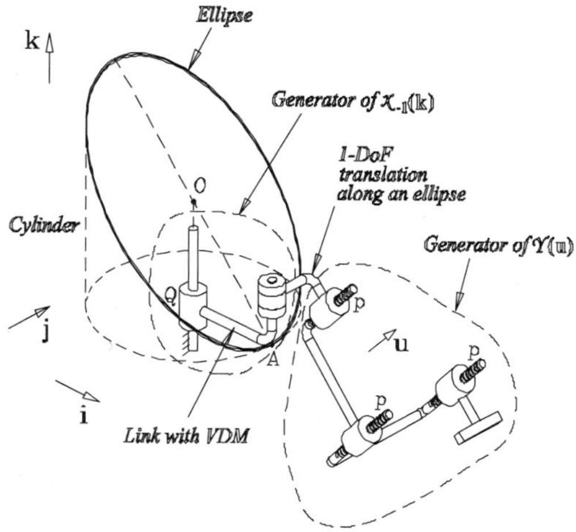

Fig. 6 The C̲R̲-*HHH* overconstrained chain with a *VDM*.

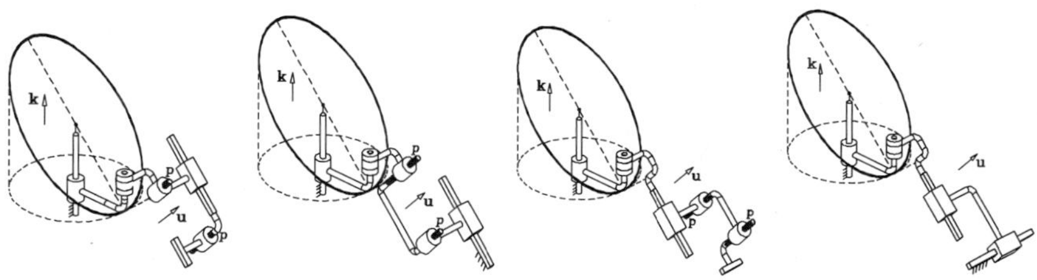

(a) C̲R̲-*HPH* (b) C̲R̲-*HHP* (c) C̲R̲-*PHH* (d) C̲R̲-*PP*

Fig. 7 Overconstrained *VDM* generators.

The set of the end-effector feasible displacements is $\mathcal{X}_{-1}(\mathbf{k}) \cap \mathcal{Y}(\mathbf{u}, p)$. From $\mathcal{X}(\mathbf{k}) \cap \mathcal{Y}(\mathbf{u}, p) = \mathcal{T}(\perp\mathbf{u})$, $\forall$ pitch p, we derive: $\mathcal{X}_{-1}(\mathbf{k}) \cap \mathcal{Y}(\mathbf{u}, p) = \mathcal{T}_{-1}(\perp\mathbf{u})$; $\mathcal{T}_{-1}(\perp\mathbf{u})$ denotes a 1D submanifold of the 2D group $\mathcal{T}(\perp\mathbf{u})$ of planar translations perpendicular to **u** [14]. Therefore, the mechanism is movable with one DoF and its mobility can be established by using the product closure in two or several displacement subgroups. This chain belongs to the exceptional category [9]. Furthermore, the end-effector motion is a 1-DoF translational motion, which is, in other words, translation along a curve, which is an ellipse in the present issue. The rigid link connecting C̲ and R̲ in the C̲R̲-*HHH* mechanism undergoes a *VDM*.

The HHH generator of $\mathcal{Y}(\mathbf{u}, p)$ group can be replaced by any one of its kinematic equivalences. Hence, C̲R̲-*HPH*, C̲R̲-*HHP*, C̲R̲-*PHH* are *VDM* generators. However, in $\mathcal{Y}(\mathbf{u}, p)$ generators including two *P*s, the *H* pair is idle (inactive) and, consequently, can be removed. These results are summarized in Table 4 and are graphically displayed in Fig. 7.

Table 4 Overconstrained *VDM* generators.

Type	Overconstrained VDM generators	Amount
$\underline{\mathrm{CR}}$-$p\mathscr{Y}$	$\underline{\mathrm{CR}}$-HHH $\underline{\mathrm{CR}}$-HPH $\underline{\mathrm{CR}}$-HHP $\underline{\mathrm{CR}}$-PHH	5
	$\underline{\mathrm{CR}}$-PP	

5 Conclusions

A set of *VDM*s is a special 1D submanifold of a 2D group of cylindrical motions, which is a subgroup of a 4D group of Schoenflies motions. Basic mechanical means to generate a *VDM* are revealed. Isoconstrained *VDM* mechanical generators are obtained through the parallel arrangement of a C pair and a generator of planar-spherical motion. Subfamilies of *VDM* generators result from the numerous ways to produce the same *Pl-Sph* kinematic bond. Moreover, overconstrained *VDM* generators with exceptional mobility are synthesized. In a future work, combining these kinematic chains will lead to more mechanical generators of a *VDM*.

Acknowledgements The authors are very thankful to the National Science Council for supporting this research under grants NSC 99-2221-E-151-016 and NSC 100-2221-E-151-022.

References

1. Darboux, G.: Sur le déplacement d'une figure invariable. Comptes Rendus de L'Académie Des Sciences **XCII**, 118–121 (1881)
2. Darboux, G.: Sur le déplacement d'une figure invariable. Ann. Sci. Ec. Norm. Super. 3e Série **7**, 323–326 (1890)
3. Koenigs, G.: Leçons de cinématique. avec des notes par M.G. Darboux, et par MM. E. Cosserat, F. Cosserat, Paris (1897) 499 p
4. Veldkamp, G.R.: Canonical systems and instantaneous invariants in spatial kinematics. J. Mech. **2**(3), 329–388 (1967)
5. Bottema, O., Roth, B.: Theoretical Kinematics. North-Holland, Amsterdam (1979)
6. Selig, J.M., Husty, M.: Half-turns and line symmetric motions. Mech. Mach. Theory **46**(2), 156–167 (2011)
7. Hervé, J.M.: The planar-spherical kinematic bond: implementation in parallel mechanisms. http://www.parallemic.org/Reviews/review013.html (2003)
8. Lee, C.-C., Hervé, J.M.: Uncoupled 6-dof tripods via group theory. In: Kecskeméthy A., Müller A. (eds.) Computational Kinematics, pp. 201–208. Springer (2009)
9. Hervé, J.M.: Analyse structurelle des mécanismes par groupe des déplacements. Mech. Mach. Theory **13**(4), 437–450 (1978)
10. Angeles, J.: Spatial Kinematic Chains. Springer-Verlag, Berlin, (1982)
11. Hervé, J.M.: Conjugation in the displacement group and mobility in mechanisms. Trans. Can. Soc. Mech. Eng. **33**(2), 3–14 (2009)
12. Lee, C.-C., Hervé, J.M.: Isoconstrained parallel generators of Schoenflies motion. ASME Trans. J. Mech. Rob. **3**(2), 021006 (2011)
13. Lee, C.-C., Hervé, J.M.: Cartesian parallel manipulators with pseudo-planar limbs. ASME Trans. J. Mech. Des. **129**(12), 1256–1264 (2007)
14. Lee, C.-C., Hervé, J.M.: Generators of the product of two Schoenflies motion groups. Euro. J. Mech. A/Solids **29**(1), 97–108 (2010)

Kinematic Analysis of a Planar Tensegrity Mechanism for Wave Energy Harvesting

Rafael E. Vasquez, Carl D. Crane III and Julio C. Correa

Abstract Tensegrity systems have been used in several disciplines such as architecture, biology, aerospace, mechanics and robotics during the last fifty years. However, just a few references in literature have stated the possibility of using them in ocean or energy-related applications. This work addresses the kinematic analysis of a planar 3–dof tensegrity mechanism for ocean wave energy harvesting. A planar tensegrity mechanism is proposed based on the "X-frame" morphology developed by Kenneth Snelson in 1960s. A geometric approach is used to solve the forward and reverse displacement problems. The theory of screws is used to perform the forward and reverse velocity analyses of the device. The result of shows that tensegrity systems could play an important roll in the expansion of clean energy technologies that help the world's sustainable development.

Key words: Planar mechanisms, tensegrity, wave energy

1 Introduction

Due to sustainability concerns, a world race started several years ago to incentivize the research, development and utilization of renewable energy sources [10]. The ocean represents an enormous potential energy source [12]; however, its exploitation is still incipient compared to other well-established power harvesting technologies such as wind and solar energies.

Rafael E. Vasquez · Julio C. Correa
Department of Mechanical Engineering, Universidad Pontificia Bolivariana, Medellín, Colombia, e-mail: {rafael.vasquez, julio.correa}@upb.edu.co

Carl D. Crane III
Department of Mechanical and Aerospace Engineering, University of Florida, Gainesville, FL 32611 USA, e-mail: carl.crane@gmail.com

J. Lenarčič, M. Husty (eds.), *Latest Advances in Robot Kinematics*,
DOI 10.1007/978-94-007-4620-6_14,

Wave energy constitutes the most noticeable form of ocean energy, maybe because of its impressive capabilities [6]. The U.S. Department of Energy developed the Marine and Hydrokinetic Technology Database [16], as a resource for the marine/hydrokinetic industry. There are more than 160 different devices for ocean energy harvesting registered, with about 40% corresponding to wave energy.

The word *tensegrity* is a combination of the words *tension* and *integrity*. Tensegrity systems were introduced in the 20th century by Füller [7], Emmerich [5] and Snelson [15]. These systems are formed by a combination of rigid elements (struts) under compression, and elastic elements (ties) under tension [9].

Tensegrity systems have been used in several disciplines such as architecture, biology, aerospace, mechanics and robotics during the last fifty years [14]. Applications in sciences and engineering include, among others, development of structural domes and bridges, deployable systems for space applications, description and modeling of living organisms and biological systems, and applications in robotics.

Just a few references in literature have stated the possibility of using tensegrity systems in ocean applications. Scruggs and Skelton [13] suggested their suitability to harvest energy from ocean waves. Jensen et al. [8] proposed tensegrity structures to develop wave compliant structures for aquaculture.

Vasquez [17] compared the dynamic behavior of a tensegrity mechanism with a direct drive heaving under the influence of linear ocean waves, showing that a tensegrity configuration allows to harvest 10% more power from ocean waves than a purely heaving system. This work presents the position and velocity analysis for a planar three dof tensegrity mechanism that is required for the design stage and integration of the mechanism with electrical generators.

2 Tensegrity Mechanism

Several planar tensegrity-based mechanism have been proposed, see [1, 2] for instance. The proposed 3–dof tensegrity mechanism is based a two-dimensional morphology proposed by Kenneth Snelson in [15]. The mechanism comprises four members in tension and two members in compression. The members in compression are replaced by two bars connected by prismatic joints which represent electrical generators. Two of the ties have a very high modulus of elasticity (the base and the top platform) with respect to the other two ties. Therefore, the deformations of the base and the top platform are negligible and the lateral ties are the two deformable members under tension that are necessary to keep the tensegrity configuration.

3 Position Analysis

The kinematic diagram of the proposed tensegrity mechanism is shown in Fig. 1.

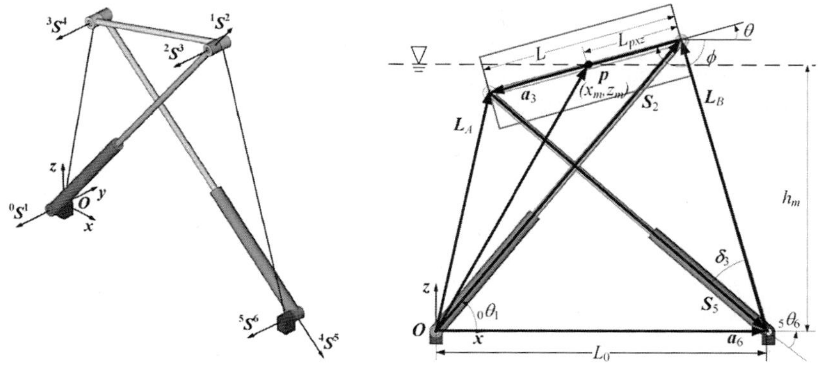

(a) (b)

Fig. 1 Kinematic diagram of the mechanism. (a) Joint axes. (b) Vector diagram.

3.1 Forward Position Analysis

- Given: the constant mechanism parameters a_3, a_6 and L_{pxz} (position of point **p** on the top platform); and the joint variables, ${}_1S_2$, ${}_4S_5$ and ${}_0\theta_1$.
- Find: the position and the orientation of the top platform, x_m, z_m and ϕ.

Let us define the following vectors

$$\mathbf{S}_2 = {}_1S_2 \cos({}_0\theta_1)\mathbf{i} + {}_1S_2 \sin({}_0\theta_1)\mathbf{k}, \tag{1}$$

$$\mathbf{a}_6 = a_6\mathbf{i}. \tag{2}$$

From Fig. 1, the following vector loop equation can be written

$$\mathbf{L}_B = \mathbf{S}_2 - \mathbf{a}_6. \tag{3}$$

Equation (3) permits to evaluate L_B and θ_B. The angle δ_3 can be computed as

$$\delta_3 = \arccos\left(\frac{L_B^2 + {}_4S_5^2 - a_3^2}{2L_B\,{}_4S_5}\right). \tag{4}$$

Hence,

$${}_5\theta_6 = \theta_B + \delta_3 + \pi. \tag{5}$$

Then, the vector $\mathbf{S}_5$ is given by

$$\mathbf{S}_5 = {}_4S_5 \cos({}_5\theta_6)\mathbf{i} + {}_4S_5 \sin({}_5\theta_6)\mathbf{k}. \tag{6}$$

From Fig. 1, the other two vector loop equations can be written as

$$\mathbf{L}_A = \mathbf{a}_6 - \mathbf{S}_5, \tag{7}$$

$$\mathbf{a}_3 = \mathbf{L}_A - \mathbf{S}_2. \tag{8}$$

Equation (8) permits to evaluate the orientation ϕ. There are two solutions for δ_3, however, only one in the first quadrant is taken for this mechanism's configuration.

Finally, the position of the point $\mathbf{p} = [\, x_m \; 0 \; z_m \,]^T$ can be computed as

$$\mathbf{p} = \mathbf{S}_2 + \frac{L_{pxz}}{L}\mathbf{a}_3. \tag{9}$$

The length of the springs can be computed in terms of the position and orientation ($\phi = \theta + \pi$) of the top platform, when $L_{pxz} = L/2$, as follows

$$L_A = \left[(x_m - L/2\cos\theta)^2 + (z_m - L/2\sin\theta)^2 \right]^{\frac{1}{2}}, \tag{10}$$

$$L_B = \left[(x_m + L/2\cos\theta - L_0)^2 + (z_m + L/2\sin\theta)^2 \right]^{\frac{1}{2}}. \tag{11}$$

3.2 Reverse Position Analysis

- Given: the constant mechanism parameters a_3, a_6 and L_{pxz}; and the position and the orientation of the top platform, x_m, z_m and ϕ.
- Find: the joint variables, ${}_1S_2$, ${}_4S_5$ and ${}_0\theta_1$.

Let us define the following vector

$$\mathbf{a}_3 = a_3 \cos(\phi)\mathbf{i} + a_3 \sin(\phi)\mathbf{k}. \tag{12}$$

From Fig. 1, the following vector loop equations can be written

$$\mathbf{S}_2 = \mathbf{p} - L_{pxz}\mathbf{a}_3, \tag{13}$$

$$\mathbf{L}_A = \mathbf{S}_2 + \mathbf{a}_3, \tag{14}$$

$$\mathbf{S}_5 = \mathbf{a}_6 - \mathbf{L}_A. \tag{15}$$

Equation (13) permits to evaluate ${}_1S_2$ and ${}_0\theta_1$, and (15) permits to evaluate ${}_4S_5$.

4 Velocity Analysis

The velocity state is defined as a set of parameters from which the velocity of any body/point of the linkage can be determined relative to a reference body [4, 11]. Since the presented planar mechanism is the first approximation to harvest ocean energy from ocean waves, the theory of screws [3] is used in this analysis to illustrate the methodology that can be used in further works that include spatial tensegrity configurations.

The directions of the unit vectors ${}^i\mathbf{s}^{i+1}$ along each axis and the coordinates of one point $\mathbf{r}_i$ on each joint axis are known. Then, Plücker coordinates of the lines along revolute and prismatic joints are given, respectively, by

$$\{{}^i\mathbf{s}^{i+1}; {}^i\mathbf{s}_{OL}^{i+1}\} = \{{}^i\mathbf{s}^{i+1}; \mathbf{r}_i \times {}^i\mathbf{s}^{i+1}\}, \tag{16}$$

$$\{{}^i\mathbf{s}^{i+1}; {}^i\mathbf{s}_{OL}^{i+1}\} = \{\mathbf{0}; \mathbf{r}_i \times {}^i\mathbf{s}^{i+1}\}. \tag{17}$$

Using (16) and (17), the Plücker coordinates of the lines along the joint axes are given by

$$\begin{aligned}
{}^0\$^1 &= \{{}^0\mathbf{s}^1; {}^0\mathbf{s}_{OL}^1\} = \{\mathbf{s}_1; \mathbf{0} \times \mathbf{s}_1\}, & {}^1\$^2 &= \{\mathbf{0}; {}^1\mathbf{s}_{OL}^2\} = \{\mathbf{0}; \mathbf{s}_2\},\\
{}^2\$^3 &= \{{}^2\mathbf{s}^3; {}^2\mathbf{s}_{OL}^3\} = \{\mathbf{s}_3; \mathbf{S}_2 \times \mathbf{s}_3\}, & {}^3\$^4 &= \{{}^3\mathbf{s}^4; {}^3\mathbf{s}_{OL}^4\} = \{\mathbf{s}_4; \mathbf{L}_A \times \mathbf{s}_4\},\\
{}^4\$^5 &= \{\mathbf{0}; {}^4\mathbf{s}_{OL}^5\} = \{\mathbf{0}; \mathbf{s}_5\}, & {}^5\$^6 &= \{{}^5\mathbf{s}^6; {}^5\mathbf{s}_{OL}^6\} = \{\mathbf{s}_6; \mathbf{a}_6 \times \mathbf{s}_6\}.
\end{aligned}$$

All the unit vectors are known and are given by:

$$\mathbf{s}_1 = \mathbf{s}_3 = \mathbf{s}_4 = \mathbf{s}_6 = [0\ -1\ 0]^T, \quad \mathbf{s}_2 = \frac{\mathbf{S}_2}{|\mathbf{S}_2|}, \quad \mathbf{s}_5 = \frac{\mathbf{S}_5}{|\mathbf{S}_5|}.$$

Since the mechanism is a closed-loop kinematic chain, bodies 0 and 6 are the same (i.e. the ground of the mechanism).

4.1 Forward Velocity Analysis

- Given: the constant mechanism parameters a_3, a_6 and L_{pxz}; the joint variables, ${}_1S_2$, ${}_4S_5$ and ${}_0\theta_1$; and the velocities of the joint variables ${}_1v_2$, ${}_4v_5$ and ${}_0\omega_1$.
- Find: the velocity state of the top platform $\left[{}^0\boldsymbol{\omega}^3\ {}^0\mathbf{v}_O^3\right]^T$ and $\dot{x}_m$, $\dot{z}_m$ and $\dot{\phi}$.

The closed-loop velocity equation can be written in screw form as follows [4]:

$${}_0\omega_1{}^0\$^1 +_1 v_2{}^1\$^2 +_2 \omega_3{}^2\$^3 +_3 \omega_4{}^3\$^4 +_4 v_5{}^4\$^5 +_5 \omega_6{}^5\$^6 = 0. \tag{18}$$

Equation (18) can be written as a 3x3 system in matrix form as follows

$$\begin{bmatrix} -1 & -1 & -1 \\ s_{2_z} & L_{A_z} & 0 \\ -s_{2_x} & -L_{A_x} & -a_{6_x} \end{bmatrix} \begin{bmatrix} {}_2\omega_3 \\ {}_3\omega_4 \\ {}_5\omega_6 \end{bmatrix} = -{}_0\omega_1 \begin{bmatrix} -1 \\ 0 \\ 0 \end{bmatrix} - {}_1v_2 \begin{bmatrix} 0 \\ s_{2_x} \\ s_{2_z} \end{bmatrix} - {}_4v_5 \begin{bmatrix} 0 \\ s_{5_x} \\ s_{5_z} \end{bmatrix} \tag{19}$$

The solution of (19) gives the magnitudes of the angular velocities between consecutive bodies. Then, the velocity state of the top platform can be computed as

$$\begin{bmatrix} {}^0\boldsymbol{\omega}^3 \\ {}^0\mathbf{v}_O^3 \end{bmatrix} ={}_0 \omega_1{}^0\$^1 +_1 v_2{}^1\$^2 +_2 \omega_3{}^2\$^3. \tag{20}$$

Now the velocity of any point $\mathbf{p}$ on the top platform, whose position is represented by $\mathbf{r}_{O\rightarrow P}$, is given in terms of the velocity state by

$$ {}^{0}\mathbf{v}_{p}^{3} = {}^{0}\mathbf{v}_{O}^{3} + {}^{0}\boldsymbol{\omega}^{3} \times \mathbf{r}_{O\rightarrow P}. \tag{21} $$

Equations (19), (20) and (21) complete the forward velocity analysis.

4.2 Reverse Velocity Analysis

- Given: the constant mechanism parameters a_3, a_6 and L_{pxz}; the joint variables, ${}_1S_2$, ${}_4S_5$ and ${}_0\theta_1$; and the velocity state of the top platform $\left[{}^{0}\boldsymbol{\omega}^{3}\ {}^{0}\mathbf{v}_{O}^{3}\right]^{T}$ in terms of $\dot{x}_m$, $\dot{z}_m$ and $\dot{\phi}$.
- Find: the velocities of the joint variables ${}_1v_2$, ${}_4v_5$ and ${}_0\omega_1$.

Since the velocity of the point $\mathbf{p}$ (${}^{0}\mathbf{v}_{p}^{3} = \dot{x}_m\mathbf{i} + \dot{z}_m\mathbf{k}$), and the angular velocity of the top platform (${}^{0}\boldsymbol{\omega}^{3} = \dot{\phi}\mathbf{j}$) are known, the element associated with linear velocities in the velocity state can be computed using (21) as

$$ {}^{0}\mathbf{v}_{O}^{3} = {}^{0}\mathbf{v}_{p}^{3} - {}^{0}\boldsymbol{\omega}^{3} \times \mathbf{r}_{O\rightarrow P}. \tag{22} $$

Substituting (20) into (18) yields

$$ \begin{bmatrix} {}^{0}\boldsymbol{\omega}^{3} \\ {}^{0}\mathbf{v}_{O}^{3} \end{bmatrix} = -{}_3\omega_4\,{}^{3}\$^{4} - {}_4v_5\,{}^{4}\$^{5} - {}_5\omega_0\,{}^{5}\$^{0}. \tag{23} $$

Since the velocity state of the top platform is known, the joint velocities of the mechanism can be computed from (20) and (23) as follows

$$ \begin{bmatrix} -1 & 0 & -1 \\ 0 & s_{2_x} & s_{2_z} \\ 0 & -s_{2_z} & -s_{2_x} \end{bmatrix} \begin{bmatrix} {}_0\omega_1 \\ {}_1v_2 \\ {}_2\omega_3 \end{bmatrix} = \begin{bmatrix} {}_0\omega_3 \\ {}_0v_{3_x} \\ {}_0v_{3_z} \end{bmatrix}, \tag{24} $$

$$ \begin{bmatrix} -1 & 0 & -1 \\ L_{A_z} & s_{5_x} & 0 \\ -L_{A_x} & s_{5_z} & 0 \end{bmatrix} \begin{bmatrix} {}_3\omega_4 \\ {}_4v_5 \\ {}_5\omega_6 \end{bmatrix} = - \begin{bmatrix} {}_0\omega_3 \\ {}_0v_{3_x} \\ {}_0v_{3_z} \end{bmatrix}. \tag{25} $$

Equations (24) and (25) complete the reverse velocity analysis.

5 Numerical Example

The constant mechanism parameters are defined as $L_0 = 6$ m, $L_{pxz} = 1/2L$ m and $h_m = 6$ m. Vasquez [17] performed the solution of the differential equation of motion for the tensegrity mechanism under the influence of linear ocean waves.

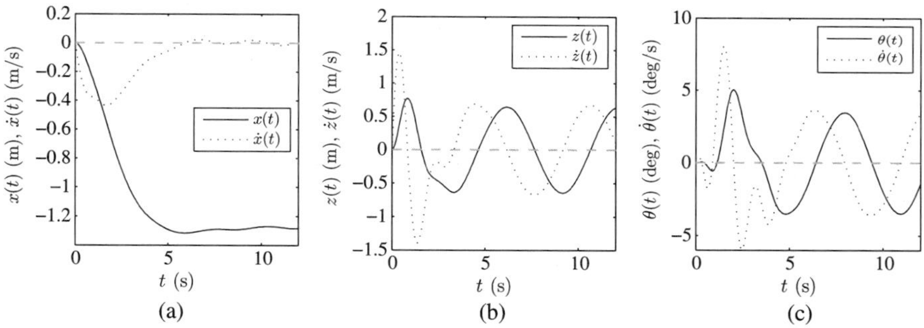

Fig. 2 Motion of the top platform: (a) Surge, (b) Heave, (c) Pitch [17].

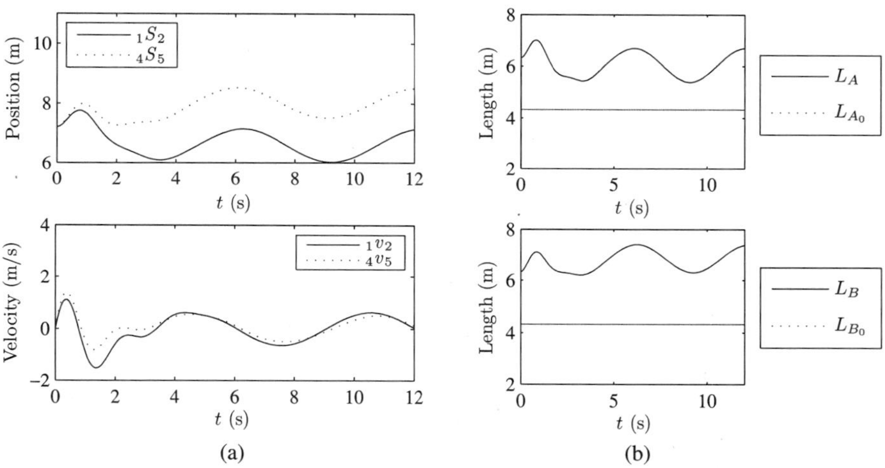

Fig. 3 Reverse kinematic analysis: (a) Motion of generators, (b) Motion of springs.

The solution for the position and velocity of point **p** and the orientation and angular velocity of the top platform are taken from [17] and represent the inputs for the kinematic analysis, Fig. 2.

Equations derived in Sections 3.2 and 4.2 are used to perform the reverse analysis, see Fig. 3. Fig. 3a shows the motion of the electrical generators under the influence of ocean waves. Fig. 3b shows that the length of the springs is always greater than their free length, hence, the tensegrity configuration is preserved.

6 Conclusions

This work addressed the kinematic analysis of a tensegrity mechanism for ocean wave energy harvesting. A planar tensegrity morphology was selected for the system, and the position analyses was performed using a geometric approach. The velocity analysis was performed using theory of screws.

The interaction between ocean waves, a multi-degree-of-freedom linkage and electrical generators poses challenging problems in terms of mathematical modeling and simulation. Nonetheless, the ideas presented in this document will be useful for the analysis and testing of more advanced and complex energy harvesting devices.

This research constitutes an interesting approach to show how the extensive knowledge acquired in the analysis and design of mechanisms can be used in new applications that contribute to the world's sustainable development.

References

1. Arsenault, M., Gosselin, C.: Kinematic, static, and dynamic analysis of a planar one-degree-of-freedom tensegrity mechanism. J. Mech. Des. **127**(6), 1152–1160 (2005)
2. Arsenault, M., Gosselin, C.: Kinematic, static and dynamic analysis of a planar 2-DOF tensegrity mechanism. Mech. Mach. Theory **41**(9), 1072–1089 (2006)
3. Ball, S.R.S.: A Treatise on the Theory of Screws. Cambridge (1900)
4. Crane, C.D. III, Rico, J.M., Duffy, J.: Screw theory and its application to spatial robot manipulators. Tech. rep., CIMAR, University of Florida (2009)
5. Emmerich, D.: Construction de réseaux autotendants, French patent no. 1,377,290 (1964)
6. Falcão, A.: Wave energy utilization: A review of the technologies. Renew. Sustain. Energy Rev. **14**(3), 899–918 (2010)
7. Füller, R.: Tensile-integrity structures, US Patent 3,063,521 (1962)
8. Jensen, O., Wroldsen, A., Lader, P., Fredheim, A., Heide, M.: Finite element analysis of tensegrity structures in offshore aquaculture installations. Aquac. Eng. **36**(3), 272–284 (2007)
9. Motro, R.: Tensegrity systems: The state of the art. Int. J. Space Struct. **7**(2), 75–83 (1992)
10. Redondo-Gil, C., Esquibel, L., Alonso Sanchez, A., Zapico, P.: European strategic energy technology plan. In: Proceedings of ICREPQ09 (2009)
11. Rico, J.M., Gallardo, J., Duffy, J.: Screw theory and higher order kinematic analysis of open serial and closed chains. Mech. Mach. Theory **34**(4), 559–586 (1999)
12. Scruggs, J., Jacob, P.: ENGINEERING: Harvesting ocean wave energy. Science **323**(5918), 1176–1178 (2009)
13. Scruggs, J., Skelton, R.: Regenerative tensegrity structures for energy harvesting applications. In: Proceedings of the 45th IEEE CDC (2006)
14. Skelton, R., de Oliveira, M.C.: Tensegrity Systems. Springer (2009)
15. Snelson, K.: Continuous tension, discontinuous compression structures, US Patent 3,169,611 (1965)
16. U.S. Department of Energy: Marine and hydrokinetic technology database. http://www1.eere.energy.gov/water/index.html
17. Vasquez, R.E.: Analysis of a tensegrity system for ocean wave energy harvesting. Ph.D. thesis, Department of Mechanical and Aerospace Engineering, University of Florida (2011)

Motion Planning for Parallel Robots with Non-holonomic Joints

Krzysztof Tchoń, Janusz Jakubiak, Patrick Grosch and Federico Thomas

Abstract Designing a robot manipulator with fewer actuators than the dimension of its configuration space – to reduce bulk, weight and cost – becomes feasible by introducing mechanical elements that lead to non-holonomic constraints. Unfortunately, the mechanical advantages of these non-holonomic designs are usually darkened by the complexity of their control. This paper deals with motion planning for parallel robots with non-holonomic joints shedding new light on their control strategies. As a case study, the motion planning problem is solved for a 3-ŬPU parallel robot, where Ŭ stands for a non-holonomic joint whose instantaneous kinematics are equivalent to that of a universal joint. It is thus shown how the three prismatic actuators can maneuver to reach any six-degree-of-freedom pose of the moving platform. The motion planning has been addressed as a control problem in the control system representation of the robot's kinematics and a motion planning algorithm has been devised based on a Jacobian inversion of the end-point map of the representation. Performance of the algorithm is illustrated with numeric computations.

Key words: Parallel non-holonomic manipulator, Jacobian motion planning

1 Introduction

The joints of standard robots, either serial or parallel, implement lower kinematic pairs. An alternative to these joints are non-holonomic joints, a mechanical concept probably used for the first time in [8], which can be implemented using convex

Krzysztof Tchoń · Janusz Jakubiak
Institute of Computer Engineering, Control and Robotics, Wrocław University of Technology, Wrocław, Poland, e-mail: krzysztof.tchon|janusz.jakubiak@pwr.wroc.pl

Patrick Grosch · Federico Thomas
Institut de Robòtica i Informàtica Industrial (CSIC-UPC), Barcelona, Spain, e-mail: pgrosch|fthomas@iri.upc.edu

J. Lenarčič, M. Husty (eds.), *Latest Advances in Robot Kinematics*,
DOI 10.1007/978-94-007-4620-6_15,

bodies rolling on spherical surfaces. Two kinds of contacts have been considered: marble rolling, when the convex body can freely roll in contact with the sphere without slipping [4], and rubber rolling, when the convex body satisfies additionally a no-twist condition [9].

In the practical implementations of non-holonomic joints, the rolling convex body is usually a disk implementing a marble rolling contact with the sphere. If the disk rolls upon the interior surface of a spherical shell, the resulting joint is said to implement the Suslov constraint [15]. Alternatively, if the contact is performed on the outer surface of the spherical shell, the resulting joint is said to implement the Veselova constraint [3], the kind of non-holonomic joint used throughout this paper. Lower-mobility spatial parallel robots have become an active research topic in the field of parallel robot during the last decade because of their simple structure, low price and easy control. The dimension of the space of admissible velocities for the end-effector of this kind of parallel robots is lower than six and, if singular configurations are excluded, equal to the dimension of the tangent space of the reachable manifold. The substitution of a standard joint in a lower-mobility parallel robot by a non-holonomic joint with equivalent instantaneous kinematics has dramatic consequences: while the dimension of the space of admissible velocities for the end-effector remains the same, the dimension of the reachable space is increased. To the best of our knowledge, this idea was first used by Ben-Horin and Thomas in [1], where a three-legged parallel robot is proposed whose each leg is connected to the base through non-holonomic joints. The kinetostatics of this architecture was analyzed by Grosch et al. in [6], who proved that this robot was able to *locally* move its moving platform – excluding singular configurations – in a six-dimensional configuration space. In this paper we go a step further by presenting a solution to the motion planning problem for this robot which can be adapted to other designs.

It is worth to mention that the use of non-holonomic devices in the design of robot manipulators has some tradition. For example, in [13], Stammers et al. present a robot wrist that can attain any orientation with two motors only. This is achieved by means of a friction drive, using rollers on a spherical ball to which the end effector is fixed, and by fixing the two motors to the arm. In [12], Peshkin et al. present a passive spherical robot which can display programmable constraints. The device is based on a non-holonomic element involving a sphere and three reorientable rollers. In [11], Nakamura et al. describe an n-joint serial manipulator which can reach any pose in its n-dimensional configuration space with only two actuators. The joints of this manipulator are coupled by $(n-1)$ non-holonomic devices, based on spheres and rollers, so that its control is equivalent to maneuvering a car with n-trailers. Considerable effort has been made to clarify different aspects of non-holonomic mechanical systems [2]. A challenge in control of these systems results from a limited applicability of the feedback control, discovered by Brockett [5] and Lizárraga [10]. In this paper, the motion planning problem for the parallel non-holonomic robot will be addressed using the endogenous configuration space approach [14], specified in [7] to the class of mechanical systems including the parallel non-holonomic robot used as case study in this paper. The motion planning problem for the parallel non-holonomic robot will be decomposed into two steps:

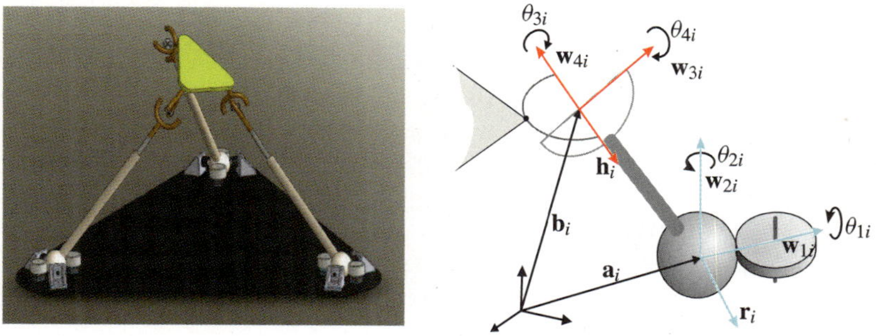

Fig. 1 The 3-ŬPU parallel robot used as a case study and notations associated with leg i.

first the control system representing the robot's kinematics is subject to a feedback transformation, and afterwards the end-point map of the obtained system is inverted.

The remainder of the paper is organized as follows. Section 2 summarizes the main characteristics of the non-holonomic parallel robot used as the case study, and its instantaneous kinematics. Section 3 introduces the motion planning algorithm. Its performance is illustrated in Section 4 by a numeric example. Section 5 concludes the paper.

2 Instantaneous Kinematics of the 3-ŬPU Parallel Robot

The parallel robot manipulator appearing in Fig. 1 (left) can be thought of as a 3-UPU lower mobility parallel robot in which each universal joint attached to the base is substituted by a non-holonomic joint (see [1] for details). Let us consider the leg number i, $i = 1, 2, 3$. According to Fig. 1, we shall introduce the following notations:

- $\mathbf{a}_i$ and $\mathbf{b}_i$ are the position vectors, in the base reference frame, of the centers of the sphere and of the universal joint, respectively.
- l_i is the length of leg i, that is, $\|\mathbf{b}_i - \mathbf{a}_i\|$. Then, $\mathbf{g}_i = (\mathbf{b}_i - \mathbf{a}_i)/l_i$ is the unit vector in the direction of leg i.
- $\{\mathbf{w}_{1i}, \mathbf{w}_{2i}, \mathbf{r}_i\}$ defines a right-handed reference frame with origin at the center of the sphere. $\mathbf{w}_{1i}$ is aligned with $\mathbf{a}_i$ and $\mathbf{w}_{2i}$ is parallel to the roller axis.
- $\{\mathbf{w}_{3i}, \mathbf{w}_{4i}, \mathbf{h}_i\}$ defines a right-handed reference frame with origin at the center of the universal joint. $\mathbf{w}_{3i}$ and $\mathbf{w}_{4i}$ are defined by the two revolute axis constituting the universal joint.
- θ_{ji} is the joint variable denoting the rotation angle about the axis defined by $\mathbf{w}_{ji}$.
- $\mathbf{s}_i = \mathbf{h}_i \times \mathbf{r}_i - [\mathbf{g}_i \cdot (\mathbf{h}_i \times \mathbf{r}_i)]\mathbf{g}_i$ is the component of $\mathbf{h}_i \times \mathbf{r}_i$ perpendicular to $\mathbf{g}_i$.

Then, it can be proved that (see [6] for details):

$$\begin{pmatrix} \mathbf{1}_{3\times3} \\ \mathbf{0}_{3\times3} \end{pmatrix} \dot{\mathbf{l}} = \begin{bmatrix} \mathbf{G}_{3\times3} & \mathbf{K}_{3\times3} \\ \mathbf{S}_{3\times3} & \mathbf{J}_{3\times3} \end{bmatrix} \begin{pmatrix} \dot{\mathbf{p}} \\ \boldsymbol{\omega} \end{pmatrix}, \tag{1}$$

where $\dot{\mathbf{l}} = (\dot{l}_1, \dot{l}_2, \dot{l}_3)^T$ is the vector of velocities in the actuators, $\binom{\dot{\mathbf{p}}}{\omega}$ is the vector of linear and angular velocities of the moving platform, $\mathbf{1}_{3\times 3}$ and $\mathbf{0}_{3\times 3}$ are the 3×3 identity and the zero matrix, respectively, and $\mathbf{G} = \mathbf{G}(\mathbf{p}, \mathbf{R})$, $\mathbf{K} = \mathbf{K}(\mathbf{p}, \mathbf{R})$, $\mathbf{S} = \mathbf{S}(\mathbf{p}, \mathbf{R})$, $\mathbf{J} = \mathbf{J}(\mathbf{p}, \mathbf{R})$ are 3×3 matrices dependent on the end-effector pose (position and orientation) $(\mathbf{p}, \mathbf{R}) \in \mathbb{R}^3 \times SO(3)$ whose entries are defined as

$$\begin{aligned} \mathbf{K}^T[i,:] &= (\mathbf{b}_i - \mathbf{p}) \times \mathbf{g}_i, & \mathbf{G}^T[i,:] &= \mathbf{g}_i \\ \mathbf{J}^T[i,:] &= (\mathbf{b}_i - \mathbf{p}) \times \mathbf{s}_i - l_i(\mathbf{r}_i \cdot \mathbf{g}_i)\mathbf{h}_i, & \mathbf{S}^T[i,:] &= \mathbf{s}_i, \end{aligned} \tag{2}$$

where $\mathbf{A}[i,:]$ denotes the i-th row of a matrix $\mathbf{A}$.

3 Motion Planning

As a starting point we shall adopt the kinematics representation (1) of the parallel non-holonomic robot. Assuming invertibility of the whole block matrix standing on the right hand side of (1) and taking $\mathbf{u} = \dot{\mathbf{l}}$ as a control variable, the kinematics model is converted to the driftless control system

$$\dot{\mathbf{p}} = \mathbf{E}(\mathbf{p}, \mathbf{R})\mathbf{u}, \quad \dot{\mathbf{R}} = [\mathbf{F}(\mathbf{p}, \mathbf{R})\mathbf{u}]\mathbf{R}, \tag{3}$$

used in [7], where $[\ \ ] : \mathbb{R}^3 \to so(3)$ denotes the standard Lie algebras isomorphism of $\mathbb{R}^3$ with the cross product and the space of skew symmetric 3×3 matrices with the matrix commutator, so that $[\mathbf{v} \times \mathbf{w}] = [\mathbf{v}][\mathbf{w}] - [\mathbf{w}][\mathbf{v}]$, and

$$\begin{bmatrix} \mathbf{E}(\mathbf{p}, \mathbf{R}) \\ \mathbf{F}(\mathbf{p}, \mathbf{R}) \end{bmatrix} = \begin{bmatrix} \mathbf{G} & \mathbf{K} \\ \mathbf{S} & \mathbf{J} \end{bmatrix}^{-1} \Bigg|_{3 \text{ first columns}} . \tag{4}$$

Given the control system (3), the motion planning problem for the parallel non-holonomic robot can be stated in the following way: compute a control function $\mathbf{u}(t)$ steering the system from an initial end effector pose $(\mathbf{p}_0, \mathbf{R}_0)$ to the desired one $(\mathbf{p}_d, \mathbf{R}_d)$ within a prescribed time T. More formally, setting $\mathbf{p}(t) = \mathbf{p}_{\mathbf{p}_0,\mathbf{R}_0,t}(\mathbf{u}(\cdot))$, $\mathbf{R}(t) = \mathbf{R}_{\mathbf{p}_0,\mathbf{R}_0,t}(\mathbf{u}(\cdot))$ to be the trajectory of (3) starting at $(\mathbf{p}_0, \mathbf{R}_0)$ and driven by the control $\mathbf{u}(t)$, this means that at time T the end-point map of (3) assumes the prescribed values $\mathbf{p}(T) = \mathbf{p}_d$ and $\mathbf{R}(T) = \mathbf{R}_d$.

Due to the complexity of the matrix entries on the right hand side of (1), the analytic form of (3) is not very enlightening. To make it more tractable, two regularity assumptions will be made. First, the matrix $\mathbf{G}$ will be assumed invertible, resulting in the following form of the system (4)

$$\begin{bmatrix} \mathbf{E}(p, R) \\ \mathbf{F}(p, R) \end{bmatrix} = \begin{bmatrix} \mathbf{G}^{-1} + \mathbf{G}^{-1}\mathbf{K}(\mathbf{J} - \mathbf{S}\mathbf{G}^{-1}\mathbf{K})^{-1}\mathbf{S}\mathbf{G}^{-1} \\ -(\mathbf{J} - \mathbf{S}\mathbf{G}^{-1}\mathbf{K})^{-1}\mathbf{S}\mathbf{G}^{-1} \end{bmatrix} . \tag{5}$$

The second assumption is the invertibility of $\mathbf{S}$. Under this assumption the feedback

$$\mathbf{u} = \mathbf{G}\mathbf{S}^{-1}(\mathbf{J} - \mathbf{S}\mathbf{G}^{-1}\mathbf{K})\mathbf{v}, \tag{6}$$

where $\mathbf{v} \in \mathbb{R}^3$ is a new control, makes the control system (3) equivalent to

$$\dot{\mathbf{p}} = \mathbf{S}^{-1}\mathbf{J}\mathbf{v}, \quad \dot{\mathbf{R}} = -[\mathbf{v}]\mathbf{R}. \tag{7}$$

Thanks to the regularity assumptions, the solution of the motion planning problem may be obtained in two steps: first a control $\mathbf{v}(t)$ solving the motion planning problem for the system (7) is found, and then the original control $\mathbf{u}(t)$ is computed using (6). The first step can be accomplished in accordance with the guidelines presented in [7], that will be concisely recalled below. Let $\mathbf{v}_\theta(t)$ be a family of control functions smoothly dependent on a parameter $\theta \in \mathbb{R}$, and $\mathbf{p}_t(\theta) = \mathbf{p}_{\mathbf{p}_0,\mathbf{R}_0,t}(\mathbf{v}_\theta(\cdot))$, $\mathbf{R}_t(\theta) = \mathbf{R}_{\mathbf{p}_0,\mathbf{R}_0,t}(\mathbf{v}_\theta(\cdot))$ denote the trajectory of the system (7) initialized at $(\mathbf{p}_0, \mathbf{R}_0)$ and subject to the control $\mathbf{v}_\theta(t)$. The derivation of the motion planning algorithm for the system (7) relies on an assumption that there exists a control family $\mathbf{v}_\theta(t)$, such that the error

$$\mathbf{e}(\theta) = \left(\mathbf{p}_T(\theta) - \mathbf{p}_d, \log(\mathbf{R}_T(\theta))\mathbf{R}_d^T)\right), \tag{8}$$

decreases to zero exponentially along with θ with a prescribed decay rate $\gamma > 0$,

$$\frac{d\mathbf{e}(\theta)}{d\theta} = -\gamma\mathbf{e}(\theta). \tag{9}$$

The logarithm of the rotation matrix in (8) is defined as $\log\mathbf{R} = \frac{\alpha}{2\sin\alpha}(\mathbf{R} - \mathbf{R}^T)$, where $\cos\alpha = \frac{1}{2}(tr\mathbf{R} - 1)$ and the angle of rotation $0 \leq \alpha < \pi$.

To proceed, a pair of auxiliary variables will be introduced, denoted by $\mathbf{w}_t(\theta)$, $\mathbf{s}_t(\theta)$, satisfying the following dependencies

$$\mathbf{w}_t(\theta) = \frac{\partial\mathbf{p}_t(\theta)}{\partial\theta}, \quad [\mathbf{s}_t(\theta)] = \frac{\partial\mathbf{R}_t(\theta)}{\partial\theta}\mathbf{R}_t^T(\theta). \tag{10}$$

The differentiation with respect to θ of the matrices on the r.h.s of the system (7) results in a collection of differential equations (for details, see [7, proof of theorem 2.1])

$$\begin{pmatrix} \dot{\mathbf{w}}_t(\theta) \\ \dot{\mathbf{s}}_t(\theta) \end{pmatrix} = \begin{bmatrix} \mathbf{A}_{11\theta}(t) & \mathbf{A}_{12\theta}(t) \\ \mathbf{0} & -[\mathbf{v}_\theta(t)] \end{bmatrix} \begin{pmatrix} \mathbf{w}_t(\theta) \\ \mathbf{s}_t(\theta) \end{pmatrix} + \begin{bmatrix} \mathbf{B}_{1\theta}(t) \\ -\mathbf{1}_3 \end{bmatrix} \frac{d\,\mathbf{v}_\theta(t)}{d\theta}, \tag{11}$$

where the entries of the matrices $\mathbf{A}_{11\theta}(t)$, $\mathbf{A}_{12\theta}(t)$ and $\mathbf{B}_{1\theta}(t)$ have been computed on the basis of the data provided in [6]. The assumption that $\mathbf{p}_0(\theta) = \mathbf{p}_0$ and $\mathbf{R}_0(\theta) = \mathbf{R}_0$ yields the initial conditions for (11) $\mathbf{w}_0(\theta) = \mathbf{0}$ and $\mathbf{s}_0(\theta) = \mathbf{0}$. With these initial conditions the solution of (11) at T can be represented as

$$\begin{pmatrix} \mathbf{w}_T(\theta) \\ \mathbf{s}_T(\theta) \end{pmatrix} = \int_0^T \Phi_\theta(T, t)\mathbf{B}_\theta(t)\frac{d\,\mathbf{v}_\theta(t)}{d\theta}dt, \tag{12}$$

where the fundamental matrix $\Phi_\theta(T,t)$ satisfies the evolution equation $\frac{\partial\Phi_\theta(t,s)}{\partial t} = \mathbf{A}_\theta(t)\Phi_\theta(t,s)$, $\Phi_\theta(s,s) = \mathbf{1}_6$, and $\mathbf{A}_\theta(t) = \begin{bmatrix} \mathbf{A}_{11\theta}(t) & \mathbf{A}_{12\theta}(t) \\ \mathbf{0} & -[\mathbf{v}_\theta(t)] \end{bmatrix}$, $\mathbf{B}_\theta(t) = \begin{bmatrix} \mathbf{B}_{1\theta}(t) \\ -\mathbf{1}_3 \end{bmatrix}$.

The integral operator in (12) can be regarded as a Jacobian operator of the parallel non-holonomic robot [14]. Now, it has been proved in [7] that the error vanishing formula (9) is tantamount to the integral equation

$$\int_0^T \Phi_\theta(T,t)\mathbf{B}_\theta(t)\frac{d\,\mathbf{v}_\theta(t)}{d\theta}dt = -\gamma \begin{pmatrix} \mathbf{p}_T(\theta) - \mathbf{p}_d \\ \mathbf{r}_T(\theta) \end{pmatrix}, \tag{13}$$

where $[\mathbf{r}_T(\theta)] = \log(\mathbf{R}_T(\theta)\mathbf{R}_d^T)$. This being so, the motion planning algorithm for the parallel non-holonomic robot is obtained by solving the equation (13) using a generalized inverse of the Jacobian. If the Moore-Penrose pseudo inverse is chosen, the resulting differential equation for the control function $\mathbf{v}_\theta(t)$ takes the form

$$\frac{d\mathbf{v}_\theta(t)}{d\theta} = -\gamma\mathbf{B}_\theta^T(t)\Phi_\theta^T(T,t)\mathbf{D}_\theta^{-1}\begin{pmatrix} \mathbf{p}_T(\theta) - \mathbf{p}_d \\ \mathbf{r}_T(\theta) \end{pmatrix}. \tag{14}$$

The matrix $\mathbf{D}_\theta = \int_0^T \Phi_\theta(T,t)\mathbf{B}_\theta(t)\mathbf{B}_\theta^T(t)\Phi_\theta^T(T,t)dt$, is the Gram matrix of the system (11). Given the system (14), the solution of the motion planning problem is computed as the limit $\mathbf{v}(t) = \lim_{\theta\to+\infty}\mathbf{v}_\theta(t)$. The system (7) subject to the control $\mathbf{v}(t)$ produces a trajectory $(\mathbf{p}(t), \mathbf{R}(t))$. A suitable substitutions to the feedback equation (6) defines the control $\mathbf{u}(t)$ solving the motion planning problem for the parallel non-holonomic robot.

4 Computations

Since the motion planning algorithm (14) operates in an infinite dimensional space of control functions, its computer implementation needs to be preceded by the introduction of a finite dimensional space of controls. This is done in a standard way, by representing the control function by its truncated orthogonal expansion [14]. In this paper the truncated Fourier series is exploited, so each control $v_i(t)$, $i = 1, 2, 3$, will consist of a constant term and up to h harmonics,

$$v_i(t) = \lambda_{i,0} + \sum_{k=1}^{h}\left(\lambda_{i,2k-1}\sin\frac{2\pi}{T}kt + \lambda_{i,2k}\cos\frac{2\pi}{T}kt\right), \tag{15}$$

so the control is finitely parametrized by $\Lambda = (\lambda_{1,0}, \ldots, \lambda_{1,2h}, \ldots, \lambda_{3,0}, \ldots, \lambda_{3,2h})^T \in \mathbb{R}^{6h+3}$. In the finite dimensional case the control family takes the form $\mathbf{v}_\theta(t) = \mathbf{P}(t)\Lambda(\theta)$, where the block matrix $\mathbf{P}(t)$ aggregates the basic harmonic functions. Consequently, the differential equation (14) underlying the motion planning algorithm determines the control coefficients Λ

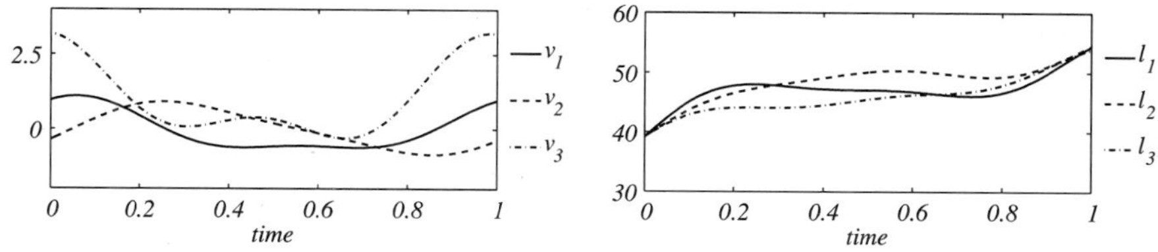

Fig. 2 Solution of the motion planning problem: controls $\mathbf{v}(t)$ and leg lengths $\mathbf{l}(t)$.

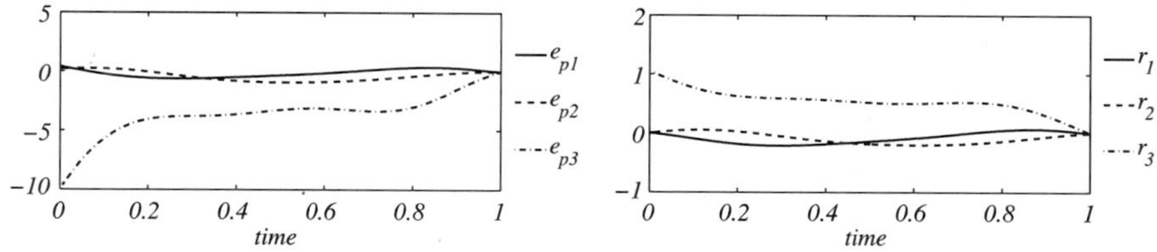

Fig. 3 Relative position $\mathbf{e}_p(t)$ and orientation $\mathbf{r}(t)$ trajectories.

$$\frac{d\Lambda_\theta}{d\theta} = -\gamma \mathbf{J}^{\#}_{\mathbf{p}_0,\mathbf{R}_0,T}(\Lambda_\theta)\begin{pmatrix}\mathbf{p}_T(\theta) - \mathbf{p}_d\\ \mathbf{r}_T(\theta)\end{pmatrix}, \tag{16}$$

where $\mathbf{J}^{\#}_{\mathbf{p}_0,\mathbf{R}_0,T}(\Lambda_\theta)$ denotes the Moore-Penrose pseudo inverse of the $6 \times (6h+3)$ Jacobian matrix

$$\mathbf{J}_{\mathbf{p}_0,\mathbf{R}_0,T}(\Lambda_\theta) = \int_0^T \Phi_\theta(T,t)\mathbf{B}_\theta(t)\mathbf{P}(t)dt$$

of the parallel non-holonomic robot. The differential equation (16) should be integrated numerically in accordance with a suitable integration scheme. In the sequel the simplest Euler scheme will be applied leading to the following difference equation for Λ_θ, where $\theta = 0, 1, \ldots$

$$\Lambda_{\theta+1} = \Lambda_\theta - \gamma \mathbf{J}^{\#}_{\mathbf{p}_0,\mathbf{R}_0,T}(\Lambda_\theta)\begin{pmatrix}\mathbf{p}_T(\theta) - \mathbf{p}_d\\ \mathbf{r}_T(\theta)\end{pmatrix}. \tag{17}$$

Performance of the motion planning algorithm will be illustrated with a numeric example. The initial position of the platform is $\mathbf{p}_0 = (0, 0, 25)^T$, while its orientation $\mathbf{R}_0 = RPY(0, 0, -\pi/6)$ corresponds to the Roll-Pitch-Yaw angles $(0, 0, -\pi/6)$. The desired end effector position and orientation $\mathbf{p}_d = (-0.4, -0.2, 35)^T$ and $\mathbf{R}_d = RPY(0, 0, -\pi/2)$. The initial values of control parameters have been set to 0, except for $\lambda_{11} = \lambda_{21} = \lambda_{32} = 0.5$, $\lambda_{30} = 1$. The planning time horizon $T = 1$. The algorithm has been stopped when the total error $\mathcal{E}(\theta) = \sqrt{||\mathbf{p}_T(\theta) - \mathbf{p}_d||^2 + ||\mathbf{r}_T(\theta)||^2}$ drops below 10^{-3}. In the computations the number h of harmonics is set to two. Results of computations are shown in Figures 2 and 3. In Figure 3 the relative trajectories are shown, defined as $\mathbf{e}_p(t) = \mathbf{p}(t) - \mathbf{p}_d$ and $[\mathbf{r}(t)] = \log(\mathbf{R}(t)\mathbf{R}_d^T)$.

5 Conclusion

This paper provides a motion planning algorithm of the parallel non-holonomic robot. The algorithm's synthesis has been based on an application of the endogenous configuration space approach preceded by a feedback transformation of the system (3). Presented results provide a novel motion planning algorithm and essentially extend the applicability of the endogenous configuration space approach.

Acknowledgements The research of the first two authors was supported by a statutory grant from Wrocław University of Technology.

References

1. Ben-Horin, P., Thomas, F.: A Nonholonomic 3-Motor Parallel Robot. In: Advances in Robot Kinematics. Springer Verlag, Berlin (2008)
2. Bloch, A.M.: Nonholonomic Mechanics and Control. Springer-Verlag, New York (2003)
3. Borisov, A.P., Mamaev, I.S.: Chaplygin's ball. The Suslov problem and Veselova problem. Integrability and realization of constraints. In: Non-holonomic Dynamical Systems, pp. 118–130. Inst. Computer Sci, Moscow (2002)
4. Borisov, A.V., Mamaev, I.S.: The rolling body motion of a rigid body on a plane and a sphere. Hierarchy of dynamics. Regul. Chaotic Dyn. **7**(2), 177–200 (2002)
5. Brockett, R.W.: Asymptotic stability and feedback stabilization. Differential Geometric Control Theory, pp. 171–181. Birkhäuser, Boston (1983)
6. Grosch, P., Di Gregorio, R., Thomas, F.: Generation of under-actuated manipulators with nonholonomic joints from ordinary manipulators. ASME J. Mech. Robot. **2**(1), (2010)
7. Jakubiak, J., Tchoń, K., Magiera, W.: Motion planning in velocity affine mechanical systems. Int. J. Control **83**(9), 1965–1974 (2010)
8. Kharlamov, A.P., Kharlamov, M.P.: Nonholonomic joint. Mekh. Tverd. Tela **27**, 1–7 (1995)
9. Koiller, J., Ehlers, K.: Rubber rolling over a sphere. Regul. Chaotic Dyn. **12**(2), (2007)
10. Lizárraga, D.A.: Obstructions to the existence of universal stabilizers for smooth control systems. Math. Control Signals Syst. **16**, 255–277 (2004)
11. Nakamura, Y., Chung, W., Sordalen, O.J.: Design and control of the nonholonomic manipulator. IEEE Trans. Robot. Autom. **1**(17), 48–59 (2001)
12. Peshkin, M., Colgate, J.E., Moore, C.: Passive robots and haptic displays based on nonholonomic elements. In: Proc. 1996 IEEE ICRA, pp. 551–556 (1996)
13. Stammers, C.W., Prest, P.H., Mobley, C.G.: A friction drive robot wrist: electronic and control requirements. Mechatronics **2**(4), 391–401 (1992)
14. Tchoń, K., Jakubiak, J.: Endogenous configuration space approach to mobile manipulators: a derivation and performance assessment of Jacobian inverse kinematics algorithms. Int. J. Control **76**(9), 1387–1419 (2003)
15. Vagner, V.: Geometrical interpretation of the motion of nonholonomic dynamical systems. Proc. Seminar on Vector and Tensor Analysis **5**, 301–327 (1941)

Synthesis of a Family of Regular Deployable Polyhedral Mechanisms (DPMs)

Guowu Wei and Jian S. Dai

Abstract This paper for the first time presents the synthesis of a family of overconstrained regular deployable polyhedral mechanisms (DPMs). The mechanisms are developed based on a novel plane-symmetric eight-bar linkage with exact straight-line motion. By implanting the plane-symmetric eight-bar linkages into the regular polyhedron bases, the synthesis of a family of overconstrained regular DPMs are presented in this paper and the constraint matrix of the mechanisms is constructed using the screw-loop equation method verifying the mobility of the mechanisms. The synthesis method presented in this paper can used to synthesize more DPMs and the proposed DPMs have potential applications in the fields of machines, deployable robots, architectural applications and space technologies.

Key words: Deployable polyhedral mechanisms (DPMs), exact straight-line motion, eight-bar linkage, overconstrained mechanisms, constraint matrix

1 Introduction

Polyhedral mechanisms, the mechanisms developed by implanting elementary kinematic chains into faces, edges and vertices of the polyhedrons, have drawn research interest starting from the pioneering work of Verheyen [1] on the expandable polyhedral structures named "Jitterbug transformers" by following Fuller's [2] introduction of a geometrical structure which he called the "Jitterbug", as a set of eight identical regular triangles connected to one another by the vertices. However, probably it was the showpiece of a mobile octahedron which was named "Heureka-

Guowu Wei
King's College London, London, UK, e-mail: guowu.wei@kcl.ac.uk

Jian S. Dai
Tianjin University, Tianjin, China, King's College London, London, UK, e-mail: jian.dai@kcl.ac.uk

J. Lenarčič, M. Husty (eds.), *Latest Advances in Robot Kinematics*,
DOI 10.1007/978-94-007-4620-6_16,

polyhedron" [3] shown at the Heureka Exposition in Zurich that aroused more interest for the research of polyhedron mechanisms from kinematicians and structure researchers. After the Heureka-polyhedron, the most notable contribution in the field was brought out by Wohlhart starting from the "Turning Tower" by treating the Heureka octahedron and the Brusels folding as its special cases [4] to the recent development of new polyhedral star linkages [5] obtained by arranging stretching spatial "star modules" into the faces of regular and irregular polyhedrons. Agrawal et al. [6] proposed a simple approach of constructing expanding polyhedrons based on prismatic joints which preserve their shape because of the rigidity of the vertices. Similar to Wohlhart's work, by identifying various types of closed kinematic chains and implanting them into the face of polyhedrons, several expandable polyhedral mechanisms have also been presented by different researchers [7, 8]. Over the same period, a fancy toy "Hoberman Switch Pitch" was brought out based on the geared expanding structure [9].

In this paper, based on a plane-symmetric eight-bar linkage with exact straight-line motion [10], a family of overconstrained regular deployable polyhedral mechanisms are synthesized and for the first time presented.

2 A Plane-symmetric Eight-bar Linkage with Exact Straight-Line Motion

A plane-symmetric eight-bar linkage with exact straight-line motion has been proposed by the authors [10] as illustrated in Fig. 1. The linkage contains four links 1, 3, 5 and 7 with two parallel joint axes at each end and four vertexes V_0, V_2, V_4 and V_6 in isosceles triangle shape with two revolute joint axes being spread at both ends of the vertexes by φ_1 and φ_2. Vertices V_0 and V_4 are identical and so are vertices V_2 and V_6. The eight-bar linkage is a linkage of mobility two. In order to define the configuration of the linkage, two inputs are required. To facilitate the study, two of the vertex links V_0 and V_4 are picked up one (V_0) as base and the other (V_4) as platform. Thus, the base and platform vertexes are connected by two identical limbs each of which consists of two links, one isosceles triangular vertex and four joints. It is found that in the case when lengths of links 1, 3, 5 and 7 are the same and the dimensions of vertices V_0 and V_4, and V_2 and V_6 are respectively identical, the linkage becomes a plane-symmetric linkage with its two limbs symmetric to a plane π which passes the centre points of V_0 and V_4 and is perpendicular to vertices V_0 and V_4. If the two joints connected to the base are assigned to be actuated joints. With the symmetric inputs, i.e. $\theta_{11} = \theta_{21}$ (see Fig. 1), vertex V_4, as the platform of the linkage, performs exact straight-line motion with respect to the base. According to the detailed analysis in [10], the angle β between this straight-line traced by the trajectories of point P of V_4 and the base (a plane which is coplanar with vertex V_0) only dependents on the value of angle φ_1 and φ_2 as $\beta = \arctan(1/(\cot\gamma\sin(\varphi_1/2) + \cos(\varphi_1/2)\csc\gamma\tan(\varphi_2/2)))$, with γ being the angle between two adjacent normals (see $\boldsymbol{n}_0$, $\boldsymbol{n}_2$, $\boldsymbol{n}_4$ and $\boldsymbol{n}_6$ in Fig. 1) of the vertices. As

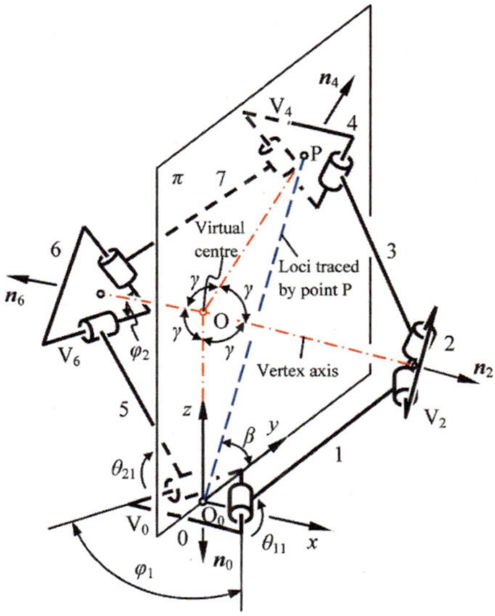

Fig. 1 A plane-symmetric eight-bar linkage.

indicated in [10], the orientations of the normals $\boldsymbol{n}_0$, $\boldsymbol{n}_2$, $\boldsymbol{n}_4$ and $\boldsymbol{n}_6$ maintain constant and intersect at a common centre O which is referred to as *virtual centre* of the eight-bar linkage as shown in Fig. 1, thus in every configuration the angle γ can be represented by the structure parameters as

$$\gamma = \arccos\left(\frac{\tan(\varphi_1/2)(1-\cos\varphi_2)}{\sin\varphi_2/2}\right). \tag{1}$$

It should be pointed out that since all the four links 1, 3, 5 and 7 are of the same length and vertices V_0 and V_4, and V_2 and V_6 are identical, the linkage is also symmetric to another plane that passes through the centre points of V_2 and V_6 and is perpendicular to vertices V_2 and V_6. Therefore this eight-bar linkage is referred to as spatial *plane-symmetric eight-bar linkage*.

Based on the eight-bar linkage, a family of regular deployable polyhedral mechanisms can be synthesized in the following sections.

3 Synthesis of the Deployable Polyhedral Mechanisms (DPMs)

It is well known that there are only five regular convex polyhedrons and they are named Platonic polyhedrons. Based on the five Platonic polyhedrons and the proposed plane-symmetric eight-bar linkage, a family of deployable polyhedral mechanisms (DPMs) can be constructed and the synthesis of the regular deployable polyhedral mechanisms (DPMs) starts from the synthesis of a deployable tetrahedral mechanism by implanting a group of eight-bar linkages into a regular tetrahedron

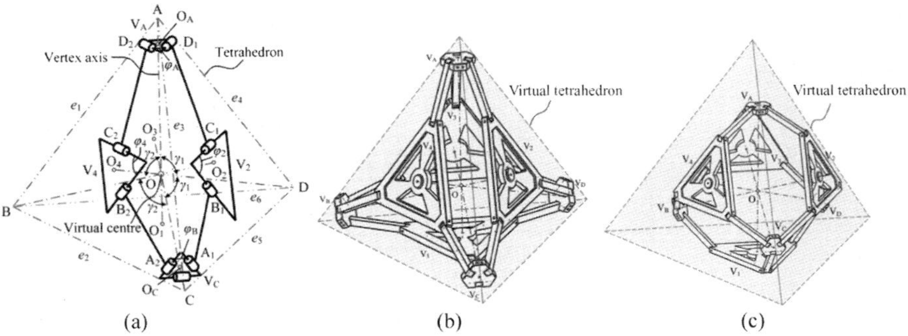

Fig. 2 Synthesis of a deployable tetrahedral mechanisms.

base. The method used for synthesizing the deployable tetrahedral mechanism can then be applied to the synthesis of the whole family of regular DPMs and can also be used for synthesis of the other series of DPMs based on the semi-regular and irregular polyhedrons.

As illustrated in Fig. 2a, a regular tetrahedron ABCD is given with its six edges denoted by e_1 to e_6. O_1, O_2, O_3 and O_4 are the centres of the four equilateral triangular faces such that AO_1, BO_2, CO_3 and DO_4 are all perpendicular to the faces they pass through and therefore O is the centroid of the tetrahedron. AO_1, BO_2, CO_3 and DO_4 are heights of the tetrahedron. The central angle is denoted as α.

In the tetrahedron, as shown in Fig. 2a, an eight-bar linkage is implanted along edge e_3 in such a way that two identical vertices V_A and V_B of the eight-bar linkage are perpendicular to AO_1 and CO_3 such that AO_1 and CO_3 become vertex axes [10] of vertices V_A and V_B. The other two identical vertices of the eight-bar linkage, i.e. vertices V_2 and V_4 are placed on faces 2 and 4 in such an arrangement that BO_2 passes through the centre of V_2 and is perpendicular to V_2, and DO_4 passes through centre of V_4 and is perpendicular to V_4. The revolute joints of the two chains in the eight-bar linkage, i.e. A_1, B_1, C_1, D_1 and A_2, B_2, C_2, D_2 are arranged in such a configuration that joints A_1 and B_1 are parallel to edge e_4, joints C_1 and D_1 are parallel to edge e_5, joints A_2 and B_2 are parallel to edge e_1, and joints C_2 and D_2 are parallel to edge e_2. The lengths of the two link groups, i.e. links A_1B_1 and C_1D_1, and A_2B_2 and C_2D_2 are arbitrary allocated with the condition that $A_1B_1 = C_1D_1 = A_2B_2 = C_2D_2$. Arranged in this way, it can be found that the four angles φ_A, φ_B, φ_2 and φ_4 of the isosceles triangle vertices are identical and they all equal 60°. Substituting $\varphi_A = \varphi_B = \varphi_2 = \varphi_4 = 60°$ into Eq. (1) it has $\gamma_1 = \gamma_2 = 70.53°$, and according to the geometric property of the tetrahedron, it has $\alpha = 70.53°$, thus $\gamma_1 = \gamma_2 = \alpha$ holds.

Then, imaging that the linkage is fixed at point O, given symmetric inputs at any pair of joints in any of the four vertices, the four vertices perform radially reciprocating motions along their vertex axes and the centre point O of the tetrahedron become the *virtual centre* of the eight-bar linkage.

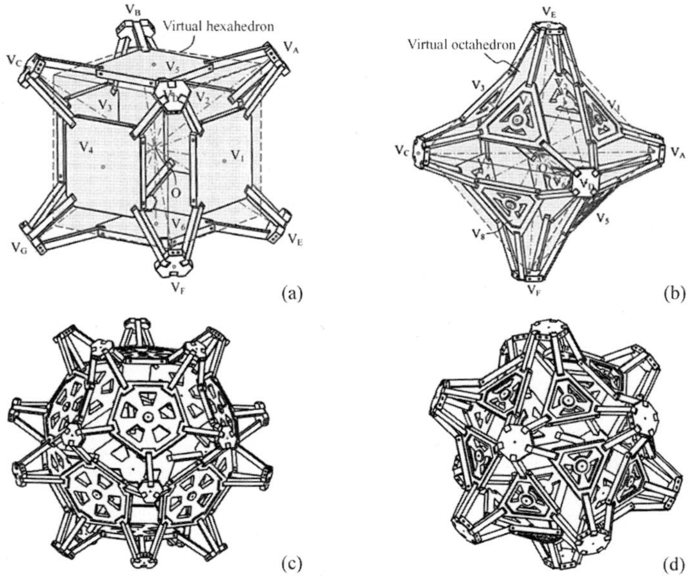

Fig. 3 Synthesis of a deployable tetrahedral mechanisms.

Further, taking the same procedure, integrating four more plane-symmetric eight-bar linkages into the tetrahedron base along edges e_1, e_2, e_4 and e_5, and carrying out a detailed structure design, a deployable tetrahedral mechanism can be generated in Figs. 2b and c. Once the deployable tetrahedral mechanism is constructed, all the virtual centres of the five eight-bar linkages coincident at one common point, i.e. the virtual centre of the mechanism. Therefore, this synthesis method is referred to as *virtual-centre-based* (VCB) synthesis method.

Figures 2b and c show a deployable tetrahedral mechanism that is synthesized based on plane-symmetric eight-bar linkages, the mechanism is a highly overconstrained mechanism of mobility one. The mechanism can perform a radially reciprocating motion in such a manner that vertices V_A, V_B, V_C, V_D move radially towards the virtual centre O, vertices V_1, V_2, V_3 and V_4 move radially away from the virtual centre O, and vice versa. It should be pointed out that, in every work configuration, the four vertices V_1, V_2, V_3 and V_4 locate on the faces of a virtual tetrahedron.

Subsequently, based on the same principle, the whole family of regular PDMs can be synthesized in Fig. 3. They are all overconstrained mechanisms of mobility one. In all the mechanisms, each edge of the virtual polyhedrons contains a plane-symmetric eight-bar linkage presented in Section 2 with various vertex angles φ_1 and φ_2.

Further, the mobility of the mechanisms can be verified in the next section through the constraint matrix using the screw-loop method proposed in [11].

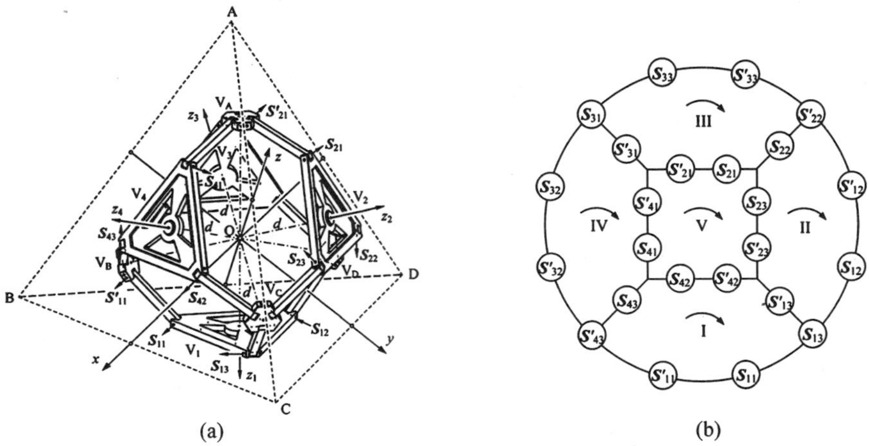

Fig. 4 Synthesis of a deployable tetrahedral mechanisms.

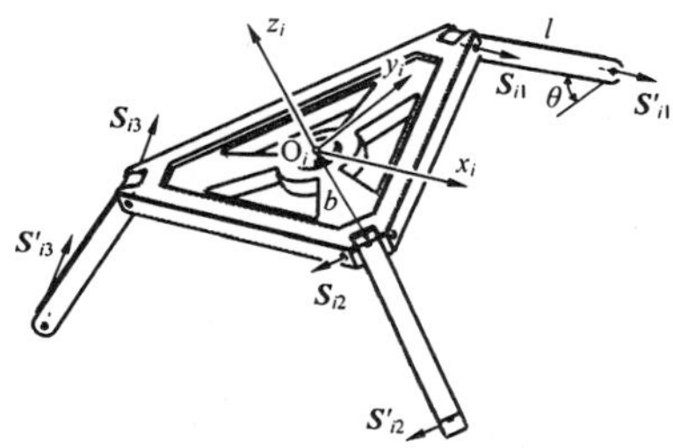

Fig. 5 Synthesis of a deployable tetrahedral mechanisms.

4 Constraint Matrix and Mobility of the Regular Deployable Polyhedral Mechanisms (DPMs)

Mobility of the regular DPMs can be analyzed through the screw-loop equation which is evolved from the mechanical network stemmed from Kirchhoff's circulation law. Taking the deployable tetrahedral mechanism as an example, in order to facilitate the analysis, as illustrated in Fig. 4a, a general configuration of the mechanism is picked up and a reference coordinate system is established with its origin locating at the virtual centre of the mechanism and x-axis passing through the middle points of edges AD and BC, y-axis passing through the middle points of edges AB and CD, and z-axis passing through the middle points of edges AC and BD. In vertices V_1, V_2, V_3 and V_4, local coordinate system O_i-$x_iy_iz_i$ are established with origin O_i locating at the centre of the ith vertex ($i = 1,2,3,4$) as shown in Fig. 5. In the local coordinate system, z_i-axis is collinear with OO_i, and x_i-axis is parallel to one of the sides of the equilateral triangle vertex. The lengths of the binary links are all

l, the distance between the centre of the vertex and the joint axis is b, and the distances between the virtual centre of the deployable tetrahedral mechanism and the centres of vertices V_1, V_2, V_3 and V_4 are all d. From Fig. 5, the joint screws in every individual vertex can be obtained in the corresponding local coordinate system as

$$\begin{cases} \boldsymbol{S}_{i1} = [1\,0\,0\,0\,0\,b]^{\mathrm{T}},\ \boldsymbol{S}_{i2} = \left[-1/2\ -\sqrt{3}/2\,0\,0\,0\ -b\right]^{\mathrm{T}} \\ \boldsymbol{S}_{i3} = \left[-1/2\ \sqrt{3}/2\,0\,0\,0\ -b\right]^{\mathrm{T}},\ \boldsymbol{S}'_{i1} = [1\,0\,0\,0\ l\sin\theta\ -b-l\cos\theta]^{\mathrm{T}}, \\ \boldsymbol{S}'_{i2} = \left[-1/2\ -\sqrt{3}/2\ 0\ -\sqrt{3}l\sin\theta/2\ l\sin\theta/2\ -b-l\cos\theta\right]^{\mathrm{T}}, \\ \boldsymbol{S}'_{i3} = \left[-1/2\ \sqrt{3}/2\ 0\ \sqrt{3}l\sin\theta/2\ l\sin\theta/2\ -b-l\cos\theta\right]^{\mathrm{T}}. \end{cases} \tag{2}$$

In the above equation, the first subscript $i = 1,2,3$ and 4 indicates the number of vertex. The joint screws in the local coordinate system of each individual vertex can be transformed to the reference coordinate system through a screw transformation matrix $\mathbf{T}_i = \begin{bmatrix} \mathbf{R}_i & \mathbf{0} \\ \widetilde{\boldsymbol{p}}_i\mathbf{R}_i & \mathbf{R}_i \end{bmatrix}$, with $\mathbf{R}_i$ being the rotation transformation matrix and $\widetilde{\boldsymbol{p}}_i$ being a skew-symmetric matrix derived from $\boldsymbol{p}_i$ presenting the displacement of point O_i in the reference coordinate system. From Fig. 4a and according to the geometry of a tetrahedron, $\mathbf{R}_i$ and $\boldsymbol{p}_i$ ($i = 1,2,3,4$) can be obtained as

$$\begin{aligned} \mathbf{R}_1 &= \begin{bmatrix} \sqrt{2}/2 & \sqrt{6}/6 & \sqrt{3}/3 \\ 0 & -\sqrt{6}/3 & \sqrt{3}/3 \\ \sqrt{2}/2 & -\sqrt{6}/6 & -\sqrt{3}/3 \end{bmatrix}, \quad \mathbf{R}_2 = \begin{bmatrix} -\sqrt{2}/2 & -\sqrt{6}/6 & -\sqrt{3}/3 \\ 0 & -\sqrt{6}/3 & \sqrt{3}/3 \\ -\sqrt{2}/2 & \sqrt{6}/6 & \sqrt{3}/3 \end{bmatrix}, \\ \mathbf{R}_3 &= \begin{bmatrix} \sqrt{2}/2 & -\sqrt{6}/6 & -\sqrt{3}/3 \\ -\sqrt{2}/2 & -\sqrt{6}/6 & -\sqrt{3}/3 \\ 0 & \sqrt{6}/3 & -\sqrt{3}/3 \end{bmatrix}, \quad \mathbf{R}_4 = \begin{bmatrix} 0 & -\sqrt{6}/3 & \sqrt{3}/3 \\ \sqrt{2}/2 & -\sqrt{6}/6 & \sqrt{3}/3 \\ \sqrt{2}/2 & \sqrt{6}/6 & \sqrt{3}/3 \end{bmatrix}, \end{aligned} \tag{3}$$

and

$$\begin{aligned} \boldsymbol{p}_1 &= d\left[\sqrt{3}/3\ \sqrt{3}/3\ -\sqrt{3}/3\right]^{\mathrm{T}}, \quad \boldsymbol{p}_2 = d\left[-\sqrt{3}/3\ \sqrt{3}/3\ \sqrt{3}/3\right]^{\mathrm{T}}, \\ \boldsymbol{p}_3 &= d\left[-\sqrt{3}/3\ -\sqrt{3}/3\ -\sqrt{3}/3\right]^{\mathrm{T}}, \quad \boldsymbol{p}_4 = d\left[\sqrt{3}/3\ -\sqrt{3}/3\ \sqrt{3}/3\right]^{\mathrm{T}}. \end{aligned} \tag{4}$$

Thus, through the screw transformation matrix $\mathbf{T}_i$, all the joints screws in the mechanism can be obtained in the reference coordinate system.

According to Euler's formula for independent loop of mechanical graph, the mechanism contains five independent loops such that the constraint graph of the mechanism can be obtained in Fig. 4b. Based on the constraint graph, the constraint matrix [11] of the deployable tetrahedral mechanism can be obtained as

$$\mathbf{M}_c = \begin{bmatrix} S_{11} & S'_{11} & 0 & 0 & S_{13} & S'_{13} & 0 & 0 & 0 & 0 & 0 & 0 & 0 & 0 & 0 & 0 & 0 & 0 & 0 & 0 & S_{42} & S'_{42} & S_{43} & S'_{43} \\ 0 & 0 & S_{12} & S'_{12} & -S_{13} & -S'_{13} & 0 & 0 & S_{22} & S'_{22} & S_{23} & S'_{23} & 0 & 0 & 0 & 0 & 0 & 0 & 0 & 0 & 0 & 0 & 0 & 0 \\ 0 & 0 & 0 & 0 & 0 & 0 & S_{21} & S'_{21} & -S_{22} & -S'_{22} & 0 & 0 & S_{31} & S'_{31} & 0 & 0 & S_{33} & S'_{33} & 0 & 0 & 0 & 0 & 0 & 0 \\ 0 & 0 & 0 & 0 & 0 & 0 & 0 & 0 & 0 & 0 & 0 & 0 & -S_{31} & -S'_{31} & S_{32} & S'_{32} & 0 & 0 & S_{41} & S'_{41} & 0 & 0 & -S_{43} & -S'_{43} \\ 0 & 0 & 0 & 0 & 0 & 0 & -S_{21} & -S'_{21} & 0 & 0 & -S_{23} & -S'_{23} & 0 & 0 & 0 & 0 & 0 & 0 & -S_{41} & -S'_{41} & -S_{42} & -S'_{42} & 0 & 0 \end{bmatrix}. \tag{5}$$

This is a 30×24 matrix with $\boldsymbol{0} = \begin{bmatrix} 0\,0\,0\,0\,0\,0 \end{bmatrix}^{\mathrm{T}}$ and through computation, the mobility of the mechanism can be given [11] as

$$m = n_c - rank\,(\mathbf{M}_c) = 24 - 23 = 1, \tag{6}$$

where m denotes the mobility of the mechanism and n_c is the number of joints.

The above analysis proves that the mobility of the deployable tetrahedral mechanism is actually one and it is an overconstrained mechanism. The mobility of the whole proposed family of DPMs can be verified with the same approach.

5 Conclusions

Base on the novel plane-symmetric eight-bar linkage with exact straight-line motion, this paper investigated the synthesis of a family of regular deployable polyhedral mechanisms (DPMs), the principle of the synthesis was introduced based on the construction of a deployable tetrahedral mechanism and the mobility of the mechanisms has been verified utilizing the screw-loop equation method. The regular DPMs presented in this paper have potential applications in the fields of machines, deployable robots, architectural applications and space technologies.

Acknowledgements The authors gratefully acknowledge the support from the EU FP7 project TOMSY under Grant No. 270436 and the EU FP7 project ECHORD DEXDEB under Grant No. 231143.

References

1. Verheyen, H.F.: The complete set of Jitterbug transformers and the analysis of their motion. Comput. Math. Appl. **17**(1–3), 203–250 (1989)
2. Fuller, R.B.: Synergetics: Exploration in the geometry of thinking. Macmillan, NY (1975)
3. Stachel, H.: The Heureka-Polyhedron. In: Fejes Toth, G. (ed.) Intuitive Geometry. Colloq. Math. Soc. Janos Bolyai, vol. 63, pp. 447–459, North-Holland, Amsterdam (1994)
4. Wohlhart, K.: Heureka octahedron and Brussels folding cube as special cases of the turning tower. In: Proc. the Sixth IFToMM International Symposium on Linkages and Computer Aided Design Methods, Bucharest, Romania (1993)
5. Wohlhart, K.: New polyhedral star linkages. In: Proc. the 10th International Conference on the Theory of Machines and Mechanisms, Liberec, Czech Republic (2008)
6. Agrawal, S.K., Kumar, S., Yim, M.: Polyhedral single degree-of freedom expanding structures: Design and prototypes. ASME, J. Mech. Des. **124**(9), 473–478 (2002)
7. Gosselin, C.M., Gagnon-Lachance, D.: Expandable polyhedral mechanisms based on polygonal one-degree-of-freedom faces. J. Mech. Eng. Sci. **220**, 1011–1018 (2006)
8. Kiper, G.: Fulleroid-like linkages. In: Proc. EUCOMES 8. Springer (2009)
9. Hoberman, C.: Geared expanding structures. US patent: US007464503B2 (2004)
10. Wei, G., Dai, J.S.: Geometry and kinematics of a plane-symmetric spatial eight-bar linkage with exact straight-line motion. In: DETC2011-48281, Proc. ASME 2011 Int. Design Eng. Tech. Conferences, 28–31 Aug. 2011, Washington, DC, USA (2011)
11. Wei, G., Ding, X., Dai, J.S.: Mobility and geometric analysis of the Hoberman Switch-Pitch ball and its variant. Trans. ASME, J. Mech. Robot. **2**(3), 031010 (2010)

Type Synthesis of Binary Actuated Parallel Mechanisms

D. Schütz, R.J. Ellwood, A. Raatz and J. Hesselbach

Abstract Binary actuators have two stable positions which are defined by mechanical end positions. These actuators can be used within parallel structures, to create binary robots. These robots feature high repeatability in their discrete configurations. The discrete nature of these actuators simplify the robots as they no longer need complex feedback controllers. The result is simple cost effective robots that are suitable for positioning tasks which require only a few destination points. A drawback of these discrete robots is that the conditions of a specific task must be fulfilled by the geometric parameters of the binary robot's kinematic structure. This contribution focuses on the realization of a type synthesis method based on the combination of simple binary robot structures. The presented method allows a suitable parallel structure whose end effector is able to reach the given number of positions of the defined task to be synthesized.

Key words: Type synthesis, binary robots, binary mechanisms

1 Introduction

The simple and robust design of binary actuators allow two stable states to be reached. Dictated by the mechanical end positions of these actuators, the two resulting positions can be reached with a high repeatability and do not require extensive feedback control [1]. As a result of this simplified control, a programmable logic controller (PLC) is more than adequate to control the actuation of the resulting binary robot. In the field of binary robotics, many research groups have focused their efforts on the integration of a large number of n binary actuators inside of a kinematic structure. This results in a discrete workspace with $j = 2^n$ possible end

Daniel Schütz · Robert John Ellwood · Annika Raatz · Jürgen Hesselbach
Institute of Machine Tools and Production Technology, Braunschweig, Germany,
e-mail: {d.schuetz, j.ellwood, a.raatz, j.hesselbach}@tu-bs.de

J. Lenarčič, M. Husty (eds.), *Latest Advances in Robot Kinematics*,
DOI 10.1007/978-94-007-4620-6_17,

effector poses $\mathbf{x}_j$ [4, 8, 12]. If the main part of these discrete poses lie inside of a small area, then the end effector of the robot can be moved within the robot's quasi-continuous workspace [1]. To achieve the large number of points required for a quasi-continuous workspace, identical modules are placed in series. The parameters of the basis module are then found using a geometric synthesis, which takes the workspace of the resulting structure into consideration. As the parameters of all modules within the structure are based on the basis module, this approach allows the desired structure to be quickly found. Chirikjian *et al.* have worked on this topic and the resulting approaches can be found in several of their publications [2, 3, 7].

This contribution focuses on parallel robots where the minimum required number of binary actuators to fulfill a specific task is sought. In other publications it has been shown that a given task can be accomplished with a suitable structure if a geometric synthesis leads to a parameter set where the reachable end effector poses can be adapted to the desired poses of the specific task. In [9] a geometric synthesis approach has been presented for a simple fivebar with an R$\underline{\text{P}}$R$\underline{\text{P}}$R structure. After the installation of the mechanical structure, deviations occur between the actual poses $\mathbf{x}_{j,a}$ and the computed nominal poses $\mathbf{x}_{j,n}$. A possible cause of these differences are due to production and assembly errors which are an inevitable result of production methods. As there is no chance to correct the actuated joints' coordinates $\mathbf{q}$ within the mathematical model of the structure inside the robot control, a method of mechanical calibration has been introduced [9, 10]. This technique is based on the adjustment of the mechanical structure with additional components which allow for the physical correction of the identified parameter values.

This publication looks at the first steps of the adapted task configuration of a binary robot, being the type synthesis of the kinematic chains. A type synthesis method for both planar and spatial binary robots is then presented. This method is then implemented to find the structure for an exemplary sorting task in which 50 given points within a plane need to be reached. Performance criteria such as velocity, force transmission, and stiffness are not taken into account and should be the topic of future work.

Given a set of desired points, a simple structure which has a minimal number of binary actuators is sought. This type synthesis method starts by finding the needed degrees of freedom (DOF) of the kinematic structure based on the desired number of end effector poses. For more complex tasks, a method that allows the extension of typical parallel structures is shown. Furthermore this approach considers an output link which is suitable as the end effector. After presenting the methods for binary robot type synthesis, the article points out the required steps to install and calibrate the mechanical structure.

2 Identification of the Necessary Degrees of Freedom

A structure with n integrated binary actuators is able to position its end effector in j poses $\mathbf{x}_j$ with

$$j = 2^n. \tag{1}$$

Solving for n, it is then possible to find the minimum number of binary actuators for the end effector poses $\mathbf{x}_j$ for a given task.

$$n \geq \log_2 \cdot j \quad \text{with} \quad n \in \mathbb{N} \tag{2}$$

The number of binary actuators defines the DOF of the robot structure F as the condition

$$F \stackrel{!}{=} n. \tag{3}$$

This condition assures that the robot can switch between all the end effector poses. If F is less than n, the robot is overdetermined and a change of the actuator configuration $\mathbf{q}$ is locked. In the case that the DOF of the structure is larger than the number of actuators, it is not possible to say whether the structure is in the desired position as there are not enough constraints on the system. Equation (2) can be used for the 50 point example to show that the structure must have at least six binary actuators and six DOF respectively.

The need of a planar kinematic structure with $F \leq 3$ or a spatial structure with $F \leq 6$ where one to three respectively six binary actuators have to be integrated, can be accomplished using already published methods for type synthesis of parallel robots, e.g. [5]. If new structures are required with more than three or six DOF respectively then the following technique supports the synthesis of suitable structures.

3 Extension of Kinematic Structures

Looking at the geometric considerations of a robot, one of the most cost efficient robots for a specific task is one in which the number of poses within its discrete workspace is equal to the number of task poses. As there are several options to arrange a structure with the desired DOF, additional restrictions are required to allow a suitable solution to be found. With the condition of a closed kinematic chain, the simplest planar parallel structure with DOF F is a single loop chain. The DOF can be calculated using the well-known Grübler formula

$$F = b(l-1) - \sum_{i=1}^{g} (b - f_i). \tag{4}$$

In Equation (4), the DOF b of the space in which the structure is designed is either three for planar structures or six for spatial structures. The notation l stands for the number of links and g represents the number of joints. The DOF permitted by the ith joint is declared as f_i [6].

The identification of the number of needed links and joints for a planar parallel structure with one single loop can be determined by transposing Eq. (4) into

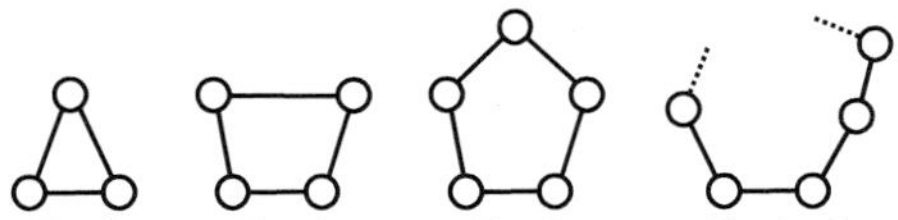

Fig. 1 Planar single loop kinematic chains.

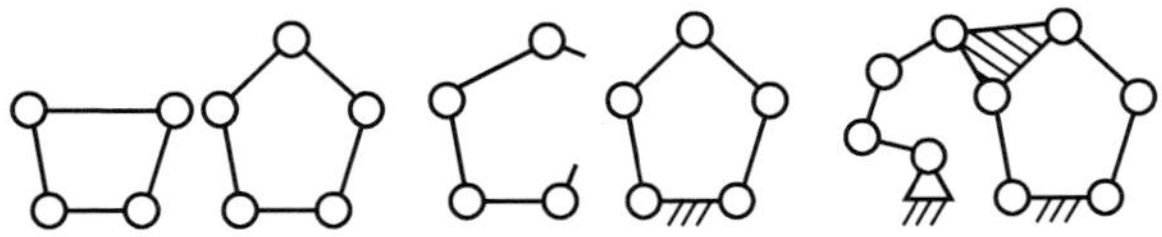

Fig. 2 Connection of an additional branch to a basis structure.

$$l = \frac{F + \sum_{i=1}^{g}(b - f_i)}{b} + 1. \tag{5}$$

In the planar case where $b = 3$, the number of DOF of each joint is one, $f_i = 1$, as two links are connected by one joint. Thus a single loop of a closed kinematic chain consisting of $g = l$ joints and links, allows Eq. (5) to be reduced to Eq. (6). Figure 1 illustrates single loop kinematic chains sorted by their DOF.

$$l = F + 3. \tag{6}$$

After the parallel structure is chosen, the selection of the joint type (revolute or prismatic joint) as well as which joints should remain passive and which should become binary actuators can be handled. To achieve a stiff structure without additional guides while offering high dynamics, the binary actuators should be located closed to the base. Extending this to planar single loop chains, the number of binary actuators is limited to a maximum of two actuators (cf. Fig. 1).

Tsai has shown in [11] the extension of a single loop chain to a multi loop chain and can be applied for binary parallel robots. Through the addition of branches, it is possible to extend a given planar basis structure. As illustrated in Fig. 2, one of the links within the branch structure has been split. One end of this open chain is connected to the basis structure, while the other end is fixed to the base frame.

The result of this procedure is a parallel structure with a closed kinematic chain with multiple loops, where the number of loops depends on the number of additional branches. The advantage of this method is the opportunity to extend the basis structure until the DOF of the entire kinematic chain is in accordance with Eq. (3). This is due to the fact that the sum of the DOF of the basis structure F_{basis} and the DOF of each pth additional branch $F_{branch,p}$ represents the total DOF of the resulting structure

$$F_{total} = F_{basis} + \sum_{p=1}^{c} (F_{branch})_p, \tag{7}$$

where c denotes the number of branches.

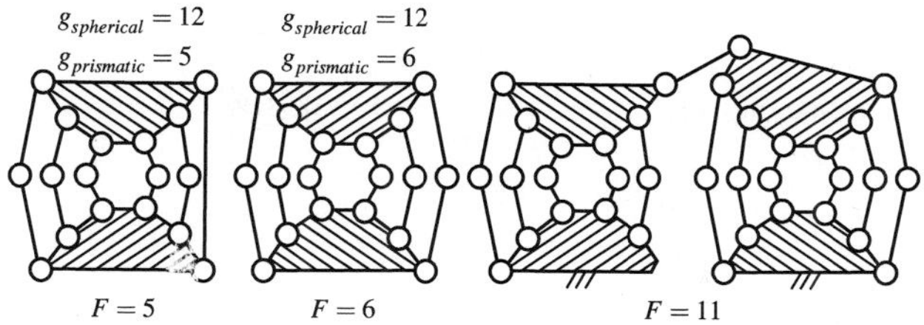

Fig. 3 Connection of spatial kinematic chains of a six DOF Stewart–Gough platform and an additional structure with five DOF to create a structure with eleven DOF.

The principle of this method can be transfered from single loop basis structures and branches to multiple loop basis structures with additional structures. Thus Equation (7) can be written as

$$F_{total} = F_{basis} + \sum_{s=1}^{d} (F_{additional})_s \,. \tag{8}$$

Here d denotes the number of additional structures, where the sth structure has $F_{additional,s}$ DOF.

The advantage of this simple technique is that a type synthesis of spatial parallel structures for robots with binary actuators can be accomplished in the same way. The identification of the necessary DOF F_{total} can also be achieved by Eq. (2) and (3). In contrast the simplification of the Grübler formula cannot be used for spatial structures as there are different types of joints which feature more than one DOF, $f_i \geq 1$.

The adapted method is a general approach to design a closed kinematic chain of a parallel structure using the minimum number of binary actuators. On the basis of already presented approaches for type synthesis of parallel robots, it is possible to derive spatial kinematic chains with up to six DOF. These kinematic chains can be used as a basis structure which allow a connection of additional structures to gain the desired DOF expressed in Eq. (8). Therefore one link of an additional closed kinematic chain must be split up in two pieces. These links are combined with one or two links of the basis structure, allowing a new structure to be realized. For instance in Fig. 3 the well-known Stewart-Gough platform is extended with a similar structure that features five DOF, which results in a new structure that allows the integration of eleven binary prismatic actuators. This enables the end effector to reach $2^{11} = 2048$ poses.

4 End Effector Location

The main objective of type synthesis of parallel structures with the integration of n desired binary actuators is the positioning of the end effector in the 2^n poses.

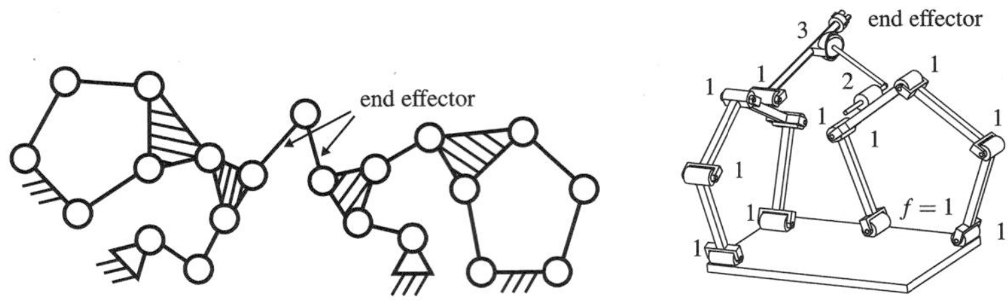

(a) Planar kinematic chain with six DOF. (b) Spatial parallel mechanism with four DOF

Fig. 4 Connection of two independent loops with a transmission linkage.

Therefore the pose of one link of the structure must be changed for every possible actuator configuration $\mathbf{q}_j$. Two techniques can be utilized to fulfill this condition. The first technique is based on the combination of affixing one end of the additional structure to the base to allow the localization of binary actuators near the base frame. The link on the other side of the open chain must be combined with the link which acts as the output of the basis structure.

If the second end of an additional structure is also connected to the base frame or directly to a binary actuator, then the total DOF of the structure is still governed by Eq. (8). Due to independent loops within such a structure, no one link can reach all of the 2^n actuator configurations. Another technique to ensure that an output link can reach the desired poses is the integration of an additional closed kinematic chain which has zero DOF. This chain is connected to two output links of independent loops of the basis structure. As the DOF of the kth transmission linkage is $F_{transmission,k} = 0$, the total DOF of the entire structure is not changed. Equation (8) can be extended with the DOF of u transmission linkages

$$F_{total} = F_{basis} + \sum_{s=1}^{d} (F_{additional})_s + \sum_{k=1}^{u} (F_{transmission})_k \,. \tag{9}$$

Extending this to the 50 point example, it has been shown that the structure will need six DOF. This can be realized by joining two structures with three DOF, such as the one illustrated in Fig. 2. As depicted in Fig. 4(a), an end effector link has been incorporated using a single loop kinematic chain with zero DOF. This structure with six actuators is thus able to reach $2^6 = 64$ poses. For completeness, it is also possible to use this concept with spatial mechanisms, as illustrated in Fig. 4(b). This structure is based on two planar fivebars and a transmission linkage with one revolute, one spherical, as well as a combination of revolute and prismatic joint. The structure allows the integration of four binary actuators at the base, resulting in a discrete workspace of $2^4 = 16$ end effector poses.

5 Geometric Synthesis and Mechanical Calibration

The design process of parallel structures with binary actuators consists of several additional steps to ensure that the resulting structure can reach each of the desired poses with its end effector. After the structure and the joint types have been found using type synthesis, a geometric synthesis can be used to find the structure's parameters. There are many different approaches to compute these parameter values and are dependent on the desired workspace as well as the task conditions. The discrete nature of binary robots makes the geometric synthesis an important step in ensuring the robots capability to reach the desired points. Simple parallel robots with only a few binary actuators can be synthesized on the basis of geometrical constraints, as shown for an R$\underline{P}$R$\underline{P}$R structure [9]. If a geometric synthesis has been established for simple structures, it can be used as a module in more complex structures. Thus the single modules can be synthesized independently from one another, allowing a reduction in the total effort. For instance the R$\underline{P}$R$\underline{P}$R structure can be used within the structures of Fig. 4(a) and 4(b). The knowledge gained about the parameters of the fivebar structures for a given four point constellation can be used in further steps such as the geometric synthesis of the transmission linkages.

Production and assembly errors cause deviations between the desired poses $\mathbf{x}_{j,n}$ of the task and the reachable poses $\mathbf{x}_{j,a}$ of the installed mechanical structure. Due to the discrete nature of binary robots, it is not possible to change a parameter within the kinematic model to overcome this discrepancy. In order to overcome this, some of the physical properties of the binary robot have to be changed.

As presented in [10], it is possible to incorporate additional mechanical components into the robot to allow for a mechanical calibration. Here the residual vector between the desired poses and the measured poses is calculated. The residual is then minimized using an optimization method and used to find a new mechanical parameter set. Following the mechanical adjustment of the robot, a new residual vector is calculated. This iterative process can be conducted until the residuals have reached acceptable values. After the mechanical correction has taken place, it is advantageous that these joints be locked for operation mode.

6 Conclusions

Parallel structures with a minimum number of binary actuators can be applied to positioning tasks with a discrete number of end effector poses. This article presents methods for type synthesis of these structures depending on the number of desired end effector poses. It has been shown that parallel structures based on typical type synthesis can be connected to create structures with a DOF equal to the sum of the original structures' DOF. Therefore two methods have been shown to calculate an output link which is suitable as the end effector location. The first techniques is based on the connection of one end of an opened chain to the base frame and the other to the output link of the basis structure. If a structure features independent loops, then two independent output links can be connected by a transmission linkage, which is predicated on a single loop chain with zero DOF. After type synthesis of a suitable structure, a geometric synthesis and mechanical calibration process must be applied to accomplish the needed accuracy of the regarded task.

References

1. Chirikjian, G.S.: A Binary paradigm for robotic manipulators. In: IEEE International Conference on Robotics and Automation, pp. 449–455. San Diego, USA (1994)
2. Chirikjian, G.S.: Kinematic synthesis of mechanisms and robotic manipulators with binary actuators. J. Mech. Des. **117**(4), 573–580 (1995)
3. Chirikjian, G.S., Ebert-Uphoff, I.: Numerical convolution on the Euclidean group with applications to workspace generation. IEEE Trans. Robot. Autom. **14**(1), 123–136 (1998)
4. Ebert-Uphoff, I., Chirikjian, G.S.: Generation for binary manipulators with many actuators. J. Robot. Syst. **12**(6), 383–400 (1995)
5. Frindt, M., Krefft, M., Hesselbach, J.: Structure and type synthesis of parallel manipulators. In: Schütz, D., Wahl, F.M. (eds.) Robotic Systems for Handling and Assembly, STAR, vol. 67, pp. 17–37. Springer, Berlin (2010)
6. Grübler, M.F.: Getriebelehre. Springer, Berlin (1917)
7. Kyatkin, A.B., Chirikjian, G.S.: Synthesis of binary manipulators using the Fourier transform on the Euclidean group. J. Mech. Des. **121**, 9–14 (1999)
8. Plante, J.S., Dubowsky, S.: The calibration of a parallel manipulator with binary actuation. In: Lenarčič, J., Roth, B. (eds.) Advances in Robot Kinematics, pp. 391–398. Springer, Dordrecht (2006)
9. Schütz, D., Raatz, A., Hesselbach, J.: The development of a reconfigurable parallel robot with binary actuators. In: Lenarčič, J., Stanišić, M.M. (eds.) Advances in Robot Kinematics, pp. 225–232. Springer, Dordrecht (2010)
10. Schütz, D., Raatz, A., Hesselbach, J.: A mechanical calibration approach for binary parallel robots. In: World Congress in Mechanism and Machine Science, Guanajuato, México (2011)
11. Tsai, L.W.: Robot Analysis: The Mechanics of Serial and Parallel Manipulators. Wiley, New York, USA (1999)
12. Wingert, A., Lichter, M.D., Dubowsky, S.: On the design of large degree-of-freedom digital mechatronic devices based on bistable dielectric elastomer actuators. IEEE/ASME Trans. Mechatron. **11**(4), 448–456 (2006)

Biokinematic Study of Barn Owl Head Movements for the Development of a Bio-Inspired Active Vision Robotic System

Ouriel Barzilay, Yoram Gutfreund and Alon Wolf

Abstract Active vision is a process used by human beings and most animals to improve their visual recognition and avoid ill-posed visual problems. It has been proved that, by combining motion to their visual senses and perception, active observers can solve basic visual problems more efficiently than a passive one and that complex problems can also be addressed more easily. Autonomous robotic systems acting in dynamic environments should therefore imitate this process for a better image recognition and target tracking. Visually guided robotic systems also need to actively select visual information from the environment for detailed processing through mechanisms that mimic visual attention and saccadic eye movements. Developing and verifying computational models for visual search and implementing them on a robotic system are challenging important tasks that we will address based on our planned exploration of these mechanisms in the barn owl. We intend to investigate how, by imitating the barn owl's repertoire of motor behaviors and search patterns, the autonomous agent could obtain meaningful information on the structure of the environment structure and possibly target position and motion. Towards that end, we have investigated the conspicuous head motions of barn owls and searched a kinematic characterization of the movements by means of screw theory.

Key words: Barn owl, head movements, screw theory

1 Introduction

Understanding how we perceive the world is one of the greatest challenges facing neuroscience. Historically, research into this problem has mostly focused on pas-

Ouriel Barzilay · Alon Wolf
Biorobotics and Biomechanics Lab., Dept. Mechanical Engineering, Technion I.I.T, Haifa 32000, Israel {barzilay}{alonw}@technion.ac.il

Yoram Gutfreund
Audiovisual Systems Lab., Dept. Medical Studies, Technion I.I.T, Haifa 32000, Israel

J. Lenarčič, M. Husty (eds.), *Latest Advances in Robot Kinematics*,
DOI 10.1007/978-94-007-4620-6_18,

sive perception where information is acquired passively from static sensory organs. However vast psychophysical and physiological evidence indicate that biological perception is to a large extent an *active* process.

In addition to understanding vision as a biological sensory process, the field of *active vision* opens up new opportunities for integration with the fields of computational vision and robotics which strive to capture, model, and mimic complex behaviors in artificial agents. Artificial systems which analyze natural environments face the same tasks and challenges as the brains of active animals. Therefore, identifying and understanding strategies of active vision in animals is a critical source of information for successful artificial vision systems, as well as for the implementation of mobile autonomous agents.

The novelty of the proposed research lies in the combination and integration of approaches from three different disciplines. We utilize a classical neuroethological approach to focus on a single carefully selected animal model. A biomechanical approach will be adopted to study and interpret head motions from a task-based and anatomical point of view. Finally, computational vision approaches will be employed to study and model strategies of active vision. Previous investigations of active vision behaviors have been conducted in humans [2] and animals such as insects [3] and primates [7]. In this research we aim to establish such a basic and novel line of research in one non-mammalian species – the barn owl.

Barn owls are skillful predators that employ their visual system to detect small targets in dim illumination. Their evolutionary adaptation to low light vision has almost completely eliminated their ability to move their eyes [8]: eye movements in barn owls are limited to about 3 degrees in each direction [4] and are therefore compensated by conspicuous head movements. A barn owl freely scanning the environment will occasionally perform stereotypic side-to-side *peering* motion of its head, a micro behavior believed to facilitate depth estimation [6]. In addition it will rapidly shift its head position (head saccades) to face regions of interest [5], a macro behavior that produces a noticeable scan path reminiscent of what a human eye would perform in its socket. The fact that the process of active vision in barn owls is mainly manifested in head motions provides a substantial experimental advantage, since head motions are much easier to track than eye movements.

The primary goal of this study is to provide valuable information towards the understanding of active vision strategies and mechanisms in barn owls. Our research strategy is based on the exploration of possible correlations between the animal's head movements, the dynamic visual signals acquired by its visual system, and other task or scene information. Toward that end, we have developed an experimental setup combining motion capture and a head mounted camera providing an estimation of the bird's view.

Head movements as measured by our setup were analyzed mathematically in order to identify characteristic movement patterns. We expect reaching insights on the rotational/translational nature of these movements from screw analysis. In a further stage, we intend to study the bird's visual attention and scan path by designing experiments aiming at the characterization of the perceived saliency of objects in the scene and subsequently propose saliency models in barn owls.

Based on the insights provided by the kinematic analysis and the study of the bird's visual attention, we plan to build a camera-mounted robotic platform and reproduce barn owl's typical head movements and visual search patterns. The results of the motion analysis will then be formalized into algorithms and software running on the autonomous agent for target localization and tracking. We believe this biomimetic approach to the design of attention model and active vision system may bring considerable advances in the field of scene reconstruction and mapping.

2 Methods

2.1 Experimental Setup

To investigate the bird's head movements, we measured kinematics in perching, freely moving barn owls. Head motion tracking was achieved via the Vicon™ motion capture system allowing the recording of kinematics of markers with 0.2 millimeters of accuracy at sampling frequencies reaching 2000 Hz. The owl's head kinematic recordings were performed at 120 Hz. We have designed a special head-mounted device attached to the owl's head. This tracking device is equipped with five reflectors tracked by the Vicon system, a miniature wireless CCD camera and a battery pack for the CCD camera. The CCD camera is fixed and oriented along the presumed visual axis of the owl. The device has been developed with the intent to minimize its weight and moment of inertia such that the natural behavior of the bird remains unaltered. The final version of the device weighs 14 gr. (including marker, camera and battery pack) and can be adjusted to any subject. The birds preliminarily underwent a minor surgical operation at the Technion Faculty of Medical Studies, where a small bolt was glued onto the top of their skull.

The experiments were conducted on two free perching barn owls bred at the Audiovisual Systems Laboratory at the faculty of Medical Studies. To adapt the system to record head kinematics in freely viewing owls, young barn owls were acclimatized to perch on a small rod position about 1.5 meters above the ground. Furthermore, to ensure an ordinary behavior during recording and with the device mounted on their head, the barn owls were trained to wear the device on a regular basis. In every experimental session, an external camera aimed at the perching owl captured an external view on the head motions of the bird.

2.2 Head Orientation Estimation

We use screw theory to describe the instantaneous head motion of the owl. Screw coordinates provide information on the nature of the head motion, i.e. pure rotation/pure translation or helical motion, and on the instantaneous biomechanical axis of the head motion. These data are then analyzed for a deeper understanding of the kinematic nature of owl head motions and used for clustering typical head motion behavior of the owl.

The rigid-body head orientation was obtained from computation of Euler angles from the location of three markers. Form the computed device's orientation and location, we computed the screw parameters according to methods developed in [1, pp. 77–94].

2.3 Isolation of Peering Motions

In the first part of this research, we isolated and characterized stereotypic head side-to-side translational motions. These movements are often referred to in the literature as *peering* motions. Based on the recordings of the external camera aimed at the barn owl and of the head-mounted camera, we pre-selected several segments on which the owl seemed to move its head in the sagittal plane without changing its orientation. The marker trajectories on those segments were then more closely examined in Matlab in order to refine the selection of those translational motions. Certain of the complex motions were decomposed into smaller and simpler motion primitives.

Two sessions have been conducted on two different birds in free perching and viewing. From the first experiment session (about 24 minutes of recording) were extracted 20 peering motions with duration 1.51 ± 0.85 seconds (Owl #1). In the second session (about 18 minutes of recording), 48 segments were identified as peering motions, with duration 1.46 ± 0.92 seconds (Owl #1). Sample peering motions are represented in Fig. 1.

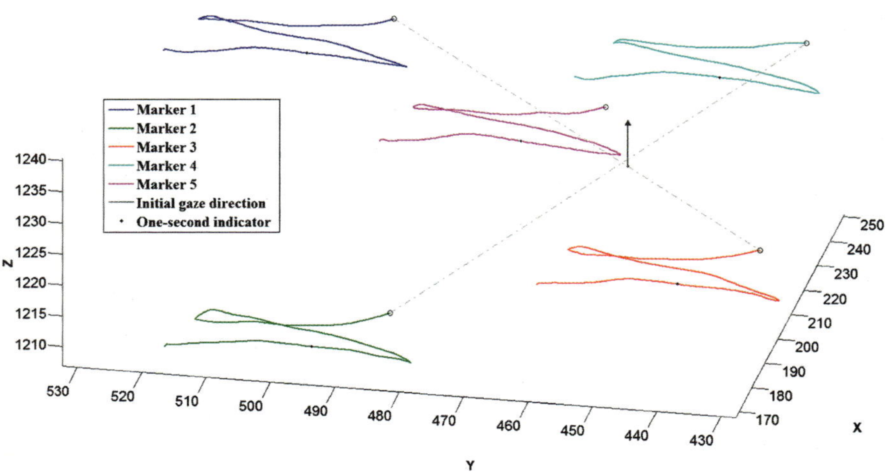

Fig. 1 Sample marker trajectories of a peering motion.

2.4 *Isolation of Non-Peering Motions*

For comparison with peering segments, we extracted motions different from peerings. However, fixations (prolonged pauses in the head motion) should not be considered, as the kinematic parameters are almost constant on these intervals. Fixations can be identified as segments on which the amplitudes of both the body's angular velocity and linear velocity are below a preset threshold. The thresholds were set as 10 deg/s for the angular velocity and 30 mm/s for the linear velocity, as suggested by [6].

The length of these reference segments was established as the approximate mean value of all the peering motions durations, 1.5 sec, for consistency of the comparison. Since the velocities may vary in a great measure on small intervals, we were reluctant to compare the average velocities to the respective thresholds. Instead, we considered the proportion of trajectory samples that were below the given threshold values (we refer to this parameter as the *fixation ratio*). We set the threshold as the average fixation ratio of the peering motions. The intervals for which less than 30% of the samples were tagged as fixation points were retained for comparison with the peering motions.

Based on these considerations, we developed an algorithm for the automatic selection of reference non-peering and non-fixation motions among the set of recorded kinematic data for each session. Intervals of 180 time samples (1.5 sec) are extracted from the recordings that are not marked as peering. From the calculated orientation and position of the solid body, the linear and angular velocity amplitudes are computed and compared to the velocity thresholds. If the segment contains too many fixation points, the operation is repeated on the recorded segment starting two seconds later. The algorithm stops when as many segments as the number of peering intervals for a given subject have been selected, unless the end of the data has previously been reached. Segments on which the owl directed its gaze towards the floor beneath it (typical sign of stress in the barn owl) were additionally detected from the orientation of the device and automatically rejected.

3 Results

From the estimation of the orientation and location of the barn owl's head, the screw axes and screw parameters were computed on the peering and non-peering segments. We computed on each segment the mean values and standard deviations of the screw parameters. These values were then averaged over the number of segments in the peering cases and reference motions. A comparison was performed by means of a Mann–Whitney U-test. This non-parametric statistical hypothesis test is adequate for assessing whether two independent samples of observations have equally large values. The results of this test are presented in Table 1, together with the mean values averaged over the number of trajectories. Table 2 shows the average standard deviations of the screw parameters for the translations and reference motions, along

Table 1 Mean screw parameters values averaged over the set of trajectories.

		Peering	Non-peering	p-value
Owl 1	Pitch	696.88 ± 386.16	782.41 ± 489.72	0.84
	Distance from SA	515.56 ± 403.12	656.95 ± 525.20	0.42
	Screw Intensity	0.49 ± 0.23	630.20 ± 2745.00	0.0385
Owl 2	Pitch	1159.24 ± 653.70	1010.93 ± 551.17	0.35
	Distance from SA	952.15 ± 974.30	880.73 ± 584.91	0.41
	Screw Intensity	0.44 ± 0.35	1.53 ± 3.43	0.0242

Table 2 Mean standard deviations of the screw parameters averaged over the set of trajectories.

		Peering	Non-peering	p-value
Owl 1	Pitch	11.12 ± 10.21	372.28 ± 318.60	$2.96E-07$
	Distance from SA	17.19 ± 21.10	496.73 ± 580.01	$1.80E-06$
	Screw Intensity	0.004 ± 0.004	8375.9 ± 36810	$6.80E-08$
Owl 2	Pitch	52.27 ± 169.42	322.58 ± 354.54	$3.84E-12$
	Distance from SA	32.06 ± 65.43	449.78 ± 504.37	$4.15E-11$
	Screw Intensity	0.003 ± 0.006	1.86 ± 0.35	$5.46E-16$

with the p-values obtained from Mann–Whitney test. The results are presented there as *mean* $\pm$ *standard deviation* over the number of recordings for each owl.

The most prominent observable fact in Tables 1, 2 is that the standard deviations of the pitch and distances from screw axes along the translational segments were typically much lower than the standard deviations on the reference motions. This suggests evidence for the constancy of those values during peerings.

From observation of Table 1, it appears that the screw intensity was smaller in average on head translations than on other motions for both subjects ($p = 0.039$ and 0.024, respectively). However, some of the reference motions exhibited high peaks in the screw magnitude values (see the high standard deviation of non-peering cases in Table 2). These peaks occurred mainly when the pitch reached 0 mm, i.e. in the case of pure rotations. After omission of the cases where the average screw intensity was higher than 10 mm^{-1}, the test results were less statistically significant but yet suggested lower screw intensity average values on translational motions ($p = 0.16$ and 0.05).

As expected, the standard deviations for the pitch, distance from the screw axes and screw intensity were always lower (with statistical significance) on translational movements than on the reference motions (Table 2). These results were hardly altered by the omission of the cases with high twist magnitudes.

4 Discussion

We have presented in this paper preliminary results of the kinematic analysis of side-to-side peering motions of the barn owl's head in free perching. The goal was to find a characterization of the movement in screw theory representation to develop in a further step a mobile robotic platform that would move its visual sensors similarly to the raptor.

The first finding was that, in a general manner, the six screw coordinates were approximately constant on peering motions: screw axes varied in a small measure in their direction and location (four parameters), and the pitch and screw intensity showed steadiness along those segments. This was not the case on the control group of reference motions, where sudden changes in the angular velocity of the barn owl's head produced notable variations in the values of the six screw parameters. The reference motions were selected as motions different from peerings and fixations, and therefore mostly contained head rotations. It was expectable to find variable angular velocities on these rotations, as the rotations performed by the barn owl are generally done as quick head saccades, shorter than the segment's overall duration (1.5 s). However, the pitch values did not appear to be lower on these motions. Only on a few cases did the pitch reach zero, as expected for pure rotational movements. This fact suggests that the head rotations of barn owls generally include a non-negligible translational element and support the findings of [5]. We believe this phenomenon finds its basis in the complex physiognomy of the barn owl's neck.

Another observable fact was the low values of the average screw intensity (correlated to the twist angle), on the peering motions. Conversely, head rotations reached high values of screw intensity. This parameter emphasizes the "screw-like" nature of a movement. In the case of peering, the dominant translational nature is thus understandably characterized by low values of screw intensities.

The previously described results should be validated on additional kinematic data. For this purpose, new experiments on barn owls will be conducted where the attention of the perching birds would be stimulated in a controlled way, such as static and dynamic salient objects in the bird's environment, auditory signals or with dead and alive preys. Analysis of the images obtained from the head-mounted camera in a controlled environment could then provide an understanding on the bird's visual attention and the followed scan path. Based on the insights provided by the head kinematic study and by the head-mounted camera image analysis, a model of visual search and attention should be implemented in a robotic agent. This agent would then be mounted on a mobile robotic platform with active gaze control.

We propose in this study a novel biokinematic approach to the challenge of active vision. This approach might additionally provide insights on the way to address the problem of visual self-location and mapping (visual SLAM) on autonomous robots. Our proposal is also a technically challenging inter-disciplinary research with a high potential to contribute to our understanding of animal and human vision.

References

1. Davidson, J.K., Hunt, K.H. (eds.): Robots and Screw Theory: Applications of Kinematics and Statics to Robotics. Oxford University Press, Oxford (2004)
2. Hayhoe, M., Ballard, D.: Eye movements in natural behavior. Trends Cogn. Sci. **9**, 188–194 (2005)
3. Katsman, I., Rivlin, E.: The mantis head camera (why the praying mantis is so good at catching its prey). In: IEEE International Conference on Image Processing, pp. 612–617 (2003)
4. du Lac, S., Knudsen, E.I.: Neural maps of head movement vector and speed in the optic tectum of the barn owl. J. Neurophysiol. **63**, 131–146 (1990)
5. Ohayon, S., Harmening, W., Wagner, H., Rivlin, E.: Through a barn owl's eyes: Interactions between scene content and visual attention. Biol. Cybern. **98**, 115–132 (2008)
6. Ohayon, S., van der Willigen, R.F., Wagner, H., Katsman, I., Rivlin, E.: On the barn owl's visual pre-attack behavior: I. Structure of head movements and motion patterns. J. Comp. Physiol. A Neuroethol. Sens. Neural. Behav. Physiol. **192**, 927–940 (2006)
7. Schafer, R.J., Moore, T.: Attention governs action in the primate frontal eye field. Neuron **56**(3), 541–551 (2007)
8. Steinbach, M.J., Money, K.E.: Eye movements of the owl. Vis. Res. **13**, 889–891 (1973)

Persistent Screw Systems of Dimension Four

Marco Carricato

Abstract The subalgebras of the Lie algebra $se(3)$ of the Euclidean group are at the basis of most families of mechanisms with special motion capabilities. Recently, it was shown that, by conveniently composing subalgebra generators, persistent screw systems (PSSs) may be obtained. PSSs are not subalgebras of $se(3)$, but they still exhibit remarkable invariant properties. For this reason, they may play an important role in both mobility analysis and mechanism design. This paper presents all generators of PSSs of dimension 4.

Key words: Screw theory, mechanism synthesis, mobility analysis

1 Introduction

Screw systems, i.e. the subspaces of the Lie algebra $se(3)$ of the Euclidean group, are fundamentals tools in mechanism theory. In 1978, Hunt [4] found a set of screw systems that guarantee 'full-cycle mobility' of mechanisms. These systems were proven to be the subalgebras of $se(3)$ and labeled *invariant* (ISSs) [3]. They exhibit remarkable properties and they are at the basis of most families of mechanisms with special motion capabilities (planar, spherical, translational, Schoenflies, etc.).

Recently, Carricato and Rico Martínez [1, 2] showed that, by conveniently composing generators of ISSs, *persistent screw systems* (PSSs) may be obtained. PSSs have noteworthy properties, even though they are not subalgebras of $se(3)$. In particular, a kinematic chain whose freedoms belong to the generator of a PSS in a reference configuration generates, in *any* other nonsingular pose, a screw subspace that is congruent to the reference one under a proper isometry. The concept of PSSs generalizes that of ISSs, with the latter emerging from the former when the mentioned isometry is the identity map. PSSs are believed to play an important role in both mobility analysis and the synthesis of mechanisms, particularly parallel robots.

Marco Carricato
University of Bologna, Bologna, Italy, e-mail: marco.carricato@unibo.it

J. Lenarčič, M. Husty (eds.), *Latest Advances in Robot Kinematics*,
DOI 10.1007/978-94-007-4620-6_19,

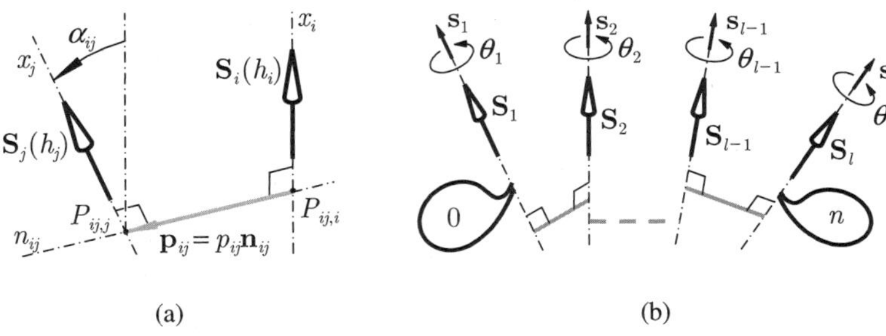

Fig. 1 Relative posture between two screws (a); a chain of adjacent screws (b).

This paper presents all generators of PSSs of dimension 4. Due to space limitation, a rigorous derivation and a proof that the given classification is exhaustive cannot be included. They will be provided in an extended version of this contribution.

In the following, n-dimensional subspaces of screws are briefly referred to as n-systems and they are denoted by capital italic letters, e.g. A. Hunt's classification of screw spaces is adopted [4], so that the locutions 'nG system' and 'nR system', with R being a Roman numeral, denote, respectively, a *general* n-system and a *special* n-system of the Rth type. ISSs are denoted by capital calligraphic letters, followed by their characteristic geometric quantities in parentheses, e.g. $\mathscr{Y}(\mathbf{u},h)$. A screw representing a relative twist between two bodies is designated by $\mathbf{S}$, followed, if necessary, by the pitch h in parentheses, e.g. $\mathbf{S}(h)$. The axis of the screw is denoted by x, with $\mathbf{s}$ being a unit vector along x. Line directions are always identified by unit vectors. The screws of a convenient basis of a system S are called *principal* and the quantities related to them are denoted by a right subscript p, so that $\mathbf{S}_{pi}$ is the ith principal screw, and h_{pi} and x_{pi} are, respectively, its pitch and its axis. The *principal reciprocal screws*, namely the principal screws of the reciprocal system of S, are denoted by a right subscript r. Given $\mathbf{S}_i$ and $\mathbf{S}_j$ (Fig. 1(a)), n_{ij} is the common perpendicular between their axes; $P_{ij,i}$ and $P_{ij,j}$ are the feet of n_{ij} on x_i and x_j; $\mathbf{n}_{ij}$ is a unit vector parallel to n_{ij} and directed from $P_{ij,i}$ to $P_{ij,j}$; p_{ij} and α_{ij} are the shortest distance and the relative angle between x_i and x_j, with α_{ij} being evaluated according to the right-hand rule about $\mathbf{n}_{ij}$ and being comprised in the interval $(-\pi/2,\pi/2]$; finally, $\mathbf{p}_{ij} = P_{ij,j} - P_{ij,i} = p_{ij}\mathbf{n}_{ij}$. Once the relative pose between the axes of two screws is assigned (by assigning the parameters p_{ij} and α_{ij}), a *connection* or a *link* is said to be laid between them and the screws are called *adjacent*.

2 Persistent Screw Systems

Let A be an n-dimensional ISS. A may be thought of as generated by l (not necessarily independent) screws $\mathbf{S}_i(h_i)$, with $i = 1\dots l$ and $l \geq n$. The screws are *composed* or *connected* if a link i is laid between $\mathbf{S}_i$ and $\mathbf{S}_{i+1}$. In this case, the pose between the axes of any pair of adjacent screws is fixed and $\mathbf{S}_i$ may be thought of as governing the *finite* displacement $\mathbf{D}_i(\theta_i)$ between the axes of $\mathbf{S}_{i-1}$ and $\mathbf{S}_{i+1}$ (Fig. 1(b)). $\mathbf{D}_i(\theta_i)$ depends on both the geometric parameters of $\mathbf{S}_i$ (i.e. the axis x_i and the pitch h_i) and the displacement variable θ_i. A represents all the admissible twists of link n, attached to the second kinematic element of $\mathbf{S}_l$, with respect to link 0, attached to the first kinematic element of $\mathbf{S}_1$. Because of the properties of ISSs, the described screw chain[1] is, in *any* nonsingular configuration, a generator of A. To lighten the notation, we will often *not* distinguish between an ISS and its generator.

Given two ISSs A^1 and A^2, respectively generated by chain $\{\mathbf{S}_1,\ldots,\mathbf{S}_{l_1}\}$ and chain $\{\mathbf{S}_{l_1+1},\ldots,\mathbf{S}_{l_1+l_2}\}$, they may be *composed* by laying a link between $\mathbf{S}_{l_1}$ and $\mathbf{S}_{l_1+1}$. The resulting generator, denoted by $A^1 \circ A^2$, generates a subspace that is the sum of A^1 and A^2, and whose elements represent the admissible twists of link l_1+l_2 with respect to link 0. Unless $A^1 \circ A^2$ is an ISS, it generates *distinct subspaces* in *distinct configurations*, depending on the instantaneous 'pose' of A^2 with respect to A^1. The concept of *composition* may be extended in an obvious way to a sequence of ISSs. Unless $A^1 \circ \ldots \circ A^h$ is an ISS, if S is the k-system spawned in the reference configuration, after a generic displacement of the chain, a different k'-system S' is obtained. According to [1], $A^1 \circ \ldots \circ A^h$ generates a *persistent screw system* (PSS) if, for arbitrary finite movements of the generator (and out of singular configurations), S' is *congruent* to S under a *proper isometry*. In other terms, S is persistent if, for an arbitrary set of displacements $\{\mathbf{D}_1(\theta_1),\ldots,\mathbf{D}_{l_1+\ldots+l_h}(\theta_{l_1+\ldots+l_h})\}$, some Euclidean displacement $\mathbf{D}$ exists such that $S' = \mathbf{D}S$. Since the subspaces generated by the generator of a PSS in *all* possible nonsingular configurations are mutually congruent, a PSS is not only *type-invariant* (in the sense that it preserves the 'qualitative' geometric pattern that gives reason for its classification as an nG or nR system), but it also preserves the 'quantitative' features defining its 'shape', such as the space dimension and the relative posture and pitches of its constituent screws. It is worth observing that the generator of a PSS does *not* depend on the architecture of the ISS generators and on the values of variables θ_i, but it only depends on the 'shape' of the connections between the constituent ISSs (i.e. links l_1, l_1+l_2, $\ldots$, $l_1+\ldots+l_{h-1}$).

Any sequence $A^1 \circ \ldots \circ A^h$ of ISSs that generates a PSS, and such that no two adjacent A^j, $j = 1\ldots h$, may be merged together to originate an ISS, is referred to as a *form* of the PSS. A form is said to be *unary* if $h = 1$, *binary* if $h = 2$, and so on. A form $A^1 \circ \ldots \circ A^h$ is said to be *conjoint* or *disjoint* depending on whether, respectively, $A^s \cap A^t \neq \mathbf{0}$ or $A^s \cap A^t = \mathbf{0}$, for *all* pairs of adjacent A^s and A^t therein; it is *partially joint* otherwise.

A fundamental theorem guarantees that the composition between two ISSs A and B, namely $C = A \circ B$, always yields a PSS [1]. Two handy methods sufficient to obtain further forms of C are the following. If $I = A \cap B$ and $\dim I = h \neq 0$, h adjacent screws in both A and B may be suitably chosen in order to form generators of I (since A and B are ISSs, I is also an ISS). Accordingly, if $\dim A = n$ and $\dim B = m$, $n-h$ adjacent screws may be chosen in A to form a generator of a subspace A^* such

[1] Though A is generated here by a chain of serially-connected screws, this is not a necessary requisite.

that $A = A^* \oplus I$, and $m - h$ screws may be chosen in B to form a generator of a subspace B^* such that $B = I \oplus B^*$.[2] $A^* \circ B$ and $A \circ B^*$ are non-binary forms of C. A dual procedure is applicable when an ISS I exists that, though it is included in neither A nor B, is comprised in $A \circ B$. In this case, h^A adjacent screws in A and h^B in B may be suitably chosen so that $h^A + h^B = h$ and so as to form a generator of I. In this case, $A \circ I \circ B$ is a non-binary form of C. The described methods (denoted as 'ISS subtraction' and 'ISS addition', respectively) are sufficient, but not necessary, to synthesize non-binary forms of PSSs.

3 The General and the First Special Four-System

According to Hunt [4], a 4-system is *general* if the principal screws $\mathbf{S}_{r1}(h_{r1})$ and $\mathbf{S}_{r2}(h_{r2})$ of the reciprocal cylindroid have distinct finite pitches; conversely, it is *special* if either $h_{r1} = h_{r2}$ (4I-system) or one or both of the principal reciprocal pitches go to infinity (4II through 4V systems).

In the 4G and 4I systems, the principal screws $\mathbf{S}_{p1}(h_{p1})$ and $\mathbf{S}_{p2}(h_{p2})$ ($h_{pi} = -h_{ri}$, $i = 1,2$) are collinear with $\mathbf{S}_{r1}$ and $\mathbf{S}_{r2}$, respectively, whereas $\mathbf{S}_{p3}$ and $\mathbf{S}_{p4}$ form a cylindrical ISS $\mathscr{C}(x_{p3})$ aligned with the nodal line of the reciprocal cylindroid.

In the 4G-system, in particular, $h_{p1} \neq h_{p2}$ and $\mathscr{C}(x_{p3})$ is the only existing ISS [4]. Since $\mathscr{C}(x_{p3})$ has dimension 2, the generation of a 4G-PSS requires the composition of no less than three ISSs (e.g. $\mathscr{C}(x_{p3})$ and two independent screws). Accordingly, there are no binary forms of 4G-PSSs. *Ternary* and *quaternary forms* do exist, but their derivation is too long to be presented here. A brief description may be found in Table 2 of [1].

In the 4I-system, $h_{p1} = h_{p2}$ and the reciprocal cylindroid degenerates into a planar pencil of equal-pitch screws. The entire 4I-system is axis-symmetric about x_{p3}, with the screws of pitch h_{p1} forming both a planar field on the $x_{p1}x_{p2}$-plane and a bundle originating in the center O of the reciprocal cylindroid. If $h_{p1} = h_{p2} = 0$, the bundle is a spherical ISS $\mathscr{S}(O)$. Since $\mathscr{S}(O)$ has dimension 3, the fundamental Theorem presented in Section 2 may be used to generate binary forms of 4I-PSSs. In particular, if $\mathscr{S}(O)$ is composed with a 1-dimensional ISS $\mathscr{H}(x_1, h_1)$ (i.e. a single screw of axis x_1 and pitch h_1), x_{p3} lies at right-angle with the perpendicular line from O to x_1 (Fig. 2(a)) and it forms with x_1 an angle $\alpha_{p3,1}$ such that

$$\tan \alpha_{p3,1} = -p_{p3,1}/h_1, \qquad (1)$$

with $p_{p3,1}$ being the distance between x_{p3} and x_1. If $h_1 = 0$ (namely, $\mathscr{H}(x_1, h_1)$ is a revolute ISS $\mathscr{R}(x_1)$) and x_1 does not pass through O, then $\alpha_{p3,1} = \pi/2$ and $p_{p3,1}$ is arbitrary (as long as different from zero). If x_1 passes through O, i.e. $p_{p3,1} = 0$, and $h_1 \neq 0$, then $\alpha_{p3,1} = 0$ and x_{p3} coincides with x_1. $\alpha_{p3,1}$ goes to zero even when

[2] A^* and B^* are, respectively, tangent spaces at the identity to quotient manifolds of the subgroups associated with A_n and B_m by the subgroup associated with I_h [5].

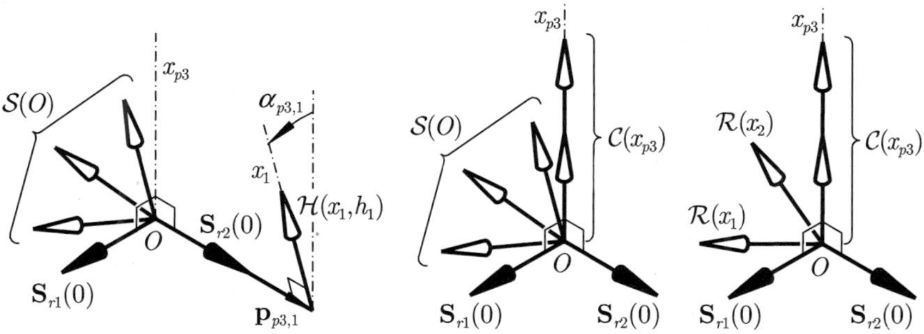

(a) (b) (c)

Fig. 2 Persistent 4*I*-systems: (a) disjoint binary form, (b) conjoint binary form and (c) disjoint ternary form.

$h_1 \to \infty$ (namely, when $\mathscr{H}(x_1, h_1)$ is a linear-translation ISS $\mathscr{P}(\mathbf{s}_1)$), in which case x_{p3} becomes the line through O parallel to $\mathbf{s}_1$.

Another persistent form may be obtained by composing $\mathscr{S}(O)$ with the ISS $\mathscr{C}(x_{p3})$, whose axis passes through O (Fig. 2(b)). Since $\mathscr{S}(O)$ and $\mathscr{C}(x_{p3})$ intersect in a 0-pitch screw aligned with x_{p3}, 'subtracting' this screw from $\mathscr{S}(O)$ yields the ternary form in Fig. 2(c).

4 The Second, the Fourth and the Fifth Special Four-Systems

The 4*II*, 4*IV* and 4*V* systems emerge when the reciprocal cylindroid becomes, respectively, a 2*II*, a 2*IV* and a 2*V* system [4]. Figure 3 shows the main geometric features of these subspaces. In the reciprocal 2*II* and 2*IV* systems, all finite-pitch screws are parallel to a single direction $\mathbf{u}_{n\infty}$ and lie on a characteristic plane Λ (with the axis x_{rh} of the screw of pitch $-h$ being located at a distance $|-h\tan\zeta_r|$ from the axis of the screw of 0-pitch); the ∞-pitch screws are, instead, parallel to a direction $\mathbf{u}_n$ forming a prescribed nonzero angle ζ_r with $\mathbf{u}_{n\infty}$ $(0 < \zeta_r \leq \pi/2)$. In particular, in the 2*II* system, all finite-pitch screws have the same pitch h_{r2} (with $h_{r2} = -h_{p2}$) and $\zeta_r = \pi/2$. In the reciprocal 2*V* system, all screws lie on the same line x_r, thus forming a cylindrical ISS (hence, $\mathbf{u}_{n\infty} = \mathbf{u}_n$ and $\zeta_r = 0$). The 4*II*, 4*IV* and 4*V* systems comprise ∞-pitch screws along all directions perpendicular to $\mathbf{u}_{n\infty}$, whereas all finite-pitch screws lie in planes perpendicular to $\mathbf{u}_n$. The 4*II* and the 4*IV* system contains cylindrical ISSs $\mathscr{C}(x_c)$ on all lines of Λ parallel to $\mathbf{u}_c = \mathbf{u}_{n\infty} \times \mathbf{u}_n / \sin\zeta_r$. The 4*V* system comprises cylindrical ISSs on all lines intersecting x_r at right-angle and all finite-pitch screws belong to this bundle. In all these special systems, the ∞-pitch screws form a planar-translation ISS $\mathscr{T}_2(\mathbf{u}_{n\infty})$; moreover, in the 4*II* system, the screws of pitch h_{p2} parallel to $\mathbf{u}_{n\infty}$ form, together with $\mathscr{T}_2(\mathbf{u}_{n\infty})$, a helicoid-planar ISS $\mathscr{Y}(\mathbf{u}_{n\infty}, h_{p2})$.

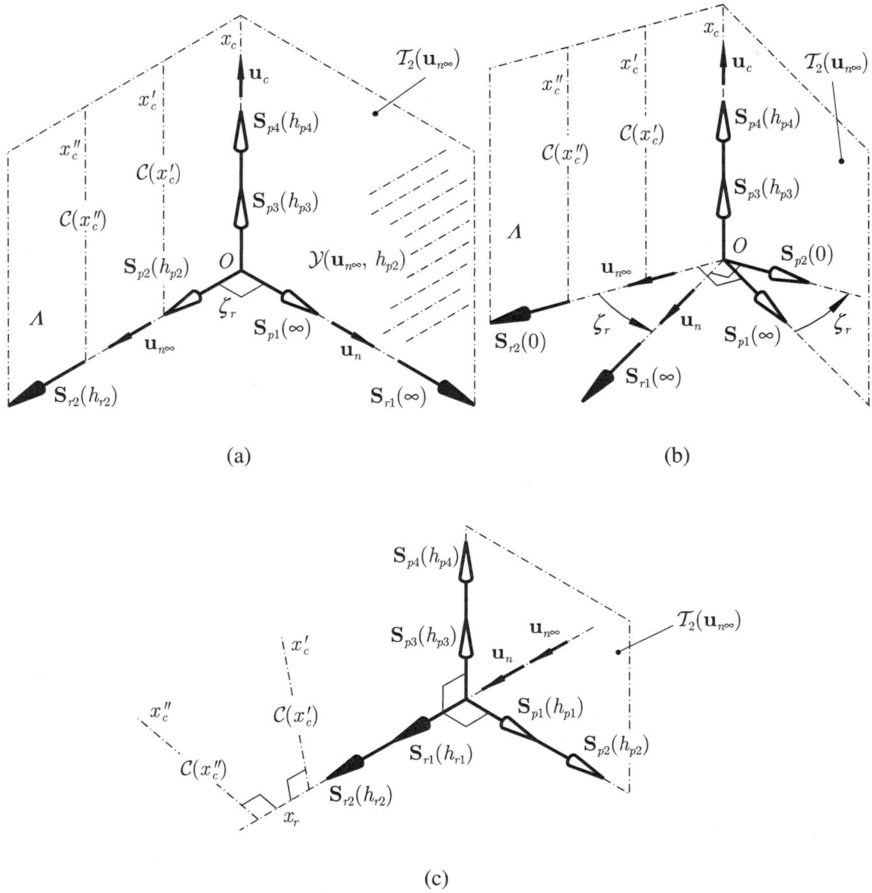

Fig. 3 The 4*II* (a), 4*IV* (b) and 4*V* (c) systems.

4.1 The Second and the Fifth Special Four-Systems

Binary forms of 4*II* and 4*V*-PSSs may be synthesized by way of the fundamental Theorem in Section 2, by combining any two ISSs among those described at the beginning of this Section, provided that their composition generates a 4-dimensional space (which discards, indeed, the composition of a planar-translation ISS with a cylindrical ISS parallel to it as well as the composition of two mutually parallel cylindrical ISSs). In particular, a 4*II*-PSS emerges by composing $\mathscr{Y}(\mathbf{u}_{n\infty}, h_{p2})$ with either a 1-dimensional ISS $\mathscr{H}(x_1, h_1)$ not parallel to $\mathbf{u}_{n\infty}$ (Fig. 4(a)) or a cylindrical ISS $\mathscr{C}(x_c)$ perpendicular to it (Fig. 4(b)). In the former case, the position of Λ with respect to the plane parallel to $\mathbf{u}_{n\infty}$ passing through x_1 is determined by the distance

$$p_{1,p2} = (h_1 - h_{p2}) \cot \alpha_{1,p2}, \tag{2}$$

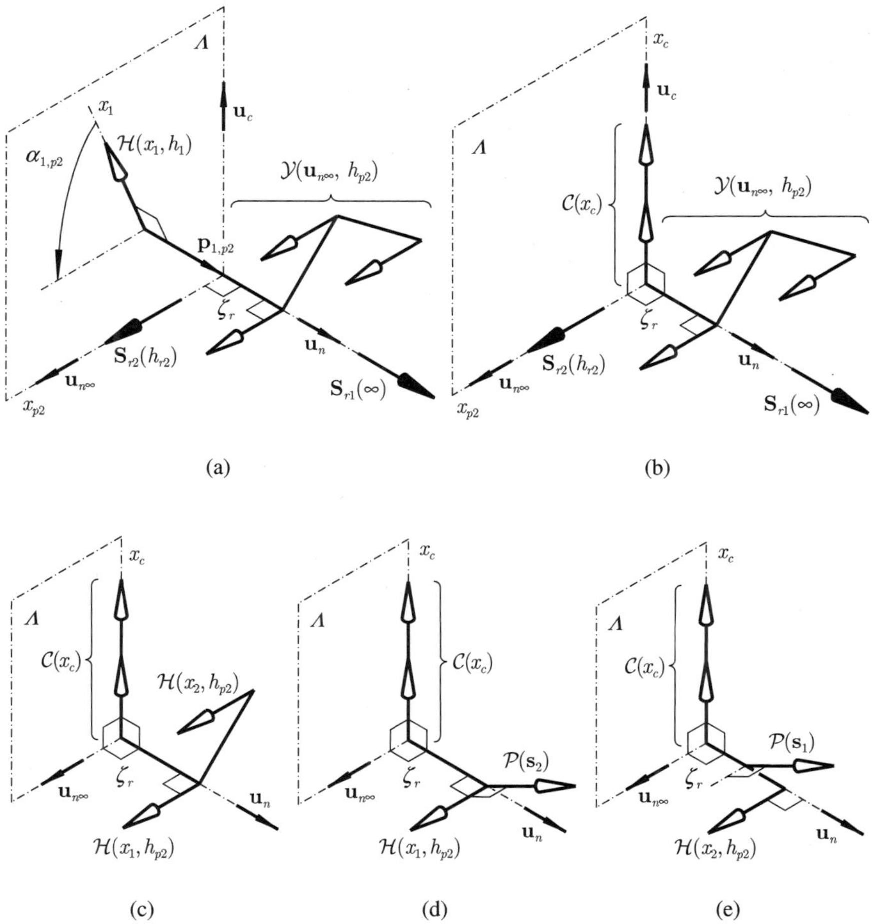

Fig. 4 Persistent $4II$-systems: (a) disjoint binary form, (b) conjoint binary form, (c–e) disjoint ternary forms.

with $\alpha_{1,p2} \neq 0$, and by the sign of $\alpha_{1,p2}$ (in Fig. 4(a), $h_1 < h_{p2}$ and $\alpha_{1,p2} > 0$). When $\mathscr{Y}(\mathbf{u}_{n\infty}, h_{p2})$ is combined with $\mathscr{C}(x_c)$ (Fig. 4(b)), Λ is the plane through x_c parallel to $\mathbf{u}_{n\infty}$ ($\alpha_{c,p2} = \pi/2$ infers $p_{c,p2} = 0$). By composing two nonparallel cylindrical ISSs $\mathscr{C}(x_c)$ and $\mathscr{C}(x_c')$ (Fig. 5(a)), a $4V$-PSS is instead obtained. In this case, x_r is the common perpendicular between x_c and x_c'.

By following the procedures recalled in Section 2, ternary forms of persistent $4II$ and $4V$ systems may be obtained by 'subtracting' or 'adding' convenient ISSs to the binary forms described above. As far as the $4II$ system is concerned, since $\mathscr{Y}(\mathbf{u}_{n\infty}, h_{p2}) \cap \mathscr{C}(x_c) = \mathscr{P}(\mathbf{s})$, with $\mathbf{s}$ parallel to x_c, ternary forms of $4II$-PSS may be obtained by subtracting an ∞-pitch screw from the kinematic generator of $\mathscr{Y}(\mathbf{u}_{n\infty}, h_{p2})$, which is thus reduced to the composition of either two parallel h_{p2}-

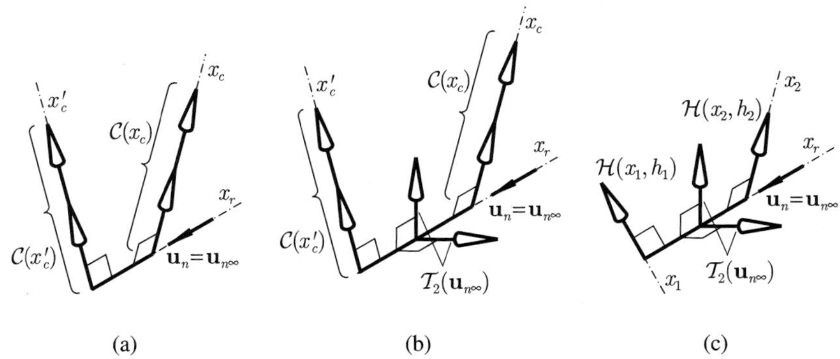

Fig. 5 Persistent 4*V*-systems: (a) disjoint binary form, (b) conjoint ternary form, (c) disjoint ternary form.

pitch screws (Fig. 4(c)) or one h_{p2}-pitch screw plus an ∞-pitch screw perpendicular to it (Figs. 4(d)-4(e)). As for the 4*V* system, since the ∞-pitch screws contained in $\mathscr{C}(x_c)$ and $\mathscr{C}(x_c')$ span a planar-translation ISS perpendicular to x_r, ternary forms of 4*V*-PSS may be obtained as $\mathscr{C}(x_c)\circ\mathscr{T}_2(\mathbf{u}_{n\infty})\circ\mathscr{C}(x_c')$ (Fig. 5(b)), $\mathscr{C}(x_c)\circ\mathscr{T}_2(\mathbf{u}_{n\infty})\circ\mathscr{H}(x_1,h_1)$, $\mathscr{C}(x_c)\circ\mathscr{P}(\mathbf{s})\circ\mathscr{C}(x_c')$, $\mathscr{H}(x_1,h_1)\circ\mathscr{T}_2(\mathbf{u}_{n\infty})\circ\mathscr{H}(x_2,h_2)$ (Fig. 5(c)), or $\mathscr{C}(x_c)\circ\mathscr{P}(\mathbf{s})\circ\mathscr{H}(x_1,h_1)$, with x_c, x_c', x_1, x_2 and $\mathbf{s}$ being all perpendicular to $\mathbf{u}_{n\infty}$.

4.2 The Fourth Special Four-System

It may be proven that there are no binary forms of 4*IV*-PSSs. However, the form $\mathscr{H}(x_1,h_1)\circ\mathscr{T}_2(\mathbf{u}_{n\infty})\circ\mathscr{H}(x_2,h_2)$ presented in Section 4.1 (Fig. 5(c)) provides a clue for the design of ternary forms. It may be readily seen, in fact, that when a planar-translation ISS $\mathscr{T}_2(\mathbf{u}_{n\infty})$ is interposed between two generally-oriented finite-pitch screws, say $\mathscr{H}(x_1,h_1)$ and $\mathscr{H}(x_2,h_2)$, the relative orientation between all screws of the kinematic generator is constant for arbitrary chain displacements. Consequently, the angle ζ_r between $\mathbf{u}_{n\infty}$ and $\mathbf{u}_n$, with $\mathbf{u}_n=\mathbf{s}_1\times\mathbf{s}_2/\sin\alpha_{12}$, is invariant and, provided that $\mathscr{H}(x_1,h_1)$ and $\mathscr{H}(x_2,h_2)$ are not parallel (i.e. $\alpha_{12}\neq 0$), the chain generates a persistent 4-system. If $\zeta_r\in(0,\pi/2)$, this is of the 4th special type (Fig. 6(a)). The geometric features of the system are determined as follows. $\mathscr{H}(x_1,h_1)$ and $\mathscr{H}(x_2,h_2)$ generate, in a generic configuration, a cylindroid *A*, whose nodal line coincides with n_{12}. If $\mathbf{S}_c$ is the screw of *A* parallel to $\mathbf{u}_c$ (namely, to the direction perpendicular to both $\mathbf{u}_{n\infty}$ and $\mathbf{u}_n$), P_c is the foot of x_c on n_{12}, and $p_{ci}'=(P_{12,i}-P_c)\cdot\mathbf{u}_n$ ($i=1,2$) (the prime is used to emphasize that p_{ci}' is not necessarily positive, so that it may not coincide with the distance defined in Section 1), it is not difficult to prove that

$$p_{ci}'=\frac{\sin\alpha_{ci}}{\sin\alpha_{12}}\left[(h_2-h_1)\sin\alpha_{cj}+p_{12}'\cos\alpha_{cj}\right], \tag{3}$$

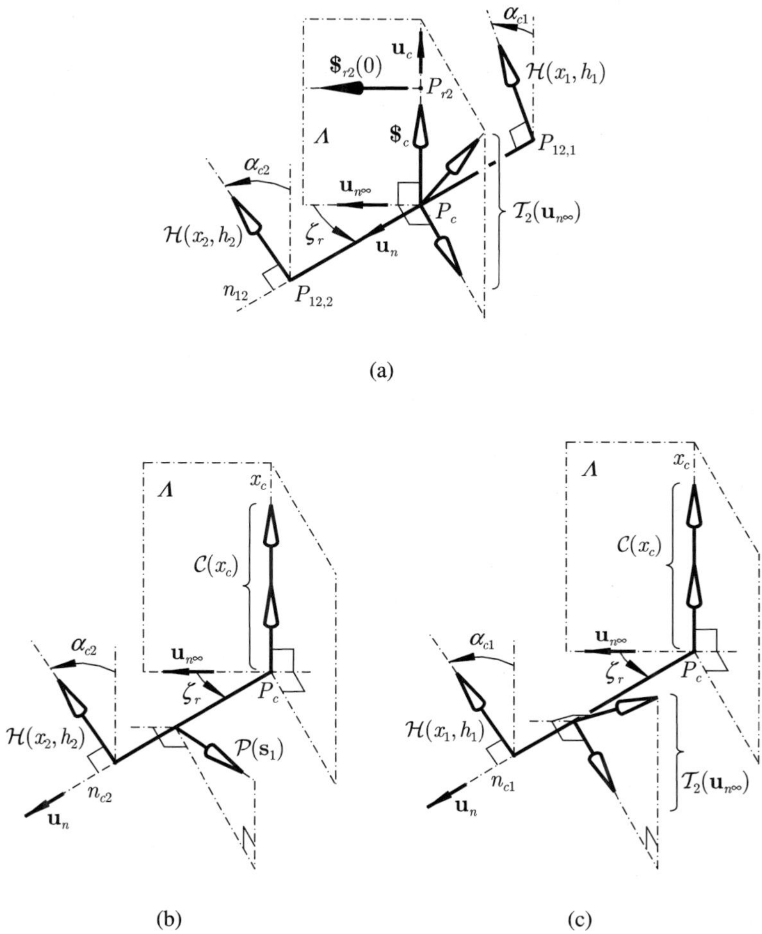

Fig. 6 Persistent 4*IV*-systems: (a) disjoint ternary form, (b) disjoint ternary form, (c) partially joint ternary form.

where $(i,j) \in \{(1,2),(2,1)\}$, α_{ci} is evaluated according to the right-hand rule about $\mathbf{u}_n$, $\alpha_{12} = \alpha_{c2} - \alpha_{c1}$ and $p'_{12} = p'_{c2} - p'_{c1}$. Since the linear combination of $\mathbf{S}_c$ with the screw of $\mathscr{T}_2(\mathbf{u}_{n\infty})$ parallel to $\mathbf{u}_c$ generates a cylindrical ISS, the identification of x_c allows one to determine the characteristic plane Λ of the 4*IV* system, which must pass through x_c and be parallel to $\mathbf{u}_{n\infty}$. The principal reciprocal screw $\mathbf{S}_{r2}(0)$ may be conveniently located on Λ by imposing the reciprocity condition between it and either $\mathscr{H}(x_1,h_1)$ or $\mathscr{H}(x_2,h_2)$. By denoting the foot of x_{r2} on x_c with P_{r2} and by letting $p'_{n12,r2} = (P_{r2} - P_c) \cdot \mathbf{u}_c$, one obtains that

$$p'_{n12,r2} = \frac{\tan \zeta_r}{\sin \alpha_{12}} \left(h_2 \cos \alpha_{c1} \sin \alpha_{c2} - h_1 \sin \alpha_{c1} \cos \alpha_{c2} + p'_{12} \cos \alpha_{c1} \cos \alpha_{c2} \right). \quad (4)$$

It is worth observing that all scalars at the right-hand side of Eqs. (3)-(4) are constant, with the exception of p'_{12}, which depends on the variables θ_2 and θ_3. Accordingly, the pose of x_c and x_{r2} depends on θ_2 and θ_3 and the reciprocal cylindroid is fixed to no link of the generator.

If either $\mathscr{H}(x_1, h_1)$ or $\mathscr{H}(x_2, h_2)$ is chosen perpendicular to $\mathbf{u}_{n\infty}$, it generates, together with the screw of $\mathscr{T}_2(\mathbf{u}_{n\infty})$ parallel to it, a cylindrical ISS and it may be replaced by a corresponding generator $\mathscr{C}(x_c)$, thus originating two further forms of $4IV$-PSS (Fig. 6(b)-6(c)).

5 The Third Special Four-System

The $4III$ system is the Schoenflies ISS $\mathscr{X}(\mathbf{u})$, which comprises screws of all pitches on all lines parallel to $\mathbf{u}$ and ∞-pitch screws along all directions in space.

6 Conclusions

In [1], Carricato and Rico Martínez presented the idea of *persistent screw systems* (PSSs), a novel concept generalizing the concept of *invariant screw systems*. The latter are the subalgebras of the Lie algebra $se(3)$ of the Euclidean group. PSSs are not subalgebras of $se(3)$, but they still exhibit remarkable invariant properties for full-cycle motions. PSSs may play an important role in both mobility analysis and robot synthesis [1, 2]. In [1], all screw systems of dimension 2 were proven to be persistent. All generators of PSSs of dimension 3 were revealed and classified in [2]. This paper presented the generators of PSSs of dimension 4. Due to space limitation, a formal derivation and the proof that the given classification is exhaustive could not be included. They will be provided in a future extended version of the contribution.

References

1. Carricato, M., Rico Martínez, J.M.: Persistent screw systems. In: Lenarčič J., Stanišič, M.M. (eds.) Advances in Robot Kinematics: Motion in Man and Machine, pp. 185–194. Springer, Dordrecht (2010)
2. Carricato, M., Rico Martínez, J.M.: Persistent screw systems of dimension three. In: Proc. of 13th World Congress in Mechanism and Machine Science, pp. 1–12, Paper No. A7_430. Guanajuato, Mexico (2011)
3. Gibson, C.G., Hunt, K.H.: Geometry of screw systems—2, classification of screw systems. Mech. Mach. Theory **25**(1), 11–27 (1990)
4. Hunt, K.H.: Kinematic Geometry of Mechanisms. Clarendon, Oxford (1978)
5. Meng, J., Liu, G., Li, Z.: A geometric theory for analysis and synthesis of sub-6 dof parallel manipulators. IEEE Trans. Robot. **23**(4), 625–649 (2007)

Simplified Voronoi Diagrams for Motion Planning of Quadratically-Solvable Gough–Stewart Platforms

Rubén Vaca, Joan Aranda and Federico Thomas

Abstract The obstacles in Configuration Space of quadratically-solvable Gough–Stewart platforms, due to both kinematic singularities and collisions, can be uniformly represented by a Boolean combination of signs of 4×4 determinants involving the homogeneous coordinates of sets of four points. This Boolean combination induces a measure of distance to obstacles in Configuration Space from which a simplified Voronoi diagram can be derived. Contrary to what happens with standard Voronoi diagrams, this diagram is no longer a strong deformation retract of free space but, as Canny proved in 1987, it is still complete for motion planning. Its main advantage is that it has lower algebraic complexity than standard Voronoi diagrams based on the Euclidean metric.

Key words: Gough–Stewart platform, pure condition, Voronoi diagrams, path planning

1 Introduction

Gough–Stewart platforms whose assembly modes can be obtained by solving only quadratic equations are said to be *quadratically-solvable*. This family of parallel platforms is defined by certain geometric constraints in the location of their leg attachments to the fixed base and/or moving platform such as coincidence and collinearity (see Fig. 1), or, in general, certain algebraic relations between the coordinates of the attachments in their local reference frames (see [1] for a non-trivial example).

Rubén Vaca · Joan Aranda
Department of Automatic Control and Computing Engineering (UPC), Edifici Vertex, 08034 Barcelona, Spain, e-mail: {ruben.hernando.vaca, joan.aranda}@upc.edu

Federico Thomas
Institut de Robòtica i Informàtica Industrial (CSIC-UPC), Llorens Artigas 4-6, 08028 Barcelona, Spain, e-mail: fthomas@iri.upc.edu

J. Lenarčič, M. Husty (eds.), *Latest Advances in Robot Kinematics*,
DOI 10.1007/978-94-007-4620-6_20,

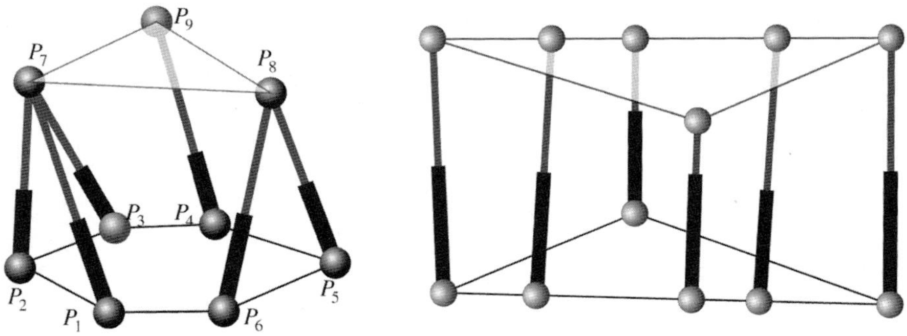

Fig. 1 Examples of two well-known quadratically-solvable Gough–Stewart platforms. Left: the 3-2-1 platform. Right: the Zhang-Song platform.

It is interesting to observe how, when all the elements of a quadratically-solvable Gough–Stewart platform and its environment are approximated by polyhedra, the obstacles in its configuration space, due to both kinematic singularities and collisions, can be uniformly represented by a Boolean combination of signs of 4×4 determinants involving the homogeneous coordinates of sets of four points. This Boolean combination induces a measure of distance to obstacles from which a simplified Voronoi diagram can be derived. A detailed explanation of all these facts requires a lot of mathematical details but, due to space limitations, we have opted for an informal style to convey the main concepts.

This paper is organized as follows. Section 2 explains how the kinematic singularities of quadratically-solvable Gough–Stewart platforms can always be interpreted as degeneracies of a set of tetrahedra. Section 3 deals with the collision detection between arbitrary polyhedra. Then, it is shown how the regions where either collisions or singularities occur can be uniformly represented by a Boolean combination of signs of 4×4 determinants. How this representation induces a measure of distance to these forbidden regions is explained in Section 4 and how a simplified Voronoi diagram is derived, in Section 5. A simple planar case is analyzed in Section 6. Finally, we conclude in Section 7.

2 Singularities of Quadratically-Solvable Parallel Robots

The singularities of quadratically-solvable Gough–Stewart obey a neat algebraic structure. The pure condition of this kind of platforms factorizes into 4×4 determinants which involve the homogeneous coordinates of sets of four attachments [3]. When the coordinates of the leg attachments in the moving platform are expressed in terms of the position and orientation of the platform, $\mathbf{x}$ and Θ respectively, three kinds of determinants arise:

- Plane-Point determinants: They involve three attachments of the fixed base and one of the moving platform. They vanish if, and only if, the point lies on the plane defined by the three points fixed to the base.
- Point-Plane determinants: They involve three attachments of the moving platform and one of the fixed base. The vanish if, and only if, the fixed point lies on the moving plane defined by the three points attached to the platform.
- Line-Line determinants: They involve two attachments of the base and two of the moving platform. They vanish if, and only if, the lines defined by the two couples of points lie on the same plane.

Hence, the singularities of quadratically-solvable platforms can be geometrically interpreted as the degeneracy of tetrahedra. For example, the pure condition of the robot in Fig. 1(left) can be expressed as:

$$[\mathbf{p}_1\mathbf{p}_2\mathbf{p}_3\mathbf{p}_7][\mathbf{p}_5\mathbf{p}_6\mathbf{p}_7\mathbf{p}_8][\mathbf{p}_4\mathbf{p}_7\mathbf{p}_8\mathbf{p}_9] = 0$$

where $\mathbf{p}_i$ stand for the homogeneous coordinates of leg attachment P_i in the world reference frame. Then, this robot is in a singularity if any of the three determinants vanish (i.e., if any of the three associated tetrahedra degenerate). Then, if we define the predicate

$$\triangle(i,j,k,l) = \begin{cases} \text{true if } [\mathbf{p}_i\mathbf{p}_j\mathbf{p}_k\mathbf{p}_l] > 0, \\ \text{false otherwise,} \end{cases} \tag{1}$$

the robot's singularity-free region is the set of configurations where

$$(\triangle(1,2,3,7) \vee \overline{\triangle(1,2,3,7)}) \wedge (\triangle(5,6,7,8) \vee \overline{\triangle(5,6,7,8)}) \wedge (\triangle(4,7,8,9) \vee \overline{\triangle(4,7,8,9)}) \tag{2}$$

is true. This boolean expression can be seen as a shallow (depth 2) AND-OR tree with an AND node at the root. The interest of expressing singularity-free regions in this apparently awkward way becomes obvious when integrating them with collision-free regions as described next.

3 Basic Contacts Between Polyhedra

The configuration space obstacle for a pair of polyhedra is the set of configurations at which the two polyhedra overlap. The overlap predicates between polyhedra are logical combinations of primitive predicates. These primitive predicates correspond to the three basic types of contacts between polyhedra [7], which in turn can be expressed using the same three kinds of determinants already used to characterize the singularities of quadratically-solvable platforms.

The formulation of polyhedral interference detection as the evaluation of a boolean formula that depends only on the features (vertices, edges, and faces) of polyhedra with convex faces was proposed by Canny [5]. This test was later generalized, and fully expressed in terms of signs of 4×4 determinants, by Thomas and Torras [9].

As a simple example of boolean formulas for overlap detection, let us consider the segment defined by P_1P_2 and the triangle defined by $P_3P_4P_5$ in 3D. Then, the segment is intersecting the triangle if

$$(\triangle(1,3,4,5) \otimes \triangle(2,3,4,5)) \wedge [(\triangle(1,2,3,4) \wedge \triangle(1,2,4,5) \wedge \triangle(1,2,5,3)) \\ \vee (\overline{\triangle(1,2,3,4)} \wedge \overline{\triangle(1,2,4,5)} \wedge \overline{\triangle(1,2,5,3)})]$$

is true, $\otimes$ being the exclusive or operator defined as $(a \otimes b) = (a \wedge \overline{b}) \vee (\overline{a} \wedge b)$.

Then, the predicates that define collision-free regions can also be represented as AND-OR trees that can be fused in a seamless way with those representing singularity-free regions through an AND node. The resulting trees can be obviously organized in different ways for higher efficiency using the properties of boolean algebra. One of the overlap predicates for polyhedra with convex faces given in [5] generate a shallow (depth 2) AND-OR predicate tree, whose root is an OR node. Nevertheless, as Canny already observed, it is computationally advantageous to make the predicate tree as deep as possible and it is also desirable for the root to be an AND node.

4 A Measure of Distance to Obstacles in Configuration Space

Following the discussion in [4], we can now observe that by letting positive real values represent logical one, and non-positive values represent logical zero, that the *min* function implements logical AND, and the *max* function implements logical OR. Thus, for example, the predicate (2) that characterizes the obstacle in configuration space due to kinematic singularities of the robot in Fig. 1 can be expressed as:

$$\begin{aligned}\rho(\mathbf{x},\Theta) = \min(&\max([\mathbf{p}_1\mathbf{p}_2\mathbf{p}_3\mathbf{p}_7], -[\mathbf{p}_1\mathbf{p}_2\mathbf{p}_3\mathbf{p}_7]),\\ &\max([\mathbf{p}_5\mathbf{p}_6\mathbf{p}_7\mathbf{p}_8], -[\mathbf{p}_5\mathbf{p}_6\mathbf{p}_7\mathbf{p}_8]),\\ &\max([\mathbf{p}_4\mathbf{p}_7\mathbf{p}_8\mathbf{p}_9], -[\mathbf{p}_4\mathbf{p}_7\mathbf{p}_8\mathbf{p}_9])).\end{aligned}$$

The above quantity can be used as a measure of distance – it is not a true metric – to the robot's kinematic singularities, because it varies continuously through configuration space. Thus, the translation from a predicate representing the singularity and collision-free regions to a measure of distance to forbidden regions is straightforward. It will be positive at configurations outside them and negative at configurations inside them. Clearly, this measure of distance decomposes the robot's configuration space into regions in which one of the involved determinants is critical in determining the value of ρ, that is, small changes in its value cause identical changes in the value of ρ. The boundaries between these regions can be seen as a Voronoi diagram. Then, searching the configuration space for singularity and collision-free paths can be reduced to a search on this diagram as explained in the next section.

5 Simplified Voronoi Diagrams

A Voronoi diagram is defined to be the set of points equidistant from two or more generators (points, segments, polygons, ...) under the appropriate metric (usually the Euclidean distance). This construction received considerable attention in the early eighties as a useful tool for motion planning (see the textbook of Schwartz and Yap [8], and the references therein, for an introduction and review of the use of Voronoi diagrams in motion planning).

The main advantage of using the Voronoi diagrams based on the measure of distance described in the previous section is that they have a lower algebraic complexity than those resulting from using the Euclidean distance. These diagrams are piecewise linear for fixed orientations of the moving platform while standard Voronoi diagrams would contain quadratic sheets. These simplified Voronoi diagrams are sometimes also called *straight skeletons* [6]. Despite this important simplification, they still have an important property: any path through free space which starts and ends on the diagram can be continuously deformed so that it lies entirely on the diagram [4]. Thus, they are complete for motion planning, i.e., searching the original space for paths can be reduced to a search on the diagram.

Now, to find a path between two points in free space, it suffices to find a path for each point onto the diagram, and to join these points with a path that lies wholly on the diagram.

6 Example

The ideas presented above are for spatial parallel robots. Nevertheless, to exemplify them, we will rely on a planar case because the corresponding configuration space is three-dimensional and hence easily representable.

Let us consider a planar robot whose base is determined by the segment P_1P_2, its moving platform by the segment P_3P_4, and its three legs by the segments P_1P_3, P_2P_3, and P_2P_4. Let us also assume that the homogeneous coordinates of these points are given by $\mathbf{p}_1 = (0,0,1)^t$, $\mathbf{p}_2 = (15,0,1)^t$, $\mathbf{p}_3 = (x,y,1)^t$, and $\mathbf{p}_4 = (x+10\cos(\phi), y+10\sin(\phi), 1)^t$. Therefore, the configuration of the moving platform is determined by (x,y,ϕ). We also introduce a boundary region determined by the square $P_5P_6P_7P_8$ whose vertex coordinates are $\mathbf{p}_5 = (-25,40,1)^t$, $\mathbf{p}_6 = (-25,-40,1)^t$, $\mathbf{p}_7 = (55,-40,1)^t$, and $\mathbf{p}_8 = (55,40,1)^t$.

In this planar case, the regions free from singularities and collisions can be represented by a Boolean combination of signs of 3×3 determinants. Actually, the region free from kinematic singularities is defined by the predicate

$$\mathscr{P}_1 = (\triangle(1,2,3) \vee \overline{\triangle(1,2,3)}) \wedge (\triangle(2,3,4) \vee \overline{\triangle(2,3,4)}),$$

and the region free from collisions between the moving platform, the base, and the four segments defining the boundary, by the predicate

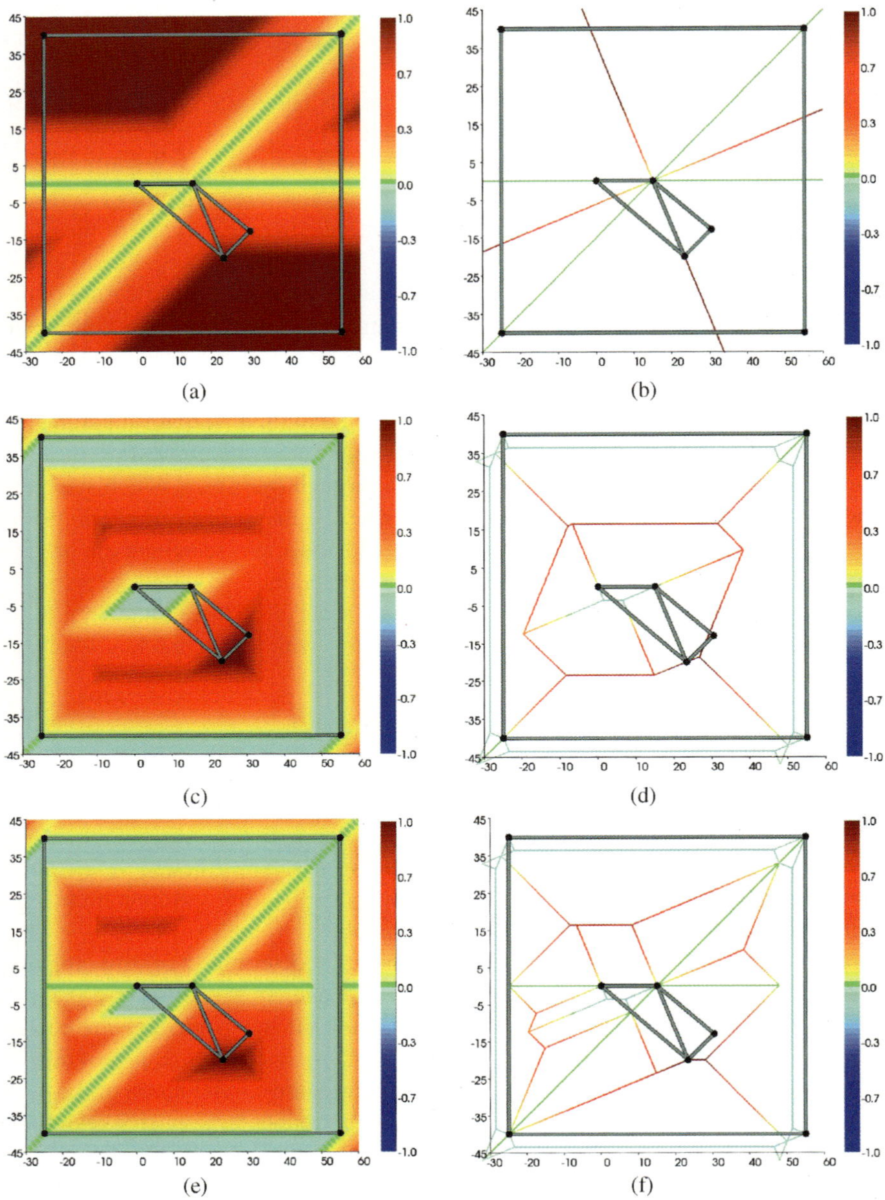

Fig. 2 Measure of distance to the kinematic singularities (a), and the corresponding Voronoi diagram (b). Measure of distance to the regions where collisions occur (c), and the corresponding Voronoi diagram (d). Measure of distance taking into account both singularities and collisions (e), and the corresponding Voronoi diagram (f). All images are slices of the robot's configuration space for $\phi = \frac{\pi}{4}$.

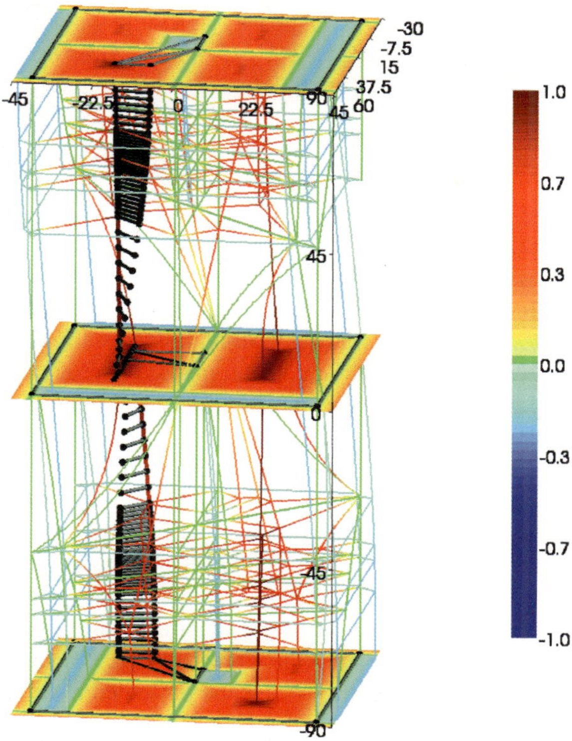

Fig. 3 Representation of the analyzed robot's configuration space in the region defined by $(x,y,\phi) = ([-30,60],[-45,45],[-\pi/2,\pi/2])$. The plot shows the measure of distance associated with $\mathscr{P}_1 \wedge \mathscr{P}_2$ for $\phi = -\pi/2, 0$, and $\pi/2$, and slices of the corresponding Voronoi diagram for values of ϕ where there is a change in its topology. The lines connecting them correspond to the edges of the Voronoi diagram.

$$\begin{aligned}\mathscr{P}_2 =&\overline{(\triangle(1,2,3)\otimes\triangle(1,2,4))\wedge(\triangle(1,3,4)\otimes\triangle(2,3,4))}\\ &\wedge\overline{(\triangle(5,6,3)\otimes\triangle(5,6,4))\wedge(\triangle(5,3,4)\otimes\triangle(6,3,4))}\\ &\wedge\overline{(\triangle(6,7,3)\otimes\triangle(6,7,4))\wedge(\triangle(6,3,4)\otimes\triangle(7,3,4))}\\ &\wedge\overline{(\triangle(7,8,3)\otimes\triangle(7,8,4))\wedge(\triangle(7,3,4)\otimes\triangle(8,3,4))}\\ &\wedge\overline{(\triangle(8,5,3)\otimes\triangle(8,5,4))\wedge(\triangle(8,3,4)\otimes\triangle(5,3,4))}\end{aligned}$$

Figs. 2a and 2b show the measure of distance associated with $\mathscr{P}_1$ and its corresponding Voronoi diagram, respectively. Figs. 2c and 2d show the same information for $\mathscr{P}_2$. These Voronoi diagrams represent the regions free from kinematic singularities and free from collisions, respectively. Figs. 2e and 2f show the measure of distance associated with $\mathscr{P}_1 \wedge \mathscr{P}_2$ and the corresponding Voronoi diagram thus integrating all the information in a single diagram. Finally, Fig. 3 shows how this diagram evolves, as ϕ varies, and a singularity and collision-free path connecting the configuration $(35, -20, \pi/2)$ and $(-10, -15, -\pi/2)$. Observe how all slices for fixed orientations are arrangements of straight line segments.

7 Conclusion

The use of Voronoi diagrams for motion planning received a lot of attention in the eighties. This interest progressively decreased in favor of algorithms based on random samplings. Nevertheless, it still remains as a fundamental tool for those applications in which an exact representation of free space is required. This paper shows how a simplified Voronoi diagram of the singularity and collision-free regions of any quadratically-solvable Gough–Stewart platform can be obtained assuming that the robot itself and its environment can be well-approximated by sets of polyhedra. The result is an elegant algorithm fully expressed in terms of 4×4 determinants.

Finally, concerning computational efficiency, it is worth noting that many basic geometric tests other than interference detection such as classification, containment, and depth priority tests can be performed by computing sets of determinants. This has motivated the search for efficient determinant computations using hardware implementations such as the triangle processor and its successor, the polygon engine [10]. This is a point that could be explored if the presented ideas should be implemented for complex Gough–Stewart platforms in cluttered environments.

References

1. Borràs, J., Thomas, F., Torras, C.: On Delta transforms. IEEE Trans. Robot. **25**(6), 1225–1236 (2009)
2. Borràs, J., Thomas, F., Torras, C.: A family of quadratically-solvable 5-SPU parallel robots. Proc. of the 2010 IEEE Int. Conf. on Robotics and Automation, pp. 4703–4708 (2010)
3. Borràs, J., Thomas, F.: On the primal and dual forms of the Stewart platform pure condition. Submitted
4. Canny, J., Donald, B.: Simplified Voronoi diagrams. Discrete Comput. Geom. **3**(3), 219–236 (1988)
5. Canny, J.: The Complexity of Robot Motion Planning. MIT Press, Cambridge, Mass. (1987)
6. Eppstein, D., Erickson, J.: Raising roofs, crashing cycles, and playing pool: Applications of a data structure for finding pairwise interactions. Discrete Comput. Geom. **22**(4), 569–592 (1999)
7. Lozano-Pérez, T.: Spatial planning: A configuration space approach. IEEE Trans. Comput. **C-32**(2), 108–120 (1983)
8. Schwartz, J., Yap, C.K.: Advances in Robotics. Lawrence Erlbaum Associates, New Jersey (1986)
9. Thomas, F., Torras, C.: A projectively invariant intersection test for polyhedra. Vis. Comput. **18**(7), 405–414 (2002)
10. Yamaguchi, F., Tsuda, S., Nagasaki, T.: Applications of the 4×4 determinant method and the polygon engine. Vis. Comput. **4**(4), 176–187 (1988)

Dynamic Capabilities of a Parallel Robot Based Routing Machine

J. Corral, Ch. Pinto, F.J. Campa and O. Altuzarra

Abstract In this paper, a study of the capabilities of a parallel manipulator with four degrees of freedom for machining application is presented. Specifically, routing tasks will be considered. To carry out this study, a process model that allows the dynamic behaviour of robot based on the phenomenon of chatter to be considered is used. The model takes into account characteristics of the process like the tool, the material to be machined and spindle speed. The dynamic stiffness of the manipulator is also taken into account. This modal parameter has a decisive influence on the dynamic behaviour. Using a single-frequency model of stability, values of the critical depth of cut and chatter frequency are obtained. Since the structural behaviour of parallel manipulators is a function of the location in the workspace, a post-processing of both critical depth of and chatter frequency is presented. From these maps, the dependence of the depth of cut and chatter frequency on the location is also derived.

Key words: Parallel machine, routing, structural dynamics, depth of cut, chatter

1 Introduction

An index of productivity in the chip removing machining process is the chip flow that depends on the speed of the tool, feed rate and depth of cut. The automation of these processes leads to a dependency on the characteristics of the machine in both productivity and the quality of the finishing. The main rough machining, semi-finishing and finishing machining require Computer Numerical Control (CNC), the architecture of which ensures precision at cutting edge. This is usually achieved by very robust machines in which the large amount of mass helps to provide stiffness.

J. Corral · Ch. Pinto · F.J. Campa · O. Altuzarra
Department of Mechanical Engineering, University of the Basque Country, 48013 Bilbao, Spain,
e-mail: {j.corral,charles.pinto,fran.campa,oscar.altuzarra}@ehu.es

J. Lenarčič, M. Husty (eds.), *Latest Advances in Robot Kinematics*,
DOI 10.1007/978-94-007-4620-6_21,

However, some finishing tasks are more efficiently met in terms of accuracy and surface quality by lighter machines with good rigidity, size and mass. This applies, for example, the routing, certain finishing milling, small planning and reaming. Performing these operations with a heavy and robust machining centres leads to inefficiency because, on the one hand, the actual capabilities of the machines are wasted and, on the other hand, the energy consumption associated with moving a heavy machine is comparatively high [11]. That is why, in certain sectors, the use of robots and manipulators with good dynamics, good dexterity of the tool to ensure accessibility to cavities and profiles, and less rigid than traditional machines but big enough to fit entrusted machining tasks is a trending topic [6]. These requirements fit properly in the most advantageous features of parallel machines and, more specifically, in those of lower mobility manipulators because some degrees of freedom may be either not necessary or inherent to the type of the tool itself such as for drilling or milling. Therefore, in this paper, a characterization based on the dynamic structural performance of a parallel kinematic manipulator for routing operations is proposed. The routing is used in the machining of thin components and defines the final shape of the piece. That is why dimensional accuracy requirements and surface finish qualities determined by the dynamic stiffness of the machine have to be ensured. critical in that, an unexpected problem in the routing, can damage a 90% processed piece and the costs it involves.

2 Routing Process Characterization and Modelling

The routing is an operation common in the manufacture of structural components of reduced thickness of aluminium, titanium and composites for the aviation sector. The process to obtain these workpieces starts with a block of material which is firstly shaped by removing up to 90–95% of material by high speed milling. In the final stages of the machining, the framework of the block itself supports the endpiece. Routing is one of the last machining operations in which a milling cutter, diameter of which is between 10 and 16 mm, penetrates the piece to slot it along the outline. Specifically, full immersion milling with a four-edged, 12 mm diameter carbide end mill will be considered. Because of the aforementioned milling, the radial depth of cut equals the tool diameter, 12 mm. The axial depth of cut is defined by the thickness of the part, which in this type of components is small (from 0.5 mm to 5 mm). So the case study of this work will be based on an operation with an axial depth of 1 mm. Workpiece material will be 7075-T6 Aluminium, typical of the aeronautical industry. This material is machined at high cutting speeds, so the limitation on the spindle speed is given by the reduction capabilities of the spindle. A small screw with a range from 2,500 to 4,000 rpm will be considered and the advance per edge will be set at 0.05 mm. Figure 1 represents a sketch of the routing operation. The horizontal arrow represents the direction of tool advance with clockwise rotation. In this scheme, the milling cut, with a radial depth a_e equal to the tool diameter d, and axial depth of cut a_p, defined by the thickness of part is shown.

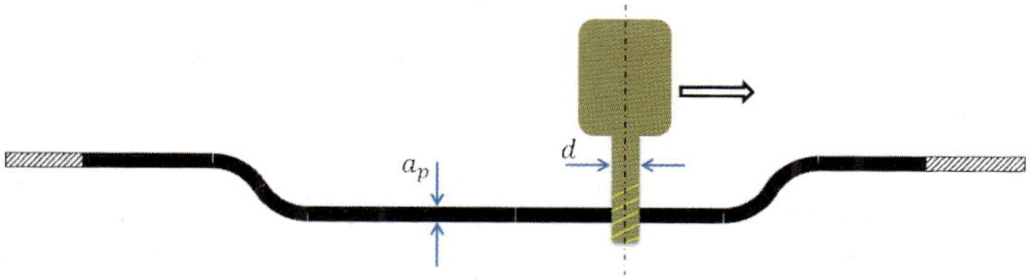

Fig. 1 Routing process sketch.

These two parameters will determine the dynamic characterization of the process. While the radial depth of cut remains constant ($a_e = d$), the axial depth of cut will be established to ensure routing conditions, in the most efficient way, in terms of geometric accuracy and surface finish from the dynamics point of view. The dynamic characterization determines whether the dynamic stiffness of the manipulator allows a satisfying behaviour against vibrations during the process. The parameter to evaluate the dynamic stiffness is the critical depth of cut of the stability lobes diagram (Fig. 2(b)) of the manipulator for a routing operation with a given tool and workpiece material. As the modal parameters of the manipulator depend heavily on its position along the workspace, several stability diagrams have to be calculated inside the workspace. Moreover, the dynamic behaviour of the manipulator is not symmetrical with respect to the tool axis, so a different stability diagram is obtained for each feed direction. Finally, as the maximum stable depth of cut of each stability diagram depends on the spindle speed, to simplify the overall dynamic behaviour characterization, the authors have decided to use the critical depth of cut, taking into account that it is a conservative approach. Regenerative chatter in milling is a common problem that has been extensively studied during the last 60 years [7]. The works of [5, 9, 10] provided the theoretical basis to study the problem. Later, Budak and Altintas [1], proposed a relatively simple yet effective model to predict the stability lobes diagram. That model was improved with the multifrequency solution. In the last decade, the works of Insperger, Butcher or Baily [2, 4] have proposed alternative algorithms that improved previous models in terms of accuracy and computational time. From the point of view of the designer of a machine tool, stability models can be used to evaluate the dynamic behaviour of a designed machine tool or even to redesign features of a machine, as it is shown in the work of [11], who used the lobes diagrams to find a balance between the dynamic behaviour and the mass of a machine tool. The stability model developed is based on the well-known stability model with a single-frequency solution proposed by Budak and Altintas [1]. As the end mill commonly used to perform routing operations has a cutting edge lead angle of $\kappa = 90°$, the self-excitation in the Z direction, that is, the tool axis direction, is avoided, so a two-dimensional stability model has been used. The basics of the model are summarized in the following paragraphs. First, the dynamic chip thickness, which is the origin of the regeneration mechanism, is considered as dependent on the dynamic displacements of the manipulator in X and Y axes. Hence, the dynamic chip thickness that a flute j located at an angular position ϕ_j cuts, can be obtained as:

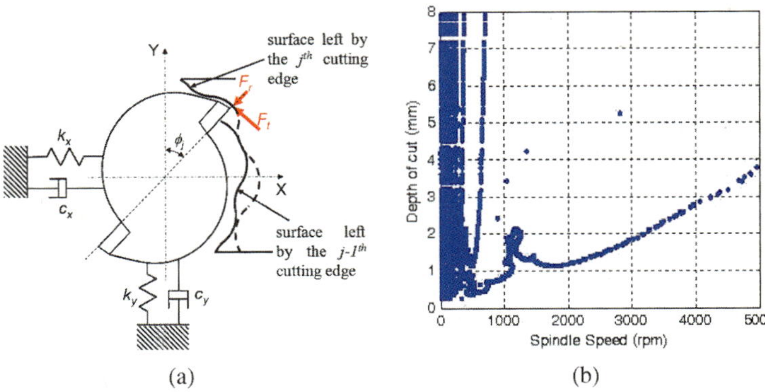

Fig. 2 (a) Axis and forces considered in the model and (b) lobes diagram.

$$h_d(t) = [\Delta x(t) \cdot \sin\phi_j(t) + \Delta y(t) \cdot \cos\phi_j(t)] \cdot g(t) \tag{1}$$

where $g(t)$ is a unity window function that takes into account if the flute is inside the cutting area or not. The dynamic forces model is a linear mechanistic model where the tangential and radial cutting forces depend on the corresponding shearing cutting coefficients. These forces are then projected to the XY axis of the reference system of the tool.

$$\begin{Bmatrix} F_t(t) \\ F_r(t) \end{Bmatrix} = k_t \cdot a_p \cdot \begin{Bmatrix} 1 \\ k_r \end{Bmatrix} \cdot h(\phi_j(t)) \cdot \begin{Bmatrix} F_x(\phi_j) \\ F_y(\phi_j) \end{Bmatrix} = \begin{bmatrix} -\cos\phi_j & -\sin\phi_j \\ \sin\phi_j & -\cos\phi_j \end{bmatrix} \cdot \begin{Bmatrix} F_t(t) \\ F_r(t) \end{Bmatrix} \tag{2}$$

Introducing Eq. (1) into Eq. (2) the dynamic forces and dynamic displacements are related by the directional factors matrix $\mathbf{A}(t)$.

$$\begin{Bmatrix} F_x(t) \\ F_y(t) \end{Bmatrix} = a_p \cdot k_t \cdot \mathbf{A}(t) \cdot \begin{Bmatrix} \Delta x(t) \\ \Delta y(t) \end{Bmatrix} \tag{3}$$

Following the single-frequency approach, where the chatter appears only at one frequency near the natural frequency, the dynamic displacements depend on the modal parameters of the manipulator which are included in the frequency response matrix $\mathbf{G}(\omega_c)$, and also on the current and precedent position of the manipulator.

$$\begin{Bmatrix} \Delta x(t) \\ \Delta y(t) \end{Bmatrix} = (1 - e^{-i\omega_c T}) \begin{bmatrix} G_{xx}(\omega_c) & G_{xy}(\omega_c) \\ G_{yx}(\omega_c) & G_{yy}(\omega_c) \end{bmatrix} \cdot \begin{Bmatrix} F_x \\ F_y \end{Bmatrix} \cdot e^{i\omega_c T} \tag{4}$$

Introducing Eq. (4) into Eq. (3), an eigenvalue problem of second order is obtained, which can be solved after averaging the directional factors matrix by its average term of the Fourier series expansion. The eigenvalue problem is shown in Eq. (5).

$$det[\mathbf{I} + \Lambda \mathbf{A}_0 \mathbf{G}(\omega_c)] = 0 \longrightarrow \Lambda = -\frac{z}{4\pi} \cdot a_p \cdot k_t (1 - e^{-i\omega_c T}) \tag{5}$$

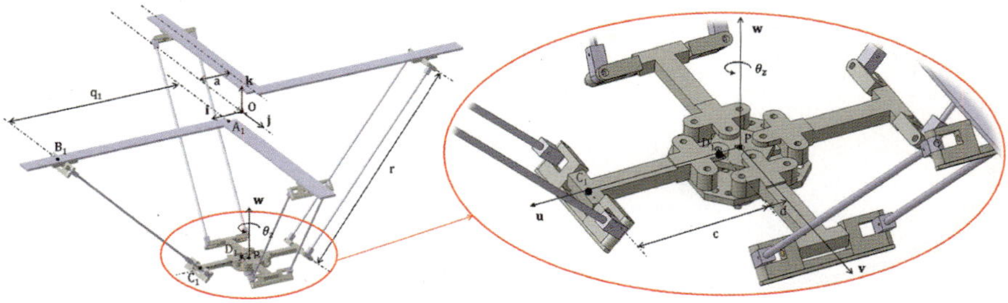

Fig. 3 ARABA I kinematic skeleton.

Solving the eigenvalue problem for several values of the chatter frequency, the limiting depth of cut, a_p, and the corresponding spindle speed, N, are calculated as:

$$a_p = -\frac{2\pi}{z \cdot k_t} \cdot real(\Lambda) \cdot \left[1 + \left(\frac{real(\Lambda)}{imag(\Lambda)} \right)^2 \right]; \quad N = \frac{60 \cdot \omega_c}{z \cdot (\varepsilon + 2\pi k)} \tag{6}$$

where ε is the phase difference between the surface left by the precedent tooth and the current tooth.

$$\varepsilon = \pi - 2 \cdot atan\left(\frac{real(\Lambda)}{imag(\Lambda))} \right) \tag{7}$$

Plotting the depth of cut values against the corresponding spindle speed values calculated in Eq. (6), the stability lobes diagram is completed.

3 Case Study

The ARABA I parallel manipulator has 4 dof, three translations and one rotation about an axis contained in a vertical plane. Next, the main characteristics to understand properly the structural behaviour will be presented, although more details can be found in [8]. The manipulator has been designed to assist in machining tasks and, therefore, the requirements of precision and stiffness typical of this kind of task. The architecture of the manipulator is presented in Fig. 3 and consists of a fixed platform and a mobile platform. The linear guides represent the fixed platform and they are not perfectly aligned but have an offset $2a$. Connecting both platforms four identical legs are arranged which consist of several elements. In this case, the basic kinematic chain of each leg is *PRPaRR*, i.e., the connection to the fixed component, represented by the guides, is performed through a prismatic joint, then a rotational joint B_i, a parallelogram joint, and two mutually perpendicular rotational joints in C_i and D_i, respectively.

From Kinematic Problem resolution, the volume corresponding to the manipulator ARABA I workspace is obtained (Fig. 4(a)).

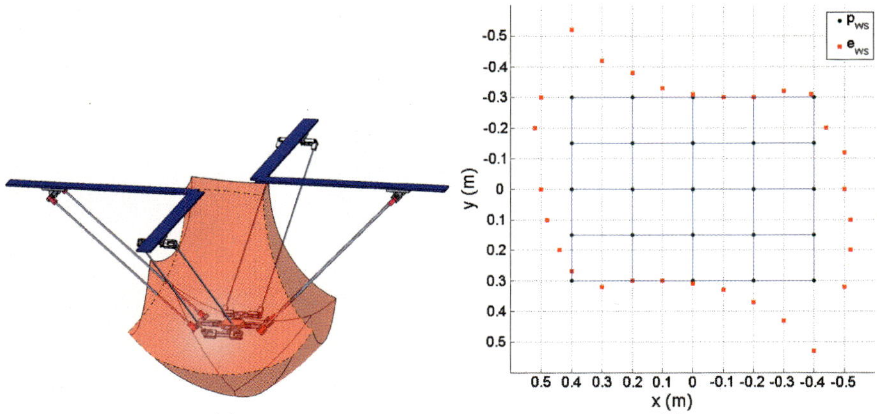

(a) (b)

Fig. 4 (a) Workspace and (b) workplane.

In this work, an initial study in which a constant orientation of the platform is maintained will be carried out. Moreover, the workspace is discretized into planes of constant height so that a second variable is fixed (z coordinate constant). In each plane a homogeneous distribution of measuring points is provided and a representative point density depending on the size of the working plane is chosen. Figure 4(b) shows the section of maximal workspace by a z constant plane in which measurements have been performed. Specifically, this is the $z = -700$ mm height plane which, henceforth, is referred to as Z_{700} plane. The distribution of points of Fig. 4(b) shows the boundary of the workspace experimentally obtained, e_{ws}, and the points that define the practical workspace, p_{ws}, where the study is carried out. The workspace in which the analysis has been performed and which is free of singularities will be referred to as *operational workspace*, W_{op}.

The dynamic structural behaviour has been carried out by analysing the dynamic stiffness of the manipulator. For parallel kinematics manipulators dynamic structural behaviour is a function of the location in their workspace [3]. Due to this dependency, vibratory analyses have to be done in the whole workspace. The results of these analyses have led the values of the natural frequencies in the Z_{700} workplane. Shape modes associated to each natural frequency have been derived in order to detected possible crossings between modes, consequence of the location dependent behaviour. For each mode shape the values of the dynamic stiffness have been derived. These values are used in the stability model for the performance evaluation used in this work and presented in the previous section. A detailed explanation of the natural frequencies of the model as well as figures and description of the modes can be found in [3].

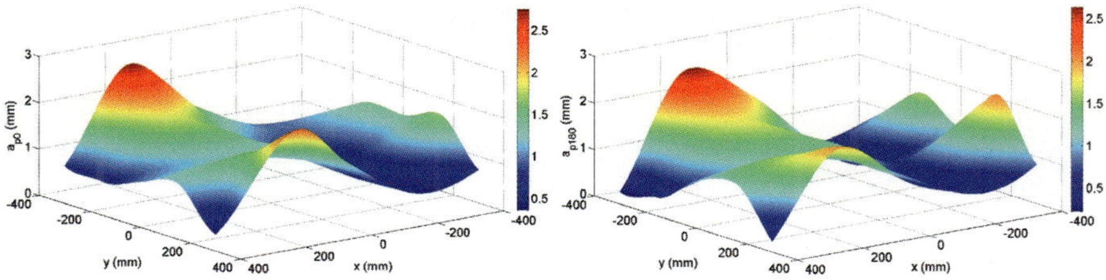

(a) (b)

Fig. 5 X-direction critical a_p maps: (a) $X(+)$ and (b) $X(-)$.

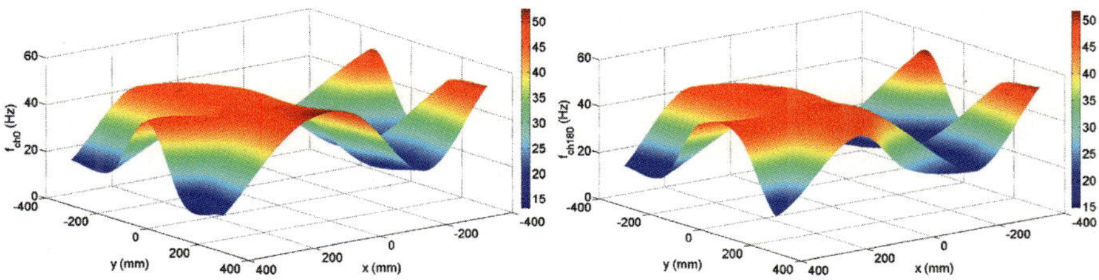

(a) (b)

Fig. 6 Chatter frequency for X-direction advance: (a) positive and (b) negative senses.

4 Results

In the maps of Fig. 5 the values of the critical depth of cut as a function of the position in the plane, $a_{p_{crit}}(x,y)$ are presented. These maps represent the critical depth of cut for an advance of the tool along the X-direction. The map on the left corresponds to the positive X-direction advance and, on the right, the negative one. As shown in the maps, a critical depth of cut is obtained for each point in the work plane. If the manipulator exceeds this operational parameter during the machining task, chatter instability phenomenon will appear resulting in poor finishing and even machine collapse. For each point, the chatter frequency, the one at the chatter occurs, is also different. Associated to the chatter instability, the frequency of chatter is also obtained from the model. This is the frequency at which the chatter occurs. That is, when the machine removes more material than that permitted by the critical depth of cut, chatter happens at a given frequency. This fact is useful in order to detect which shape modes are meaningful and have some influence in the dynamic capabilities of the machine. In Fig. 6 chatter frequency maps are presented for X-direction of the tool advance. In this work, the *critical depth of cut maps* has been obtained. In these maps, optimal regions of the workplane can be selected for a certain depth of cut. And so, these maps prove that dynamic capabilities of the manipulator are function of the tool tip location in the workspace. A second postprocessing parameter has been obtained from the model: the *frequency of chatter*. In future works, this chatter frequency could be used to find which are the most influential shape modes for the parallel machine and, in this way, find design criteria to improve the dynamic capabilities of the routing machine.

Acknowledgements The authors wish to acknowledge the financial support received from the Spanish Government through the Ministerio de Educación y Ciencia (Project DPI2011-22955) and the regional Government of the Basque Country through the Departamento de Educación, Universidades e Investigación (Project IT445-10).

References

1. Altintas, Y., Budak, E.: Analytical prediction of stability lobes in milling. CIRP Ann. – Manuf. Technol. **44**(1), 357–362 (1995)
2. Butcher, E.A., Ma, H., Bueler, E., Averina, V., Szabo, Z.: Stability of linear time-periodic delay-differential equations via Chebyshev polynomials. Int. J. Numer. Methods Eng. **59**(7), 895–922 (2004)
3. Corral, J.: Caracterización del comportamiento estructural de manipuladores de cinemática paralela. Ph.D. thesis, University of the Basque Country (2011)
4. Insperger, T., Mann, B., Stépán, G., Bayly, P.: Stability of up-milling and down-milling, Part 1: Alternative analytical method. Int. J. Mach. Tools Manuf. **43**(1), 25–34 (2003)
5. Merrit, H.: Theory of self-excited machine tool chatter. Trans. ASME J. Eng. Ind. **87**, 447–454 (1965)
6. Pan, Z., Zhang, H., Zhu, Z., Wang, J.: Chatter analysis of robotic machining process. J. Mater. Process. Technol. **173**(3), 301–309 (2006)
7. Quintana, G., Ciurana, J.: Chatter in machining processes: A review. Int. J. Mach. Tools Manuf. **51**(5), 363–376 (2011)
8. Salgado, O.: Síntesis, análisis y diseño de manipuladores paralelos de baja movilidad. Ph.D. thesis, University of the Basque Country (2008)
9. Tlusty, J., Polacek, M.: The stability of machine tools against self excited vibrations in machining. In: International Research in Production Engineering, vol. 1, pp. 465–476. ASME (1963)
10. Tobias, S.A., Fishwick, W.: A theory of regenerative chatter. Engineer (Lond.) **42**(1), 463–466 (1958)
11. Zulaika, J., Campa, F., de Lacalle, L.L.: An integrated process-machine approach for designing productive and lightweight milling machines. Int. J. Mach. Tools Manuf. **51**(7–8), 591–604 (2011)

Kinematic Synthesis of Multi-Fingered Robotic Hands for Finite and Infinitesimal Tasks

E. Simo-Serra, A. Perez-Gracia, H. Moon and N. Robson

Abstract In this paper we present a novel method of designing multi-fingered robotic hands using tasks composed of both finite and infinitesimal motion. The method is based on representing the robotic hands as a kinematic chain with a tree topology. We represent finite motion using Clifford algebra and infinitesimal motion using Lie algebra to perform finite dimensional kinematic synthesis of the multi-fingered mechanism. This allows tasks to be defined not only by displacements, but also by the velocity and acceleration at different positions for the design of robotic hands. The additional information enables an increased local approximation of the task at critical positions, as well as contact and curvature specifications. An example task is provided using an experimental motion capture system and we present the design of a robotic hand for the task using a hybrid Genetic Algorithm/Levenberg-Marquadt solver.

Key words: Kinematic synthesis, multi-fingered grippers, Clifford and Lie algebra

1 Introduction

The design of end-effector robotic tools has traditionally taken place in an application-oriented fashion within the framework of the mechanical design theory [7]. Among the rich variety of robotic end-effectors, those generally defined as robotic hands are considered suited not only for grasping, but also for dexterous manipu-

E. Simo-Serra
Institut de Robòtica i Informàtica Industrial (CSIC-UPC), Barcelona, Spain, e-mail: esimo@iri.upc.edu

A. Perez-Gracia
Idaho State university, Pocatello, Idaho 83209, USA, e-mail: perealba@isu.edu

H. Moon · N. Robson
Texas A&M University, College Station, TX 77843, USA, e-mail: hsmoon@neo.tamu.edu, ninarobson@tamu.edu

J. Lenarčič, M. Husty (eds.), *Latest Advances in Robot Kinematics*,
DOI 10.1007/978-94-007-4620-6_22,

lation. We can define a multi-fingered robotic hand as an end-effector in which the base, or palm, spans several serial chains in a tree-like structure.

There are a great variety of designs for robotic hands. Some designs mimic the human hand and exhibit a high number of degrees of freedom [1, 15]; others are designed for specific applications [4] and may or not be anthropomorphic [5].

Most of the designs have been oriented either towards maximum anthropomorphism or towards optimizing grasping, manipulability or workspace size. A good review on the efforts toward kinematic hand design can be found in [6]. As robotic hands become more common in industrial applications and human environments, it makes sense to think that their design will become more task-oriented. Soto Martell and Gini [14] expose the need for a task-based design process for robotic hands.

The use of kinematic synthesis for the design of the multi-fingered robotic hand has been applied to individual fingers, see [2]. We believe that the reason why dimensional synthesis has been scarcely applied to robotic hand design is because of the lack of a method that takes a multi-fingered task as the input and outputs a multi-fingered design.

In this paper, we extend the work presented in [13] by combining it with the results on kinematic synthesis for infinitesimal positions [8, 9] and expressing the kinematics using the Clifford algebra of dual quaternions [10]. Note that mechanical linkages are traditionally synthesized by specifying a task, consisting of a number of positions that the end-effector has to move through, with the goal of determining the design parameters, i.e. fixed and moving pivot locations, as well as the size of the linkage. The difference between the traditional design of mechanical linkages and the current design with contact direction, used in this research, is basically in the task, which consists not only of positions, but velocities and accelerations compatible with contact and curvature specifications between the end-effector/fingers and the object to be grasped. In comparison to the traditional synthesis techniques, these velocities and accelerations yield to a more complicated system of position, velocity and acceleration design equations, as well as more complicated trajectory planning techniques.

As an example, we apply this methodology to the design of a multi-fingered hand for operating a door knob. The motivation for this design arose from an individual, who is confined to a wheel chair after an accident. He has limited movement and weakness in his hands, making it difficult for him to grasp doorknobs at his workspace. The synthesis presented here is the first step towards developing assistive manipulation devices.

2 Infinitesimal Kinematics

The generic screw S for a twist can be represented as an element of the Lie algebra $se(3)$ [12],

$$\mathsf{S} = \lambda(\mathbf{s}; \mathbf{r} \times \mathbf{s} + h\mathbf{s}) = \lambda(\mathbf{s}; \mathbf{s}^0 + h\mathbf{s}) = (\boldsymbol{\omega}; \mathbf{v}) \in se(3) \tag{1}$$

where $\mathbf{s}, \mathbf{r}, \boldsymbol{\omega}, \mathbf{v} \in \mathbb{R}^3$ with $\mathbf{s} \cdot \mathbf{s} = 1$, $\mathbf{s}^0 = \mathbf{r} \times \mathbf{s}$ and $\lambda, h \in \mathbb{R}$.

The relative velocities between a pair of rigid bodies form one-dimensional subalgebras of the Lie algebra $se(3)$ [11]. The most generic subalgebra is generated by the screw or helical joint S which becomes a revolute joint $\mathsf{S}_R = (\mathbf{s}; \mathbf{s}^0)$ with $h = 0$ or a prismatic joint $\mathsf{S}_P = (0; \mathbf{s})$ with the screw axis at infinity. The binary operation of the Lie algebra is the Lie bracket, which can be expanded for screws as,

$$[\mathsf{S}_1, \mathsf{S}_2] = [(\boldsymbol{\omega}_1; \mathbf{v}_1), (\boldsymbol{\omega}_2; \mathbf{v}_2)] = (\boldsymbol{\omega}_1 \times \boldsymbol{\omega}_2; \boldsymbol{\omega}_1 \times \mathbf{v}_2 + \mathbf{v}_1 \times \boldsymbol{\omega}_2) \tag{2}$$

The velocity of the end-effector for a serial articulated chain with n joints in a given configuration can be written as [12],

$$\frac{d\mathsf{P}}{dt} = \dot{\mathsf{P}} = \sum_{i=1}^{n} \dot{\theta}_i \mathsf{S}_i \tag{3}$$

where $\dot{\mathsf{P}} = (\boldsymbol{\omega}; \mathbf{v})$ with $\boldsymbol{\omega}$ being the angular velocities and $\mathbf{v}$ being the Cartesian velocities. The screws S_i represent the infinitesimal screws of each joint.

The infinitesimal screws can be transformed to an instantaneous position from a reference position using the Clifford algebra conjugation action,

$$\mathsf{S}_i^k = \left(e^{\frac{\Delta\hat{\theta}_{i-1}^k}{2}\mathsf{S}_{i-1}^k}\right) \mathsf{S}_i \left(e^{\frac{\Delta\hat{\theta}_{i-1}^k}{2}\mathsf{S}_{i-1}^k}\right)^* = \left(\prod_{j=1}^{i-1} e^{\frac{\Delta\hat{\theta}_j^k}{2}\mathsf{S}_j}\right) \mathsf{S}_i \left(\prod_{j=1}^{i-1} e^{\frac{\Delta\hat{\theta}_j^k}{2}\mathsf{S}_j}\right)^* \tag{4}$$

where $\Delta\hat{\theta}_i^k = \hat{\theta}_i^k - \hat{\theta}_i^r$ with $\hat{\theta}_i^r$ is the joint parameter in the reference configuration and S_i^k is the i-th screw in a serial chain at position k.

The velocity of a joint j in a chain is written as the derivative of a finite screw [3],

$$\frac{d\mathsf{S}_j}{dt} = \dot{\mathsf{S}}_j = \sum_{i=1}^{j-1} \dot{\theta}_i [\mathsf{S}_i, \mathsf{S}_j] \tag{5}$$

Cross terms and the non-commutation of the derivation operator must be taken into account as seen by differentiating each velocity component of (3) using the chain rule. This can be expanded to obtain the acceleration of the end-effector,

$$\frac{d^2\mathsf{P}}{dt^2} = \ddot{\mathsf{P}} = \frac{d}{dt}\dot{\mathsf{P}} = \sum_{i=1}^{n}\left(\ddot{\theta}_i \mathsf{S}_i + \dot{\theta}_i \dot{\mathsf{S}}_i\right) = \sum_{i=1}^{n} \ddot{\theta}_i \mathsf{S}_i + \sum_{i=1}^{n-1} \dot{\theta}_i \sum_{j=i+1}^{n} \dot{\theta}_j [\mathsf{S}_i, \mathsf{S}_j] \tag{6}$$

where $\ddot{\mathsf{P}} = (\boldsymbol{\alpha}; \mathbf{a})$ with $\boldsymbol{\alpha}$ being the angular accelerations and $\mathbf{a}$ being the Cartesian accelerations.

The approach is general in the sense that the chain rule can be successively applied to obtain higher derivatives if necessary.

3 Design Equations for Tree Topologies

Tree topologies can be seen as many different serial chains that share a number of common joints. The equations can be written as for serial chains, but the task definition will vary with the topology. The finite motion of a joint can be expressed using the exponential map of a screw S. This can be expressed using the unit element of the Clifford even subalgebra of the projective space $Cl^{+}(0, 3, 1)$ or dual quaternion,

$$e^{\frac{\hat{\theta}}{2}\mathsf{S}} = \left(\cos\frac{\theta}{2} - \frac{d}{2}\sin\frac{\theta}{2}\epsilon\right) + \left(\sin\frac{\theta}{2} + \frac{d}{2}\cos\frac{\theta}{2}\epsilon\right)\mathsf{S} = \cos\frac{\hat{\theta}}{2} + \sin\frac{\hat{\theta}}{2}\mathsf{S}. \quad (7)$$

where ϵ is the dual unit such that $\epsilon^2 = 0$.

For a serial chain with n joints, the forward kinematics of a serial chain can be written relative to a reference configuration of the serial chain,

$$\hat{Q}(\boldsymbol{\Delta}\hat{\boldsymbol{\theta}}) = \prod_{i=1}^{n} e^{\frac{\Delta\hat{\theta}_i}{2}\mathsf{S}_i} = \prod_{i=1}^{n}\left(\cos\frac{\Delta\hat{\theta}_i}{2} + \sin\frac{\Delta\hat{\theta}_i}{2}\mathsf{S}_i\right) \quad (8)$$

where $\Delta\hat{\theta}_i = \hat{\theta}_i - \hat{\theta}_0$ with $\hat{\theta}_0$ being the joint parameters of the reference configuration.

For a task composed of finite positions, the relative forward kinematics can be compared to the relative motion from the reference configuration to each position $\hat{P}_{1k} = \hat{P}_k\hat{P}_1^{-1}$ [10],

$$\hat{P}_{1k} = \prod_{i=1}^{n} e^{\frac{\Delta\hat{\theta}_i^k}{2}\mathsf{S}_i}, \quad k = 2, \ldots, m_p \quad (9)$$

where m_p is the number of positions considered and $\Delta\hat{\theta}_i^k = \hat{\theta}_i^k - \theta_i^1$, with $k = 1$ being the reference configuration.

For a task with velocities, we can use (3) to write,

$$\dot{\mathsf{P}}_k = \sum_{i=1}^{n} \dot{\theta}_i^k \mathsf{S}_i^k, \quad k = 1, \ldots, m_v \quad (10)$$

where $\dot{\mathsf{P}}_k$ is the absolute velocity information for a given position k in the form $(\boldsymbol{\omega}; \mathbf{v})$. The instantaneous joint screw axis S_i^k can be calculated from (4).

The same procedure can be applied to acceleration to obtain from (6),

$$\ddot{\mathsf{P}}_k = \sum_{i=1}^{n} \ddot{\theta}_i^k \mathsf{S}_i^k + \sum_{i=1}^{n-1} \dot{\theta}_i^k \sum_{j=i+1}^{n} \dot{\theta}_j^k [\mathsf{S}_i^k, \mathsf{S}_j^k], \quad k = 1, \ldots, m_a \quad (11)$$

where $\ddot{\mathsf{P}}_k$ is the absolute acceleration for a given position k in the form $(\boldsymbol{\alpha}; \mathbf{a})$.

The velocity and acceleration equations can be seen as additional pose information that reduce the number of poses needed. Counting the number of independent unknowns n_x and independent equations n_f we obtain,

$$n_x = n_s + n_j(m_p + m_v + m_a - 1) \tag{12}$$
$$n_f = n_c + n_d(m_p + m_v + m_a - 1) \tag{13}$$

where n_s is the number of independent structural parameters, n_j the number of joint degrees of freedom, n_c the number of independent constraints and n_d the degrees of freedom of the end-effector motion. The number of positions, velocities and accelerations are given by m_p, m_v and m_a respectively. If we consider $m = m_p + m_v + m_a$ we obtain the familiar formula [10],

$$m = \frac{n_s - n_c}{n_d - n_j} + 1 \tag{14}$$

4 Experimental Set up and Task Specification

Since the first step in our synthesis technique is related to choosing a specific task, the kinematic task selected for the design is the operation of a standard door knob. In order to define this kinematic task, the door knob grasping and turning movement was performed, from a start to end spatial locations. During the movement, the subject emulates the opening of the door motion with an apparatus shown in Fig. 1a. The upper limb kinematics at specific points of interest are captured by a 3D Motion Capture System (Vicon, OMG Plc., UK), available in our Human Interactive Robotics Lab at Texas A&M University.

Three infrared cameras track the position of each marker relative to a predefined global coordinate frame, with a sampling rate of 100 Hz. Five moving frames are defined at the: elbow, wrist, and tip of thumb, tip of index and tip of middle fingers, respectively. Fig. 1b shows the marker attachment. To only synthesize the motion of the forearm, the positions chosen were transformed from absolute positions $\hat{P}_i$ to

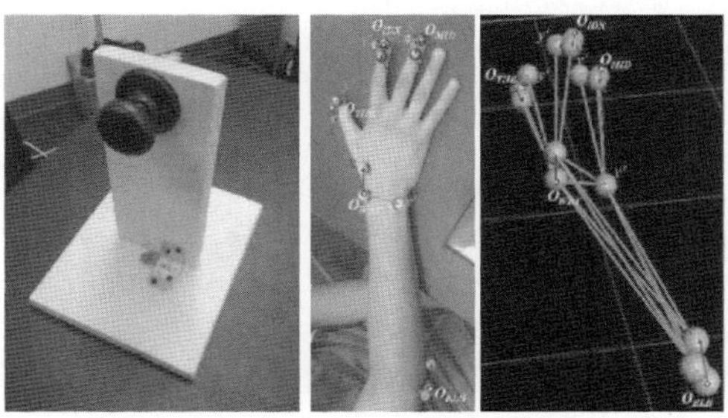
(a) (b)

Fig. 1 (a) The experimental apparatus for emulating the door knob task and (b) the kinematic model configured in the 3D Motion Capture System.

positions local to the elbow $\hat{P}_{ij} = \hat{P}_j^{-1}\hat{P}_i$ with $\hat{P}_J$ is the position of the elbow. For velocities this can be written as $\dot{\mathsf{P}}_{ij} = \dot{\mathsf{P}}_i - \dot{\mathsf{P}}_j$.

The obtained positions and velocities were then used as a task for the kinematic synthesis of the multi-fingered robotic hand. The kinematic specification consists of set of three spatial displacements defined by $\hat{P}_k = (\boldsymbol{\psi}_k, \mathbf{d}_k)$, $k = 1, 2, 3$, and the associated angular and linear velocities $\dot{\mathsf{P}}_k = (\boldsymbol{\omega}_k; \mathbf{v}_k)$, $k = 1, 2$ in two of the positions for each fingertip.

5 Solving Numerically

The solver used is an updated version of the kinematic synthesis solver for tree structures [13] updated to support the design equations (10) and (11). The solver is composed of a Genetic Algorithm (GA) paired with a Levenberg–Marquadt (LM) local optimizer. For more details on the solving approach see [13].

The full equation system for a multi-fingered robot with b fingers formed by n revolute joints can be defined directly by manipulation of the design equations,

$$\mathbf{F}(\mathsf{S}, \boldsymbol{\Delta}\hat{\boldsymbol{\theta}}, \dot{\boldsymbol{\theta}}, \ddot{\boldsymbol{\theta}}) = \begin{pmatrix} \hat{P}^c_{1k} - \prod_{i=1}^{n_c} e^{\frac{\Delta\hat{\theta}^k_i}{2}\mathsf{S}_{i,c}}, & \begin{matrix} k = 2, \ldots, m_p \\ c = 1, \ldots, b \end{matrix} \\ \dot{\mathsf{P}}^c_k - \sum_{i=1}^{n_c} \dot{\theta}^k_i \mathsf{S}^k_{i,c}, & \begin{matrix} k = 1, \ldots, m_v \\ c = 1, \ldots, b \end{matrix} \\ \ddot{\mathsf{P}}^c_k - \left(\sum_{i=1}^{n_c} \ddot{\theta}^k_i \mathsf{S}^k_{i,c} + \sum_{i=1}^{n_c - 1} \dot{\theta}^k_i \sum_{j=i+1}^{n_c} \dot{\theta}^k_j [\mathsf{S}^k_{i,c}, \mathsf{S}^k_{j,c}] \right), & \begin{matrix} k = 1, \ldots, m_a \\ c = 1, \ldots, b \end{matrix} \end{pmatrix} \tag{15}$$

where b is the number of branches or fingers, n_c is the number of joints for a branch c, n is the total number of joints in the structure and $S^k_{i,c}$ are the instantaneous axis of the joint i in the branch c for the frame k calculated by (4). A valid mechanism is said to be found when $\mathbf{F}(\mathsf{S}, \boldsymbol{\Delta}\hat{\boldsymbol{\theta}}, \dot{\boldsymbol{\theta}}, \ddot{\boldsymbol{\theta}}) = \mathbf{0}$.

6 Results

The kinematic structure of the hand consists of a three degree of freedom palm+wrist complex (RRR), and three fingers, each of which is modeled as a two degree of freedom RR kinematic chain as seen in Fig. 2a. This structure was chosen for the kinematic synthesis of the task as it has fewer degrees of freedom than the human hand while having an non-fractional number of required samples. As this paper does not deal with structural synthesis, a pre-determined topology is used. No additional constraints were placed on the structure. For this kinematic structure with 9 revolute joints, a total of $m = 5$ samples are needed as obtained from (14).

The experimental task consists of many hundreds of frames of which $m_p = 3$ were selected. For two of them velocity information was also used providing

(a) Topology (b) Full System (c) Index Finger (d) Middle Finger (e) Thumb

Fig. 2 The topology used and overview of a solution mechanism found.

$m_v = 2$. Due to the nature of the door-knob opening task, the accelerations, related to curvature constraints (i.e. sliding motion of the fingers along the door-knob) are fairly small in comparison to the other task specifications and were not taken into account. Therefore, our task consists of positions, prescribed at the point where the fingers need to grasp the door knob, and velocities, describing the contact of the fingers with the door knob. The resulting equation system has 90 unknowns and 102 equations of which only 72 are independent. A set of 50 solutions was obtained taking an average of 18.06 minutes per solution and needing an average of 13.76 generations per solution with a population of 100 entities.

A selected solution mechanism of the doorknob task can be seen in Fig. 2. The thick joint axes are connected by thinner lines at the intersections of the common normals of the joints and the joint axes. The origin is represented by using a square and the end-effectors are represented by using spheres. This solution shown is more compact than the human hand as all the joints except one are grouped together and is able to perform the same task as the human hand. It was observed that generally the solutions have a similarity between the index and middle finger, while the thumb has a different shape, which is similar to how the human hand is designed.

7 Conclusions

This paper presents a novel dimensional synthesis methodology for articulated systems with a tree structure using additional constraints such as velocity and acceleration. It also presents a new design for numerically solving kinematic systems using these new constraints, adding upon the previous work of numerically solving tree structures with kinematic synthesis. The addition of velocity, acceleration and other derivatives at the positions allows a better local approximation of the task motion, as well tasks with contact and curvature specifications.

For the selected knob-operating task, a tree-like robot with three two-jointed fingers has been designed. Kinematic synthesis is just one step in the design process, one that allows you to create innovative candidates fitted for the kinematic tasks under consideration. Enough solutions have been found to suggest that a method to analyze and rank those needs to be a part of the design process. These results show that the dimensional synthesis of robotic multi-fingered hands is possible. In addition, multi-fingered hands appear not to be redundant when a task involving several fingers is to be performed.

Acknowledgements This work is partially supported by the Spanish Ministry of Science and Innovation under project DPI2010-18449.

References

1. Borst, C., Fischer, M., Haidacher, S., Liu, H., Hirzinger, G.: DLR Hand II: Experiments and experiences with an anthropomorphic hand. In: ICRA (2003)
2. Ceccarelli, M., Nava Rodriguez, N., Carbone, G.: Design and tests of a three finger hand with 1-DOF articulated fingers. Robotica **24**, 183–196 (2006)
3. Cervantes-Sánchez, J.J., Rico-Martínez, J.M., González-Montiel, G., González-Galván, E.J.: The differential calculus of screws: Theory, geometrical interpretation, and applications. Proc. Inst. Mech. Eng., Part C, J. Mech. Eng. Sci. **223**(6), 1449–1468 (2009)
4. Chalon, M., et al.: Dexhand: A space qualified multi-fingered robotic hand. In: ICRA (2011)
5. Dai, J.S., Wang, D., Cui, L.: Orientation and workspace analysis of the multifingered metamorphic hand-metahand. IEEE Trans. Robot. **25**, 942–947 (2009)
6. Grebenstein, M., Chalon, M., Hirzinger, G., Siegwart, R.: A method for hand kinematics design. 7 billion perfect hands. In: ICABB (2010)
7. Mason, M., Srinivasa, S., Vazquez, A., Rodriguez, A.: Generality and simple hands. Int. J. Robot. Res. (2010)
8. Patarinsky Robson, N., McCarthy, J.M.: Kinematic synthesis with contact direction and curvature constraints on the workpiece. In: IDETC (2007)
9. Patarinsky Robson, N., McCarthy, J.M.: Synthesis of a spatial SS serial chain for a prescribed acceleration task. In: World Congress in Mechanism and Machine Science (IFToMM), Besancon, France (2007)
10. Perez-Gracia, A., McCarthy, J.M.: Kinematic synthesis of spatial serial chains using Clifford algebra exponentials. Proc. Inst. Mech. Eng., Part C, J. Mech. Eng. Sci. **220**(7), 953–968 (2006)
11. Rico, J.M., Gallardo, J., Ravani, B.: Lie algebra and the mobility of kinematic chains. J. Robot. Syst. **20**(8), 477–499 (2003)
12. Selig, J.M.: Geometric Fundamentals of Robotics (Monographs in Computer Science). Springer Verlag (2004)
13. Simo-Serra, E., Moreno-Noguer, F., Perez-Gracia, A.: Design of non-anthropomorphic robotic hands for anthropomorphic tasks. In: IDETC (2011)
14. Soto Martell, J., Gini, G.: Robotic hands: Design review and proposal of new design process. World Academy of Science, Engineering and Technology **26** (2007)
15. Ueda, J., Kondo, M., Ogasawara, T.: The multifingered naist hand system for robot in-hand manipulation. Mech. Mach. Theory **45**(2), 224–238 (2010)

Inverse Kinematics Solver for Android Faces with Elastic Skin

Emarc Magtanong, Akihiko Yamaguchi, Kentaro Takemura, Jun Takamatsu and Tsukasa Ogasawara

Abstract The ability of androids to display facial expressions is a key factor towards more natural human-robot interaction. However, controlling the facial expressions of such robots with elastic facial skin is difficult due to the complexity of modeling the skin deformation. We propose a method to solve the inverse kinematics of android faces to control the android's facial expression using target feature points. In our method, we use an artificial neural network to model the forward kinematics and minimizing a weighted squared error function for solving the inverse kinematics. We then implement an inverse kinematics solver and evaluate our method using an actual android.

Key words: Android, artificial neural networks, facial expressions, inverse kinematics, human-robot interaction

1 Introduction

One of the main goals in android research is to design robots that are able to interact with humans in a natural manner. To achieve this objective, efforts have been made to incorporate the ability to display facial expressions on android robots [3].

Commonly, the android's face is controlled by directly adjusting the actuator displacements. However, controlling *feature points* on the android face is more suitable for making facial expressions since it directly adjusts the appearance of the face [6]. A feature point is defined as a specific point on the android's face that moves when displacing the facial actuators. On the other hand, there are only a few methods to control the android's face using feature points because solving the inverse kinematics, which is the relationship between the feature point positions and actuators dis-

Emarc Magtanong · Akihiko Yamaguchi · Kentaro Takemura · Jun Takamatsu · Tsukasa Ogasawara
Robotics Laboratory, Nara Institute of Science and Technology, Nara, Japan,
e-mail: {emarc-m, akihiko-y}@is.naist.jp

J. Lenarčič, M. Husty (eds.), *Latest Advances in Robot Kinematics*,
DOI 10.1007/978-94-007-4620-6_23,

placements, is difficult due to the elastic facial skin of androids. This paper presents a method for solving the inverse kinematics of such an android face using a machine learning technique.

The problem in solving for the inverse kinematics of android faces with elastic skin is that the facial skin surface is *deformable*. This causes the feature points to move with each other when displacing the actuators. Specifically, there is *coupling* between the feature points. Therefore, it is difficult to formulate an analytic solution to the inverse kinematics since the feature points are not fixed on a rigid link. Furthermore, specifying target feature points for the inverse kinematics is complicated because of the coupling problem.

In this paper, we propose three ideas to solve the inverse kinematics of android faces. Initially, the forward kinematics of the face is modeled using an artificial neural network. The forward kinematics model determines the feature point positions given the actuator displacements. Artificial neural networks are employed because of its capability to learn the complex forward kinematics of android faces. Next, using an iterative minimization technique for an error function, we compute the actuator displacements that satisfies the specified target feature point positions. Also, a weighting method is introduced for computing the difference between the target and the computed feature point positions to address the problem resulting from the feature points being coupled. Lastly, we propose a face segmentation technique to group the facial feature points and the actuators. Segmenting the face reduces the complexity of modeling the forward kinematics and solving the inverse kinematics.

The proposed inverse kinematics solver is validated by conducting several experiments using an actual android. The experimental results demonstrate the ability of the proposed inverse kinematic solver to control the android's facial expressions using target feature points.

There have been several research done to control the facial expressions of android robots. A method used in [4] retargets captured human facial expressions from video to an android by converting 2D feature point positions to actuator displacements using partial least squares regression. Another method in [7] retargets human facial motion capture data to actuator displacements of an android by interpolating weights of blendshape models. Unlike these methods, our proposed inverse kinematics solver will provide a proper solution even for infeasible target feature point positions.

The rest of this paper is organized as follows. Section 2 proposes the method to solve the inverse kinematics of the android's face and the face segmentation technique. Section 3 describes the experimental results. Finally, Section 4 concludes the paper.

2 Inverse Kinematics Solver for an Android Face

To model the forward kinematics, we employ an artificial neural network (ANN). Then, the actuator displacements are computed using an iterative minimization of the difference between the target and the feature point positions computed from the ANN. Here, a weighting method for the feature points is introduced to handle the coupling problem. Also, we describe a technique to segment the face of the android.

2.1 Forward Kinematics Using Neural Network

To model the forward kinematics, we use a multilayer feedforward ANN that is composed of an input, an output, and a hidden layer. Concretely, the ANN is defined as,

$$\mathbf{x} = ANN(\mathbf{u};\Theta), \tag{1}$$

where Θ denotes the parameters of the neural network and is optimized during training. The vector $\mathbf{u} = (u_1, u_2, \ldots, u_{N_\mathrm{a}})$ represents the input vector of actuator displacements and $\mathbf{x} = (x_1, y_1, z_1, \ldots, x_{N_\mathrm{f}}, y_{N_\mathrm{f}}, z_{N_\mathrm{f}})$ defines the output vector of feature point positions, where N_a and N_f denotes the number of actuators and feature points respectively.

For training the ANN, sets of actuator displacements and feature point positions, $\mathscr{D} = \{\mathbf{u}_n, \mathbf{x}_n | n = 1, 2, \ldots\}$, are used. In training the ANN, a backpropagation algorithm is used to optimize the parameter Θ of the ANN. To avoid overfitting, an early stopping technique is applied during the training [2]. For the experimental section, we implement this ANN using MATLAB's Neural Network Toolbox [1]. The number of neurons in the hidden layer is determined through initial experimentation.

2.2 Solving for the Inverse Kinematics

This section discusses our proposed solution for the inverse kinematics of the android's face based on the forward model learned by the ANN. We aim to address the difficulty of obtaining an analytic solution for the inverse kinematics from the forward kinematics ANN and solve the coupling problem when specifying target feature points. To consider these problems, the inverse kinematics is formulated as the minimization of the weighted squared error of the feature point positions with respect to the actuator displacements. That is,

$$\begin{aligned} \min_{\mathbf{u}} \quad & \sum_{i=1}^{3N_\mathrm{f}} w_i \left[ANN(\mathbf{u})_{[i]} - x^*_{[i]} \right]^2, \\ \text{such that,} \quad & u_{\min j} \leq u_j \leq u_{\max j}. \end{aligned} \tag{2}$$

In Eq. (2), $x^*_{[i]}$ denotes an i-th element of the target feature point position, the subscript $[i]$ denotes the i-th element of the vector, and $(w_1, w_2, \ldots, w_{3N_\mathrm{f}})$ are the weights for the feature points. The weights handle the coupling problem by emphasizing the error contribution of each feature point.

2.3 Face Segmentation

The inverse kinematics solver described in Sections 2.1 and 2.2 is applicable to any number of feature points and actuators. We refer to the inverse kinematics solver applied to all feature points and actuators as the *full face* inverse kinematics solver. However, we can improve the precision of the inverse kinematics solver by *segmenting* the feature points and actuators. Segmentation means that the feature points and actuators are grouped to be modeled separately using independent forward kinematics ANNs. Specifically, the set of all feature points $\mathscr{F}_{\mathrm{w}}$ and the set of all actuators $\mathscr{A}_{\mathrm{w}}$ are separated into their subsets: $\{\mathscr{F}_m | m = 1, \ldots, N_{\mathrm{g}}\}$ and $\{\mathscr{A}_m | m = 1, \ldots, N_{\mathrm{g}}\}$, where N_{g} denotes the number of segments.

To determine the segmentation, we measure the effect of each actuator to each feature point position. Wherein each facial actuator is independently displaced several times and the feature point positions are recorded. We then compute the *effect index* defined as,

$$\delta_{j,i} = \frac{1}{N_{\mathrm{s}}} \sum_{n=1}^{N_{\mathrm{s}}} \left\| \begin{pmatrix} x_{i,n} \\ y_{i,n} \\ z_{i,n} \end{pmatrix} - \begin{pmatrix} x_{i,0} \\ y_{i,0} \\ z_{i,0} \end{pmatrix} \right\| \tag{3}$$

where N_{s} denotes the number of samples per each actuator and $(x_{i,n}, y_{i,n}, z_{i,n})$ and $(x_{i,0}, y_{i,0}, z_{i,0})$ denote the current and the neutral (i.e., all facial actuator displacements are at minimum) feature point positions respectively. The effect index indicates the effect of an actuator j to the position of a feature point i.

Using Eq. (3) and a threshold, we can define a subset of feature points $\mathscr{F}_m$ which is affected by an actuator subset $\mathscr{A}_m \in \mathscr{A}_{\mathrm{w}}$. Concretely,

$$\mathscr{F}_m = \{i |^{\forall} i \in \mathscr{F}_{\mathrm{w}}, \delta_{j,i} > threshold, ^{\forall} j \in \mathscr{A}_m\}. \tag{4}$$

To segment properly, the feature point subsets $\mathscr{F}_1, \ldots, \mathscr{F}_{N_{\mathrm{g}}}$ should not overlap with each other; the same requirement goes for the actuator subsets $\mathscr{A}_1, \ldots, \mathscr{A}_{N_{\mathrm{g}}}$. Choosing actuator subsets that satisfy these requirements, independent inverse kinematics solvers can be created for each group of feature points and actuators. The resulting inverse kinematics solver is referred to as the *segmented face* inverse kinematics solver.

3 Experiments of Controlling an Android Face

3.1 Capturing Feature Point Positions

The android used for the evaluation of the proposed method is an Actroid-SIT android from Kokoro Co. The android's facial expression is controlled using 11 actuators. For a detailed explanation about the android and the experimental setup, please refer to [5].

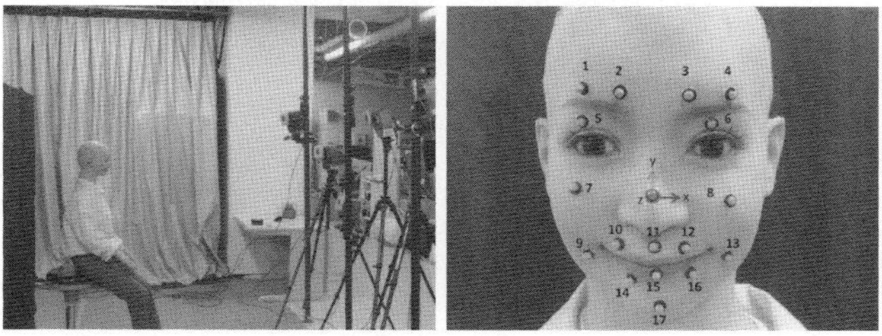

(a) Motion capture camera setup. (b) Feature point layout.

Fig. 1 Motion capture setup used for capturing feature point positions.

The android's feature point positions are captured using a motion capture system for gathering training data for training the forward kinematics ANN in Section 2.1 (see Fig. 1). On the face, 17 feature point markers are placed where significant feature point movement occurs [7]. For the training data of the forward kinematics ANN, several actuator configurations such as independent, combination, and random actuator displacements are recorded. Feature point positions are recorded while keeping the actuator displacements stationary. Additional random actuator configurations are captured for testing the generalization of the forward kinematics ANN.

Using the defined feature points shown in Fig. 1b, the forward kinematics ANN for the full face inverse kinematics solver has 11 actuators as input and 51 (17 feature points $\times$ 3 dimensions) feature point dimensions as output.

3.2 Grouping Feature Points and Actuators

To segment the feature points and actuators of the android, the method mentioned in Section 2.3 is applied. The threshold value for $\delta_{j,i}$ is manually selected through experimentation. Fig. 2 shows the feature points that are affected by each actuator. This figure indicates that the feature points and actuators can be segmented into two parts: the upper face and the lips. The upper face segment has 4 actuator displacements as input and 24 dimensional feature point positions as output, while the lips segment has 7 actuator displacements as input and 27 dimensional feature point as output.

3.3 Evaluation of the Forward Kinematics Model

Since the forward kinematics of the android's face is modeled using an ANN, the generalization of the model should be evaluated. The generalization measures the accuracy of the trained ANN when the input data are data not used during training. This tests if the training data is overfitted by the ANN.

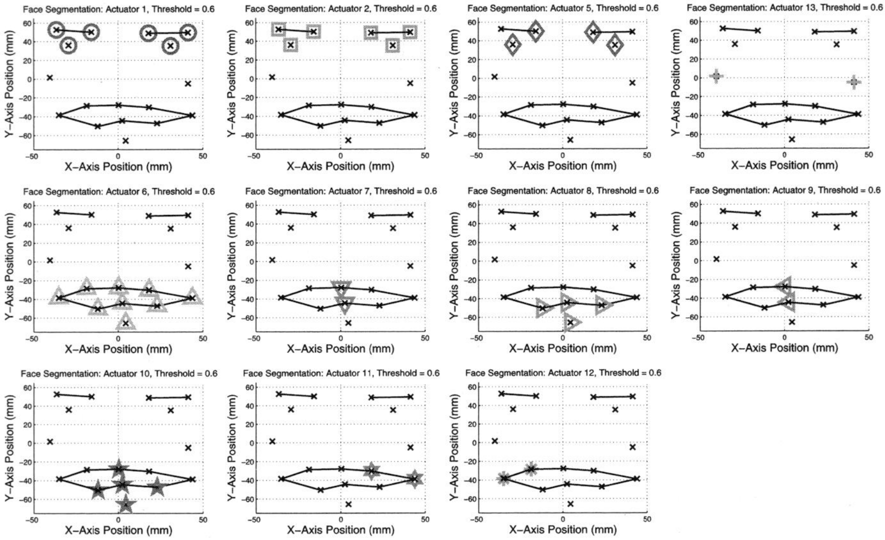

Fig. 2 Face segmentation of the android robot at *threshold* = 0.6.

Table 1 Average norm error of the forward kinematics artificial neural network.

Model Type	Mean (mm)	STD (mm)
Segmented Face	0.54	0.30
Full Face	1.15	0.62

The norm errors of the feature points are computed between the output feature point positions of the forward kinematics ANN and actual feature points positions captured by the motion capture system. This is done over 500 samples of random actuator displacements and the results are averaged. In this experiment, the full face forward kinematics ANN and the segmented face forward kinematics ANN are compared. As shown in Table 1 the error is small compared to the average displacement range of the feature points which is 13.32 mm. This suggests that the forward kinematics ANN has good generalization. However, it should be noted that the segmented face has better generalization compared to the full face forward kinematics ANN. The reason for this is that the complexity of modeling the forward kinematics is decreased by reducing the dimensions for each segmented ANN.

3.4 Evaluation of the Inverse Kinematics

Next, we evaluate the inverse kinematics solvers using two cases; also, the full face inverse kinematics solver and the segmented face inverse kinematics solver are compared. First, the target feature point positions captured from random actuator dis-

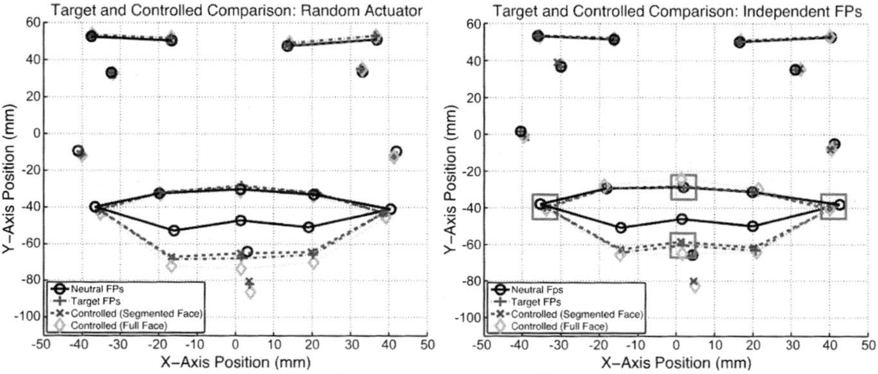

(a) Target feature points captured from random actuator displacements. The Z-axis is omitted.

(b) Independently moved feature points as target feature points. The Z-axis is omitted.

Fig. 3 Plot comparison of target (target FPs) and controlled feature points.

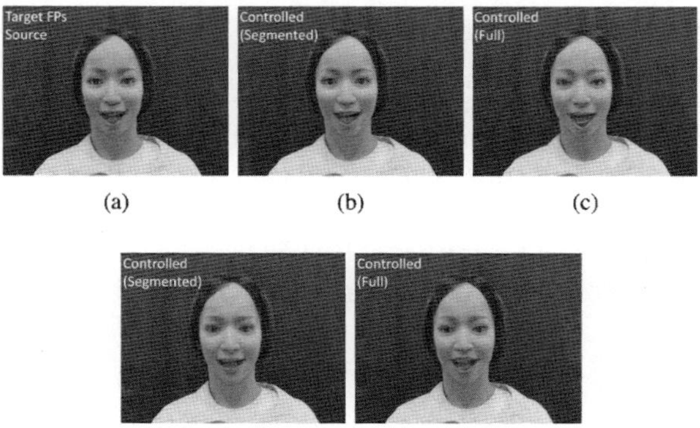

(a) (b) (c)

(d) (e)

Fig. 4 Images (b)–(c) are controlled using target feature points from random actuator displacements (a) and correspond to plots in Fig. 3a. Images (d)–(e) are controlled using independently moved target feature points and correspond to plots in Fig. 3b.

placements is used as input for the inverse kinematics solver. The weights of each feature point are set to 1 since such target feature points are assured to be feasible. The resulting feature points are shown in Fig. 3a. Inspecting the plot, observe that the target and the controlled feature point positions are close to each other indicating that the inverse kinematics solver can estimate the actuator displacements.

For the second evaluation, independently displaced feature points are set as target feature points for the inverse kinematics solver. This verifies if the weighting method proposed in Eq. (2) solves the coupling problem. The weight of the moved feature points are assigned as 1 and others as 0.01 to emphasize the moved feature points during minimization of the error function in Eq. (2). The results in Fig. 3b show that the controlled feature points are close to the target feature points. This signifies that the inverse kinematics solver is able to handle the coupling of the feature points.

Furthermore, we can observe from the plots in Fig. 3 that the segmented face inverse kinematics solver achieves better results than the full face inverse kinematics solver. The difference can also be seen by visually inspecting the controlled android face shown in Figs. 4b–e. This is because the full face inverse kinematics solver considers all the feature points and actuator displacements using a single ANN which makes modeling the forward kinematics and the minimization process more complex. This proves that the proposed face segmentation in Section 2.3 increased the precision of the inverse kinematics solver.

4 Conclusion

This paper presented a method to solve the inverse kinematics of androids with elastic faces. We addressed the problem of solving the inverse kinematics of such android faces, that is, the complexity of modeling the deformable face and the coupling of the feature points. Our proposed method employed an artificial neural network to model the forward kinematics. Then, the inverse kinematics was solved by using an iterative minimization technique, where a weighted squared error is introduced to handle the coupling of the feature points. This solution to the coupling enables the input of infeasible target feature point positions. Lastly, a face segmentation technique for grouping the feature points and the actuators was proposed to improve the accuracy of the inverse kinematics solver. Experimental results showed that the proposed inverse kinematics solver can control the android's facial expression using target feature points.

References

1. Beale, M., Demuth, H.: Neural network toolbox. For use with MATLAB, User's guide. The MathWorks, Natick (1998)
2. Bishop, C.: Pattern recognition and machine learning, vol. 4. Springer, New York (2006)
3. Hashimoto, T., Hitramatsu, S., Tsuji, T., Kobayashi, H.: Development of the face robot SAYA for rich facial expressions. In: Int. Joint Conf. on SICE-ICASE, pp. 5423–5428. IEEE (2006)
4. Jaeckel, P., Campbell, N., Melhuish, C.: Facial behaviour mapping from video footage to a robot head. Robot. Auton. Syst. **56**(12, 31), 1042–1049 (2009)
5. Magtanong, E., Yamaguchi, A., Takemura, K., Takamatsu, J., Ogasawara, T.: Inverse kinematics solver for an android face using neural network. In: 29th Annual Conf. of the Robotics Society of Japan, pp. 1Q3–1 (2011)
6. Tolani, D., Goswami, A., Badler, N.: Real-time inverse kinematics techniques for anthropomorphic limbs. Graph. Models **62**(5), 353–388 (2000)
7. Wilbers, F., Ishi, C., Ishiguro, H.: A blendshape model for mapping facial motions to an android. In: Int. Conf. on Intelligent Robots and Systems, pp. 542–547. IEEE (2007)

Decomposing Envelopes of Rational Hypersurfaces

Tino Schulz and Bert Jüttler

Abstract The envelope of a family of real, rational hypersurfaces is defined by an implicit equation in the parameter space. This equation can be decomposed into factors that are mapped to varieties of different dimension. The factorization can be found using solely gcd computations and polynomial divisions. The decomposition is used to derive some general results about envelopes, which also contribute to the analysis of self-intersections.

Key words: Envelopes, singularities, self-intersection

1 Introduction

Envelopes of curves and surfaces are a classical topic of differential geometry and kinematics [2, 7]. Due to their importance in various applications, computational techniques for dealing with envelopes have attracted the interest of researchers from several fields. These include robotics (collision detection and avoidance) and gearing theory (design of matching pairs of gear teeth surfaces), geometrical optics (caustics), NC-machining (offset curves for tool path generation) and Computer Aided Geometric Design (sweeps, convolutions, Minkowski sums). See e.g. [1, 5, 6, 8–10] and the references cited therein.

In this paper, we generalize the approach presented in [11], which is restricted to the curve case, to envelopes of general families of hypersurfaces. More precisely, we will focus on the fact that envelopes are essentially *singularities* of the mapping that

Tino Schulz
GALAAD, INRIA Méditerranée, 2004 Route des Lucioles, 06902 Sophia-Antipolis, France, e-mail: tino.schulz@inria.fr

Bert Jüttler
Institute of Applied Geometry, Johannes Kepler University, Altenberger Str. 69, 4040 Linz, Austria, e-mail: bert.juettler@jku.at

J. Lenarčič, M. Husty (eds.), *Latest Advances in Robot Kinematics*,
DOI 10.1007/978-94-007-4620-6_24,

describes the defining family of hypersurfaces. We will identify and compute the parts of the envelope possessing different dimension, and we derive some general results of envelopes.

2 Envelopes of Rational Hypersurfaces

We recall basic properties of envelopes and show how to identify their parts possessing different dimensions. Consider a rational mapping

$$\mathbf{x}(\mathbf{t}) = \left(x_1(\mathbf{t})/x_0(\mathbf{t}), \ldots, x_n(\mathbf{t})/x_0(\mathbf{t})\right)^\top \tag{1}$$

where the $x_i(\mathbf{t})$ $(i=0\ldots n)$ are real, n-variate polynomials. Here

$$\mathbf{t} = (t_1, \ldots, t_n)^\top \in I_1 \times \cdots \times I_n = \mathbf{I} \subset \mathbb{R}^n \tag{2}$$

with closed real intervals I_i $(i = 1\ldots n)$. Moreover, we assume that $x_0 \neq 0$ for $\mathbf{t} \in \mathbf{I}$, $\gcd(x_0, \ldots, x_n) = 1$ and that the image $\mathbf{x}(\mathbf{I})$ is not completely contained in any hypersurface. Though we are mostly interested in real properties of $\mathbf{x}$, it will be necessary to consider the complex extension of $\mathbf{x} : \mathbb{C}^n \to \mathbb{C}^n$. When not explicitly stated differently we will make use of a complex variable $\mathbf{s} \in \mathbb{C}^n$ and examine $\mathbf{x}(\mathbf{s})$ in the remainder of this paper.

If we pick any index j and use t_j as a time-like parameter $t_j = \tau$, then the mapping $\mathbf{x}$ defines a family of rational hypersurfaces. For each value of τ, the corresponding hypersurface is obtained by varying the remaining $n-1$ parameters t_i $(i \neq j)$.

The envelope of this family of hypersurfaces is defined by the property that it is tangent to almost every member of the family. With respect to the mapping $\mathbf{x}$, we can characterize the envelope as the image of those points where the Jacobian $\mathbf{J}$ is singular, i.e., *envelopes* are essentially *singularities*. Consequently, the envelope is independent of the choice of the index j. A short computation confirms that

$$\det \mathbf{J} = \frac{1}{x_0} \begin{vmatrix} x_0 & 0 & \cdots & 0 \\ x_1 & \frac{\partial_1 x_1}{x_0} - \frac{\partial_1 x_0}{x_0^2} x_1 & \cdots & \frac{\partial_n x_1}{x_0} - \frac{\partial_n x_0}{x_0^2} x_1 \\ \vdots & \vdots & \ddots & \vdots \\ x_n & \frac{\partial_1 x_n}{x_0} - \frac{\partial_1 x_0}{x_0^2} x_n & \cdots & \frac{\partial_n x_n}{x_0} - \frac{\partial_n x_0}{x_0^2} x_n \end{vmatrix} = \frac{1}{x_0^{n+1}} \underbrace{\begin{vmatrix} x_0 & \partial_1 x_0 & \cdots & \partial_n x_0 \\ x_1 & \partial_1 x_1 & \cdots & \partial_n x_1 \\ \vdots & \vdots & \ddots & \vdots \\ x_n & \partial_1 x_n & \cdots & \partial_n x_n \end{vmatrix}}_{=h} \tag{3}$$

where ∂_i denotes the differentiation with respect to the i-th variable. The determinant defining the polynomial h is obtained by adding $(\partial_i x_0)/x_0^2$ times the first column to the $(i+1)$-th one for $i = 1, \ldots, n$, and then factoring out the common denominators.

Since the points $\mathbf{s} \in \mathbb{C}^n$ satisfying $h(\mathbf{s}) = 0$ are mapped to the envelope, we call h the *envelope function*. The zero set of h consists of one or several (possibly complex) hypersurfaces, i.e. surfaces whose dimension is exactly $n-1$.

3 Decomposing Envelopes

The envelope function possesses a factorization in $\mathbb{C}[\mathbf{s}]$ into irreducible and relatively prime polynomials h_j, $j = 1, \dots, M$ with certain multiplicities. By a suitable ordering we can guarantee that the first N factors ($N \leq M$) do not divide x_0, while the remaining ones do. After eliminating all factors shared with x_0 and reducing the multiplicities of the remaining factors to 1 we obtain the *reduced envelope function*

$$\hat{h}(\mathbf{s}) = \prod_{j=1,\dots,N} h_j(\mathbf{s}). \tag{4}$$

Instead of using factorization techniques, the reduced envelope function can be found via suitable gcd computations, cf. [3]. Clearly, $\hat{h}$ is squarefree and satisfies $(\nabla_{\mathbf{s}}\hat{h})(\mathbf{s}) \neq \mathbf{0}$ almost everywhere, where $\nabla_{\mathbf{s}}$ denotes the gradient with respect to $\mathbf{s}$.

The zero sets of the factors h_j are mapped to components that might possess different dimensions. In order to identify those factors h_j whose zero sets are mapped to varieties of a certain dimension, we consider the restriction of the differential of $\mathbf{x}$ at a point $\mathbf{s}$ to the tangent spaces of these zero sets. If the rank of this restriction is equal to r for almost all points satisfying $h_j(\mathbf{s}) = 0$, then this algebraic variety is mapped to a variety of dimension r.

The differential of $\mathbf{x}$ at $\mathbf{s}$ is the linear mapping defined by the Jacobian $\mathbf{J}(\mathbf{s})$. We consider the augmented Jacobian $\mathbf{J}^+(\mathbf{s})$ which is obtained by adding the row vector $\left((\nabla_{\mathbf{s}}\hat{h})(\mathbf{s})\right)^\top$ to $\mathbf{J}(\mathbf{s})$. The augmented Jacobian thus has $n+1$ rows and n columns.

The dimension of the kernel of $\mathbf{J}^+$ equals $n - r$, where $r = \operatorname{rk}\mathbf{J}^+$. For all points $\mathbf{s}_0 \in \mathbb{C}^n$ satisfying $(\nabla_{\mathbf{s}}\hat{h})(\mathbf{s}_0) \neq \mathbf{0}$, the hypersurface $\hat{h}(\mathbf{s}) = \hat{h}(\mathbf{s}_0)$ possesses a well-defined tangent space at $\mathbf{s}_0$ and the kernel of $\mathbf{J}^+(\mathbf{s}_0)$ is contained in it, due to the additional row in the augmented Jacobian. Consequently, the augmented Jacobian $\mathbf{J}^+(\mathbf{s})$ – and hence also the Jacobian $\mathbf{J}(\mathbf{s})$ – maps this tangent space into a space of dimension $r - 1$.

Thus, for almost all points satisfying $h_j(\mathbf{s}) = 0$ for a particular index j, the dimension of the image of this hypersurface under $\mathbf{x}$, i.e., of the associated component of the envelope, is equal to $\operatorname{rk}\mathbf{J}^+(\mathbf{s}) - 1$. This property is inherited by the matrix $\mathbf{V} = (x_0)^2\mathbf{J}^+$, which has polynomial entries. The vanishing of all i-th order minors of $\mathbf{V}$ (the determinants of all its $(i \times i)$-submatrices) is a necessary and sufficient condition for $\operatorname{rk}\mathbf{V} \leq i - 1$.

This observation leads us to formulate the following procedure for decomposing $\hat{h}$ into factors whose zero sets are mapped into components of different dimensions:

- Let $g_{n+1} = \hat{h}$. Further, let g_i be the greatest common divisor of $\hat{h}$ and of all i-th order minors of $\mathbf{V}$ ($i = 1 \dots n$). Obviously, g_i divides g_{i+1}, and the zero set of g_i is mapped into components of maximum dimension $i - 2$.
- Further, let $f_i = g_{i+2}/g_{i+1}$, ($i = 0 \dots n-1$). The zero set of f_i is mapped into components of dimension i. This gives the decomposition

$$\hat{h}(\mathbf{s}) = \prod_{i=0\dots n-1} f_i(\mathbf{s}). \tag{5}$$

The polynomial f_{n-1} is called the *proper envelope function*. A factor h_j of $\hat{h}$ is called *proper*, if and only if it is also a factor of f_{n-1}, otherwise it is said to be *improper*.

We summarize these observations in

Theorem 1. *The rational mapping* $\mathbf{x}$ *maps the zero set of the polynomial* f_i *into a component of the envelope. The smallest algebraic variety containing this component is of dimension* i*.*

Let $\mathscr{D}_i$ be the image of the zero set of f_i under $\mathbf{x}$ ($i = 0,\dots,n-1$). The sets $\mathscr{D}_i$ are images of real algebraic hypersurfaces under a real rational mapping and are of complex dimension i. Their real dimension might be lower.

The *proper part* $\mathscr{D}_{n-1}$ of the envelope is of particular interest. There exists a real, squarefree polynomial q such that

$$\mathscr{D}_{n-1} \subseteq V(q) = \{\mathbf{p} \in \mathbb{C}^n : q(\mathbf{p}) = 0\}. \tag{6}$$

If the degree of q is minimal, then $q = 0$ is the implicit equation of the envelope. The following example illustrates these facts.

Example 1. Let $n = 3$ and consider

$$\mathbf{x}(s,t,u) = \left(\frac{(s+t)(st+1)(u-1)}{1+s^2}, \frac{4u}{1+u^2}, \frac{s^2(1+t^2)(1-u^2)^2}{(1+s^2)(1+u^2)} \right)^\top. \tag{7}$$

The reduced envelope function is $\hat{h} = (t+I)(t-I)(1+st)(u+1)(u-1)s$, where $I^2 = -1$. A short computation gives

$$f_0 = u-1, \quad f_1 = (t+I)(t-I)(1+st)(u+1) \quad \text{and} \quad f_2 = s. \tag{8}$$

By applying $\mathbf{x}$ to the zero sets of the polynomials f_i we obtain that

- $\mathscr{D}_0$ is the point $(0,2,0)^\top$,
- $\mathscr{D}_1$ consists of an ellipse, a line, and two complex conjugate ellipses and
- $\mathscr{D}_2$ is a certain subset of the xy-plane.

Consequently, we get $q(x,y,z) = z$. The numerator of $q \circ \mathbf{x}$ includes all those factors of $\hat{h}$ that are mapped on the proper part of the envelope. Note that f_2 and two additional factors appear in $q \circ \mathbf{x}$ with multiplicity 2. This will be investigated in the next section.

The computation of the exact implicit equation of an envelope is rather expensive in general. Although several methods exist [3], their complexity usually restricts their practical application to planar or low-degree problems. Techniques for *approximate implicitization* are a valuable alternative, see [4, 11].

4 Using the Decomposition

In this section, we will use the factorization (5) to derive several properties of the envelope. The first result generalizes Theorem 1 from [11].

Theorem 2. *Let q be the implicit equation of the proper part of the envelope as defined in* (6). *There exists a real, n-variate, polynomial $\tilde{\lambda} : \mathbb{C}^n \to \mathbb{C}$, such that*

$$(q \circ \mathbf{x}) \cdot (x_0)^d = \tilde{\lambda} \cdot (f_{n-1})^2, \tag{9}$$

where d is the degree of q.

Proof. Since $\mathscr{D}_{n-1} \subset V(q)$ consists of an $n-1$-dimensional family of points $\mathbf{x}(\mathbf{s})$ fulfilling $f_{n-1}(\mathbf{s}) = 0$, we can conclude that f_{n-1} is a factor of the numerator of $q \circ \mathbf{x}$. Additionally we note that if $f_{n-1}(\mathbf{s}) = 0$, then

$$\nabla_{\mathbf{s}} (q \circ \mathbf{x})(\mathbf{s}) = \mathbf{J}(\mathbf{s})^\top (\nabla_x q \circ \mathbf{x})(\mathbf{s}) = \mathbf{0}, \tag{10}$$

because $\mathbf{J}(\mathbf{s})$ spans the tangent space of the envelope. This implies that $(f_{n-1})^2$ is a factor of the numerator of $q \circ \mathbf{x}$ since f_{n-1} is squarefree. □

Theorem 2 implies that $(f_{n-1})^2$ is a factor of the composition $q \circ \mathbf{x}$. Now we study the remaining factors of multiplicity 2:

Corollary 1. *If $\tilde{\lambda}$ is not squarefree, then its factors of multiplicity greater than one are also factors of $\hat{h}$.*

Proof. If $\tilde{\lambda}$ is not squarefree then there exist polynomials $\nu, \mu : \mathbb{C}^n \to \mathbb{C}$ such that $\tilde{\lambda} = \nu\mu$, where ν is squarefree and μ has only factors of multiplicity greater than one. For every $\mathbf{s} \in \mathbb{C}^n$ with $\mu(\mathbf{s}) = 0 \neq x_0(\mathbf{s})$ we get that

$$(q \circ \mathbf{x})(\mathbf{s}) = \nu(\mathbf{s})\mu(\mathbf{s})\left(f_{n-1}(\mathbf{s})\right)^2 = 0 \tag{11}$$

which implies

$$\nabla_{\mathbf{s}} (q \circ \mathbf{x})(\mathbf{s}) = \mathbf{J}(\mathbf{s})^\top (\nabla_x q \circ \mathbf{x})(\mathbf{s}) = \mathbf{0}, \tag{12}$$

since $\nabla_{\mathbf{s}}\mu(\mathbf{s}) = \mathbf{0}$. The rightmost identity of equation (12) can only be fulfilled for a $n-1$-dimensional family of points if $\mathbf{J}$ is singular. Thus every factor of the square-free representation of μ must also be a factor of $\hat{h}$. □

Consequently, the factors of $\tilde{\lambda}$ with a multiplicity greater than one correspond to those factors of $\hat{h}$ that are "singularly" mapped on the proper part of the envelope. Note that $\tilde{\lambda}$ might contain factors of x_0 which we eliminate by setting

$$\lambda = \tilde{\lambda} / \gcd(\tilde{\lambda}, x_0). \tag{13}$$

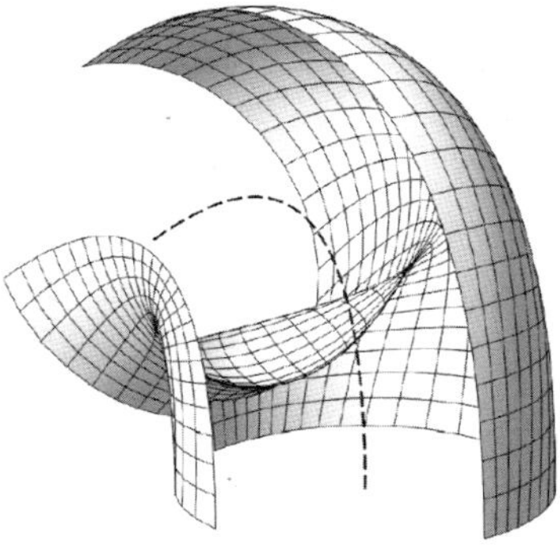

Fig. 1 Example 2: A part of the envelope, which is the offset of a parabola (dashed curve), and its self-intersections.

5 Self-intersections and "Undercuts"

According to Theorem 2, $q \circ \mathbf{x} = 0$ holds also for every point on the zero set of λ. The factors of λ which are not factors of f_{n-1} characterize additional intersections of the family $\mathbf{x}$ with the proper part of the envelope:

Corollary 2. *Let* $\mathbf{s}' \in \mathbb{C}^n$ *such that* $f_{n-1}(\mathbf{s}') \neq 0 \neq x_0(\mathbf{s}')$. *Assume there exists* $\mathbf{s} \in \mathbb{C}^n$ *satisfying* $f_{n-1}(\mathbf{s}) = 0$ *and* $\mathbf{x}(\mathbf{s}) = \mathbf{x}(\mathbf{s}')$. *Then* $\lambda(\mathbf{s}') = 0$.

Proof. We directly obtain $0 = (q \circ \mathbf{x})(\mathbf{s}) = (q \circ \mathbf{x})(\mathbf{s}') = \lambda(\mathbf{s}') \cdot f_{n-1}(\mathbf{s}')$, which implies $\lambda(\mathbf{s}') = 0$. □

Consequently, $\lambda(\mathbf{s}') = 0$ is a necessary condition for the point $\mathbf{x}(s')$ to be located on the proper part of the envelope, and therefore to create an "undercut". This interesting observation may be used for the trimming of offsets and for eliminating the undercut of envelope surfaces. We explain this by an example.

Example 2. Consider the rational mapping

$$\mathbf{x}(s,t,u) = \frac{1}{x_0(s,t)} \begin{pmatrix} (1+s^2)\left(u+ut^2+2t\right) \\ 2s(1+t)(1-t) \\ s^2+t^2-u^2-s^2t^2-s^2u^2-t^2u^2-s^2t^2u^2-1 \end{pmatrix}, \tag{14}$$

with $x_0 = \left(1+s^2\right)\left(1+t^2\right)$. It describes as a sphere of radius 1 whose center is moving along a parabola in the xz-plane, where u is the time-like parameter.

The proper envelope function is $f_{n-1} = (1-s^2-t^2-s^2t^2)u + (1+t)s^2$ and the proper part of the envelope is the offset surface of distance 1 of the parabola, see Fig. 1. It is a pipe surface with the implicit equation

$$\begin{aligned} q(x,y,z) = \; & 16(x^2+y^2)^2(x^2+y^2+z^2) - 2z(3x^2-36y^2-20z^2) \\ & +8z(5x^4-4y^4+x^2y^2+4x^2z^2-4y^2z^2)+28x^2+65y^2+9z^2 \\ & -47x^4-56y^4+16z^4-76x^2y^2-24y^2z^2-40z-25, \end{aligned} \tag{15}$$

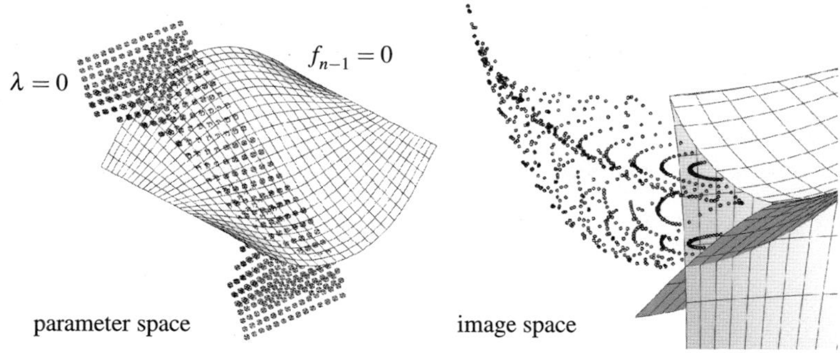

Fig. 2 Example 2: The zero sets of f_{n-1} (shows as surface) and λ (shown as point cloud) in the parameter space (left) are mapped onto the envelope $\mathscr{D}_2$ in the image space (right). In particular, the zero set of λ s mapped to the undercut region.

and it possesses a certain region of self intersection. If one thinks of **x** as describing a moving cutting tool which moves along some path, then the part of the envelope that is bounded by its singularities would be cut away. Thus, in situations like in this example, this part is referred to as *undercut*. In several applications (e.g. offset trimming), it is an important task to determine it.

Let $\lambda(s,t,u)$ be defined as in section 4, i.e. take $\tilde{\lambda} = (q \circ \mathbf{x}) \cdot (x_0)^6/(f_{n-1}^2)$ and remove common factors with x_0. It is a rather complicated polynomial of tri-degree $(6,6,4)$ which describes a surface with two sheets that are almost parallel. Figure 2 shows the zero sets of λ and f_{n-1} in parameter space (which are visualized by a point cloud and by parameter lines, respectively) and their images under **x**. The image of $\lambda = 0$ is the undercut region, and the curves defined by $\lambda = f_{n-1} = 0$ are mapped to the self-intersection curves of the envelope.

The additional components which are defined by λ also appear in the problem of sorting assembly modes in robot kinematics. In that context they are referred to as *characteristic (hyper-)surfaces*, see [12].

6 Conclusion

We have shown how to decompose the defining equations of an envelope into polynomial factors that are mapped onto varieties of different dimension. The proposed method is algorithmically simple and constructs an explicit decomposition only using gcd computations and polynomial division.

We then deduced some general properties of envelopes, generalizing existing results for curves. In particular, we addressed some aspects which are closely related to the analysis of self-intersections and "undercuts".

Future work could be devoted to a more detailed investigation on the properties of the factorization described in Theorem 2, to its application in the determination of undercut regions, and to the use of approximate implicitization techniques for envelope surfaces.

Acknowledgements The first author was supported by the Marie-Curie Network SAGA (FP7, GA no. 214584), and by the Doctoral Program "Computational Mathematics" (W1214) at Johannes Kepler University, Linz. The authors thank the anonymous referees for their useful comments, in particular for their contributions that led to a correct version of Theorem 1.

References

1. Abdel-Malek, K., Yang, J., Blackmore, D., Joy, K.: Swept volumes: fundation, perspectives, and applications. Int. J. Shape Model. **12**(1), 87–127 (2006)
2. Bottema, O., Roth, B.: Theoretical Kinematics. Dover Publications (1990)
3. Cox, D., Little, J., O'Shea, D.: Ideals, Varieties, and Algorithms. Springer-Verlag, New York (2007)
4. Dokken, T., Thomassen, J.: Overview of approximate implicitization. In: Topics in Algebraic Geometry and Geometric Modeling, vol. 334, pp. 169–184. American Mathematical Society (2003)
5. Flaquer, J., Garate, G., Pargada, M.: Envelopes of moving quadric surfaces. Comput. Aided Geom. Design **9**(4), 299–312 (1992)
6. Kim, Y., Varadhan, G., Lin, M., Manocha, D.: Fast swept volume approximation of complex polyhedral models. Comput. Aided Des. **36**(11), 1013–1027 (2004)
7. Kreyszig, E.: Differential Geometry. Dover (1991)
8. Peternell, M., Pottmann, H., Steiner, T., Zhao, H.: Swept volumes. Comput. Aided Des. Appl. **2**, 599–608 (2005)
9. Pottmann, H., Peternell, M.: Envelopes-computational theory and applications. In: Spring Conf. on Computer Graphics, pp. 3–23. Comenius Univ., Bratislava (2000)
10. Rabl, M., Jüttler, B., Gonzalez-Vega, L.: Exact envelope computation for moving surfaces with quadratic support functions. In: Advances in Robot Kinematics: Analysis and Design, pp. 283–290. Springer (2008)
11. Schulz, T., Jüttler, B.: Envelope computation in the plane by approximate implicitization. Appl. Algebra Engrg. Comm. Comput. **22**, 265–288 (2011)
12. Wenger, P.: A new general formalism for the kinematic analysis of all nonredundant manipulators. In: Robotics and Automation, pp. 442–447. IEEE (1992)

Influence of Pulley Kinematics on Cable-Driven Parallel Robots

Andreas Pott

Abstract In this paper the modeling of a pulley mechanism for cable-driven parallel robots is presented. In many works, the proximal anchor points of the robots are simplified to be ideal points. Real cables achieve reasonable life time only when a minimum bending radius is exceeded. Therefore, pulley mechanisms have to be used which in turn require the extension of the kinematic modeling. In this paper a kinematic model for a pulley mechanism of the winches is revisited. Then we derive a corrected structure equation and compare the different results from the extended model with the estimation from the simplified standard model with respect to kinematics transformation, workspace, and force distribution.

Key words: Cable-driven parallel robot, pulley mechanism, kinematics, statics, stiffness, workspace

1 Introduction

A cable-driven parallel robot is a special type of parallel kinematic machine where the rigid struts are replaced by light-weight cables. Therefore, the inertia of the robot is largely reduced allowing for application in large-scale [2], ultra-fast [5], and heavy duty applications [1]. Due to their advantages cable robots attracted increasing attention during the last year. Although some fundamental issues of cable robots are still open, researchers have started to address practical issues related to construction and control of prototypes. The results presented in this paper were driven by the development of the cable robot IPAnema [7], which targets at application in large-scale handling and assembly.

Cables are very flexible and versatile construction elements that are used in applications such as bridges, cable-cars, and elevators. Nevertheless, there are important

Andreas Pott
Fraunhofer IPA, Stuttgart, Germany, e-mail: andreas.pott@ipa.fraunhofer.de

J. Lenarčič, M. Husty (eds.), *Latest Advances in Robot Kinematics*,
DOI 10.1007/978-94-007-4620-6_25,

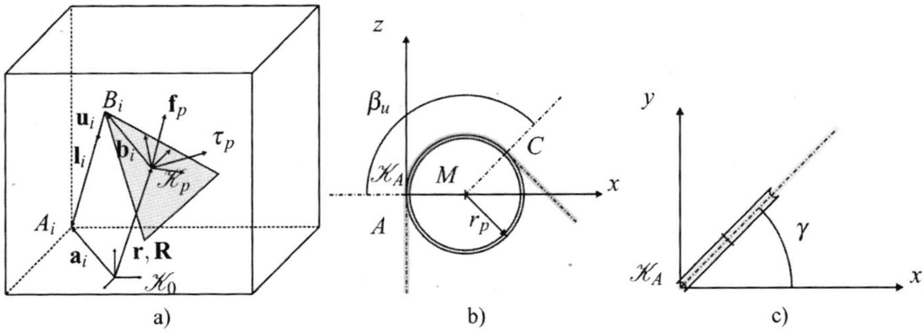

Fig. 1 Geometry and kinematics of a cable-driven parallel robot: (a) simplified model, (b,c) definition of coordinate frame $\mathcal{K}_A$ and variables for pulley kinematics.

design rules when using cables that have to be taken into account such as minimum feasible bending radius. Therefore, one has to integrate elements such as pulleys to allow for acceptable durability as well as safety.

Only few authors have addressed the influence of guiding pulleys on the kinematics of cable-driven parallel robots. Bruckmann [3] derived an inverse kinematic algorithm to cope with pulleys. The influence of pulleys was also taken into account for the dynamic simulation of cable robots [6].

This paper aims at studying the influence of a pulley in the winch on the properties of a cable robot. Therefore, a kinematic modeling is presented and the equations for inverse kinematics and statics are derived taking into account the effect of a pulley as guiding element in the winches of the robot. The rest of the paper is organized as follows. In Sec. 2 the basic equations for the modeling of pulley mechanisms are presented while in Sec. 3 the method to determine the workspace properties is briefly explained. The results from the comparison are discussed in Sec. 4 where the paper closes with the conclusions.

2 Kinematics for Pulley Mechanism

For better reference, the kinematic foundation of cable robots are briefly reviewed. We refer to the well-known approach as *standard model* and we extend it by the guiding pulley in this section. Fig. 1a shows the kinematic structure of a general spatial cable robot, where the vectors $\mathbf{a}_i$ denote the proximal anchor points on the robot base, the vectors $\mathbf{b}_i$ are the relative positions of the distal anchor points on the movable platform, and $\mathbf{l}_i$ denote the vector of the cables. The length of the cables is abbreviated by $l_i = ||\mathbf{l}_i||_2$. Applying a vector loop, the closure-constraint reads

$$\mathbf{a}_i - \mathbf{r} - \mathbf{R}\,\mathbf{b}_i - \mathbf{l}_i = \mathbf{0} \quad \text{for} \quad i = 1, \ldots, m, \tag{1}$$

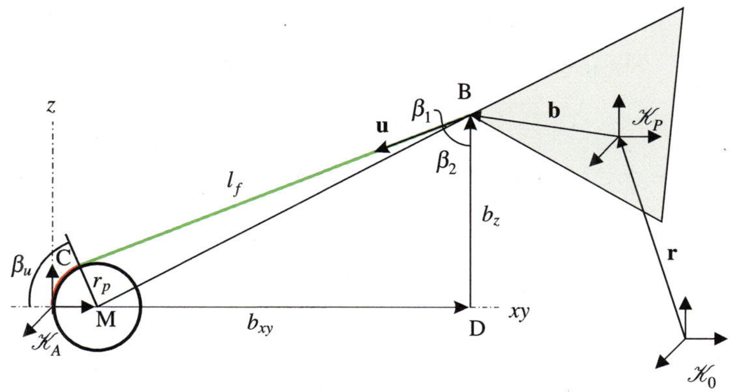

Fig. 2 Kinematics of one cable with a guiding pulley.

where the vector **r** is the Cartesian position of the platform and the rotation matrix **R** represents the orientation of the platform.

The parameters and coordinate frame used to exactly define the geometry of a guiding pulley are depicted in Fig. 1b. In the rest of this section we omit the index i for the reference points, frames, angles, and lengths for the sake of clarity. In this paper we propose to express the pulley kinematics based on a local coordinate frame $\mathscr{K}_A$ which largely simplifies the kinematic equation and represents a natural concept of arranging the winches in space. The pulley kinematics realizes a two degree-of-freedom motion. The first revolute joint is aligned with the z-axes of frame $\mathscr{K}_A$. The second joint is the pulley itself and its joint axis is initially aligned with the y-axis of frame $\mathscr{K}_A$. The center of second rotation is initially located in point M. The distance between the two screw joint axis is the effective radius r_p and it is assumed that the two joint axis are perpendicular to each other.

The fixed point A in the origin of the coordinate frame $\mathscr{K}_A$ is the characteristic point of the pulley kinematics and considered to be a design parameter **a**. The cable hits the pulley at point A and wraps around the pulley with an effective radius r_p, i.e. the radius that applies to the neutral fibre in the center of the cable. In the following considerations we assume that r_p is the effective radius, i.e. the radius resulting from both the actual radius of the pulley and the radius of the cable. Note that this holds true only if the geometric profile of the pulley and the radius of the cable perfectly match. The cable leaves the pulley at point C and the angle between point A and C is denoted by β_u. In Fig. 1c one can see the rotated pulley, where the rotation angle is denoted by γ and is taken in positive direction around the z-axis of frame $\mathscr{K}_A$. The definition of the angles β_u and γ with respect to $\mathscr{K}_A$ is crucial for both the formulation of the kinematic codes and the consideration of collisions between the cable and the pulley mechanism. For $\gamma = 0$ the pulley is located in the xz-plane of frame $\mathscr{K}_A$. Therefore, in this position β_u is measured in positive direction around the y-axis of $\mathscr{K}_A$. The orientation of $\mathscr{K}_A$ w.r.t. to $\mathscr{K}_0$ is expressed by the rotation matrix $\mathbf{R}_A$ and assumed to be given.

The kinematic equations are now derived in $\mathcal{K}_A$ as follows. From Fig. 2, the corrected cable length taking into account the pulley radius becomes

$$l = \beta_u r_p + l_f, \tag{2}$$

where β_u is the angle around the pulley, r_p is the pulley radius, and l_f is the free cable length from point C to point B. Considering the two right-angled triangles we receive

$$(b_{xy} - r_p)^2 + b_z^2 = \overline{MB}^2 = l_f^2 + r_p^2, \tag{3}$$

where $b_{xy} = \sqrt{b_x^2 + b_y^2}$ and b_z are the coordinates of the point B with respect to frame $\mathcal{K}_A$ in cylinder coordinates. To solve the inverse kinematics we need the angle β_u which is computed as follows: Considering the tetragon CMDB, we find two angles to be right-angles. Therefore, we conclude that the enclosed angle $\beta_1 + \beta_2$ at point B equals the sought complementary angle β_u at point M. Using elementary trigonometric functions yields

$$\beta_u = \beta_1 + \beta_2 = \arccos \frac{l_f}{\sqrt{(b_{xy} - r_p)^2 + b_z^2}} + \arccos \frac{b_z}{\sqrt{(b_{xy} - r_p)^2 + b_z^2}}. \tag{4}$$

Thus, we receive a closed-form solution for the cable length l. Further reduction in the computational costs can be achieved using the addition theorem for arccos. It is worthwhile to mention that one can set up similar formulas using either arctan or arcsin where both formulations require a distinction of cases when b_z changes its sign. Nevertheless, using arctan and the respective addition theorem gives a very compact expression which is only valid for positive b_z. The advantage of the presented formula is that one can get the symbolic derivative for the first-order kinematics without additional efforts.

A unique solution for the rotation of the first joint can easily be obtained using the four-quadrant arcus tangens $\gamma = \arctan 2(b_y, b_x))$. To calculate the normal vector $\mathbf{u}$ along the cable in $\mathcal{K}_0$ we rotate a negative unit vector $-\mathbf{e}_z$ along the z-axis with the following transformation matrices

$$\mathbf{u} = -\mathbf{R}_A\, \mathbf{R}_z(\gamma)\, \mathbf{R}_y(\beta_u)\mathbf{e}_z, \tag{5}$$

where $\mathbf{R}_y(\beta_u)$ and $\mathbf{R}_z(\gamma)$ are the elementary rotation matrix around the y- and z-axis, respectively.

Considering the force and torque equilibrium for the platform leads to the well-known structure equations of the standard model (see e.g. [9])

$$\mathbf{A}^{\mathrm{T}}\, \mathbf{f} + \mathbf{w} = \mathbf{0}, \tag{6}$$

where $\mathbf{A}^{\mathrm{T}}$ is the pose-dependent structure matrix, $\mathbf{f}$ is the vector of the positive cable forces, and $\mathbf{w}$ is the applied wrench at the platform. When considering a pulley model for the robot, the type of the equation is maintained where we have to use a different unit vector for the direction of the cables as given by Eq. (5). Thus, the columns of the structure matrix become $[\mathbf{A}_p^{\mathrm{T}}]_i = [\mathbf{u}_i^{\mathrm{T}}, (\mathbf{b}_i \times \mathbf{u}_i)^{\mathrm{T}}]$.

3 Workspace

To compare the results from the workspace calculation with and without pulleys we use a triangulation of the workspace's hull [7]. Although the triangulation lacks the verified nature of interval computations [4] it can be computed with a very high accuracy at moderate computational times of some seconds. Here we use a method for the determination of the border of the workspace based on discrete investigation of single points and on a line search. In this approach the workspace is represented by triangulation that allows for simple but accurate determination of the volume and surface. Let $(\mathbf{v}_a, \mathbf{v}_b, \mathbf{v}_c)_i$ be the vertices of the triangles of the border of the workspace $\mathscr{W}$ and $\widehat{\mathbf{m}}$ be the projection center of the workspace. Note, $\widehat{\mathbf{m}}$ is a parameter in the algorithm for workspace determination describing the point where the projection of a unit sphere is started. It is straightforward to calculate the surface $S(\mathscr{W})$ and the volume $V(\mathscr{W})$ of the workspace as follows

$$S(\mathscr{W}) = \frac{1}{2}\sum^{T} \|(\mathbf{v}_a - \mathbf{v}_b) \times (\mathbf{v}_a - \mathbf{v}_c)\|_2 \tag{7}$$

$$V(\mathscr{W}) = \frac{1}{6}\sum^{T} ((\mathbf{v}_a - \widehat{\mathbf{m}}) \times (\mathbf{v}_b - \widehat{\mathbf{m}})) \cdot (\mathbf{v}_c - \widehat{\mathbf{m}}) \tag{8}$$

4 Computational Results

The geometrical parameters of the robot used for this study are given in Table 1. The parameters represent the scale of the archetype IPAnema 1, but no dot exactly match the values of the prototype. For this robot, the diameter of the cable is $r_C = 0.002$ m

Table 1 Geometrical parameters of the investigated robot given as platform vectors **b** and base vectors **a**.

cable i	base vector $\mathbf{a}_i$ [m]	platform vector $\mathbf{b}_i$ [m]
1	$[-2.0, 1.5, 2.0]^{\mathrm{T}}$	$[-0.06, 0.06, 0.0]^{\mathrm{T}}$
2	$[2.0, 1.5, 2.0]^{\mathrm{T}}$	$[0.06, 0.06, 0.0]^{\mathrm{T}}$
3	$[2.0, -1.5, 2.0]^{\mathrm{T}}$	$[0.06, -0.06, 0.0]^{\mathrm{T}}$
4	$[-2.0, -1.5, 2.0]^{\mathrm{T}}$	$[-0.06, -0.06, 0.0]^{\mathrm{T}}$
5	$[-2.0, 1.5, 0.0]^{\mathrm{T}}$	$[-0.06, 0.06, 0.0]^{\mathrm{T}}$
6	$[2.0, 1.5, 0.0]^{\mathrm{T}}$	$[0.06, 0.06, 0.0]^{\mathrm{T}}$
7	$[2.0, -1.5, 0.0]^{\mathrm{T}}$	$[0.06, -0.06, 0.0]^{\mathrm{T}}$
8	$[-2.0, -1.5, 0.0]^{\mathrm{T}}$	$[-0.06, -0.06, 0.0]^{\mathrm{T}}$

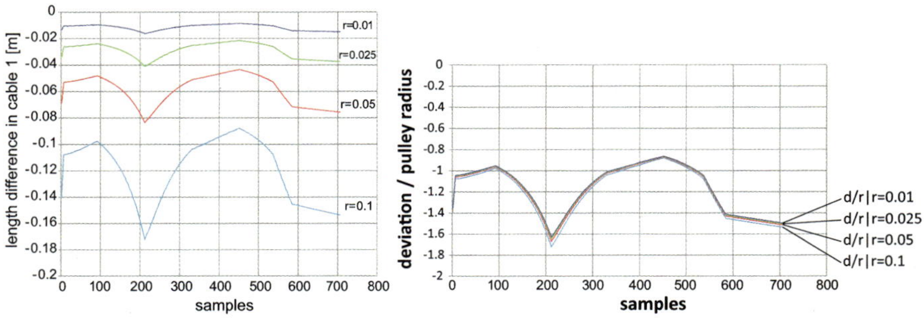

Fig. 3 Difference between the cable length computed from the standard kinematic model and the pulley model for different radii r_p of the pulley in cable 1. The left diagram shows the absolute difference where the right diagram shows the ratio between difference and radius of the pulley.

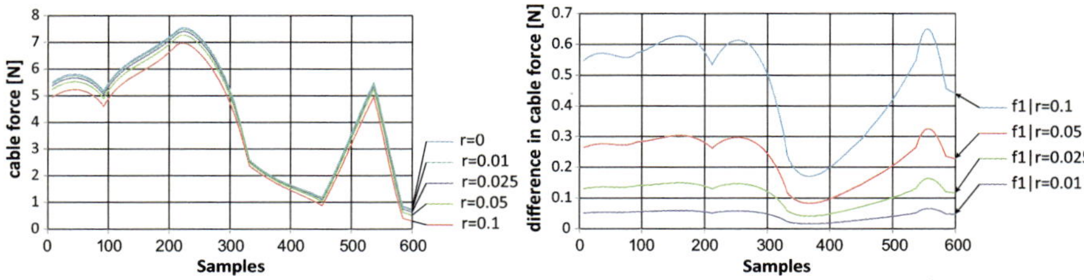

Fig. 4 Difference between the standard kinematic model and the pulley model for the forces f_1. The left diagram shows the absolute cable forces in cable 1 for a pulley radius of $r_p = \{0;0.01;0.025;0.05;0.1\}$ where the right diagram shows the difference between the standard model and different radii of the pulley.

and the effective radius of the pulleys is $r_p = 0.05$ m. All local frames $\mathscr{K}_{A,i}$ of the proximal anchor points A_i and thus the orientation of the winches were parallel aligned with the world frame $\mathscr{K}_0$.

The difference between the standard model and the pulley model for inverse kinematics is depicted in Fig. 3, where the diagrams show the difference between both inverse kinematic codes along a quadratic trajectory with 2 m edge length for different radii r_p of the pulley. One can easily see that the cable length computed from the extended formula is always longer than the standard model. This is clear since the distance around the pulley must be longer. In the right diagram one can see the relation between the radius of the pulley and the differences between the kinematic models. For the considered interval of pulley radii $r_p \in [0.1;0.01]$ m the ratio is almost constant. Thus, the dependency between the additional length of the cable and the radius of the pulley seems to be linear in this range.

We analyze the difference in the force distribution that arises from the static model taking into account the guiding pulleys. To calculate the force distribution, the following closed-form formula is used [8]

$$\mathbf{f} = \mathbf{f}_m - \mathbf{A}^{+\mathrm{T}}(\mathbf{w} + \mathbf{A}^{\mathrm{T}}\mathbf{f}_m), \tag{9}$$

Table 2 Comparison of the workspace volume and surface for different radius of the pulley r_p.

pulley radius r_p [m]	volume V [m^3]	relative volume [%]	surface S [m^2]	relative surface [%]
0.0	5.81	100.0	17.92	100.0
0.001	5.81	99.9	17.87	99.7
0.01	5.80	99.8	17.89	99.8
0.025	5.79	99.7	17.87	99.7
0.05	5.77	99.3	17.75	99.0
0.15	5.68	97.8	17.50	97.6
0.25	5.57	95.8	17.17	95.8
0.35	5.42	93.2	16.76	93.5
0.40	5.32	91.6	16.49	92.0

where $\mathbf{f}_m = \frac{1}{2}(\mathbf{f}_{\min} + \mathbf{f}_{\max})$ is the mean feasible force and $\mathbf{w}$ is the applied wrench. For the example we used $f_{\min} = 1\,\mathrm{N}$, $f_{\max} = 10\,\mathrm{N}$, and an external wrench $\mathbf{w} = \mathbf{0}$. Note that as long as no external wrench is applied only the ratio between $f_{\min}$ and $f_{\max}$ influences the results of workspace and force distribution. Fig. 4 shows the comparison for the forces f_1 in cable 1 when moving along a trajectory for different radii r_p of the pulley. The differences are again in the range of some percentage and the magnitude of the difference seems to be linear for typical sizes of the pulley. For the practical use in force control the influence seems to be less important since the error in the cable forces is the scale of the measurement error caused from typical force sensors.

In order to study the influence of pulleys on the workspace we compute the hull of the workspace and use the performance criteria surface and volume to compare the results for different radii of the pulleys. Force limits and external wrench were chosen as given above. To check for existence of the workspace for a pose, Eq. (9) was evaluated and the determined force was compared to the force limits as given in the previous section. The parameters of the workspace algorithms were set as follows: The iterations depth for the recursive refinement of the hill were chosen to be six leading to 16386 computed vertices and 32768 triangles. The accuracy for the line search was $\varepsilon = 10^{-4}$ m such that the first four digits of the performance indices shall be meaningful. The computational results from the study are given in Table 2. In this evaluation we used even larger radii for the pulleys than before. The relative error of the workspace volume is less than 2% for realistic values of the pulley's radius.

5 Conclusions

In this paper we presented the modeling of a cable-driven parallel robot taking into account the effects of pulleys in the robot's winches. Using an extended modeling for kinematics and statics the differences of the workspace between the simplified and extended model where studied. For the robot at hand it turns out that the difference of the volume and surface of the workspace is in the range of 1% for typical pulleys, i.e. with a radius of 2.5% of the robots shortest edge length. Considering other unconsidered uncertainties the influence on the workspace may be neglected in many cases. The comparison of the inverse kinematic codes and thus the expected accuracy of the robot unveils more important differences. The deviations between standard and pulley model are found to be almost linear in the considered range for the pulley radius. The shortening of the cables caused by the pulleys may significantly increase the inner tension in the robot and thus disturb the force equilibrium of the mobile platform.

Our future research aims at extending the study of pulleys to stiffness, interference, and singularities. Furthermore, we are working towards deriving a real-time capable kinematic code for the forward kinematics of the pulley model. The presented model may serve as a basis especially because it allows for a simpler Jacobian. Anyway, more sophisticated methods are needed to meet the required real-time constraints for use in the controller.

Acknowledgements This work was partially supported by the Fraunhofer-Gesellschaft Internal Programs under Grant No. WISA 823 244.

References

1. Albus, J.S., Bostelman, R.V., Dagalakis, N.G.: The NIST RoboCrane. J. Robot. Syst. **10**(5), 709–724 (1993)
2. Baoyan, D., Yuanying, Q., Fushun, Z., Zi, B.: Analysis and experiment of the feed cable-suspended structure for super antenna. In: IEEE/ASME International Conference on Advanced Intelligent Mechatronics, pp. 329–334 (2008)
3. Bruckmann, T., Mikelsons, L., Brandt, T., Hiller, M., Schramm, D.: Wire robots, Part I – Kinematics, analysis and design. In: Parallel Manipulators – New Developments, ARS Robotic Books, I-Tech Education and Publishing, Vienna, Austria (2008)
4. Gouttefarde, M., Daney, D., Merlet, J.P.: Interval-analysis-based determination of the wrench-feasible workspace of parallel cable-driven robots. IEEE Trans. Robot. **27**(1), 1–13 (2011)
5. Kawamura, S., Choe, W., Tanaka, S., Pandian, S.R.: Development of an ultrahigh speed robot Falcon using wire drive system. In: IEEE International Conference on Robotics and Automation, pp. 1764–1850 (1995)
6. Miermeister, P., Pott, A.: Modelling and real-time dynamic simulation of the cable-driven parallel robot IPAnema. In: Pisla D., Ceccarelli M., Husty M., Corves B. (eds.) New Trends in Mechanism Science, pp. 353–360. Springer, Dordrecht (2010)
7. Pott, A.: Forward kinematics and workspace determination of a wire robot for industrial applications. In: Lenarčič J., Wenger P. (eds.), Advances in Robot Kinematics: Analysis and Design, pp. 451–458. Springer-Verlag, Dordrecht (2008)
8. Pott, A., Bruckmann, T., Mikelsons, L.: Closed-form force distribution for parallel wire robots. In: Computational Kinematics, pp. 25–34. Springer-Verlag, Duisburg, Germany (2009)
9. Verhoeven, R.: Analysis of the workspace of tendon-based Stewart platforms. Ph.D. thesis, University of Duisburg-Essen, Duisburg (2004)

Fast Approximate Implicitization of Envelope Curves Using Chebyshev Polynomials

Oliver J.D. Barrowclough, Bert Jüttler and Tino Schulz

Abstract Consider a rational family of planar rational curves in a certain region of interest. We are interested in finding an approximation to the implicit representation of the envelope. Since exact implicitization methods tend to be very costly, we employ an adaptation of approximate implicitization to envelope computation. Moreover, by utilizing an orthogonal basis in the construction process, the computational times can be shortened and the numerical condition improved. We provide an example to illustrate the performance of our approach.

Key words: Implicitization, approximation, envelopes, Chebyshev polynomials

1 Introduction

In geometric applications there are two basic standards for representing curves, namely the *parametric* and the *implicit* descriptions. Both descriptions feature specific advantages and disadvantages that complement each other. For instance, parametric curves allow the simple generation of point samples, while implicit forms support the decision of point location queries. In many applications, such as intersection computations, it is an advantage if both representations are available, and

Oliver J.D. Barrowclough
SINTEF ICT, Applied Mathematics, P.O. Box 124,Blindern, 0314 Oslo, Norway,
e-mail: oliver.barrowclough@sintef.no

Bert Jüttler
Johannes Kepler University, Institute of Applied Geometry, Altenberger Str. 69, 4040 Linz, Austria, e-mail: bert.juettler@jku.at

Tino Schulz
GALAAD, INRIA Méditerranée, 2004 Route des Lucioles, 06902 Sophia-Antipolis, France,
e-mail: tino.schulz@inria.fr

J. Lenarčič, M. Husty (eds.), *Latest Advances in Robot Kinematics*,
DOI 10.1007/978-94-007-4620-6_26,

conversion algorithms are therefore of substantial practical interest. The conversion processes are called *parametrization* and *implicitization*, respectively.

A rational curve may always be implicitized, whereas the opposite is not true [10]. Several techniques for *exact* implicitization exist, e.g. Gröbner bases, moving curves/surfaces, or methods based on resultants, (see e.g. [5]). However, due to their computational complexity, their practical use is often restricted to low-degree curves. Moreover, the variety obtained by exact implicitization may contain unexpected branches and self-intersections.

A valid alternative to exact methods is *approximate implicitization*; cf. [3, 4]. Instead of the exact variety, a low degree approximation is used to represent the shape of the geometric object in a certain region of interest. This technique can be implemented using floating point numbers and thus it offers faster computation, while having very high convergence rates. As shown in [2], the speed-up may be increased even further by using an orthogonal basis in the construction process.

These features make approximate implicitization a promising candidate for an efficient computation of *envelope curves*. Envelopes are used in different contexts in mathematics and applications, ranging from gearing theory and geometrical optics to NC-machining and Computer-Aided Design. In robotics, envelopes are ubiquitous, appearing for instance as singularities or boundaries. The theory of envelopes is covered by the classical literature, and is continuously extended, due to their practical importance [1, 6–8].

Approximate implicitization has recently been adapted to the computation of envelopes in [9]. As shown there, the idea is feasible and most properties of the original method can be preserved, such as the possibility of obtaining the exact solution. However, the convergence behaviour for higher degrees has not previously been studied and the computations are still fairly expensive, needing integrals of products of high degree polynomials.

The present paper uses the latest results from approximate implicitization to obtain a fast and efficient algorithm for approximating the envelope. This will make the use of implicit methods more attractive and moreover allow us to study the convergence behaviour experimentally. The paper is organized as follows: In Section two we will recall the basics of envelopes of planar curves. After that, the third section shows how approximate implicitization can be used to compute envelopes and derives a fast and efficient algorithm. The performance of our approach is illustrated with an example and discussion in Section four.

2 Envelopes of Rational Families of Curves

Consider the family of rational curves

$$\mathbf{p}(s,t) = \big(x(s,t)/w(s,t), y(s,t)/w(s,t)\big)^T, \quad (s,t) \in I \times J \tag{1}$$

where x, y and w are bivariate polynomials of bidegree (n_1, n_2) with $\gcd(x, y, w) = 1$ and $I, J \subset \mathbb{R}$ are closed intervals. We assume that $w(s,t) \neq 0$ for all $(s,t) \in I \times J$. Either s or t can be thought of as the time-like parameters, and the remaining parameter t or s is then used to parameterize the curves forming the family.

The *envelope* of the mapping $\mathbf{p}$ consists of those points where its Jacobian $\mathbf{J}(s,t)$ becomes singular. We observe that $\det \mathbf{J}(s,t) = h(s,t)/w(s,t)^3$, where

$$h(s,t) = \det \begin{pmatrix} x(s,t) & \partial_s x(s,t) & \partial_t x(s,t) \\ y(s,t) & \partial_s y(s,t) & \partial_t y(s,t) \\ w(s,t) & \partial_s w(s,t) & \partial_t w(s,t) \end{pmatrix}. \tag{2}$$

The function h is called the *envelope function*, since its zero set determines those points in the parameter domain which are mapped to the envelope. Unfortunately, certain parts of the zero set of h may degenerate under the mapping $\mathbf{p}$.

The earlier paper [9] describes how these *improper factors* can be removed from h. This can be done via some gcd computation and gives the *reduced envelope function* $\tilde{h}$. The image of the zero set of $\tilde{h}$ under $\mathbf{p}$ is called *proper part* of the envelope.

Let $q : \mathbb{C}^2 \to \mathbb{C}$ be the polynomial which defines *the implicit equation of the proper part of the envelope* of $\mathbf{p}$. According to Theorem 1 of [9], there exists a real polynomial $\lambda(s,t)$ such that

$$(q \circ \mathbf{p})(s,t) w(s,t)^d = \lambda(s,t) \tilde{h}(s,t)^2. \tag{3}$$

Equation (3) is linear with respect to the coefficients of q and λ. Let

$$q(\mathbf{x}) = \mathbf{c}_q^T \beta(\mathbf{x}) \quad \text{and} \quad \lambda(s,t) = \mathbf{c}_\lambda^T \alpha(s,t), \tag{4}$$

where $\beta(\mathbf{x}) = (\beta_k(\mathbf{x}))_{k=1}^M$ and $\alpha(s,t) = (\alpha_i(s)\alpha_j(t))_{(i,j)=(0,0)}^{(k_1,k_2)}$ are bases of polynomials in $\mathbf{x}$ and s,t of total degree d and bidegree (k_1, k_2) respectively, where $M = \binom{d+2}{2}$. The coefficients of q and λ with respect to these bases form a vector $\mathbf{c} = (\mathbf{c}_q^T, \mathbf{c}_\lambda^T)^T$. We formulate the *problem of approximate envelope implicitization:* Find the coefficients $\mathbf{c}$ which solve the weighted least squares minimization problem

$$\min_{\|\mathbf{c}\|_2 = 1} \int_{I \times J} \omega(s,t) \left((q \circ \mathbf{p})(s,t) w(s,t)^d - \lambda(s,t) h(s,t)^2 \right)^2 \mathrm{d}(s,t), \tag{5}$$

for a nonnegative weight function ω, and chosen degrees d, k_1 and k_2.

It is important to mention that we use h instead of the exact $\tilde{h}$, since our algorithm uses floating point computations which do not support exact gcd computations. While an exact solution of this simplified problem might produce additional branches, the effect on our low degree approximation will be negligible.

The result of the minimization (5) depends both on the choice of bases of q and λ and on the weight function ω. The standard choice of a triangular Bernstein basis for q and a tensor-product Bernstein basis for λ has been used for the approximations in this paper and also in [9]. However, as a major difference to the approach in

[9] where $\omega \equiv 1$, here we use a tensor product Chebyshev weight function on the domain $I \times J$, for the reasons described in the next section.

3 Fast Approximate Implicitization of Envelope Curves

The direct method for finding an approximate implicitization of envelope curves by evaluating high degree integrals is simple, but computationally costly. In addition, the resulting symmetric positive semi-definite matrix can be rather ill conditioned, leading to inaccurate null space computations when using floating point arithmetic. This is similar to the case of approximate implicitization of parametric curves presented in [2]. In that paper, an approach using orthogonal polynomials is presented which greatly improves both the conditioning and the computation time of the problem. In this section we give the details of how to implement the approach to approximate implicitization of envelope curves using Chebyshev polynomials.

3.1 Approximate Implicitization Using Chebyshev Polynomials

As described previously, the method works by minimization of the integral (5). Such a problem is aided by expressing the function in a basis orthonormal with respect to the chosen weight function ω. The objective function is expressible in any tensor product polynomial basis of bidegree

$$(L_1, L_2) = (\max(dn_1, k_1 + 2\deg_s(h)), \max(dn_2, k_2 + 2\deg_t(h))).$$

Thus, choosing an orthonormal basis (e.g., tensor-product Chebyshev polynomials), $\mathbf{T}(s,t) = (T_i(s)T_j(t))_{i=0,j=0}^{L_1,L_2}$ written in vector form and using (4), we can write

$$(q \circ \mathbf{p})(s,t)w(s,t)^d - \lambda(s,t)h(s,t)^2 = \mathbf{T}(s,t)^T(\mathbf{D}_q\mathbf{c}_q + \mathbf{D}_\lambda\mathbf{c}_\lambda), \tag{6}$$

where the matrices $\mathbf{D}_q$ and $\mathbf{D}_\lambda$ contain coefficients in $\mathbf{T}$. Now, defining a matrix

$$\mathbf{D} = (\mathbf{D}_q, \mathbf{D}_\lambda), \tag{7}$$

we claim that the singular vector corresponding to the smallest singular value of $\mathbf{D}$ solves the minimization problem (5). To see this, we prove the following Theorem:

Theorem 1. *Let the matrix* $\mathbf{D}$ *be defined as in (7). Then we have*

$$\min_{\|\mathbf{c}\|_2=1} \int_{I\times J} \omega((q\circ\mathbf{p})w^d - \lambda h^2)^2 = \min_{\|\mathbf{c}\|_2=1} \|\mathbf{D}\mathbf{c}\|_2^2.$$

Proof. By (6) and (7) we have

$$\int_{I\times J}\omega((q\circ\mathbf{p})w^d-\lambda h^2)^2=\int_{I\times J}\omega(\mathbf{c}^T\mathbf{D}^T\mathbf{T})(\mathbf{T}^T\mathbf{D}\mathbf{c})=$$
$$\mathbf{c}^T\mathbf{D}^T\left(\int_{I\times J}\omega\mathbf{T}\mathbf{T}^T\right)\mathbf{D}\mathbf{c}=\mathbf{c}^T\mathbf{D}^T\mathbf{D}\mathbf{c}=\|\mathbf{D}\mathbf{c}\|_2^2.\quad\square$$

Since we have $\min_{\|\mathbf{c}\|_2=1}\|\mathbf{D}\mathbf{c}\|_2=\sigma_{\min}$, where $\sigma_{\min}$ is the smallest singular value of $\mathbf{D}$, the corresponding right singular vector solves the problem. The problem is, however, better conditioned and can be implemented in a more efficient way than the weak approach [2].

3.2 Implementation of the Chebyshev Method

The choice of using Chebyshev polynomials for the orthogonal basis is made mainly for computational reasons; the coefficients can be generated via a fast algorithm. This utilizes an existing method outlined for univariate polynomials in [11], which exploits the discrete orthogonality of Chebyshev polynomials at Chebyshev points.

Here we briefly describe the algorithm for efficient generation of tensor-product Chebyshev coefficients. The univariate Chebyshev points of degree L in the interval $[0,1]$ are given by:

$$t_{j,L}=(1-\cos(j\pi/L))/2,\quad j=0,\dots,L.$$

The Chebyshev coefficients of any tensor product polynomial f of bidegree no higher than (L_1,L_2) can then be generated by the following procedure [11]:

- Construct a matrix $\mathbf{f}=(f(t_{i,L_1},t_{j,L_2}))_{i=0,j=0}^{L_1,L_2}$ of values of the function f at the tensor-product Chebyshev points,
- Extend $\mathbf{f}$ to its even counterpart $\hat{\mathbf{f}}$:

$$\hat{f}_{i,j}=f_{i,j},\ i=0,\dots,L_1,\ j=0,\dots,L_2,$$
$$\hat{f}_{L_1+i,j}=f_{L_1-i,j},\ i=1,\dots,L_1-1,\ j=0,\dots,L_2,$$
$$\hat{f}_{i,L_2+j}=f_{L_1-i,j},\ i=0,\dots,L_1,\ j=1,\dots,L_2-1,$$
$$\hat{f}_{L_1+i,L_2+j}=f_{L_1-i,L_2-j},\ i=1,\dots,L_1-1,\ j=1,\dots,L_2-1.$$

- perform a bivariate fast Fourier transform (FFT) to get $\tilde{\mathbf{f}}=\mathrm{FFT}(\hat{\mathbf{f}})$,
- extract the first (L_1+1,L_2+1) coefficients of $\tilde{\mathbf{f}}$ to get $\mathbf{g}=(\tilde{f}_{i,j})_{i,j=0}^{L_1,L_2}$. The matrix $\mathbf{g}$ then contains the tensor product Chebyshev coefficients of f.

The algorithm for approximate implicitization proceeds by applying the above procedure to the functions $\{w^d(\beta_k\circ\mathbf{p})\}_{k=1}^M$ and $\{-h^2\alpha_l\}_{l=1}^{L_1L_2}$, and arranging the coefficients in matrices $\mathbf{D}_q$ and $\mathbf{D}_\lambda$ according to the definition (7). The efficiency of the method is due to it being based on point sampling and FFT. Moreover, the

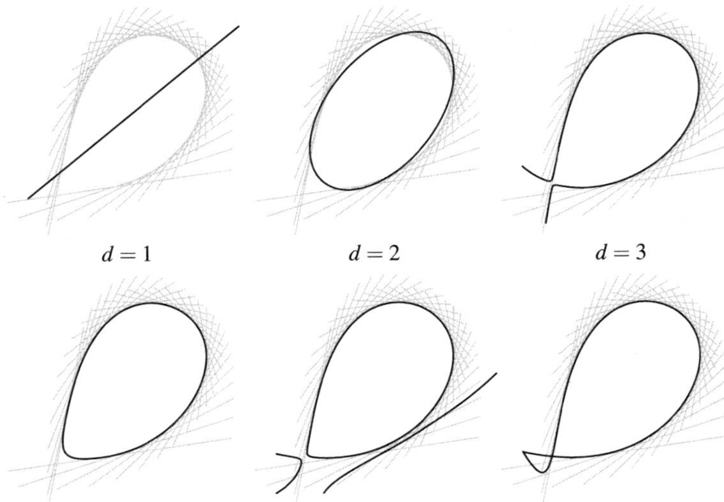

Fig. 1 Approximations of the envelope of a family of lines for degrees $d = 1,\ldots,6$.

sampling can be done entirely in parallel making the method highly suitable for implementation on heterogeneous architectures.

4 Numerical Results

In this section we present an example of the method along with both computation times and estimations for the convergence rates. In order to generate reliable data, we have chosen a degree six family of lines which has a rational envelope. We can thereby use a parametrization of the envelope to compute the algebraic error of the approximations. The family of lines is pictured in Figure 1, along with the envelope approximations up to the exact implicitization at degree six. For these approximations we take $k_1 = \max(0, dn_1 - 2\deg_s(h))$ and $k_2 = \max(0, dn_2 - 2\deg_t(h))$, since this is also the minimum needed for the exact solution.

It can be seen that with increased degree the approximations converge quickly. It is possible that with higher degrees, extra branches may appear in the region of interest. For example, the approximation of degree five has an extra branch close to the envelope curve. However, such artifacts could be avoided using a suitable collection of low-degree approximations (see [9] for an adaptive algorithm).

In Table 1 we show the computation times for the above approximations. The algorithm has been implemented in the Python programming language using the NumPy library for the built in FFT and singular value decomposition (SVD) algorithms. The results are computed on a 3.4Ghz Intel Core i7-2600 with 8GB RAM.

Table 1 Computation times and number of matrix coefficients for the examples in Figure 1.

Degree d	1	2	3	4	5	6
# coefficients	196	975	2964	7000	14136	25641
Time (s)	0.02	0.04	0.11	0.23	0.45	0.80

Table 2 Maximum algebraic error $\varepsilon_{d,i}$, of the approximations of the example in Figure 1, together with approximate convergence rates $r_{d,i}$.

		Implicit Degree d							
		1		2		3		4	
		$\varepsilon_{d,i}$	$r_{d,i}$	$\varepsilon_{d,i}$	$r_{d,i}$	$\varepsilon_{d,i}$	$r_{d,i}$	$\varepsilon_{d,i}$	r_i
Diameter 2^{-i}	1	1.69e-1	–	6.23e-3	–	1.16e-4	–	3.96e-6	–
	1/2	1.67e-1	0.02	3.46e-4	4.170	2.62e-6	5.467	2.66e-10	13.86
	1/4	3.04e-2	2.458	1.52e-5	4.511	1.50e-9	10.77	1.34e-14	14.27
	1/8	6.52e-3	2.223	5.02e-7	4.915	2.87e-12	9.028	n/a	n/a
	1/16	1.41e-3	2.213	1.58e-8	4.989	5.63e-15	8.993	n/a	n/a

Instead of increasing the polynomial degree d, one may also improve the quality of the approximations by subdivision; the envelope is then approximated by a piecewise implicit representation. It is thus of interest to see how the approximation improves as the region $\Omega = I \times J$ is reduced.

Consider a region $\Omega_i = I_i \times J_i$, of diameter 2^{-i} centered on a point (s_0, t_0) in

$$\mathscr{H} = \{(s,t) \in \Omega : h(s,t) = 0\}.$$

For an approximation $q_{d,i}$ of degree d to over the region Ω_i, we define the maximum algebraic error to be

$$\varepsilon_{d,i} = \max_{(s,t) \in \mathscr{H} \cap \Omega_i} |q_{d,i}(\mathbf{p}(s,t))|,$$

where the coefficients $\mathbf{c}_{q_{d,i}}$ of $q_{d,i}$ have been renormalized to $\|\mathbf{c}_{q_{d,i}}\| = 1$, in order to give meaningful results. Given two approximations $q_{d,i}$, and $q_{d,i+1}$, on subsequent subdivision regions Ω_i and Ω_{i+1}, we define the convergence rate to be $r_{d,i} = \log_2(\varepsilon_{d,i}/\varepsilon_{d,i+1})$. Table 2 shows values of $\varepsilon_{d,i}$ and $r_{d,i}$ for four successive subdivisions of the example in Figure 1 and degrees d up to four. Values of $\varepsilon_{d,i}$ below machine precision have been omitted.

As can be seen from Table 2, the error $\varepsilon_{d,i}$ decreases both with increased degree and increased levels of subdivision. The values of $r_{d,i}$, suggest that the convergence rates for $d = 1,2,3$ and 4 are approximately two, five, nine and 14 respectively. This corresponds directly to the number of degrees of freedom in approximating with lines, conics, cubics and quartics and is hence as high a convergence as we can expect, supporting our choices for the degrees (k_1, k_2). The results in Table 2 are typical of rational examples we have tested.

It should be noted that in general, envelope curves are not rational. Thus, this example, whilst showing that high convergence rates are attainable, cannot conclude that this is always the case. However, from studying additional examples, our experience shows that convergence behaviour is good in the general setting.

5 Conclusion

We have presented a new implementation of approximate implicitization of envelope curves using Chebyshev polynomials. We have detailed the computation times and convergence behaviour of a specific example, thereby demonstrating the feasibility of our approach. This paper also motivates theoretical work on convergence rates as a direction for future research.

Acknowledgements The research leading to these results has received funding from the European Community's Seventh Framework Programme FP7/2007-2013 under grant agreement n° PITN-GA-2008-214584 (SAGA), and from the Research Council of Norway (IS-TOPP). It was also supported by the Doctoral Program "Computational Mathematics" (W1214) at the Johannes Kepler University of Linz.

References

1. Abdel-Malek, K., Yang, J., Blackmore, D., Joy, K.: Swept volumes: Foundation, perspectives, and applications. Int. J. Shape Model. **12**(1), 87–127 (2006)
2. Barrowclough, O., Dokken, T.: Approximate implicitization using linear algebra. J. Appl. Math. (2012). doi:10.1155/2012/293746
3. Dokken, T.: Approximate implicitization. In: Mathematical Methods for Curves and Surfaces, Oslo 2000, pp. 81–102 (2001)
4. Dokken, T., Thomassen, J.: Weak approximate implicitization. In: IEEE International Conference on Shape Modeling and Applications, SMI 2006, 2006, pp. 204–214 (2006)
5. Hoffmann, C.: Implicit curves and surfaces in CAGD. IEEE Comput. Graph. Appl. **13**(1), 79–88 (1993)
6. Kim, Y., Varadhan, G., Lin, M., Manocha, D.: Fast swept volume approximation of complex polyhedral models. Comput. Aided Des. **36**(11), 1013–1027 (2004)
7. Peternell, M., Pottmann, H., Steiner, T., Zhao, H.: Swept volumes. Comput-Aided Des. Appl. **2**, 599–608 (2005)
8. Rabl, M., Jüttler, B., Gonzalez-Vega, L.: Exact envelope computation for moving surfaces with quadratic support functions. In: Lenarčič, J., Wenger, P. (eds.) Adv. in Robot Kinematics: Analysis and Design, pp. 283–290. Springer (2008)
9. Schulz, T., Jüttler, B.: Envelope computation in the plane by approximate implicitization. Appl. Algebra Eng. Commun. Comput. **22**, 265–288 (2011)
10. Sendra, J., Winkler, F., Perez-Diaz, S.: Rational Algebraic Curves. Springer (2007)
11. Trefethen, L.N.: Spectral Methods in MATLAB. SIAM, Philadelphia, USA (2000)

Construction of Overconstrained Linkages by Factorization of Rational Motions

Gábor Hegedüs, Josef Schicho and Hans-Peter Schröcker

Abstract We prove that for any sufficiently generic rational curve C of degree n in the group of Euclidean displacements, there exists an overconstrained spatial linkage with revolute joints whose linkage graph is the 1-skeleton of the n-dimensional hypercube such that the constrained motion of one of the links is exactly C. The synthesizing algorithm is based on the factorization of polynomials over the dual quaternions. The linkage contains $n!$ open nR chains, so that low degree examples include Bennett's mechanisms and are related to overconstrained 5R and 6R chains.

Key words: Dual quaternions, rational motion, factorization, overconstrained mechanism, Bennett linkage, 6R chain

1 Introduction

The research on this paper started with an attempt to understand the geometry of Bennett linkages from the point of view of dual quaternions. The group of Euclidean displacements can be embedded as an open subset of the Study quadric in the projectivization of the dual quaternions, regarded as a real vector space of dimension 8. Rotation subgroups and composition of those get then an algebraic meaning, which was exploited in [3] to devise an algorithm for the synthesis of a Bennett linkage such that the coupler assumes three pre-assigned positions in workspace. The key observation there was that the coupler curve is the intersection of a unique 2-plane

Gábor Hegedüs · Josef Schicho
Johann Radon Institute for Computational and Applied Mathematics Austrian Academy of Sciences (RICAM), 4040 Linz, Austria, e-mail: {gabor.hegedues, josef.schicho}@oeaw.ac.at

Hans-Peter Schröcker
University Innsbruck, Unit Geometry and CAD, 6020 Innsbruck, Austria, e-mail: hans-peter.schroecker@uibk.ac.at

J. Lenarčič, M. Husty (eds.), *Latest Advances in Robot Kinematics*,
DOI 10.1007/978-94-007-4620-6_27,

with the Study quadric. But one can also translate the synthesis problem entirely into the language of dual quaternions. Then the problem becomes equivalent to the factorization of a left quadratic polynomial into two linear ones.

Factorizations of left polynomials over the quaternions have been studied by in [8] by Niven, who was interested in the number of such factorizations. Gordon and Motzkin proved in [4] that the number of roots is either infinite or at most equal to the degree of the left polynomial. ([4] studies more generally polynomials over central simple algebras over commutative fields.) The more recent paper [7] gives an explicit solution formula for quadratic polynomials. It is not difficult to extend these results to dual quaternions.

Once the relation between the closure conditions of closed 4R linkages and the factorizations of left quadratic polynomials became clear, it also became clear that this relation holds for sufficiently generic left polynomials of arbitrary degree. This opens various possibilities, and we present some of them in this paper. Our general theory is capable of explaining some recent results on the synthesis of Bennett mechanisms and their limit cases. Moreover, it gives rise to a simple construction of overconstrained 5R and 6R linkages. To the best of our knowledge, many new types can be found among the latter class.

We continue this paper by recalling some well-known facts about dual quaternions and Euclidean displacements in Section 2. In Section 3 we present a factorization algorithm for generic polynomial parametrizations of curves on the Study quadric. Some of its remarkable implications are then explained at hand of examples in Section 4.

2 Preliminaries

In this section, we give a well-known and classical description of the group of Euclidean displacement by dual quaternions. More complete reference is [6].

We denote by SE_3 the group of Euclidean displacements, i.e., the group of maps from $\mathbb{R}^3$ to itself that preserve distances and orientation. It is well-known that SE_3 is a semidirect product of the translation subgroup T and the orthogonal group SO_3, which may be identified with the stabilizer of a single point.

We denote by $\mathbb{D} := \mathbb{R} + \varepsilon\mathbb{R}$ the ring of dual numbers, with multiplication defined by $\varepsilon^2 = 0$. The algebra $\mathbb{H}$ is the non-commutative algebra of quaternions, and $\mathbb{DH}$ is the algebra of quaternions with coefficients in $\mathbb{D}$. Every dual quaternion has a primal and a dual part (both quaternions in $\mathbb{H}$), a scalar part in $\mathbb{D}$ and a vectorial part in $\mathbb{D}^3$. The conjugated dual quaternion $\overline{h}$ of h is obtained by multiplying the vectorial part of h by -1. The dual number $h\overline{h}$ is called the norm of h, and the dual number $h+\overline{h}$ is called the trace of h.

By projectivizing $\mathbb{DH}$ as a real 8-dimensional vectorspace, we get $\mathbb{P}^7$. The condition that the norm of h is strictly real, i.e. its dual part is zero, is a homogeneous quadratic equation. Its zero set is called S, the Study quadric. The linear 3-space represented by all dual quaternions with zero primal part is denoted by E; we have

$E \subset S$. The complement $S - E$ can be identified with SE_3: the primal part describes SO_3, the translations are the dual quaternions with primal part ± 1 and strictly vectorial dual part. More precisely, the group isomorphism is given by sending $h = p + \varepsilon q$ to the map $\mathbb{R}^3 \to \mathbb{R}^3$, $v \mapsto \frac{pv\overline{p}+q\overline{p}}{p\overline{p}}$; it is well-known and also easy to check that the image of this map is strictly vectorial, the image is in SE_3, and the above construction is indeed a group homomorphism. The fact that it is onto follows from the fact that both groups are connected of the same dimension 6.

A nonzero dual quaternion represents a rotation if and only if its norm and trace are strictly real and its primal vectorial part is nonzero. It represents a translation if and only if its norm and trace are strictly real and its primal vectorial part is zero. The 1-parameter rotation subgroups with fixed axis and the 1-parameter translation subgroups with fixed parallel direction can be geometrically characterized as the lines on S through the identity element 1 (the translations are characterized as those lines that meet the exceptional 3-plane E).

For $n > 0$, an open linkage with n rotational joints can be described algebraically as follows. Let $h_1, \dots, h_n$ be rotations, that is, of vanishing the dual scalar part. From now on, we view them as elements of $\mathbb{P}^7$, i.e., for each i, h_i is a homogeneous coordinate vector of dimension 8. The group parametrized by $(t - h_i)_{t \in \mathbb{P}^1}$ – the parameter t determines the rotation angle – is the group of the $i+1$-th link relative to the i-th link. The position of the last link with respect to the first link is then given by a product $(t_1 - h_1)(t_2 - h_2) \cdots (t_n - h_n)$, with $t_1, \dots, t_n \in \mathbb{P}^1$.

3 Rational Curves in the Study Quadric and Open Linkages

In this section we prove a factorization theorem for curves in the Study quadric. It guarantees existence of a number of open chains with revolute joints such that the last link moves along the prescribed curve. Combining these open chains produces overconstrained mechanisms.

Let $C \subset S$ be a rational curve of degree $n > 0$. Then there exists a parametrization of C by a polynomial $(a_0 t^n + a_1 t^{n-1} + a_2 t^{n-1} + \cdots + a_n)_{t \in \mathbb{P}^1}$, where $a_0, \dots, a_n \in \mathbb{DH}$. If the curve passes through 1, then we may assume $a_0 = 1$ (the polynomial is *monic*). Note that this assumption is no loss of generality.

Conversely, let $\mathbb{DH}[t]$ be the set of left polynomials with coefficients in $\mathbb{DH}$. This set can be given a ring structure by the convention that t commutes with the coefficients. Let $P \in \mathbb{DH}[t]$ be a polynomial of degree $n > 0$. We call the map $f_P : \mathbb{P}^1 \to \mathbb{P}^7, t \mapsto P(t)$ the map associated to P. The image is a rational curve of degree at most n.

If $Q \in \mathbb{R}[t]$, $Q \neq 0$, then the maps f_P and f_{PQ} are equal. Conversely, if P has a factor in $\mathbb{R}[t]$ of positive degree, we can divide by it without changing the associated map. Note that, in general, the set of right factors is different from the set of left factors; but a polynomial in $\mathbb{R}[t]$ or in $\mathbb{D}[t]$ is a left factor if and only if it is a right factor, because it is in the center of $\mathbb{DH}[t]$. We are only interested in polynomials

$P \in \mathbb{DH}[t]$ such that the rational curve $f_P(\mathbb{P}^1)$ lies on the Study quadric S. This is the case if and only if $P\overline{P}$ is strictly real, that is, $P\overline{P} \in \mathbb{R}[t]$.

Proposition 1. *Let $P \in \mathbb{DH}[t]$ be a monic polynomial of degree $n > 0$ without strictly real factors such that $P\overline{P} \in \mathbb{R}[t]$. If there is a factorization $P(t) = (t-h_1)\cdots(t-h_n)$ with $h_i \in \mathbb{DH}$ representing rotations, then the polynomial $P\overline{P} \in \mathbb{R}[t]$ has no real roots.*

Proof. Assume $P = (t-h_1)\cdots(t-h_n)$ and let $M_i := (t-h_i)(t-\overline{h}_i)$ for $i = 1,\dots,n$. While for general h_i, M_i is a polynomial in $\mathbb{D}[t]$, M_i is in $\mathbb{R}[t]$ if h_i represents a rotation. Moreover, M_i has no real roots and the same is true for $P\overline{P} = (t-h_1)\cdots(t-h_n)(t-\overline{h}_n)\cdots(t-\overline{h}_1) = M_1\cdots M_n$. □

Theorem 1. *Let $P \in \mathbb{DH}[t]$ be a monic polynomial of degree $n > 0$ such that $P\overline{P} \in \mathbb{R}[t]$. Assume that the primal part of P has no strictly real factors. Then there is a factorization $P(t) = (t-h_1)\cdots(t-h_n)$ with $h_i \in \mathbb{DH}$ representing rotations.*

We will sketch a constructive proof of this theorem. It can be used to actually compute a factorization. Moreover, it will show that this factorization is not unique. We continue with three technical lemmas whose proofs are omitted due space restrictions. The proof of Theorem 1 is constructive so that it serves, at the same time, as description of an algorithm to actually compute factorizations.

Lemma 1 (polynomial division). *Let $P_1, P_2 \in \mathbb{DH}[t]$, and assume P_2 is monic. Then there is a unique representation $P_1 = QP_2 + R$ with $\deg(R) < \deg(P_2)$. Moreover, if $h \in \mathbb{DH}$ such that $P_2(h) = 0$, then $P_1(h) = R(h)$.*

Note that the last statement is not trivial because h need not commute with the coefficients of P_2.

Lemma 2. *Let $P \in \mathbb{DH}[t]$ and $h \in \mathbb{DH}$. Then $(t-h)$ is a right factor of P if and only if $P(h) = 0$.*

Lemma 3. *Let $P \in \mathbb{DH}[t]$ such that $P\overline{P} \in \mathbb{R}[t]$. Let $M \in \mathbb{R}[t]$ be an irreducible monic polynomial of degree two that divides $P\overline{P}$ but not the primal part of P. Then there exists a unique $h \in \mathbb{DH}$ such that $P(h) = M(h) = 0$.*

Proof (of Theorem 1). We proceed by induction on n. For $n = 0$, the statement is trivial. Assume $n \geq 1$. Since the primal part of P has no strictly real factors, P has no strictly real factors. Consequently $P\overline{P}$ has no real linear factors. Let M be one of the irreducible quadratic factors of $P\overline{P}$. By Lemma 3, there is a unique h such that $M(h) = P(h) = 0$. By Lemma 2, there exists $P' \in \mathbb{DH}[t]$ such that $P'(t-h) = P$. Obviously, P' is monic of degree $n-1$. Moreover, we have $P\overline{P} = P'(t-h)(t-\overline{h})\overline{P}' = P'\overline{P}'M$, hence $P'\overline{P}'$ is strictly real. Also, P' cannot have a strictly real factor, since this factor would also be a left factor and then it would also divide P. Similarly, the primal part of P' cannot have strictly real factor, as this factor would also divide the primal part of P. By induction hypothesis, we obtain $P' = (t-h_1)\cdots(t-h_{n-1})$ and so $P = (t-h_1)\cdots(t-h_{n-1})(t-h)$. □

This proof is constructive. The algebraic calculations necessary for actually computing a factorization are elementary with exception of the initial factorization of the real polynomial $P\overline{P}$ of degree $2n$. We illustrate this at hand of a simple example.

Example 1. The polynomial

$$P(t) = t^2 + t(-1 + \mathbf{i} - \mathbf{j} - 2\mathbf{k}) - 1 + \mathbf{i} + 2\mathbf{k} + \varepsilon(t(-\mathbf{i} + \mathbf{j} - 2\mathbf{k}) - 2 - \mathbf{j} - \mathbf{k})$$

satisfies the requirements of Theorem 1. (We assume that $(1, \mathbf{i}, \mathbf{j}, \mathbf{k})$ is the standard basis of the quaternion algebra $\mathbb{H}$.) In particular,

$$P\overline{P} = t^4 - 2t^3 + 5t^2 - 4t + 6 = (t^2 + 2)(t^2 - 2t + 3) \in \mathbb{R}[t].$$

We let $M_1 = t^2 + 2$ and, using Lemma 1, compute the rotation quaternion $h_1 = (1 - \varepsilon)\mathbf{j} + (1 + \varepsilon)\mathbf{k}$ that satisfies $M_1(h_1) = P(h_1) = 0$. By right division, we obtain the factorizations $P = (t - k_1)(t - h_1)$ where $k_1 = -1 + (1 - \varepsilon)\mathbf{i} - (1 + \varepsilon)\mathbf{k}$. $\diamond$

Now we are going to translate Theorem 1 into the language of kinematics. Some foregoing considerations allow to take into account "joints of higher degree". Let $h \in \mathbb{DH}$ be a dual quaternion representing a rotation. The parametrization $(t - h)_{t \in \mathbb{P}^1}$ of the rotation group defined by h is called a linear parametrization. More generally, let $R_1, R_2 \in \mathbb{R}[t]$ such that R_1 is monic, $\deg(R_1) = n$, $\deg(R_2) < n$, without common factor. Then the parametrization $(R_1(t) - h_1 R_2(t))_{t \in \mathbb{P}^1}$ is called a rational parametrization of degree n. Higher degree parametrizations of rotation groups may arise as the product of linear parametrizations, namely when the two axes coincide. Let us call two dual quaternions h_1, h_2 representing rotations "compatible" if the axes coincide; this is the case if and only if $h_1 h_2 = h_2 h_1$.

Corollary 1. *Let $C \subset S$ be a rational curve of degree n in the Study quadric, disjoint from the infinite 3-space E and passing through 1. Then C can be obtained as movement of the last link of an open kR-linkage, with $k \leq n$. Moreover, the rotations in the k joints have a simultaneous rational parametrization, and the sum of the degree of these parametrizations is equal to n.*

Proof. Let $P \in \mathbb{DH}[t]$ be monic polynomial of degree n such that f_P is a parametrization of C. Then $P\overline{P} \in \mathbb{R}[t]$. We may assume that P has no strictly real divisor, because otherwise we could factor it out and get a parametrization of strictly smaller degree. But also the primal part of P has no real factor: If it has a real zero t_0, then $f_P(t_0) \in E$, and this contradicts our assumption; if it has an irreducible quadratic real factor, then there is a pair of conjugate complex numbers z_0, $\overline{z}_0$ such that $q := f_P(z_0) = f_P(\overline{z}_0) \in E$, and q would also be in C, because C is an algebraic variety. By Theorem 1, there is a factorization $P = (t - h_1) \cdots (t - h_n)$ with $h_1, \ldots, h_n$ representing rotations. If $h_i, \ldots, h_{i+m-1}$ are compatible with each other, then the product $(t - h_i) \cdots (t - h_{i+m-1})$ can be written as $R_1 - h_i R_2$ for suitable $R_1, R_2 \in \mathbb{R}[t]$ such that R_1 is monic, $\deg(R_1) = m$, $\deg(R_2) < m$, because every dual quaternion compatible with h_i is a real linear combination of 1 and h_i.

Remark 1. Theorem 1 is almost a converse to Proposition 1. There are polynomials P without real factors such that $P\overline{P} \in \mathbb{R}[t]$ where the proposition and the theorem do not say anything. In this case, there is an irreducible polynomial $R \in \mathbb{R}[t]$ dividing the primal part of P but not the dual part. Since R then also divides the primal part of $\overline{P}$ and $P\overline{P} \in \mathbb{R}[t]$, the factor R appears twice in the factorization of $P\overline{P}$. For instance, if $P = t^2 + 1 + \varepsilon\mathbf{i}$, then it can be shown that P is not the product of two linear rotation polynomials. On the other hand, $P = t^2 + 1 + \varepsilon\mathbf{j}t + \varepsilon\mathbf{i} = (t - \mathbf{k})(t - \mathbf{k} + \varepsilon\mathbf{j})$ is a product of two rotation polynomials. A systematic analysis would be good, but it may be more difficult.

Remark 2. The proof of Theorem 1 can be adapted to the case when primal(P) has only simple real roots, or equivalently $P\overline{P}$ has at most double real roots. In this case, one gets a factorization into linear polynomials parametrizing rotation groups or translation groups. Together with Theorem 2 below, this observation gives an alternative explanation of some of the findings in [5].

Theorem 2. *Let $P \in \mathbb{DH}[t]$ be a monic polynomial of degree $n > 0$ without strictly real factors such that P has no real factors. Then there exist $n!$ (in general) different factorizations of the form $P(t) = (t - h_1)\cdots(t - h_n)$ with $h_i \in \mathbb{DH}$ representing a rotation. Each factorization of P corresponds to a factorization $P\overline{P} = M_1 \cdots M_n$ into irreducible quadratic polynomials over $\mathbb{R}$ where the order of the factors does matter. Moreover, we have $M_i(h_i) = 0$ for $i = 1, \dots, n$.*

Proof. Our proof of Theorem 1 can be translated into a construction of a factorization of P into linear factors over $\mathbb{DH}$. The only non-deterministic step is the choice of a quadratic factor of $P'\overline{P}'$, where P' is the left factor from which the next right linear factor is going to be constructed. The construction is also complete in the sense that every factorization can be obtained by this non-deterministic algorithm. □

We can give a precise meaning to the phrase "in general" used in Theorem 2: A necessary and sufficient condition for the existence of exactly $n!$ different factorizations is that $P\overline{P} \in \mathbb{R}[t]$ has n distinct quadratic irreducible factors.

Example 2. We continue Example 1 and divide the polynomial P by $M_2 = t^2 - 2t + 3$ instead of M_1. At first we compute the unique common zero h_2 of M_2 and P as $h_2 = \frac{1}{49}((28 - 30\varepsilon)\mathbf{i} + (7 + 3\varepsilon)\mathbf{j} - (63 + 13\varepsilon)\mathbf{k})$. Right division then yields a second factorization $P = (t - k_2)(t - h_2)$ where $k_2 = \frac{1}{49}(49 - (21 - 19\varepsilon)\mathbf{i} + (56 - 46\varepsilon)\mathbf{j} + (35 + 85\varepsilon)\mathbf{k})$. Note that $P(t)$ parametrizes a Bennett motion. The axes of the rotations k_1, h_1, h_2, and k_2 (in that order) are the axes of the underlying mechanism in the moving frame. Thus, we essentially synthesized a Bennett mechanism. ◇

We conclude this section with a description of the mechanism whose existence is guaranteed by Theorem 2.

Corollary 2. *Let $C \subset S$ be a rational curve of degree n in the Study quadric, disjoint from the infinite 3-space E and passing through 1. Then C is the coupler motion of a mechanism with revolute joints whose linkage graph (with vertices denoting links and edges denoting joints) is the 1-skeleton on the n-dimensional hypercube.*

Proof. We only have to show the statement on the linkage graph. It follows from the observation that the factorizations (open nR-chains) corresponding to permutations of $M_1, \ldots, M_n$ that differ only by a transposition of neighboring factors differ only in two consecutive revolute axes (compare also Example 3 below). □

4 Construction of Overconstrained Closed 6R Chains

Corollary 2 states the existence of an overconstrained linkage that contains many closed chains with $2n$ revolute joints. Thus, our factorization algorithm can be used to construct overconstrained 6R linkages to cubic coupler curves. We present the basic idea at hand of an example.

Example 3. We choose h_1 and h_2 as in Example 1 and let $h_3 = 1 - \mathbf{i} + \varepsilon\mathbf{k}$. In order to find the six factorizations of the rational cubic curve $C \subset S$ parametrized by

$$P = (t-h_1)(t-h_2)(t-h_3) = t^3 + t^2(-2+\mathbf{i}-\mathbf{j}-3\mathbf{k}) + \varepsilon(-\mathbf{i}-2\mathbf{k})) +$$
$$t((-2+\mathbf{i}+2\mathbf{j}+5\mathbf{k}) - \varepsilon(5+2\mathbf{i}+2\mathbf{j})) + ((3-\mathbf{i}+\mathbf{j}-\mathbf{k}) + \varepsilon(1+3\mathbf{i}+2\mathbf{j}+2\mathbf{k})),$$

we have to compute the three quadratic factors of $P\overline{P}$:

$$P\overline{P} = M_1 M_2 M_3 \quad \text{where} \quad M_1 = t^2+2,\ M_2 = t^2-2t+2,\ M_3 = t^2-2t+3.$$

The factorizations of P correspond to different orders in which the factors M_1, M_2, and M_3 enter the algorithm outlined in the proof of Theorem 1, for example:

The permutation (M_1, M_2, M_3) corresponds to $P = (t-a_1)(t-a_2)(t-a_3)$ where

$$\begin{aligned}
a_1 &= \tfrac{1}{7}(7-3\mathbf{i}+8\mathbf{j}+5\mathbf{k}) + \tfrac{1}{49}\varepsilon(19\mathbf{i}-46\mathbf{j}+85\mathbf{k}),\\
a_2 &= \tfrac{1}{161}(161-57\mathbf{i}-44\mathbf{j}+144\mathbf{k}) + \tfrac{1}{25921}\varepsilon(21456\mathbf{i}+3852\mathbf{j}+9670\mathbf{k}),\\
a_3 &= \tfrac{1}{23}(-5\mathbf{i}+3\mathbf{j}+32\mathbf{k}) + \tfrac{1}{529}\varepsilon(-114\mathbf{i}+418\mathbf{j}-57\mathbf{k}).
\end{aligned}$$

The permutation (M_2, M_1, M_3) corresponds to $P = (t-b_1)(t-b_2)(t-b_3)$ where

$$b_1 = \tfrac{1}{7}(7-3\mathbf{i}+8\mathbf{j}+5\mathbf{k}) + \tfrac{1}{49}\varepsilon(19\mathbf{i}-46\mathbf{j}+85\mathbf{k}) = a_1,$$
$$b_2 = \tfrac{1}{7}(-4\mathbf{i}-\mathbf{j}+9\mathbf{k}) + \tfrac{1}{49}\varepsilon(30\mathbf{i}-3\mathbf{j}+13\mathbf{k}), \quad b_3 = 1+\mathbf{k}+\varepsilon\mathbf{j}.$$

The permutation (M_2, M_3, M_1) corresponds to $P = (t-c_1)(t-c_2)(t-c_3)$ where

$$c_1 = \mathbf{j}+\mathbf{k}-\varepsilon(\mathbf{j}-\mathbf{k}), \quad c_2 = 1-\mathbf{i}+\mathbf{k}+\varepsilon(\mathbf{i}+\mathbf{k}), \quad c_3 = 1+\mathbf{k}+\varepsilon\mathbf{j} = b_3.$$

The equalities $a_1 = b_1$ and $b_3 = c_3$ arise because the corresponding permutations of M_1, M_2, and M_3 differ only by a neighbor transposition. ◇

Remark 3. If we start with compatible h_1 and h_2 or h_2 and h_3, one of the open 3R chains becomes a 2R chain with a link parametrization of degree two. We can combine it with any other 3R chain to obtain an overconstrained 5R linkage.

The presented method can produce overconstrained 6R chains with no obvious relations between the Denavit-Hartenberg parameters. A comparison of a random 6R chain generated by our algorithm with examples known from literature produced no match. Thus, our family contains new examples. Known mechanisms that can be synthesized by our method include Yu-Baker's syncopated double Bennett hybrid linkage [2] and Bricard's double collapsable octahedral linkage and its variants [1].

5 Conclusion

We presented an algorithm to factor a generic rational curve C of degree n on the Study quadric in $n!$ different ways into the product of n rotations. Every factorization corresponds to an open nR chain, combining all open chain results in a highly overconstrained mechanism, as described in Corollary 2. For $n = 2$, the curve C corresponds to coupler motion of a Bennett linkage (or its limiting case). For $n = 3$ we obtain six open 3R chains that can be combined to overconstrained 6R chains with the same coupler motion. Due to space restrictions, we only explained the algorithmic aspects in the proof of the central Theorem 1. A complete proof as well as further investigations and details are left to a forthcoming publication.

Acknowledgements This research was supported by the Austrian Science Fund (FWF): I 408-N13 and DK W 1214-N15.

References

1. Baker, J.E.: On Bricard's doubly collapsible octahedron and its planar, spherical and skew counterparts. J. Franklin Inst. **332**(6), 657–679 (2005)
2. Baker, J.E.: On syncopated and augmented six-revolute linkages. Mech. Mach. Theory **44**(10), 1840–1847 (2009)
3. Brunnthaler, K., Schröcker, H.P., Husty, M.: A new method for the synthesis of Bennett mechanisms. In: Proceedings of CK 2005, International Workshop on Computational Kinematics. Cassino (2005)
4. Gordon, B., Motzkin, T.S.: On the zeros of polynomials over division rings. Trans. Am. Math. Soc. **116**, 218–226 (1965)
5. Hamann, M.: Line-symmetric motions with respect to reguli. Mech. Mach. Theory **46**(7), 960–974 (2011)
6. Hao, K.: Dual number method, rank of a screw system and generation of lie sub-algebras. Mech. Mach. Theory **33**(7), 1063–1084 (1998)
7. Huang, L., So, W.: Quadratic formulas for quaternions. Appl. Math. Lett. **15**(5), 533–540 (2002)
8. Niven, I.: Equations in quaternions. Am. Math. Mon. **48**(10), 654–661 (1941)

Bond Theory and Closed 5R Linkages

Gábor Hegedüs, Josef Schicho and Hans-Peter Schröcker

Abstract We present bond theory as a new means for the analysis of overconstrained closed linkages with revolute joints. Intuitively, bonds are special points in the complex configuration curve. They exhibit discrete properties which can be visualized in bond diagrams and allow to read off directly certain properties such as the degrees of relative motions, or the special geometry of consecutive revolute axes. As an application we sketch a classification of overconstrained 5R linkages.

Key words: Dual quaternions, bond theory, overconstrained revolute chain, overconstrained 5R chain

1 Introduction

Many people have used dual quaternions as a tool for the analysis of overconstrained linkages [3, 6]. In this paper, we introduce a new technique, based on dual quaternions, to facilitate this analysis, which we call the *theory of bonds*. Intuitively, bonds are points in the configuration curve with complex coefficients where something degenerate happens. It turns out that these points exhibit discrete properties which we call the bond structure. In a certain way, a bond connects two links of a closed kinematic chain. This accounts for our choice of the name "bond". It must not be confused with the concept of "kinematic bond" as described in [1, Chapter 5].

The bonds have several useful properties. From the bond structure we can obtain geometric information. For instance, one can "read off" the degree of the coupler

Gábor Hegedüs · Josef Schicho
Johann Radon Institute for Computational and Applied Mathematics (RICAM), Austrian Academy of Sciences, 4040 Linz, Austria, e-mail: {gabor.hegedues, josef.schicho}@oeaw.ac.at

Hans-Peter Schröcker
University Innsbruck, Unit Geometry and CAD, 6020 Innsbruck, Austria,
e-mail: hans-peter.schroecker@uibk.ac.at

J. Lenarčič, M. Husty (eds.), *Latest Advances in Robot Kinematics*,
DOI 10.1007/978-94-007-4620-6_28,

curve between two arbitrary links (Theorem 3). Also, the existence of bonds with a particular structure has geometric implications, such as three consecutive axes fulfilling the Bennett conditions.

As an application of bond theory, we sketch a proof of the well-known classification of 5R linkages. Goldberg [5] gave a construction for such linkages. In [12], Wohlhart describes Goldberg's construction in full generality. Several authors then tried to prove that every 5R linkage which is neither planar nor spherical is a Goldberg linkage, and achieved partial results in this direction [2, 4, 9, 10]. The complete classification was given in [8] using the computer algebra system Mathematica; the proof involves big terms which makes it difficult to check it for humans. In this paper, we outline a complete proof without computer assistance. Due to space restrictions, some details must be left to a forthcoming publication.

This research can be considered as a continuation of [7], which introduces the factorization of left polynomials over the dual quaternions as a technique for linkage synthesis. It is to a large extend independent of [7] except for the well-known description of linkages by dual quaternions and an auxiliary result used in the proof of Theorem 4. We freely use concepts and notation of [7] and we strongly suggest to consult Section 2 of that article before continuing with the remainder of this paper.

2 Basic Facts from Bond Theory

In the following, $h_1,\dots,h_n$ will always denote rotation quaternions (unit dual quaternions with vanishing dual scalar part) such that $h_i^2=-1$ for each i. As appropriate, we identify them with elements of the real vectors space $\mathbb{R}^8$ or with points of the Study quadric $S\subset\mathbb{P}^7$. By $[n]$ we denote the sequence of integers $(1,\dots,n)$.

Definition 1. Let $h_1,h_2,\dots,h_n$ be rotation quaternions and I be an injective subsequence of $[n]$. Then the *coupling space* L_I is the linear span of all products $\prod_{i=1}^k h_{j_i}$, where $(j_1<\dots<j_k)$ is any monotonic increasing subsequence of the sequence I. The empty product is explicitly allowed and has the value 1.

Instead of $L_{\{i,j\}}$ we simply write L_{ij}, instead of $L_{\{i,j,k\}}$ we simply write L_{ijk}. The corresponding *coupling space dimensions* are denoted by l_I, l_{ij}, and l_{ijk}, respectively. Simple examples of coupling spaces are $L_\varnothing=\langle 1\rangle=\mathbb{R}$, $L_1=\langle 1,h_1\rangle$ or $L_{12}=\langle 1,h_1,h_2,h_1h_2\rangle$. The relevance of coupling spaces for the study of revolute chains follows from the observation that the end-effector positions $(t_n-h_n)\cdots(t_1-h_1)$ with $(t_0,\dots,t_n)\in(\mathbb{P}^1)^n$ of an open nR-chain lie in the coupling space.

Proposition 1. *The triple* $(L_1,+,\cdot)$ *is a field and isomorphic to* $\mathbb{C}$.

Proof. The set L_1 is closed under addition. Since quaternions in L_1 describe rotations about one fixed axis it is also closed under multiplication and inversion. This already implies that L_1 is a field. Clearly, $L_1=\langle 1,h_1\rangle$ where h_1 is the half-turn (rotation with rotation angle π) in L_1. Because of $(h_1)^2=-1$, L_1 is isomorphic to $\mathbb{C}$. □

Proposition 2. *Let $h_1, h_2, \ldots, h_n$ be rotation quaternions and $I \subset [n] \neq \varnothing$ a subsequence. Then the dimension of L_I is even (and hence in $\{2,4,6,8\}$).*

Proof. Let $i_k = \max\{I\}$. By Proposition 1, L_I is a vector space over the field $\mathbb{R}[h_{i_k}] \simeq \mathbb{C}$. Its real dimension is even. □

The following proposition relates coupling spaces and coupling dimensions to properties of the involved revolute axes. It will be very useful in the classification of overconstrained 5R linkages. We call two rotation quaternions *compatible* if they have the same revolute axis.

Proposition 3. *For rotation quaternions $h_1, \ldots, h_6$ the following statements hold:*

1. *$l_{12} = 2$ if and only if h_1 and h_2 are compatible and $l_{12} = 4$ otherwise.*
2. *The three-dimensional projective space defined by L_{12} is contained in the Study quadric S if the axes of h_1 and h_2 are parallel or meet in a point.*
3. *If the projective space defined by L_{12} is not contained in the Study quadric, then $h_3 \in L_{12}$ if and only if h_3 is compatible with h_1 or h_2.*
4. *If the projective space defined by L_{12} is not contained in the Study quadric, the vector spaces L_{12} and L_{34} are equal if and only if h_1 is compatible with h_3 and h_2 is compatible with h_4.*
5. *If h_i is not compatible with h_{i+1} for $i = 1,2$ and $\dim L_{123} = 4$, the axes of the rotations h_1, h_2, h_3 are parallel or meet in one point.*
6. *If h_i is not compatible with h_{i+1} for $i = 1,2$ and $\dim L_{123} = 6$, the axes of the rotations h_1, h_2, h_3 satisfy the Bennett conditions: The normal feet of h_1 and h_3 on h_2 coincide and the normal distances $d_{i,i+1}$ and angles $\alpha_{i,i+1}$ between consecutive axes are related by $d_{12}/\sin\alpha_{12} = d_{23}/\sin\alpha_{23}$.*
7. *If $L_{123} = L_{456}$ and $\dim L_{123} = 6$, then h_1 and h_4 or h_3 and h_6 are compatible.*

Most statements of Proposition 3 admit simple proofs or can easily be inferred from well-known facts, for example [11, Ch. 9–11]. Therefore, we only prove Proposition 3.6.

Proof. If $h_3 \in L_{12}$, it is either compatible with h_1 or h_2. The latter is excluded by assumption, the former satisfies the Bennett conditions. Hence, we can assume $h_3 \notin L_{12}$ and the vectors 1, h_1, h_2, h_3, h_1h_2 are linearly independent. As an L_1-vectorspace, L_{123} is generated by $1, h_2, h_3, h_2h_3$. Assume that these vectors form a basis of L_{123}. Then $w + xh_2 + yh_3 + zh_2h_3 = 0$ with $w,x,y,z \in L_1$ would imply $w = x = y = z = 0$ so that $l_{123} = 8$. This contradicts our assumption. Hence, there is a non-trivial linear relation

$$x + yh_2 + zh_3 = h_2h_3 \tag{1}$$

with unique $x,y,z \in L_1$. It is no loss of generality to assume $h_3^2 = -1$ (otherwise we can replace h_3 by a suitable linear combination of 1 and h_3 with real coefficients). By multiplying (1) from the right with h_3, we obtain $xh_3 + yh_2h_3 - z = -h_2$. Comparing coefficients with (1) then yields $y^2 = -1$, $z = xy$, and $x = zy$. Without loss of generality, we may assume $y = -h_1$ (otherwise we replace h_1 by a suitable linear combination of 1 and h_1 with real coefficients). Then we can also write $x = a + bh_1$ and $z = b - ah_1$ for some $a, b \in \mathbb{R}$. If $a = 0$, (1) becomes $(h_2 - b)h_3 = h_1(b - h_2)$, there is a rotation around h_2 that transforms h_1 to h_3 and the claim follows. If $a \neq 0$, we set $h_2' := a^{-1}(h_2 - b)$ (another rotation about the same axis) and find

$$h_1h_2' + h_1h_3 + h_2'h_3 = a^{-1}(h_1h_2 - bh_1 + ah_1h_3 + h_2h_3 - bh_3)$$
$$= a^{-1}(h_1h_2 - bh_1 + ah_1h_3 + a + bh_1 - h_1h_2 + bh_3 - ah_1h_3 - bh_3) = 1.$$

It follows that $h_4 := -h_1 - h_2' - h_3$ fulfills the two equations $h_1 + h_2' = \overline{h_4} + \overline{h_3}$, $h_1h_2' = \overline{h_4}\,\overline{h_3}$. Hence, the closure equation $(t - h_4)(t - h_3)(t - h_2)(t - h_1) \in \mathbb{R}$ of Bennett's mechanisms is fulfilled (see [7]). □

From now on we consider rotation quaternions $h_1, \dots, h_n$ and assume that the set $K := \{(t_1, \dots, t_n) \in (\mathbb{P}^1)^n \colon (t_1 - h_1) \cdots (t_n - h_n) \in \mathbb{R} \setminus \{0\}\}$ is of dimension one. This means that $(h_1, \dots, h_n)$ is a closed linkage with one degree of freedom, in particular $n \in \{4, 5, 6, 7\}$. We call K the *configuration curve* of the linkage.

Definition 2. Given a closed linkage $(h_1, \dots, h_n)$ with one degree of freedom we define the set of *bonds* as

$$B := \operatorname{ZarClo}(K) \cap \{(t_1, \dots, t_n) \in (\mathbb{P}^1_{\mathbb{C}})^n \colon (t_1 - h_1) \cdots (t_n - h_n) = 0\}.$$

Here, $\operatorname{ZarClo}(K)$ – the Zariski closure of K – is the zero locus of all algebraic equations that also vanish on K.

The set of bonds B is a complex algebraic variety of dimension zero. It is defined by real equations and, thus, consists of a finite number of conjugate complex points. Since conjugate complex pairs carry no additional information, we always *identify them in the context of bonds.*

We illustrate the notion of bonds at hand of a simple example.

Example 1. Let (h_1, h_2, h_3, h_4) encode a Bennett linkage, where

$$h_1 = \mathbf{i},\ h_2 = \mathbf{j} + 9\varepsilon(\mathbf{i} - \mathbf{k}),\ h_3 = -\tfrac{1}{3}(\mathbf{i} + 2\mathbf{j} - 2\mathbf{k}) - 2\varepsilon(2\mathbf{i} - 2\mathbf{j} - \mathbf{k}),$$
$$h_4 = \tfrac{1}{3}(2\mathbf{i} + \mathbf{j} + 2\mathbf{k}) + \varepsilon(5\mathbf{i} + 4\mathbf{j} - 7\mathbf{k}).$$

The ideal of the configuration curve K is $I(K) = \langle 1 + t_3 + t_4, t_2 + t_4, t_1 + t_4 + 1\rangle$. We computed it by means of Gröbner bases. Some necessary algebraic conditions are given by Theorem 2, below. The bond set, up to conjugation of coordinates, is

$$B = \{(t_1 = \mathrm{i}, t_2 = 1 + \mathrm{i}, t_3 = \mathrm{i}, t_4 = -1 - \mathrm{i}), (t_1 = \mathrm{i} - 1, t_2 = \mathrm{i}, t_3 = \mathrm{i} - 1, t_4 = -\mathrm{i})\}.$$

Theorems 1 and 2 below state important properties of bonds. For lack of space, they remain without a proof but they can easily be verified with the bonds of Example 1.

Theorem 1. *For any bond $b \in B$ there exist indices $i, j \in [n]$ such that $t_i^2 + 1 = t_j^2 + 1 = 0$.*

A proper investigation of bonds requires to consider algebraic multiplicities. However, for the sake of simplicity, we mostly ignore this concept in this paper. This limits the scope of our results but facilitates their presentation. In order to make this clear, we will speak of "simple bonds". More precisely, a bond $b \in B$ is called *simple*, if there are exactly two indices $i, j \in [n]$ such that $t_i^2 + 1 = t_j^2 + 1 = 0$, b is a simple point of K, and the vanishing order of the functions $t_i^2 + 1, t_j^2 + 1$ is one at b.

Definition 3. Let $b \in B$ be a simple bond. Then the unordered pair of indices (i, j), such that $t_i^2 + 1 = t_j^2 + 1 = 0$, is called the *combinatorial bond* corresponding to b.

Theorem 2. *If $b \in B$ is a simple bond with combinatorial bond (i, j) and $i < j$, then $\prod_{l=i}^{j}(t_l - h_l) = \prod_{l=j}^{i+n}(t_l - h_l) = 0$ (indices are to be read modulo n).*

The linkage graph of a closed nR-chain (with vertices corresponding to links and edges corresponding to joints) is the circular graph with n vertices. The combinatorial bond structure can be visualized by drawing connecting lines between the edges that correspond to connected joints. We call these visualizations *bond diagrams*. Figure 1.a displays the combinatorial bond structure of a Bennett mechanism (Example 1). The combinatorial bonds are $(1,3)$ and $(2,4)$.

It is possible that we find more simple bonds such that $t_i^2 + 1 = t_j^2 + 1 = 0$ for a fixed index pair (i, j). We call the number of bonds corresponding to the same combinatorial bond (i, j) the *bond number* and denote it by $\mathrm{bd}(i, j)$. In bond diagrams, it can be visualized by double lines, triple lines etc.

Proposition 4. *If $\mathrm{bd}(1,2) > 0$, then $l_{12} \le 2$ (the rotations h_1 and h_2 are compatible). If $\mathrm{bd}(1,3) > 0$, then $l_{123} \le 6$ (the axes of h_1, h_2, and h_3 satisfy the Bennett conditions or belong to a planar or spherical linkage).*

Proof. By Theorem 2, $\mathrm{bd}(1,3) > 0$ implies $(t_1 - h_1)(t_2 - h_2)(t_3 - h_3) = 0$. This is a nontrivial linear relation between the eight generating vectors of L_{123}. Together with Proposition 2 this implies $l_{123} \le 6$. The first assertion is proved in similar fashion. □

We denote the links of the overconstrained nR chain by $o_1, \ldots, o_n$ and use the convention that o_i is the link between h_i and h_{i+1}.

Definition 4. For an index pair (i, j) define the function $\phi_{ij}\colon K \mapsto \mathrm{SE}_3$ by

$$\phi_{ij}(t_1, \ldots, t_n) = (t_{i+1} - h_{i+1}) \cdots (t_j - h_j).$$

By $C_{ij} := \phi_{ij}(K)$ we denote the *coupler curve* between the two links o_i and o_j and by $\deg(i, j)$ the product of the algebraic degree of C_{ij} and the mapping degree of ϕ_{ij}.

Theorem 3. *The number $\deg(i, j)$ equals the sum of $\mathrm{bd}(i, j)$ over all pairs of joints (i, j) that are separated by the two links. In particular a joint is rigid, if it is not connected to any other joint.*

The proof of this fundamental result is deferred to a later publication.

Example 2. In the bond diagram of a Bennett mechanism (Figure 1.a) the relative motion of every pair of opposite linkages is of degree two. In Figure 1.d (Goldberg linkage), we have $\deg(i,j) = 1$ for $(i,j) \in \{(1,2),(1,5),(3,4),(4,5)\}$, $\deg(i,j) = 2$ for $(i,j) \in \{(1,4),(2,3),(2,5),(3,5)\}$, and $\deg(i,j) = 3$ for $(i,j) \in \{(1,3),(2,4)\}$.

3 Classification of Closed 5R Linkages

As an example for the application of bond theory, we present a classification result which is originally due to [8].

Theorem 4. *A non-trivial overconstrained 5R linkage is a Goldberg linkage [5, 12].*

Here, we call the linkage trivial, if it is planar, spherical, a Bennett linkage plus one fixed axis, or if two consecutive rotations are compatible. We describe the linkage by the sequence of its rotation quaternions $(h_1,\dots,h_5)$.

Lemma 1. *Suppose that* $\mathrm{bd}(1,4) > 0$, $\mathrm{bd}(2,4) > 0$, $\deg(1,4) > 2$ *(see Figure 1.c), and* $l_{234} = l_{154} = 6$. *Then* h_1 *and* h_2 *or* h_1 *and* h_5 *are compatible.*

Proof. Assume that h_1 and h_5 are not compatible and consider the vector spaces $L := L_{234} \cap L_{154}$ and $L' := L_{234} \cap L_{15}$. Since L is a $\mathbb{R}[h_4]$-right vectorspace, it is of even dimension. If $\dim L = 6$, we have $L_{234} = L_{154}$ and, by Proposition 3, h_1 and h_2 are compatible. Hence we can assume $\dim L = 4$.

Because of $l_{234} = 6$, the mapping degree of ϕ_{14} is one. Since we supposed that $\deg(1,4) > 2$ and C_{14} is generated by quadratic equations, it cannot be a plane curve. However, the configuration curve C_{14} is contained in L'. Thus, $\dim L' \geq 4$ and $L = L'$. But h_1 and h_5 are not compatible, hence $\dim L_{15} = 4$, $L = L' = L_{15}$, and $L_{15} \subseteq L_{234}$. This implies $L_{154} = L_{15}L_4 \subseteq L_{234}L_4 = L_{234}$ (where for $M,N \subset \mathbb{H}$ the multiplication is defined point wise as $MN = \{mn\colon m \in M, n \in N\}$). Thus, $L_{154} = L_{234}$ and it follows from Proposition 3 that h_1 and h_2 are compatible. □

Lemma 2. *Up to relabeling, the bond diagram of a non-trivial overconstrained 5R linkage is given by Figure 1.d.*

Proof. We can assume $l_{i-1,i,i+1} > 4$ for all $i \in \{1,\dots,5\}$ because otherwise the linkage is spherical by Proposition 3. We prove the statement by a discussion of all possible cases: Assume that $\mathrm{bd}(1,3)$, $\mathrm{bd}(2,4)$, $\mathrm{bd}(3,5)$, $\mathrm{bd}(4,1)$, and $\mathrm{bd}(5,2)$ are all positive (Figure 1.c; combinatorial multiplicities greater than one are not yet excluded). By Theorem 3, $\deg(1,2) = 4 > 2$, and by Lemma 1, two neighbouring links are compatible. This case cannot occur. The same reasoning applies, if we remove all bonds between one joint pair, say $(1,4)$. The next bonds to be removed cannot be those between $(1,3)$ or $(2,4)$. Without loss of generality, we remove the bonds $(2,5)$. This case is possible (Figure 1.d) and will later be shown to correspond to

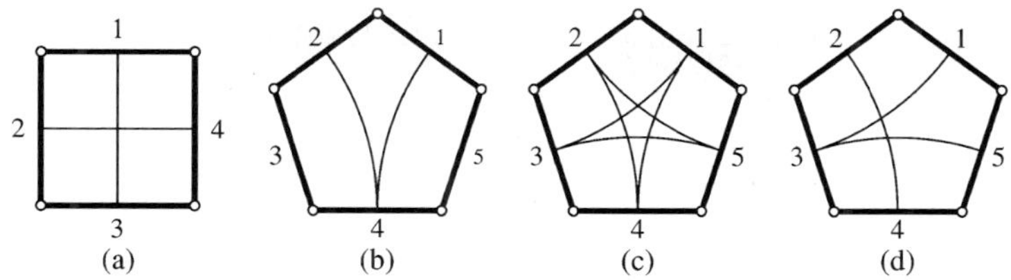

Fig. 1 Bond diagrams: Bennett linkage, two impossible 5R linkages, Goldberg linkage.

the Goldberg linkage. Now it is easy to see that removing further edges violates our assumptions. Thus, only possible multiplicities have to be discussed. However, if $\mathrm{bd}(2,4) > 1$, $\mathrm{bd}(1,3) > 1$, or $\mathrm{bd}(3,5) > 1$, then $\deg(1,3) \geq 3 > 2$ and once more Lemma 1 gives a contradiction. The discussion is complete. □

Proof (of Theorem 4). Lemma 2 gives us the bond graph of the linkage (Figure 1.d). Proposition 3 tells us that the axis triples (h_1,h_2,h_3), (h_2,h_3,h_4), and (h_3,h_4,h_5) satisfy the Bennett properties. By Theorem 3, $\deg(5,2) = \deg(5,3) = 2$. Because of $\deg(1,3) = 3$ and by an already used argument, the curve C_{13} is rational of degree three, so that the synthesis theory of [7] can be applied. (In our case, two consecutive axes of the overconstrained 6R chains of [7] coincide.) By general results of [7], the relative motion C_{13} admits parametrizations

$$(t-h_3'')(t-h_3')(t-h_2'), \quad (t-h_4')(t-h_5')(t-h_1'), \quad (t-h_3'')(t-h_6')(t-h_1')$$

with $h_i' \in L_i$ for $i = 1,\dots,6$ and $h_3'' \in L_3$ such that (h_1',h_2',h_3',h_6') and (h_3'',h_4',h_5',h_6') is a Bennett quadruple. The original 5R linkage can be constructed by composition of these two Bennett linkages, with the common axes h_3, h_6, and subsequent removal of the joint at h_6. This is exactly Goldberg's construction [5, 12]. □

4 Conclusion

This paper introduced bonds as a tool for the investigation of closed chains of revolute joints with a one-parametric mobility. We demonstrated how to read off geometric and kinematic properties directly from bond diagrams. The details of some proofs will be presented in a forthcoming paper.

There, we will also show that bonds behave well under specializations in families of linkages. Some parts of the bond structure are constant in every family, other parts may change under specialization but in a transparent way. This is particularly useful for classification purposes, because it gives strong criteria to decide whether a given bond is a special case of a given family. It motivates to use the bond structure as a classification scheme for linkages, for instance for the classification of closed 6R linkages. This is a future plan of the authors.

Acknowledgements This research was supported by the Austrian Science Fund (FWF): I 408-N13 and DK W 1214-N15.

References

1. Angeles, J.: Rational Kinematics, Springer Tracts in Natural Philosophy, vol. 34. Springer, New York (1989)
2. Baker, J.E.: On 5-revolute linkages with parallel adjacent joint axes. Mech. Mach. Theory **19**, 467–475 (1984)
3. Brunnthaler, K., Schröcker, H.P., Husty, M.: A new method for the synthesis of Bennett mechanisms. In: Proceedings of CK 2005, International Workshop on Computational Kinematics. Cassino (2005)
4. Dietmaier, P.: Simply overconstrained mechanisms with rotational joints. Habilitation thesis, Graz University of Technology (1995)
5. Goldberg, M.: New five-bar and six-bar linkages in three dimensions. Trans. ASME **65**, 649–656 (1943)
6. Hao, K.: Dual number method, rank of a screw system and generation of Lie sub-algebras. Mech. Mach. Theory **33**(7), 1063–1084 (1998)
7. Hegedüs, G., Schicho, J., Schröcker, H.P.: Construction of overconstrained linkages by factorization of rational motions. In: Lenarčič J., Husty M.L. (eds.) Advances in Robot Kinematics. Springer (2012)
8. Karger, A.: Classification of 5R closed kinematic chains with self mobility. Mech. Mach. Theory pp. 213–222 (1998)
9. Lee, C.C., Yan, H.S.: Movable spatial 5R mechanism with adjacent intersecting axis. In: Proc. Nat. Conf. Appl. Mech. Rob., pp. III-1–III-6 (1991)
10. Pamidi, P.R., Soni, H.R., Dukkipati, R.V.: Necessary and sufficient existence criteria of overconstrained five-link mechanisms with helical, cylinder, revolute and prism pairs. Trans. ASME pp. 737–743 (1973)
11. Selig, J.: Geometric Fundamentals of Robotics, 2 edn. Monographs in Computer Science. Springer (2005)
12. Wohlhart, K.: Merging two general Goldberg 5R linkages to obtain a new 6R space mechanism. Mech. Mach. Theory **26**, 659–668 (1991)

Kinematics of an Overconstrained 6R Linkage with 2-Fold Rotational Symmetry

Ketao Zhang and Jian S. Dai

Abstract This paper presents an overconstrained 6R linkage which is originated from a metamorphic 8R linkage. The 2-fold rotational symmetry reveals that the overconstrained 6R linkage is a special line-symmetric Bricard loop. The parametric constraints of the overconstrained 6R linkage are investigated in terms of Denavit–Hartenberg convention. This leads to identification of possible variation of parametric constraints for the special line-symmetric Bricard loops. The kinematics of the overconstrained 6R linkage is analyzed and motion parameters of the joint-space are derived. The resultant spherical 4R linkage of the overconstrained 6R linkage is addressed. The results of the joint-space solution are verified with a numerical example.

Key words: Overconstrained 6R linkage, 2-fold rotational symmetry, joint-space

1 Introduction

The overconstrained mechanisms which have full cycle mobility but do not satisfy the Grübler-Kutzbach mobility criterion have attracted substantial interest since the proposal of the first overconstrained mechanism, Sarrus [1], in 1853. Though very few overconstrained mechanisms have been used in practical application, a mechanism with overconstraints has the advantages to sustain great and variable loads by means of mass and compliance in heavily loaded machinery [2]. Bennett [3, 4] discovered the well-known deformable skew isogram which is the only spatial overconstrained 4R linkage and a basis of some compound mechanisms. Myard [5] presented the first known of overconstrained 5R linkage which was identified to be a special plane-symmetric form of one of the Goldberg 5R linkages [6, 7]. The mobile 4R linkages with lower kinematic pairs have been identified and the present

Ketao Zhang · Jian S. Dai
King's College London, University of London, London, UK, e-mail: {ketao.zhang,jian.dai}@kcl.ac.uk

J. Lenarčič, M. Husty (eds.), *Latest Advances in Robot Kinematics*,
DOI 10.1007/978-94-007-4620-6_29,

limited knowledge confines the systematic investigation of mobile 5R linkages to those with parallel adjacent joint-axis [2, 8].

Starting from the famous Sarrus linkage, a number of overconstrained 6R linkages have been produced. Bricard discovered three different types of mobile 6R loops during an earlier period beginning from 1897 and the other three in 1927 [9, 10]. Baker [11] went trough a thorough study of the Bricard 6R-loops and classified these linkages into five distinct kinds in terms of algebraic analysis and addressed the parametric constraints and closure equations for each category. Altmann [12] reported his workable linkage which turned out to be a special case of the Bricard line-symmetric 6R linkage. Waldron [13] described a class of overconstrained linkages obtained by combinations of other overconstrained linkages and the six-bar linkage formed from two Bennett linkages. Goldberg [6] presented asymmetrical overconstrained mobile 6R linkages by firstly attaching three selected Bennett loops in series 'back to back' and in his L-shaped manner. Schatz [14] discovered and patented the asymmetric 6R linkage which was named Turbula. This mobile 6R linkage can be derived from a special plane-symmetric Bricard loop and developed for various practical applications [15]. Recently, the overconstrained 6R linkage has been used to develop deployable structures and foldable devices [16, 17]. Zhang, Dai and Fang [18] evolved a metamorphic 8R linkage from origami fold and derived an overconstrained 6R linkage with 2-fold rotational symmetry. This paper presents the overconstrained 6R linkage originated from the metamorphic 8R linkage [18]. The geometry of the mobile 6R linkage is identified and the possible variation of parametric constraints for special line-symmetric Bricard 6R loops is revealed. The kinematics of the overconstrained 6R linkage is analyzed and the results of the joint-space solution are verified with a numerical example. The resultant spherical 4R linkage of the overconstrained 6R linkage is addressed.

2 Geometry of the Overconstrained 6R Linkage

The focus is on the overconstrained 6R linkage with 2-fold rotational symmetry in Fig. 1, and the 6R linkage is characterized by the specific geometry.

According to the derivation of the 6R linkage, any two adjacent joint axes are coplanar either with common point or parallel. The axes of joints J_1 and J_4 have the common point G and GM is the perpendicular bisector of isosceles triangle $\triangle ACG$. The axes of joint J_1 and J_2 have the common point E, and that of joint J_3 and J_4 have the common point C. The axes of joint J_2 and J_3 are parallel. For the joint axes with common point, the angle length between each two adjacent axes is determinate, that $\alpha_1 = 45^\circ$ and $\alpha_3 = 45^\circ$. Consider the 2-fold rotational symmetry, the joints J_4, J_5, and J_6 connected in series can be obtained by rotating the joints J_1, J_2, and J_3 around the symmetry axis GM. The axes of joint J_4 and J_5 have the common point F, and that of joints J_6 and J_1 have the common point A. The axes of joints J_5 and J_6 are parallel. The angle length between each two adjacent axes with common point is determinate, that $\alpha_4 = 45^\circ$ and $\alpha_6 = 45^\circ$.

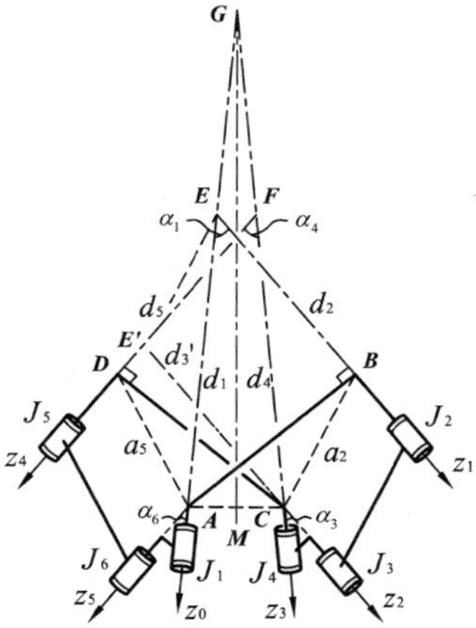

Fig. 1 The overconstrained 6R linkage with 2-fold rotational symmetry.

Considering the angle length between two intersecting adjacent joint axes, point B is the projection of common point A on axis BE of joint J_2 and the three points A, B and E form a isosceles right triangle $\triangle ABE$. Symmetrically, the three points C, D and F form a isosceles right triangle $\triangle CDF$ and CD is perpendicular to DF. The four points B, E, E' and C are coplanar and form a rectangle. This leads to identification of the parametric constraints according to design parameters.

3 Kinematics of the Overconstrained 6R Linkage

To facilitate the analysis, a Cartesian coordinate system is attached to each link in terms of the Denavit–Hartenberg convention [19] with consideration of the geometry in Fig. 1. The z_i-axis is aligned with the $(i+1)$th revolute joint axis, and the x_i-axis is defined along the common normal between the ith and $(i+1)$th joint axes, pointing from the ith to the $(i+1)$th joint axis. The axis x_0 is in the direction of the vector cross product $z_6 \times z_0$ and the origin is located at point $O_0(A)$. The origins of the other coordinate systems are $O_1(E)$, $O_2(C)$, $O_3(C)$, $O_4(F)$ and $O_5(A)$.

The parameters of a closed-loop linkage including the offset distance between two adjacent joint axes a_i, the translational distance between two incident normals of a joint axis d_i and the twist angle between two adjacent joint axes α_i are uniquely determined by the geometry of the axes. The joint variables of the revolute joints are θ_i. According to the geometry of the overconstrained 6R linkage with coplanar adjacent joint axes and 2-fold rotational symmetry, the parameters a_i, α_i and d_i are constant and given by

$$\alpha_1 = \alpha_3 = \alpha_4 = \alpha_6 = 0, \alpha_2 = \alpha_5 = l \tag{1}$$

$$\alpha_1 = \alpha_3 = \alpha_4 = \alpha_6 = 45^\circ, \alpha_2 = \alpha_5 = 0 \tag{2}$$

$$d_1 = d_4 = -\sqrt{2d}, d_2 = d_5 = d, d_3 = d_6 = 0 \tag{3}$$

in which l and d are the design parameters, and generally these two parameters are not equal. When the origins of the six coordinate systems are selected as $O_0(A)$, $O_1(E)$, $O_2(E')$, $O_3(C)$, $O_4(F)$ and $O_5(A)$, the parameters d_i become

$$d_1' = d_4' = -\sqrt{2}d, d_3' = d_5' = d, d_2' = d_6' = 0 \tag{4}$$

This shows the D-H parameters for general line-symmetric cases of Bricard loops addressed in [11] are slightly different regarding the origins of coordinate frames in Fig. 1.

The loop-closure equation of the overconstrained 6R linkage formulated by applying the homogeneous transformation [20] is derived as

$$\mathbf{H}_1 \cdot \mathbf{H}_2 \cdot \mathbf{H}_3 \cdot \mathbf{H}_4 \cdot \mathbf{H}_5 \cdot \mathbf{H}_6 = \mathbf{I} \tag{5}$$

where $\mathbf{H}_i$ $(i = 1, 2, \ldots, 6)$ is the transformation matrix indicating the change of coordinates between the ith coordinate system and the $(i-1)$th coordinate system. The loop-closure equation can be rearranged as

$$\mathbf{H}_1 \cdot \mathbf{H}_2 \cdot \mathbf{H}_3 = \mathbf{H}_6^{-1} \cdot \mathbf{H}_5^{-1} \cdot \mathbf{H}_4^{-1} \tag{6}$$

and expressed as

$$\begin{bmatrix} h_{11} & h_{12} & h_{13} & h_{14} \\ h_{21} & h_{22} & h_{23} & h_{24} \\ h_{31} & h_{32} & h_{33} & h_{34} \\ 0 & 0 & 0 & 1 \end{bmatrix} = \begin{bmatrix} h_{11}' & h_{12}' & h_{13}' & h_{14}' \\ h_{21}' & h_{22}' & h_{23}' & h_{24}' \\ h_{31}' & h_{32}' & h_{33}' & h_{34}' \\ 0 & 0 & 0 & 1 \end{bmatrix} \tag{7}$$

The elements of the matrices on the two sides are given in Appendix A. Twelve equations can be yielded from Eq. (7) since the corresponding elements of matrices on the two sides are equal.

For the overconstrained 6R linkage with one degree of freedom, the joint space of the linkage can be obtained with one input joint angle.

According to the geometry of the linkage in Fig. 1, the 2-fold rotational symmetry with symmetry axis GM leads to the following relationship for θ_i.

$$\theta_1 = \theta_4, \theta_2 = \theta_5, \theta_3 = \theta_6 \tag{8}$$

Hence, taking one of the six joint variables as an input, only two independent equations are necessary to derive the remaining motion parameters.

Three geometric equations can be generated by taking the elements equality of last column on both side of Eq. (7) and given by

$$h_{14} = h_{14}', h_{24} = h_{24}', h_{34} = h_{34}' \tag{9}$$

Expanding the equations with parameters in Appendix A, the geometric constraints can be obtained as

Table 1 Motion parameters of the overconstrained 6R linkage.

θ_2	θ_1	θ_3	θ_4	θ_5	θ_6
−2.0944	2.3253	1.5009	2.3253	−2.0944	1.5009
−1.0472	−2.3253	1.6408	−2.3253	−1.0472	1.6408
0	−0.9963	1.8302	−0.9963	0	1.8302
1.0472	0.2721	2.4768	0.2721	1.0472	2.4768
2.0944	−0.2721	0.6648	−0.2721	2.0944	0.6648

$$-\sqrt{2}lc\theta_2 s\theta_1 - c\theta_1(d - ls\theta_2) = dc(\theta_5+\theta_6) + lc\theta_6 \tag{10}$$

$$d - ls\theta_2 = dc(\theta_5+\theta_6) + ls\theta_6 \tag{11}$$

where $c\theta_i$ is a shorthand notation for $\cos\theta_i$, and $s\theta_i$ for $\sin\theta_i$.
The right side of Eqs. (10) and (11) are equal and the first fundamental equation is derived as

$$-\sqrt{2}lc\theta_2 s\theta_1 - lc\theta_1 s\theta_2) + dc\theta_1 - ls\theta_2 + d = 0 \tag{12}$$

Substituting Eq. (8) into Eq. (10), another fundamental equation is obtained as

$$d(c\theta_2 c\theta_3 - s\theta_2 s\theta_3) + l(s\theta_2 + s\theta_3) - d = 0 \tag{13}$$

Consider the motion parameters in the above two equations, θ_2 is taken as the input joint variable and the remaining can then be derived in terms of θ_2.

Introduce the trigonometric identity equations

$$s\theta_i = \frac{2x_i}{1+x_i^2}, c\theta_i = \frac{1-x_i^2}{1+x_i^2} \tag{14}$$

where $x_i = \tan(\theta_i/2)$, and substitute them into Eqs. (12) and (13), the joint variables are yielded as

$$\theta_5 = \theta_2 \tag{15}$$

$$\theta_4 = \theta_1 = \arctan\left(\frac{d - ls\theta_2}{\sqrt{2}lc\theta_2}\right) \tag{16}$$

$$\theta_6 = \theta_3 = \arctan\left(-\frac{ds(\theta_2/2) - lc(\theta_2/2)}{dc(\theta_2/2) - ls(\theta_2/2)}\right) \tag{17}$$

A numerical example is given to illustrate the motion parameters in joint-space of the overconstrained 6R linkage in Fig. 1. According to the geometry of the linkage, the motion parameters are determined by the link parameters l and d which should be fixed in the design stage. Set the two parameters as $l = 130$ and $d = 100$. Let θ_2 be the input angle in the range of $-\pi$ to π radian, a group of discrete values of the

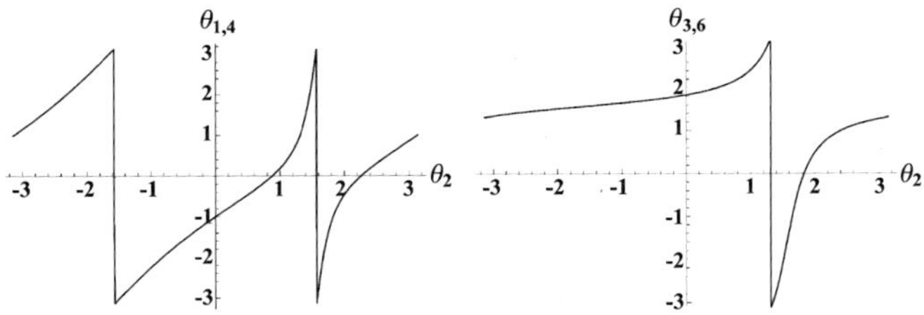

Fig. 2 The joint-space curves of the joint variables.

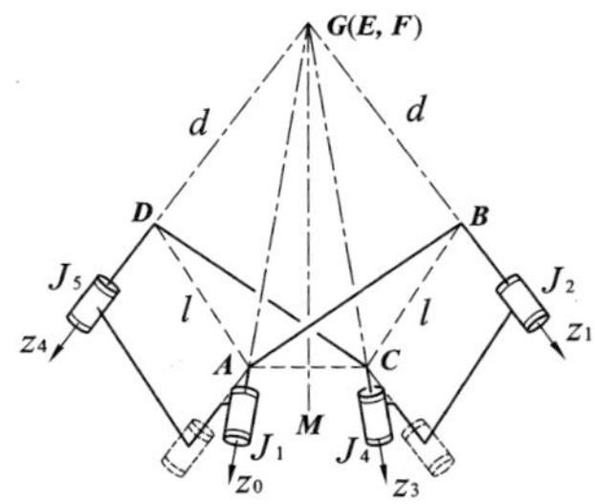

Fig. 3 The resultant spherical 4R linkage.

motion parameters are listed in Table 1. The joint-space curves of the joint variables are illustrated in Fig. 2.

4 Geometry Induced Spherical 4R Linkage

According to the expression of joint variables in Eqs. (16) and (17), the relationship between the variable joint-rotation angles is characterized by the fixed link length l and offset of links d. As aforementioned, the two design parameters l and d are not equal in general.

Specially, when the two parameters l and d are designed to be equal, the Eqs. (16) and (18) become

$$\theta_4 = \theta_1 = \arctan(-(1 - s\theta_2)/(\sqrt{2}c\theta_2)) \tag{18}$$

$$\theta_6 = \theta_3 = \pi/2 \tag{19}$$

The Eq. (19) shows that the joint-rotation angles θ_3 and θ_6 are constant. The axes of joints J_2, J_3 and J_4 become coplanar and the axes of J_5, J_6 and J_1 are coplanar symmetrically. Under such a condition, the joints J_3 and J_6 of the overconstrained 6R linkage in Fig. 1 are restricted by geometric constraints and they are not effective. This leads to the degeneration from the overconstrained 6R linkage in Fig. 1 to the spherical 4R linkage in Fig. 3.

5 Conclusions

The overconstrained 6R linkage presented in this paper is characterized by the 2-fold rotational symmetry. The geometry turns out that the mobile 6R linkage is a special case of the line-symmetric Bricard 6R linkage. The parametric constraints are derived by using the Denavit–Hartenberg convention and the possible variation of these parametric constraints for the special line symmetric Bricard 6R-loop is revealed. The motion parameters of the 6R linkage are derived in terms of algebraic kinematics analysis and the results are verified by a numerical example. The overconstrained 6R linkage degenerates to a spherical 4R linkage when the link parameters l and d are equal.

Acknowledgements The authors thank the support of European Commission – FP7 under Grant No. 270436 and help from Evangelos Emmanouil on the LaTeX file formatting.

Appendix A

$$
\begin{aligned}
h_{11} &= c\theta_1 c(\theta_2+\theta_3) - (\sqrt{2}s\theta_1 s(\theta_2+\theta_3))/2\\
h'_{11} &= c\theta_4 c(\theta_5+\theta_6) + (\sqrt{2}s\theta_4 s(\theta_5+\theta_6))/2\\
h_{12} &= (s\theta_1(c(\theta_2+\theta_3)-1))/2 - \sqrt{2}c\theta_1 s(\theta_2+\theta_3)\\
h'_{12} &= s\theta_4 c(\theta_5+\theta_6) + (\sqrt{2}c\theta_4 s(\theta_5+\theta_6))/2\\
h_{13} &= -(s\theta_1(1+c\theta_2 c\theta_3) - s\theta_2 s\theta_3)/2 - \sqrt{2}c\theta_1 s(\theta_2+\theta_3)\\
h'_{13} &= (\sqrt{2}s(\theta_5+\theta_6))/2\\
h_{14} &= lc\theta_1 c\theta_2 + \sqrt{2}s\theta_1(ls\theta_2 - d))/2\\
h'_{14} &= -lc\theta_6 + ds(\theta_5+\theta_6)\\
h_{21} &= -s\theta_1 c(\theta_2+\theta_3) + (\sqrt{2}c\theta_1 s(\theta_2+\theta_3))/2\\
h'_{21} &= s\theta_4(c(\theta_5+\theta_6)-1)) - (\sqrt{2}c\theta_4 s(\theta_5+\theta_6))/2\\
h_{22} &= (c\theta_1(c(\theta_2+\theta_3)-1))/2 + \sqrt{2}s\theta_1 s(\theta_2+\theta_3)\\
h'_{22} &= (c\theta_4(c(\theta_5+\theta_6)-1))/2 + \sqrt{2}s\theta_4 s(\theta_5+\theta_6)\\
h_{23} &= -(c\theta_1(1+c(\theta_2+\theta_3)))/2 - \sqrt{2}s\theta_1 s(\theta_2+\theta_3)\\
h'_{23} &= (1+c(\theta_5+\theta_6))/2\\
h_{24} &= -ls\theta_1 c\theta_2 + (\sqrt{2}c\theta_1(ls\theta_2 - d))/2\\
h'_{24} &= (-\sqrt{2}dc(\theta_5+\theta_6) + ls\theta_6)/2\\
h_{31} &= (\sqrt{2}s(\theta_2+\theta_3))/2\\
h'_{31} &= -(s\theta_4(1+c(\theta_5+\theta_6)) - \sqrt{2}c\theta_4 s(\theta_5+\theta_6))/2
\end{aligned}
$$

$$h_{32} = \left(c\left(\frac{\theta_2+\theta_3}{2}\right)\right)^2$$
$$h'_{32} = -c\theta_4(1+c(\theta_5+\theta_6))/2 - \sqrt{2}s\theta_4 s(\theta_5+\theta_6)$$
$$h_{33} = \left(s\left(\frac{\theta_2+\theta_3}{2}\right)\right)^2$$
$$h'_{33} = (1-c(\theta_5+\theta_6))/2$$
$$h_{34} = -\sqrt{2}(d-ls\theta_2)/2$$
$$h'_{34} = -\sqrt{2}(dc(\theta_5+\theta_6)+ls\theta_6)/2$$

References

1. Sarrus, P.T.: Note sur la transformation des mouvements rectilignes alternatifs, en mouvements circulaires; et reciproquement. Acad. Sci. **36**, 1036–1038 (1853)
2. Phillips, J.: Freedom in Machinery: Vol. 2, Screw Theory Exemplified. Cambridge University Press, Cambridge (1990)
3. Bennett, G.T.: A new mechanism. Engineering **76**, 777–778 (1903)
4. Bennett, G.T.: The skew isogram mechanism. Proc. Lond. Math. Soc., 2nd series **13**, 151–173 (1914)
5. Myard, F.E.: Contribution la géométrie. Soc. Math. Fr. **59**, 183–210 (1931)
6. Goldberg, M.: New five-bar and six-bar linkages in three dimensions. Trans. ASME **65**, 649–663 (1943)
7. Waldron, K.J.: The mobility of linkages. Doctoral Dissertation, Stanford University (1969)
8. Cui, L., Dai, J.S.: Axis constraint analysis and its resultant 6R double-centered overconstrained mechanisms. J. Mech. Robot., **3**, 031004 (2011)
9. Bricard, R.: Mémoire sur la théorie de l'octaedre articulé. J. Math. Pures Appl., Liouville **3**, 113–148 (1897)
10. Bricard, R.: Leçons de cinématique. Tome II Cinématique Appliquée, 7–12 (1927)
11. Baker, J.E.: An analysis of Bricard linkages. Mech. Mach. Theory **15**, 267–286 (1980)
12. Altmann, P.G.: Communications to Grodzinski, P. and M'ewen, E, Link Mechanisms in Modern Kinematics. Proc. Inst. Mech. Eng. **168**, 889–896 (1954)
13. Waldron, K.J.: Hybrid overconstrained linkages. J. Mech. **3**, 73–78 (1968)
14. Schatz, P.: Rhythmusforschung und Technik. Verlag Freies Geistesleben (1975)
15. Schatz Stiftung Foundation, Paul: http://www.paul-schatz.ch/en/seine-entdeckungen/dieinversina/
16. Chen, Y., You, Z.: Square deployable frames for space applications. Part 1: Theory. Proc. Inst. Mech. Eng., G J. Aerosp. Eng. **220**, 347–354 (2006)
17. Chen, Y., You, Z.: Two-fold symmetrical 6R foldable frames and their bifurcations. Int. J. Solids Struct. **46**, 4504–4514 (2009)
18. Zhang, K.T., Dai, J.S., Fang, Y.F.: Topology and constraint analysis of phase change in the metamorphic chain and its evolved mechanism. ASME J. Mech. Des. **132**, 121001 (2010)
19. Denavit, J., Hartenberg, R.S.: A kinematic notation for lower-pair mechanisms based on matrices. ASME J. Appl. Mech. **77**, 215–221 (1955)
20. Tsai, W.L.: Robot Analysis: The Mechanics of Serial and Parallel Manipulators. Wiley, New York (1999)

Design and Control of a Redundant Suspended Cable-Driven Parallel Robots

Johann Lamaury, Marc Gouttefarde, Micaël Michelin and Olivier Tempier

Abstract This paper introduces a six degree-of-freedom suspended parallel robot driven by eight cables. The determination of an optimal geometry of such a parallel cable robot together with the design of a prototype are briefly outlined. Then, based on usual kinematic modeling, a basic control strategy is presented. Since the parallel cable-driven robot presented here is redundantly actuated, this control strategy has to deal with the problem of cable tension distribution. This latter turns out to be challenging because of the under-constrained nature of the considered cable-driven robot. The extension to these robots of existing tension distribution methods is finally discussed.

Key words: Cable-driven parallel robots, actuation redundancy, tension distribution

1 Introduction

The main advantages of cable-driven parallel robots (CDPR) are fast motion, light weight, large workspaces and heavy loads capabilities, making them well adapted to tasks over important workspaces, not provided by common robot architectures. These manipulators mainly consist of a base, a mobile platform connected to the base through flexible cables and motorized winches. By means of these motors, the cable lengths are controlled, allowing the platform motion control. Thus, as cables

Johann Lamaury · Marc Gouttefarde · Olivier Tempier
Laboratoire d'Informatique, de Robotique et de Micro-électronique de Montpellier,
161 rue Ada, 34392 Montpellier Cedex 5, France,
e-mail: {johann.lamaury, olivier.tempier, marc.gouttefarde}@lirmm.fr

Micaël Michelin
TECNALIA France, MIBI, 672 rue du Mas de Verchant, 34000 Montpellier Cedex 2, France,
e-mail: micael.michelin@tecnalia.com

J. Lenarčič, M. Husty (eds.), *Latest Advances in Robot Kinematics*,
DOI 10.1007/978-94-007-4620-6_30,

can be unwound over great lengths, it is possible to design parallel robots with great workspaces.

Since the first works carried out in the eighties, many CDPR applications have been foreseen thanks to the versatility of this type of parallel robots. Crane applications [1] and rehabilitation [10] might be mentioned as examples. Our goal is to design a 6 degree-of-freedom (DOF) industrial robot able to carry out some *pick-and-place* tasks over large workspaces possibly with heavy payloads. The majority of studied fully constrained CDPR, such as FALCON robot [8] and the SEGESTA prototype [6], cannot carry out the specifications. Indeed, their structure imposes to have some cables pulling the platform from the bottom. Such cables would probably collide with some goods or workers. Suspended CDPR, e.g. the well-known NIST-ROBOCRANE [1], represent a solution as, in this configuration, all cables grow from above the mobile platform.

Additionally, actuation redundancy is used in order to significantly improve the ratio between the size of the workspace and the robot overall size, as shown in Fig. 1. In this figure, the workspace is defined as the set of feasible static equilibrium mobile platform poses where feasible means that there exists a set of non-negative cable tensions satisfying the equilibrium (cables cannot push on the platform). Fig. 1 shows that the use of two additional cables improves the suspended CDPR workspace by 225%. The 6-cable CDPR of Fig. 1(a) has the ROBOCRANE configuration, whereas the 8-cable CDPR has an original arrangement of cables outlined in the present paper.

However, suspended CDPR called a special attention from some researchers [2, 5, 16] but only few studies focused on redundant ones [12, 15]. Those architectures raise some challenging modeling and control issues such as tension distribution. In this paper, an 8-cable suspended CDPR prototype called ReelAx8 is presented, together with preliminary results and discussion on the aforementioned control issues.

2 Optimal Configuration and Design of the ReelAx8 Prototype

A method has been developed and implemented [13] in order to determine the best configuration of a suspended CDPR regarding our aforementioned needs. This *optimal* configuration consists of the cable output point positions on the base and the attachment point positions on the platform, together with the arrangement of the cables between them. The chosen optimality criterion is based on the ability to admit a shift of the center of gravity of the loaded platform, i.e. to resist to wrenches generated by off-centered payloads. Furthermore, because of the targeted industrial applications, the robot architecture needs a suspended configuration. Typically, four posts are settled at each plant corners as illustrated in Fig. 2(b). Then, eight cables are used in order to obtain a large static workspace [13]. Finally, the great number of possible configurations is strongly restrained by adding symmetry constraints in order to keep a homogeneous behavior over the whole workspace. Both the layout

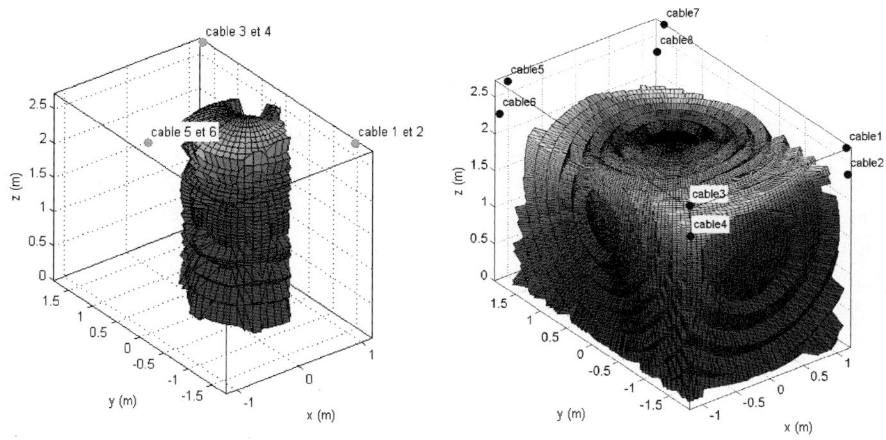

(a) 6-cables suspended CDPR (b) 8-cables suspended CDPR

Fig. 1 Workspace analysis of 6 and 8 cables suspended CDPR (constant null orientation) to fit in a rectangular room.

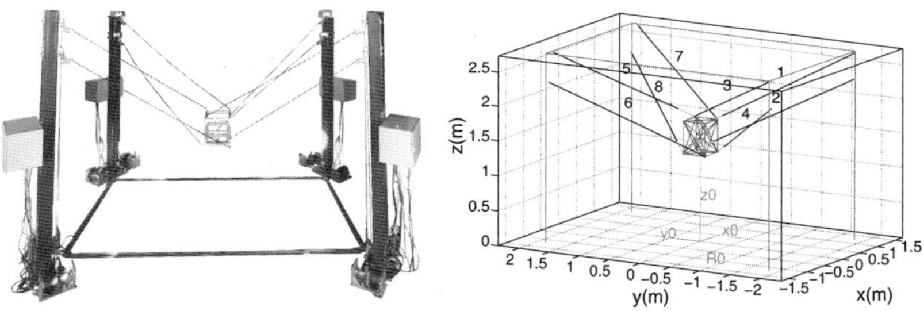

(a) ReelAx8 prototype (b) Matlab representation

Fig. 2 Optimal configuration of an 8-cable suspended CDPR.

of the drawing points and the attachment points on the platform is thus chosen to be symmetric with respect to a central vertical axis ($\mathbf{z}$ of $\mathbb{F}_b$ defined in next section). Moreover, the cable arrangement is such that if a point of the base is connected to another point on the platform, their symmetric points are also connected by a cable. It is also required that no cable interference exists across the prescribed workspace.

The resulting robot geometry, shown in Fig. 2(b), has been implemented on a prototype called ReelAx8 shown in Fig. 2(a).

In this optimal configuration, cable drawing points belong to two different horizontal planes. In the platform reference pose depicted in Fig. 2, the cables growing out from the top plane go to the top of the platform, and are not attached to their nearest platform's corner but to the next clockwise one. On the contrary, the cables growing out from the second plane go to the bottom of the platform, and are attached to their next counterclockwise platform's corner. This layout allows very good orientation capabilities and increase the robot's stiffness. ReelAx8 global dimensions are $3.8 \times 2.4 \times 3$ meters (L, l, h) and its maximum workspace is shown in Fig. 1(b). Motors are settled on the ground with idler pulleys to guide the cables along each post, which has two cable outputs at different hight equipped with pulleys and conic eyelets. The platform attachment points are embedded in spherical joints. The cable winches are installed at the foot of the poles and consist of a drum actuated by direct-drive electric motors. A ball screws allows the drum forward/backward motion so that the cables do not wound on themselves during the winding. The identification of the geometric parameters was performed with a laser tracker.

3 Kinematics

The kinematic model is based on m cables which link the base from points A_i, to points B_i of the platform, $i \in 1,\dots,m$. Vectors $\mathbf{a}_i$ and $\mathbf{b}_i$ define these points in the base frame $\mathbb{F}_b = (O, \mathbf{x}, \mathbf{y}, \mathbf{z})$ and in the platform frame $\mathbb{F}_p = (P, \mathbf{x}', \mathbf{y}', \mathbf{z}')$, located at the platform center-of-gravity (COG) P, respectively. O, the origin of the base frame, is located at the center of the robot volume. The position of the platform reference point P is defined by vector $\mathbf{p}$. The mobile platform orientation is given by the rotation matrix $\mathbf{Q}$. The cable length, denoted $\mathbf{l}_i$, is the norm of vector $\overrightarrow{A_i B_i}$. The inverse kinematics is complex when considering the mass and extensibility of the cables. In this study, massless inextensible cables are supposed. Therefore, the inverse kinematics consists in computing $\mathbf{l}_i$.

$$l_i = \left\| \mathbf{a}_i - \mathbf{p} - \mathbf{Q}\mathbf{b}_i \right\| \tag{1}$$

The expression of the wrench applied by the cables to the platform can be written as

$$\mathbf{W}\mathbf{t} = \mathbf{f} \tag{2}$$

where $\mathbf{W}$ is called the wrench matrix, $\mathbf{t} = [t_1,\dots,t_m]^T \in \Re^m$ is the cable tension vector and $\mathbf{f} = [f_1,\dots,f_n]^T \in \Re^n$. In static equilibrium, $\mathbf{f}$ is the opposite of the wrench due to the platform weight.

The challenge lies in the cable inability to transmit compressive forces, which means that all components of $\mathbf{t}$ have to remain non-negative. If $m = n$ and $\mathbf{W}$ is not singular, (2) leads to one unique solution, which does not guarantee non-negativity. Furthermore, if one or more cable have negative tensions, the platform is probably not in its current desired pose, but in another static equilibrium pose. To avoid violating this positiveness constraint, actuation redundancy can be used. Indeed, when $\mathbf{W}$ is non-square ($m > n$), (2) can be written as follows:

$$\mathbf{t} = \mathbf{W}^{+}\mathbf{f} + \mathbf{H}\lambda = \mathbf{t}_p + \mathbf{t}_n \tag{3}$$

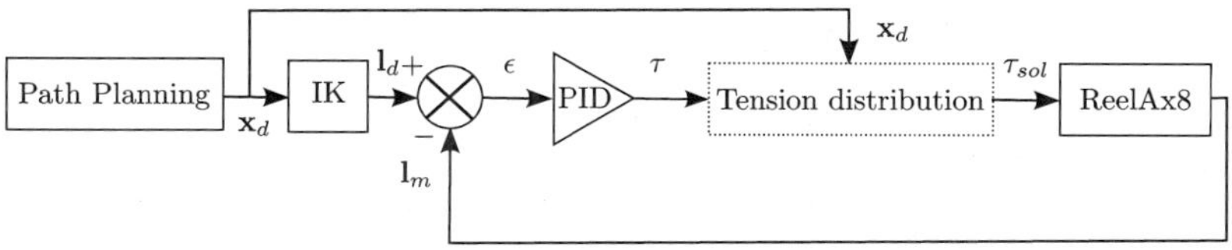

Fig. 3 ReelAx8 PID-control scheme.

where $\mathbf{W}^+$ is the Moore–Penrose pseudo-inverse of the wrench matrix and $\boldsymbol{\lambda}$ is an arbitrary m-dimensional vector. We define $\mathbf{H} = \mathbf{I} - \mathbf{W}^+\mathbf{W}$, where $\mathbf{I} \in \Re_m \times \Re_m$ is the identity matrix, as the projection matrix onto the nullspace of $\mathbf{W}$, and $\Sigma \subset \mathbb{R}^m$ as the r-dimensional affine space containing $\mathbf{t}_p$ and spanned by this nullspace. Physically, the linear combination of the columns of $\mathbf{H}$ results in internal forces which do not modify the wrench applied on the platform. $\mathbf{t}_p$ is the minimum-norm solution and $\mathbf{t}_n$ is the homogeneous solution.

4 Control

Our first closed-loop control of ReelAx8 is shown in the block diagram of Fig. 3. A Cartesian trajectory is generated and converted into the desired cable lengths $\mathbf{l}_d$, through the inverse kinematics. Thanks to a regular winding of the cables on the winch drum, the motor optical encoders supply angle information easily convertible into the current measured cable lengths $\mathbf{l}_m$. The error $\epsilon = \mathbf{l}_d - \mathbf{l}_m$ is estimated, corrected by a decentralized PID, and finally results in τ. Because this vector may contain negative tensions, a tension distribution algorithm should be used. This algorithm yields tension vector $\mathbf{t}_{\text{sol}}$ which is converted to τ_{sol}, the m-dimensional motor torque vector, as $\tau_{\text{sol}} = R\mathbf{t}_{\text{sol}}$, where R is the drum radius. Thanks to a proper winding system and tuning of the controller gains, this basic control topology is capable of providing acceptable tracking performances over an important workspace, in translation as well as in rotation. The robots is controlled with Simulink, through xPC target and embedded Matlab, currently at a 5 kHz frequency.

Close to the central vertical axis $\mathbf{z}$ of $\mathbb{F}_b$ and along the hypotenuses of the four possible three-post triangles of the $(O, \mathbf{x}, \mathbf{y})$ plane, the eight cables are taut. However, according to our experiments, as soon as the platform is located inside a triangle made by three posts, wherein the six corresponding cables can balance the weight of the mobile platform, the two cables growing out from the furthest post are often slack. In this case, the platform is probably not in the desired pose, since our control scheme is in fact asking these two cables to push on the platform.

Indeed, no effective tension distribution were implemented in real-time yet because of compatibility issues with Embedded Matlab. Hence, in our current implementation, the block "Tension distribution" shown in Fig. 3 is inactive ($\mathbf{t}_n = \mathbf{0}$ in (3)). Nevertheless, simulations allowed us to study the extension of existing tension distribution methods to the case of suspended CDPR.

5 Tension Distribution Issues

Because of the redundant nature of ReelAx8 actuation ($m = n + 2$), the nullspace of its wrench matrix is non-empty and (2) has an infinity of solutions, among which an optimal or near-optimal one is generally sought. The tension distribution problem is then to find a cable tension vector $\mathbf{t} = [t_1, \dots, t_m] \in \mathbb{R}^m$, whose components are contained between a minimum $t_{\min}$ and a maximum $t_{\max}$, i.e. such as $t_{\min} \leq t_i \leq t_{\max}$, $i \in \{1, \dots, m\}$. $t_{\max}$ may be given by the limit of cable elastic deformation, and $t_{\min}$ can be set as the lowest acceptable tension to avoid loose cables, i.e. $t_{\min} \geq 0$.

Most of works dedicated to tension distribution deal with CDPR having a non-empty force-closure workspace. Some efficient optimization methods were proposed using linear programming methods (LPM) [3, 7, 14] or quadratic programming methods (QPM). More specifically, for the redundantly actuated suspended CDPR case, Oh and Agrawal [12] proposed to find a feasible space for tension distribution described by a set of linear inequalities and to plan the robot trajectory to stay into this space. More recently, Yu et al. [15] proposed the use of a QPM in coupling basic tension optimization problem to an active stiffness control scheme.

However, LPM is subject to discontinuities in the cable tension distribution that may result in high mechanical loads, observed in simulations, and in vibrations. QPM are suffering from non-predictable worst-case runtime [4]. These procedures may be expensive in terms of computation time and usually incompatible with real-time control. Indeed, it was observed in our simulations that this type of methods often exceeds our controller loop time. That is why an efficient numerical method is required. To our knowledge, no method has been demonstrated to work in real-time on a suspended CDPR with redundant actuation.

To illustrate the effect of redundancy, Ω is defined as the set of feasible tensions, i.e. the set of vectors $\mathbf{t}$ defined as:

$$\Omega = \{\mathbf{t} \mid t_i \in [t_{\min}, t_{\max}] \; \forall i \in [1, m]\} \tag{4}$$

Ω is an intersection of halfspaces, a convex bounded polyhedron also called a polytope. More precisely, if $t_{\min}$ and $t_{\max}$ are identical over the m cables, Ω is a m-dimensional hypercube. Finally, we also define $\Lambda \subset \mathbb{R}^r$ as the convex polyhedron generated by the intersection of Σ within Ω as shown in Fig. 4. If Λ is non-empty, a solution that respects the tension limiting values exists and must be found. By computing Λ for the mobile platform along various trajectories, it was observed in simulation that for suspended CDPR the size of Λ is usually very small as Σ cut Ω close to the $\mathbf{t}_{\min}$ vertex, but Λ still exists.

To overcome the problems and disadvantages of the aforementioned LPM and QPM methods, some authors proposed to apply non-iterative algorithms. One solution, according to Lafourcade's works [9], is to find $\mathbf{t}_{\text{sol}} \in \Lambda$ the closest to an objective tension vector $\mathbf{t}_{\text{obj}} \in \Omega$. For ReelAx8, $\mathbf{t}_{\text{obj}}$ is set as a constant, equal to the coordinates of Ω COG. Then, $\mathbf{t}_n = \mathbf{H}\mathbf{t}_{\text{obj}}$ is calculated as the orthogonal projection of $\mathbf{t}_{\text{obj}}$ onto Σ. However, for suspended robots Λ is small in comparison to Ω when the platform is unloaded or carrying a light payload. In these cases, Λ is

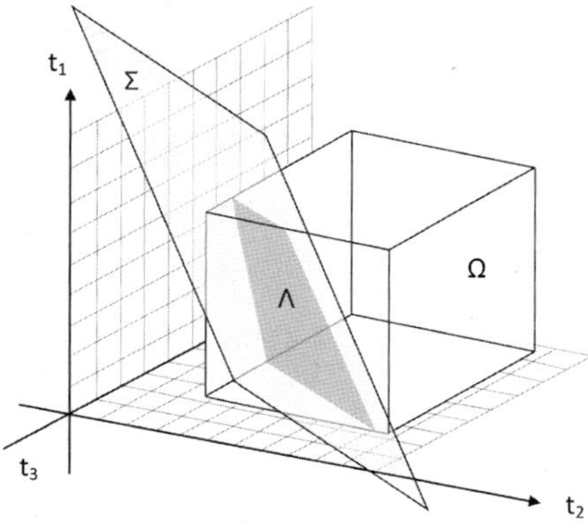

Fig. 4 Intersection Λ between Ω and Σ in the case $m = 3$ and $n = 1$.

very close from $\mathbf{t}_{\min}$. Consequently, $\mathbf{t}_{\text{sol}}$ is most of the time outside Λ, which means non-admissible. It was observed on ReelAx8: for numerous poses, one or two cables have non-admissible tensions $t_k, k \in [1,2]$. In the considered algorithm, the one or two slack cable(s) are then "saturated" at the $t_{\min}$ value and (3) is solved again with the additional constraints $t_k = t_{\min}$. If one cable is saturated, $\mathbf{t}_{\text{sol}}$ is thus moved in the direction of the saturation until it belongs to a facet of Ω, or to an edge if two cables are saturated. This algorithm appears to work well in simulation. Furthermore, $\mathbf{t}_{\text{sol}}$, defined as the minimum of the Euclidean norm $\|\mathbf{t}_{\text{obj}} - \mathbf{t}\|_2$ under the constraint (2), is a quadratic criteria that guarantees the solution continuity [9].

$\mathbf{t}_{\text{obj}}$ can also be defined as the COG of Λ. This can be computed through the determination of Λ vertices followed by a triangulation of the Λ polyhedron [11]. Finally, the COG of each simplexes is calculated together with Λ COG, defined as $\mathbf{t}_{\text{obj}}$. This method has also been implemented. However the computational time required at each loop is high because of the considered number of cables that implies many combinatorial calculations.

6 Conclusion

This paper introduced a suspended six DOF CDPR driven by eight cables called ReelAx8. It provides interesting motion capabilities over an important workspace, regarding its footprint. However, experimental tests revealed open problems such as the need of efficient tension distribution algorithms in order to improve the accuracy. Such algorithms have been tested in simulation but still remains to be apply in real-time to the prototype. This may raise issues since real-time implementation is critical, as it is the case in previous works on fully constrained CDPR which rarely report practical real-time implementations. Furthermore, due to the under-constrained nature of the robot, Λ maybe very small, e.g. if the platform is unloaded, and $\mathbf{t}_{\text{sol}} \in \Lambda$ could be hard to find. These issues are part of our future works.

Acknowledgements The authors gratefully acknowledge the financial contribution of Tecnalia. The financial support of the Région Languedoc-Roussillon (grant 115217) is also greatly acknowledged.

References

1. Albus, J., Bostelman, R., Dagalakis, N.: The NIST Robocrane. J. Robot. Syst. **10**(2), 709–724 (1993)
2. Alp, A., Agrawal, S.: Cable suspended robots: Design, planning and control. In: IEEE Int. Conf. on Robotics and Automation. Washington, DC, USA (2002)
3. Borgstrom, P.H., Jordan, B.L., Sukhatme, G.S., Batalin, M.A., Kaiser, W.J.: Rapid computation of optimally safe tension distributions for parallel cable-driven robots. IEEE Trans. Robot. **25**(6), 1271–1281 (2009)
4. Bruckmann, T., Pott, A., Hiller, M., Franitza, D.: A modular controller for redundantly actuated tendon-based Stewart platforms. In: EuCoMes. Obergurgl, Austria (2006)
5. Carricato, M., Merlet, J.P.: Geometrico-static analysis of under-constrained cable-driven parallel robots. In: Advances in Robot Kinematics. Portoroz, Slovenia (2010)
6. Fang, S., Franitza, D., Torlo, M., Bekes, F., Hiller, M.: Motion control of a tendon-based parallel manipulator using optimal tension distribution. IEEE/ASME Trans. Mechatron. **9**, 561–568 (2004)
7. Gholami, P., Aref, M., Taghirad, H.D.: On the control of the KNTU CDRPM: A cable driven redundant parallel manipulator. In: IEEE Int. Conf. on Intelligent Robots and Systems (2008)
8. Kawamura, S., Choe, W., Tanaka, S., Pandian, S.: Development of an ultrahigh speed robot FALCON using wire drive system. In: IEEE Int. Conf. of Robotics and Automation (1995)
9. Lafourcade, P.: Etude des manipulateurs parallèles à câbles, conception d'une suspension active pour soufflerie. Ph.D. thesis, Thèse de Docteur Ingénieur de l'ENSAE (2004)
10. Merlet, J.P.: MARIONET, A family of modular wire-driven parallel robots. In: Advances in Robot Kinematics. Portoroz, Slovenia (2010)
11. Mikelsons, L., Bruckmann, T., Hiller, M., Schramm, D.: A real-time capable force calculation algorithm for redundant tendon-based parallel manipulators. In: IEEE Int. Conf. on Robotics and Automation (2008)
12. Oh, S.R., Agrawal, S.K.: Cable-suspended planar parallel robots with redundant cables: Controllers with positive cable tensions. In: IEEE Int. Conf. on Robotics and Automation (2003)
13. Riehl, N.: Modélisation et conception de robots parallèles à câbles de grande dimension. Ph.D. thesis, Université de Montpellier II, France (2011)
14. Vafaei, A., Aref, M.M., Taghirad, H.D.: Integrated controller for an over constrained cable driven parallel manipulator: KNTU CDRPM. In: IEEE Int. Conf. on Robotics and Automation. Anchorage, Alaska, USA (2010)
15. Yu, K., Lee, L.F., Krovi, V.N.: Simultaneous trajectory tracking and stiffness control of cable actuated parallel manipulator. In: Int. Design Engineering Technical Conf. & Computers and Information in Engineering Conf. (2009)
16. Zi, B., Duan, B., Du, J., Bao, H.: Dynamic modeling and active control of a cable-suspended parallel robot. Mechatronics **18**(1), 1–12 (2008)

Planning Singularity-Free Force-Feasible Paths on the Stewart Platform

Oriol Bohigas, Montserrat Manubens and Lluís Ros

Abstract This paper provides a method for computing *force-feasible* paths on the Stewart platform. Given two configurations of the platform, the method attempts to connect them through a path that, at any point, allows the platform to counteract any external wrench lying inside a predefined six-dimensional region. In particular, the Jacobian matrix of the manipulator will be full rank along such path, so that the path will not traverse the forward singularity locus at any point. The path is computed by first characterizing the force-feasible C-space of the manipulator as the solution set of a system of equations, and then using a higher-dimensional continuation technique to explore this set systematically from one configuration, until the second configuration is found. Examples are included that demonstrate the performance of the method on illustrative situations.

Key words: Singularity-free path planning, higher-dimensional continuation, singularity avoidance, Stewart platform

1 Introduction

Singular configurations of the Stewart platform are well-known and widely studied in the literature [9, 11]. They yield uncontrollable motions or unresolvable end-effector forces and, thus, they are to be avoided in the vast majority of applications. On this regard, several works consider the issue of local singularity avoidance [1], but only a few tackle the more general problem of computing singularity-free paths between two distant configurations. Solutions to this problem include an algorithm based on deforming a parametrized path between the query configurations [5], a variational approach that reduces the problem to a boundary value problem [14],

Oriol Bohigas · Montserrat Manubens · Lluís Ros
Institut de Robòtica i Informàtica Industrial, Barcelona, Spain,
e-mail: {obohigas,mmanubens,lros}@iri.upc.edu

J. Lenarčič, M. Husty (eds.), *Latest Advances in Robot Kinematics*,
DOI 10.1007/978-94-007-4620-6_31,

and a numerical technique based on treating the singularity locus as a collection of obstacles [6]. All of these algorithms work well in favorable situations, but [5] and [14] mention limitations relative to proving path existence in certain cases, and [6] is computationally intensive, as it requires constructing polytope approximations of the entire singularity set before searching for the path. An important drawback of [5, 6, 14], moreover, is that the clearance of the path relative to the singularity locus is measured in terms of the determinant or the condition number of the Jacobian matrix, which, as noted in [15], lack physical significance.

In contrast to such works, this paper provides a method for planning paths that maintain a physically meaningful clearance with respect to the singularity locus. This clearance is defined following the spirit of [3, 8], as the one that results from only allowing force-feasible configurations; i.e., those on which the leg forces remain within the allowed limits, for any platform wrench lying inside a prescribed six-dimensional region (Section 2). The planning method relies on defining a system of equations whose solution manifold corresponds to the force-feasible subset of the C-space, so that maneuvering through such manifold guarantees singularity avoidance at all times (Section 3). Then, an extension of the higher-dimensional continuation strategy given in [7] is defined to explore this manifold systematically, until a path joining the start and goal configurations is found, or path non-existence is determined at a given resolution (Section 4). The approach has been implemented and validated on several experiments (Section 5).

2 Problem Statement

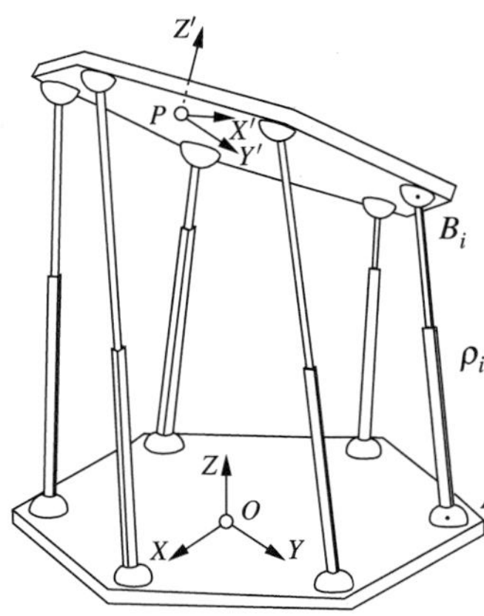

Fig. 1 The Stewart platform.

The Stewart platform consists of a moving plate, or *platform*, connected to a fixed *base* by means of six legs, where each leg is a universal-prismatic-spherical chain. The most general version of such manipulator follows the so-called 6-6 design, where the leg anchor points are all different (Fig. 1), though not necessarily coplanar [10]. The six prismatic joints are actuated, allowing to control the six degrees of freedom of the platform, and the remaining joints are passive.

Let $OXYZ$ and $PX'Y'Z'$ be fixed and moving reference frames, respectively attached to the base and the platform (Fig. 1). Any configuration of the platform can be uniquely represented by a pair $\boldsymbol{q} = (\boldsymbol{p}, \boldsymbol{R}) \in SE(3)$, where $\boldsymbol{p} = [x, y, z]^{\mathsf{T}}$ is the position vector of point P in the fixed frame, and $\boldsymbol{R}$ is a 3×3 rotation matrix providing the orientation of $PX'Y'Z'$ relative to $OXYZ$. Not all values for $\boldsymbol{R}$ and $\boldsymbol{p}$ are permitted though, because the leg lengths ρ_i need to be within the range $[\underline{\rho_i}, \overline{\rho_i}]$ of allowable values on each leg. Thus, if $\boldsymbol{a}_i$ and $\boldsymbol{b}_i$ denote the position vectors of the anchor points A_i and B_i of the ith leg (Fig. 1), expressed in $OXYZ$ and $PX'Y'Z'$ respectively, the configuration will only be *valid* if for $i = 1, \ldots, 6$ it satisfies

$$\rho_i^2 = |\boldsymbol{p} + \boldsymbol{R}\boldsymbol{b}_i - \boldsymbol{a}_i|^2, \tag{1}$$

$$\rho_i \in [\underline{\rho_i}, \overline{\rho_i}]. \tag{2}$$

Often, moreover, each configuration $\boldsymbol{q}$ must be *force feasible*, in the sense that the platform must be able to equilibrate any external wrench $\hat{\boldsymbol{w}}$ acting on it, subject to lie inside a given six-dimensional region $\mathscr{W} \subset \mathbb{R}^6$. The significance of $\mathscr{W}$ depends on the particular context of application. For example, $\mathscr{W}$ may be determined by the set of allowable inertia forces acting on the platform, or by the set of wrenches that should be suppliable to the environment. Specifically, the force-feasibility requirement on a given $\boldsymbol{q}$ implies that for any wrench $\hat{\boldsymbol{w}} \in \mathscr{W}$ there must be a vector $\boldsymbol{f} = [f_1, \ldots, f_6]^{\mathsf{T}}$ of leg forces satisfying

$$\boldsymbol{J}(\boldsymbol{q}) \cdot \boldsymbol{f} = \hat{\boldsymbol{w}}, \tag{3}$$

with

$$\boldsymbol{f} \in \mathscr{D} = [\underline{f_1}, \overline{f_1}] \times \ldots \times [\underline{f_6}, \overline{f_6}], \tag{4}$$

where $\boldsymbol{J}(\boldsymbol{q})$ is the 6×6 screw Jacobian of the manipulator at configuration $\boldsymbol{q}$, and $[\underline{f_i}, \overline{f_i}]$ is the interval of force magnitudes that can be resisted by the ith leg. In this paper, $\mathscr{W}$ will be a six-dimensional non-degenerate ellipsoid defined by

$$(\hat{\boldsymbol{w}} - \hat{\boldsymbol{w}}_0)^{\mathsf{T}} \boldsymbol{E} \, (\hat{\boldsymbol{w}} - \hat{\boldsymbol{w}}_0) \leq 1, \tag{5}$$

where $\hat{\boldsymbol{w}}_0$ is a fixed wrench and $\boldsymbol{E}$ is a constant 6×6 positive-definite symmetric matrix. Also, $\hat{\boldsymbol{w}}$, $\hat{\boldsymbol{w}}_0$, and $\boldsymbol{J}(\boldsymbol{q})$ will be assumed to be given in a frame $PXYZ$ centered in P and parallel to $OXYZ$, but any other frame could be assumed if desired.

Now, let $\mathscr{C}$ be the set of all $\boldsymbol{q} \in SE(3)$ satisfying Eqs. (1)-(4) for all $\hat{\boldsymbol{w}} \in \mathscr{W}$, which we will call the *force-feasible C-space* of the manipulator. Given two configurations in $\mathscr{C}$, $\boldsymbol{q}_1$ and $\boldsymbol{q}_2$, the goal of this paper is to provide an algorithm for computing a path on $\mathscr{C}$ connecting them, if one exists, or to determine path non-existence otherwise. To find such a path, we next define a system of equations that characterize $\mathscr{C}$.

3 Equations of the Force-Feasible C-Space

Let $\boldsymbol{f}_0$ be any vector of resultant leg forces corresponding to $\hat{\boldsymbol{w}}_0$, i.e.,

$$\boldsymbol{J}(\boldsymbol{q}) \cdot \boldsymbol{f}_0 = \hat{\boldsymbol{w}}_0. \tag{6}$$

By substitution of $\hat{\boldsymbol{w}} - \hat{\boldsymbol{w}}_0 = \boldsymbol{J}(\boldsymbol{q})(\boldsymbol{f} - \boldsymbol{f}_0)$ into Eq. (5) we realize that, for a given $\boldsymbol{q}$, the set $\mathscr{F}$ of leg forces $\boldsymbol{f}$ satisfying Eq. (3) for some $\hat{\boldsymbol{w}} \in \mathscr{W}$ is given by $(\boldsymbol{f} - \boldsymbol{f}_0)^\mathsf{T} \boldsymbol{B} (\boldsymbol{f} - \boldsymbol{f}_0) \leq 1$, where $\boldsymbol{B} = \boldsymbol{J}(\boldsymbol{q})^\mathsf{T} \boldsymbol{E}\, \boldsymbol{J}(\boldsymbol{q})$. Hence, $\mathscr{F}$ is an ellipsoid, because $\boldsymbol{E}$ is symmetric and positive-definite, but this ellipsoid will be bounded in all directions, or only in some, depending on whether $\det(\boldsymbol{J}(\boldsymbol{q})) \neq 0$ or not.

Now note that, for $\boldsymbol{q}$ to be force-feasible, it must be $\mathscr{F} \subset \mathscr{D}$, which can be checked as follows. Let $\boldsymbol{v}_i \in \mathbb{R}^6$ be a vector satisfying

$$\boldsymbol{B}^i \boldsymbol{v}_i = \boldsymbol{0} \tag{7}$$

$$\boldsymbol{v}_i^\mathsf{T} \boldsymbol{B}\, \boldsymbol{v}_i = 1 \tag{8}$$

$$v_{i,i} \geq 0 \tag{9}$$

where $v_{i,i}$ denotes the ith component of $\boldsymbol{v}_i$, and $\boldsymbol{B}^i$ stands for the matrix $\boldsymbol{B}$ with its ith row removed. If $\det(\boldsymbol{J}(\boldsymbol{q})) \neq 0$, then $\boldsymbol{B}$ and $\boldsymbol{B}^i$ are full row rank, and there is exactly one vector $\boldsymbol{v}_i$ satisfying (7)-(9). Using Lagrange multipliers, one can see that in such a case $\boldsymbol{f}_0 - \boldsymbol{v}_i$ and $\boldsymbol{f}_0 + \boldsymbol{v}_i$ are the vectors in $\mathscr{F}$ attaining the smallest and largest value along the ith coordinate. Hence, when $\det(\boldsymbol{J}(\boldsymbol{q})) \neq 0$, $\mathscr{F} \subset \mathscr{D}$ iff

$$f_{0,i} - v_{i,i} \geq \underline{f_i} \quad \text{and} \quad f_{0,i} + v_{i,i} \leq \overline{f_i}, \qquad \text{for } i = 1, \ldots, 6. \tag{10}$$

When $\det(\boldsymbol{J}(\boldsymbol{q})) = 0$, it will always be $\mathscr{F} \not\subset \mathscr{D}$, because $\mathscr{F}$ will be unbounded along some of its principal directions.

Observe that the constraints in (2), (9), and (10) are equivalent to imposing

$$(\rho_i - m_i)^2 + r_i^2 = h_i^2, \tag{11}$$

$$v_{i,i} = s_i^2, \tag{12}$$

$$f_{0,i} - v_{i,i} = t_i^2 + \underline{f_i}, \qquad f_{0,i} + v_{i,i} = -u_i^2 + \overline{f_i}, \tag{13}$$

respectively, where m_i and h_i are the midpoint and half-range of $[\underline{\rho_i}, \overline{\rho_i}]$, and r_i, s_i, t_i, and u_i are newly-defined auxiliary variables. As a result, $\mathscr{C}$ can be characterized as the set of points $\boldsymbol{q}$ that satisfy the system formed by Eqs. (1), (6), (7), (8), and (11)-(13) for some value of the remaining variables. For ease of explanation, this system will be written as $\boldsymbol{F}(\boldsymbol{x}) = \boldsymbol{0}$, where $\boldsymbol{x} \in \mathbb{R}^{n_x}$ encompasses all variables in the system, including those in $\boldsymbol{q}$.

Let $\mathscr{M}$ be the set of points $\boldsymbol{x}$ that satisfy $\boldsymbol{F}(\boldsymbol{x}) = \boldsymbol{0}$, which is a manifold of dimension $d = 6$ when no further constraints are imposed on $\boldsymbol{q} = (\boldsymbol{p}, \boldsymbol{R})$. Note that such points are in correspondence with the points $\boldsymbol{q} \in \mathscr{C}$ because any value of $\boldsymbol{q} \in \mathscr{C}$ determines the values for the remaining variables in $\boldsymbol{x} \in \mathscr{M}$. Thus, since $\boldsymbol{q}_1$ and $\boldsymbol{q}_2$ have corresponding points $\boldsymbol{x}_1$ and $\boldsymbol{x}_2$ on $\mathscr{M}$, and all paths on $\mathscr{C}$ are represented in $\mathscr{M}$, and viceversa, the original problem of computing a force-feasible path in $\mathscr{C}$ from $\boldsymbol{q}_1$ to $\boldsymbol{q}_2$ can be reduced to that of connecting $\boldsymbol{x}_1$ and $\boldsymbol{x}_2$ through a path on $\mathscr{M}$. It is not difficult to see, moreover, that for any $\boldsymbol{x} \in \mathscr{M}$ it will be $\det(\boldsymbol{J}(\boldsymbol{q})) \neq 0$, so that any path computed on $\mathscr{M}$ will be free of singular configurations. Certainly, if it were $\det(\boldsymbol{J}(\boldsymbol{q})) = 0$ for some $\boldsymbol{x} \in \mathscr{M}$, then $\boldsymbol{B}$ would be rank deficient, implying that

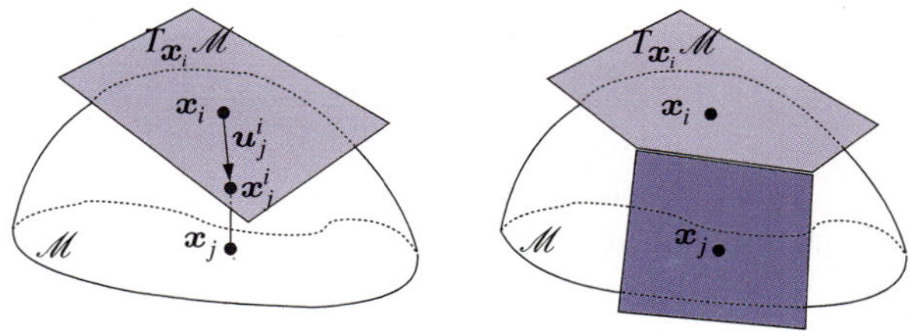

Fig. 2 The higher-dimensional continuation method applied to a 2-dimensional manifold in $\mathbb{R}^3$.

$\mathrm{Ker}(\boldsymbol{B}^i) = \mathrm{Ker}(\boldsymbol{B})$ for some i. Therefore, any $\boldsymbol{v}_i$ satisfying Eq. (7) would violate Eq. (8), which is in contradiction with the fact that $\boldsymbol{x} \in \mathscr{M}$.

4 Exploring the Force-Feasible C-Space for a Path

To determine a path on $\mathscr{M}$ connecting $\boldsymbol{x}_1$ and $\boldsymbol{x}_2$ we can gradually construct an atlas of $\mathscr{M}$, i.e., a collection of charts where each chart C_i defines a local map from a domain $\mathscr{P}_i \subset \mathbb{R}^d$ to an open set of $\mathscr{M}$ around a point $\boldsymbol{x}_i \in \mathscr{M}$, initially $\boldsymbol{x}_1$. The atlas will be computed using the higher-dimensional continuation approach proposed in [7], which defines the local map for chart C_i using $\boldsymbol{\Psi}_i$, an orthonormal basis of $T_{\boldsymbol{x}_i}\mathscr{M}$, the d-dimensional tangent space of $\mathscr{M}$ at $\boldsymbol{x}_i$. The map is defined by first selecting a vector $\boldsymbol{u}^i_j \in \mathbb{R}^d$ of parameters (Fig. 2, left), which is used to generate a point $\boldsymbol{x}^i_j \in \mathbb{R}^{n_x}$ in the neighborhood of $\boldsymbol{x}_i$, using

$$\boldsymbol{x}^i_j = \boldsymbol{x}_i + \boldsymbol{\Psi}_i \, \boldsymbol{u}^i_j. \tag{14}$$

Then, a point $\boldsymbol{x}_j \in \mathscr{M}$ corresponding to the projection of $\boldsymbol{x}^i_j$ on $\mathscr{M}$ is computed, by solving the system formed by $\boldsymbol{F}(\boldsymbol{x}_j) = \boldsymbol{0}$ and $\boldsymbol{\Psi}_i^\top (\boldsymbol{x}_j - \boldsymbol{x}^i_j) = \boldsymbol{0}$ using a Newton method initialized at $\boldsymbol{x}^i_j$. Each point $\boldsymbol{x}_j$ is the potential center of a new chart (Fig. 2, right), and Henderson introduced a method to determine how to select the chart centers to ensure a good coverage of the manifold [7]. In his approach, the domain $\mathscr{P}_i$ of chart C_i is initialized as a d-dimensional hypercube enclosing a ball $\mathscr{B}_i$ of radius r, both defined in $T_{\boldsymbol{x}_i}\mathscr{M}$, as illustrated in Fig. 3, top. A vertex of $\mathscr{P}_i$ exterior to $\mathscr{B}_i$, with position vector $\boldsymbol{s}$, is used to generate a point $\boldsymbol{x}^i_j$, using (14) with $\boldsymbol{u}^i_j = \alpha \cdot \boldsymbol{s}/\|\boldsymbol{s}\|$, where α is initialized to r. If the projection of $\boldsymbol{x}^i_j$ to $\mathscr{M}$ does not converge, or if the new chart C_j at $\boldsymbol{x}_j$ is too far or

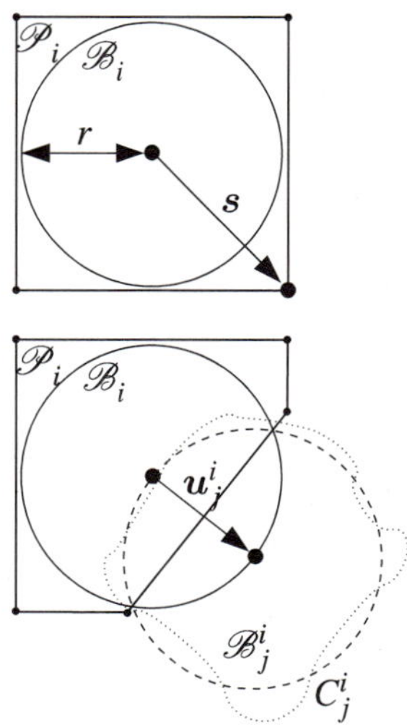

Fig. 3 Chart construction.

too different from C_i, the new chart is discarded and a new attempt of chart generation is performed with a smaller α, allowing to adapt the size of the area covered by each chart to the local curvature of the manifold. When C_j is valid, it is used to crop $\mathscr{P}_i$ from the intersection between $\mathscr{B}_i$ and C^i_j, the projection on $T_{\boldsymbol{x}_i}\mathscr{M}$ of the part of the manifold covered by C_j. This projection is approximated by a ball $\mathscr{B}^i_j$ of radius r in $T_{\boldsymbol{x}_i}\mathscr{M}$, centered at the point given by $\boldsymbol{u}^i_j$, as shown in Fig. 3, bottom. The intersection of $\mathscr{B}_i$ and $\mathscr{B}^i_j$ defines a new face for $\mathscr{P}_i$ that eliminates some of its vertices (in particular the one given by $\boldsymbol{s}$) and generates new ones. Symmetrically, the polytope $\mathscr{P}_j$ associated with C_j is cropped using C_i. When C_i is surrounded by other charts, $\mathscr{P}_i$ becomes a convex polytope included in $\mathscr{B}_i$, and C_i is considered to be *closed*, meaning that no further expansion of the atlas needs to be attempted from that chart. When all charts are closed, the connected component of $\mathscr{M}$ containing the initial point $\boldsymbol{x}_1$ gets fully covered. If a path exists from $\boldsymbol{x}_1$ to $\boldsymbol{x}_2$, $\boldsymbol{x}_2$ must be included in one of the charts of the atlas and, thus, a solution path can be determined by searching on the graph implicitly defined by the chart centers and their neighborhood relations. In practice, however, the expansion of the charts is performed according to an A* search strategy using an admissible heuristic [13], so that the path is returned as soon as it is found without computing the whole atlas, and it is guaranteed to be the shortest possible on $\mathscr{M}$. If $\boldsymbol{x}_2$ is not included in any of the charts in the end, path non-existence is established at the considered value for r.

5 Experiments

The method has been implemented in C, and run on a iMac equipped with a 2.93 GHz Intel Core i7 processor. To verify its performance on a realistic situation, the geometric parameters of the INRIA left hand have been used [10]. All legs of this manipulator admit forces in the range $[\underline{f_i}, \overline{f_i}] = [-300, 300]$, and for the experiments we have set $\hat{\boldsymbol{w}}_0 = [0,0,150,0,0,0]^\mathsf{T}$ and $\boldsymbol{E} = \boldsymbol{I}_{6\times 6}$, assuming SI units throughout. In this case, the force-feasible C-space of this manipulator is close to the workspace defined by its allowable leg lengths [10], and the algorithm solves typical planning queries in a few seconds, even when permitting the variation of all pose parameters.

To graphically illustrate the performance of the method, however, it is better to adopt the geometric parameters of the more academic manipulator in [9], where large variations of the leg lengths are allowed, leading to a very large workspace with interesting singularity surfaces. Two experiments are shown for this manipulator, assuming $[\underline{f_i}, \overline{f_i}] = [-300, 300]$ and $\boldsymbol{E} = \boldsymbol{I}_{6\times 6}$ as before, but using $\hat{\boldsymbol{w}}_0 = [0,0,1,0,0,0]^\mathsf{T}$. In a first experiment we compute a force-feasible path for the platform moving at a constant orientation, defined by the Euler angles $\phi = -2°$, $\theta = 30°$ and $\psi = -87°$ under the convention in [9], and at a constant value of z. Using the start and goal configurations defined by the positions $\boldsymbol{p}_1 = [0.4, 0, 0.1]^\mathsf{T}$ and $\boldsymbol{p}_2 = [-0.3, 0, 0.1]^\mathsf{T}$ for P, the resulting path in the XY plane is computed in

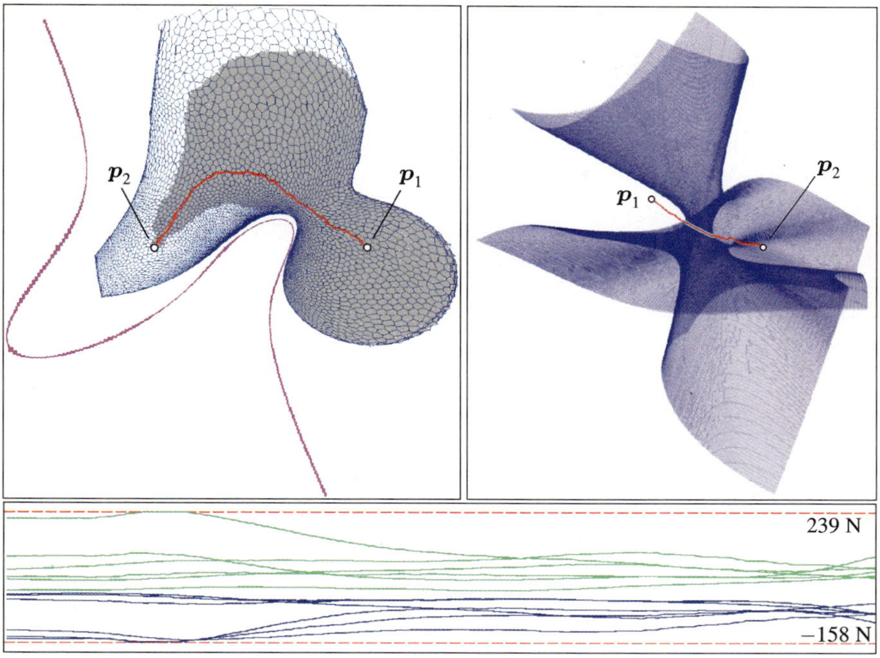

Fig. 4 Top: Two paths computed for the manipulator in [9]. **Bottom:** Plot of the maximum and minimum forces supported by each leg along the left path.

578 seconds. Fig. 4, left shows this path, together with the singularity curve to be avoided, the atlas corresponding to the whole force-feasible connected component accessible from $\boldsymbol{p}_1$ (shown as a mesh), and the region explored by the A* algorithm (shaded in grey). It can be seen that the interpolated path between $\boldsymbol{p}_1$ and $\boldsymbol{p}_2$ would go through singularities, but the computed path correctly avoids them while keeping the leg forces within the specified ranges (Fig. 4, bottom). In a second experiment, we solve the same planning query but keeping constant the orientation of the platform only, obtaining the path in XYZ shown in Fig. 4, right in 90 minutes. The singularity surface, computed using [2] and shown in the figure, is correctly avoided by the computed path. It must be noted that these are hard planning queries, since the workspace in [9] is enormous when compared to typical workspaces arising in usual platforms. Moreover, once a partial atlas is computed, all planning queries between configurations covered by such atlas can be solved in a few milliseconds.

6 Conclusions and Future Work

This paper has presented a path planning method for computing non-singular paths on the Stewart platform, based on imposing the resolvability of a six-dimensional set of wrenches at any point on the path. The method has been tested successfully on manipulators of various geometries, and computes paths in reasonable times in realistic situations. The presented approach could be generalized by requiring the path to fulfill additional constraints, like guaranteeing a certain positioning accuracy of the platform, or the avoidance of platform collisions. While the former constraints can in principle be incorporated using dual developments to those herein presented, the latter require investigating the possibility of randomizing the planner, in the spirit of [4] or [12], for example.

Acknowledgements This work has been partially funded by the Spanish Government under contract DPI2010-18449, and by a Juan de la Cierva contract supporting the second author.

References

1. Bhattacharya, S., Hatwal, H., Ghosh, A.: Comparison of an exact and an approximate method of singularity avoidance in platform type parallel manipulators. Mech. Mach. Theory **33**(7), 965–974 (1998)
2. Bohigas, O., Zlatanov, D., Ros, L., Manubens, M., Porta, J.: Numerical computation of manipulator singularities. In: Proc. of the IEEE Int. Conf. on Robotics and Automation (2012)
3. Bosscher, P., Riechel, A., Ebert-Uphoff, I.: Wrench-feasible workspace generation for cable-driven robots. IEEE Trans. Robot. **22**(5), 890–902 (2006)
4. Cortés, J., Siméon, T.: Probabilistic motion planning for parallel mechanisms. In: Proc. of the IEEE Int. Conf. on Rob. and Aut., vol. 3, pp. 4354–4359 (2003)
5. Dasgupta, B., Mruthyunjaya, T.: Singularity-free path planning for the Stewart platform manipulator. Mech. Mach. Theory **33**(6), 711–725 (1998)
6. Dash, A.K., Chen, I.M., Yeo, S.H., Yang, G.: Workspace generation and planning singularity-free path for parallel manipulators. Mech. Mach. Theory **40**(7), 776–805 (2005)
7. Henderson, M.E.: Multiple parameter continuation: Computing implicitly defined *k*-manifolds. Int. J. Bifurc. Chaos **12**(3), 451–476 (2002)
8. Hubert, J.: Manipulateurs parallèles, singularités et analyse statique. Ph.D. thesis, École Nationale Supérieure des Mines de Paris (2010)
9. Li, H., Gosselin, C., Richard, M., St-Onge, B.: Analytic form of the six-dimensional singularity locus of the general Gough-Stewart platform. ASME J. Mech. Des. **128**, 279–288 (2006)
10. Merlet, J.: Parallel robots. Springer-Verlag (2006)
11. Merlet, J.P., Gosselin, C.: Parallel Mechanisms and Robots. In: Springer Handbook of Robotics, pp. 269–285. Springer (2008)
12. Porta, J.M., Jaillet, L.: Path planning on manifolds using randomized higher-dimensional continuation. In: Hsu D., Isler V., Latombe J.C., Lin M. (eds.) Algorithmic Foundation of Robotics IX. STAR, vol. 68, pp. 337–353. Springer (2011)
13. Russell, S.J., Norvig, P.: Artificial Intelligence: A Modern Approach. Prentice Hall (2003)
14. Sen, S., Dasgupta, B., Mallik, A.K.: Variational approach for singularity-free path-planning of parallel manipulators. Mech. Mach. Theory **38**(11), 1165–1183 (2003)
15. Voglewede, P., Ebert-Uphoff, I.: Overarching framework for measuring closeness to singularities of parallel manipulators. IEEE Trans. Robot. **21**(6), 1037–1045 (2005)

Orientation Capability of a 3-RPSR Parallel Mechanism for a Movable-Die Drive Mechanism of Pipe Bender

Yukio Takeda, Satoshi Inada, Daisuke Matsuura, Kazuya Hirose and Ken Ichiryu

Abstract Kinematic analysis of a 3-RPSR parallel mechanism with six DOF, which has been applied to a movable-die drive mechanism of pipe bender, has been done. Orientation capability of this mechanism has been investigated taking into consideration singularity and motion range of spherical joints. A mechanism with a high orientation capability designed based on the analysis is shown.

Key words: Robotics, kinematics, parallel mechanism, orientation capability, singularity, pipe bender

1 Introduction

Parallel mechanisms with six degrees of freedom (DOF) have attracted attention as appropriate mechanisms for performing tasks which require high accuracy, high acceleration, high rigidity, and high power. A huge number of researches have been done for such mechanisms with six connecting chains between the base and the output link, such as those with variable link lengths (Stewart platform type), fixed length links with linear drives, and fixed length links with rotary drives. They have been applied to motion simulators, machine tools, manipulators, etc. However, due to the limited orientation capability (small range of orientation angle of the output link) of these mechanisms, their applications are quite limited. In our previous work, a six-DOF parallel mechanism with three connecting chains, named 3-RPSR mechanism with triple revolute joints on the base, has been proposed as a mechanism

Yukio Takeda · Satoshi Inada · Daisuke Matsuura
Department of Mechanical Sciences and Engineering, Tokyo Institute of Technology 2-12-1, O-okayama, Meguro-ku, Tokyo 152-8552, Japan, e-mail: takeda@mech.titech.ac.jp

Kazuya Hirose · Ken Ichiryu
Monozukiri Mechatro Research Laboratory, Kikuchi Seisakusho Co., Ltd. 2161-12, Miyama-cho, Hachioji-shi, Tokyo 192-0152, Japan

J. Lenarčič, M. Husty (eds.), *Latest Advances in Robot Kinematics*,
DOI 10.1007/978-94-007-4620-6_32,

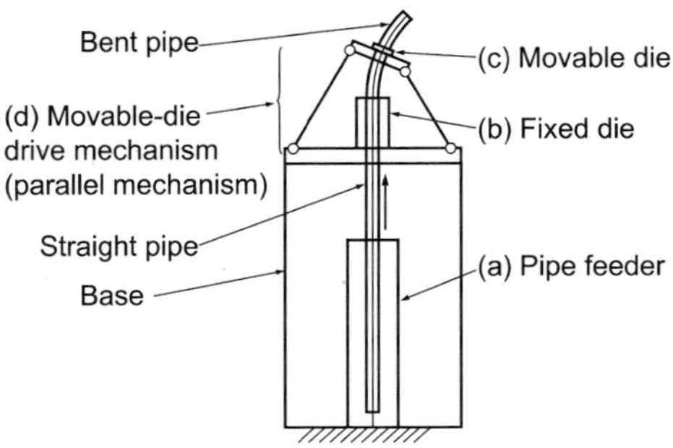

Fig. 1 Configuration of pipe bender.

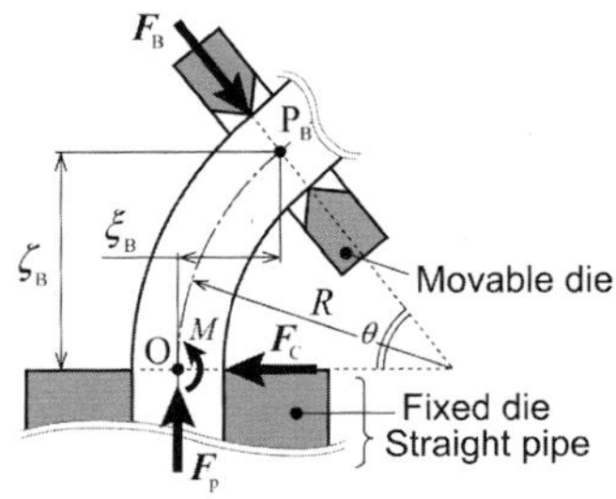

Fig. 2 Principle of penetration bending method.

which can have a good orientation capability [2]. This mechanism has been applied to a movable-die drive mechanism of pipe bender [1]. This mechanism was designed to perform a good orientation capability such as a full rotation around the center axis and inclination by 45 deg in any direction. Through its application to a pipe bender, we found that better orientation capability is required for manufacturing complex three-dimensional shaped pipes.

In the present paper, kinematic analysis of 3-RPSR mechanism has been performed to clarify the relationship between kinematic parameters and orientation capability taking into consideration singularity and motion range of joints. Then, a mechanism having a better orientation capability is shown.

2 Pipe Bender Composed of Parallel Mechanism

The basic configuration of the pipe bender is shown in Fig. 1. A straight pipe pushed upward by a pipe feeder (a) is transformed into a desired shape by being pushed through a fixed die (b) and a movable die (c). The position and orientation of the movable die are changed by the movable-die drive mechanism (d) synchronously driven with the feeding of the pipe. As the movable-die drive mechanism, we applied a parallel mechanism with six DOF.

Figure 2 shows the geometric parameters used to represent the position and orientation of the movable die relative to the fixed die, and forces applied to the pipe from two dies and pipe feeder. ξ_B, ζ_B, θ and R are the offset, angle of the movable die and the curvature radius of the bent pipe, and F_B, F_C and F_P are the forces applied to pipe by movable die, fixed die and feeder. By neglecting the friction between the pipe and two dies, bending moment, M, exerted on the pipe at point O, is written as

$$M = (F_B \cos\theta)\zeta_B + (F_B \sin\theta)\xi_B = F_B R \sin\theta. \tag{1}$$

From Eq. (1) and Fig. 1, it can be found that increasing the movable-die inclination angle, θ, is important to reduce the force amplitude, F_B, without losing the

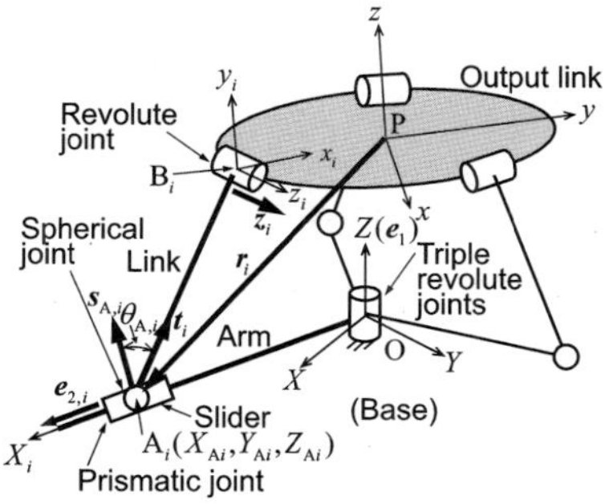

Fig. 3 3-RPSR mechanism with triple revolute joints.

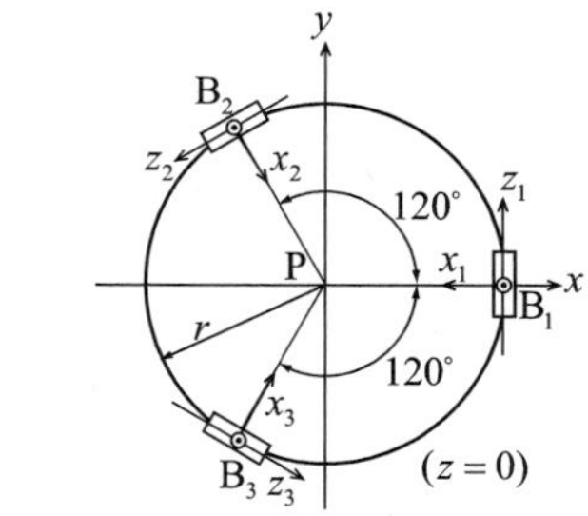

Fig. 4 Location of revolute joint on the output link.

capability to achieve large bending moment. Increasing θ is also effective to avoid the interference between the fixed die and movable die, because that will enlarge the distance between two dies.

3 3-RPSR Mechanism

3.1 Description of the Mechanism

The kinematic structure of the 3-RPSR mechanism is shown in Fig. 3. The axes of all revolute joints which connect the connecting chains with the base are coincident with the Z axis of the base coordinate system O-XYZ. The three connecting chains have the same structure and dimensions, and they are symmetrically located with respect to O-XYZ and the moving coordinate system fixed on the output link P-xyz. P represents the reference point on the output link. The revolute joints on the base and the prismatic joints are active joints. As shown in Figs. 3 and 4, a moving coordinate system B_i-$x_iy_iz_i$ is considered, while i represents the connecting chain number ($i = 1,2,3$). B_i-$x_iy_iz_i$ is fixed to the output link at point B_i, and direction of the z_i axis coincides with the axis of the revolute joint on the output link. The center of the spherical joint connecting i-th chain is denoted as A_i. The orientation of the output link is represented by the three angles θ_y, θ_z, and ψ, shown in Fig. 5. Kinematic constants are r (radius of the location circle of revolute joints on the output link, Fig. 4), l (link length $\overline{A_iB_i}$), and β_B (the angle between the axis of the prismatic joint and the base plane (XY plane), Fig. 6).

Figures 3 and 4 illustrate the definitions of important vectors for the following kinematic analysis. A unit vector $\mathbf{s}_{A,i}$ is heading the direction of the rotation center of the spherical joint. Rotation angle of the spherical joint, $\theta_{A,i}$, is defined as the angle between $\overrightarrow{A_iB_i}$ and $\mathbf{s}_{A,i}$. Vectors $\mathbf{a}_i$ and $\mathbf{r}_i$ represent the position of the spherical joint's center from the origin O and that from the output point P. Unit vectors $\mathbf{t}_i$, $\mathbf{e}_1$, $\mathbf{e}_{2,i}$, and $\mathbf{z}_i$ represent the direction of the link $\overrightarrow{A_iB_i}$, that of the revolute joint on the

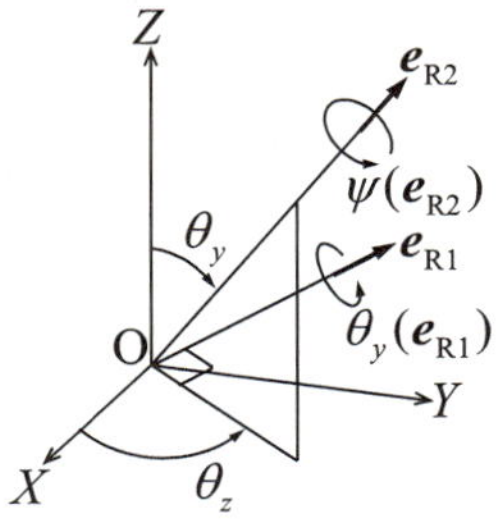

Fig. 5 Definition of orientation angles.

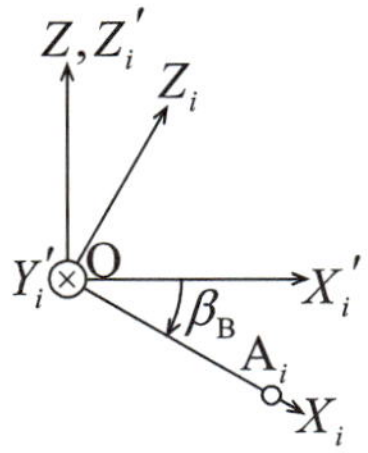

Fig. 6 Definition of β_B.

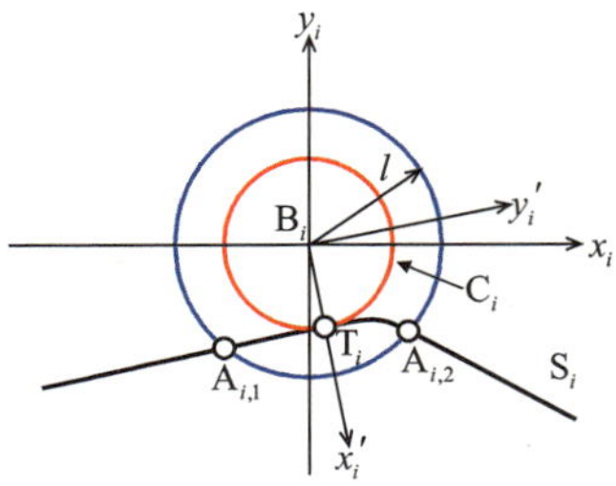

Fig. 7 Solutions in B_i-$x_i'y_i'$ coordinate system.

base, that of the prismatic joint, and that of the rotation axis of the revolute joint B_i, respectively.

Relationship among three coordinate systems O-XYZ, P-xyz and B_i-$x_iy_iz_i$ is described based on 4×4 transformation matrices $T^P_{B,i}$, $T^O_{B,i}$ and T^O_P as follows:

$$\begin{bmatrix} 1 \\ \mathbf{x}^O \end{bmatrix} = T^O_P \begin{bmatrix} 1 \\ \mathbf{x}^P \end{bmatrix} = T^O_P T^P_{B,i} \begin{bmatrix} 1 \\ \mathbf{x}^{B,i} \end{bmatrix} = T^O_{B_i} \begin{bmatrix} 1 \\ \mathbf{x}^{B,i} \end{bmatrix}, \text{ where } T^O_{B,i} = \begin{bmatrix} 1 & 0 & 0 & 0 \\ \mathbf{B}^O_i & & R^O_{B,i} & \end{bmatrix}, \quad (2)$$

$\mathbf{x}^O$, $\mathbf{x}^P$, $\mathbf{x}^{B,i}$, $\mathbf{B}^O_i = [X_{B,i} Y_{B,i} Z_{B,i}]^{\mathrm{T}}$ and $R^O_{B,i} = [a_{mn}]$ $(m,n = 1,2,3)$ are position vectors written in O-XYZ, P-xyz and B_i-$x_iy_iz_i$ coordinate systems, and position and orientation of the frame B_i-$x_iy_iz_i$ in terms of O-XYZ, respectively.

3.2 Inverse Displacement Analysis

The center of spherical joint A_i moves on the circular cone defined by a point O, Z axis and angle β_B ($Z_{A,i} \leq 0$). On the other hand, A_i lies on a circle centered on B_i and radius l. These constraints are written by the following two equations and an inequality with respect to the coordinates of $A_i(x_i, y_i, z_i)$ in B_i-$x_iy_iz_i$,

$$\left.\begin{array}{l} x_i^2 + y_i^2 = l^2 \\ \{(X_{B,i} + a_{11}x_i + a_{12}y_i)^2 + (Y_{B,i} + a_{21}x_i + a_{22}y_i)^2\} \tan^2 \beta_B = (Z_{B,i} + a_{31}x_i + a_{32}y_i)^2 \\ Z_{B,i} + a_{31}x_i + a_{32}y_i \leq 0 \end{array}\right\} \quad (3)$$

By solving these simultaneous equations with an inequality, inverse displacement analysis is performed for a given T^O_P ($T^O_{B,i}$ is calculated from T^O_P and kinematic constants). It should be noted that Eq. (3) has two solutions, but only one of them is practical, when the actual working mode is considered. As shown in Fig. 7, two solutions $A_{i,1}$ and $A_{i,2}$ on x_iy_i plane ($z_i = 0$) are illustrated as the intersections of S_i (the second equation of Eq. (3)) and a circle of radius l (the first equation of Eq. (3)), and distance between the point B_i and curvature S_i is minimized at point T_i. When Eq. (3) has a multiple root, that means the mechanism is at a singular point, and

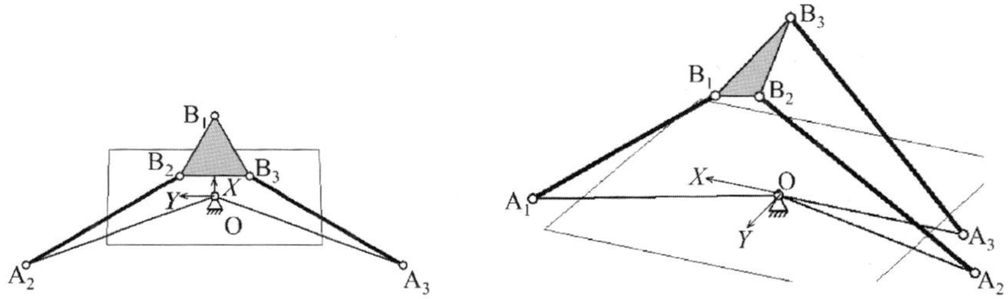

(a) Uncertain configuration (b) Stationary configuration

Fig. 8 Example of singular configurations.

$A_{i,1}$ and $A_{i,2}$ coincide with T_i. Without working mode change through the singular point T_i, the solution cannot transfer $A_{i,1(\text{or } 2)}$ to $A_{i,2(\text{or } 1)}$. Based on a coordinate system B_i-$x_i'y_i'$, in which x_i' is aligned to $\overrightarrow{B_iT_i}$ and y_i' is determined as $z_i' = x_i' \times y_i'$, two solutions are discriminated as positive or negative along the y_i' coordinate. In our study, negative solution is chosen.

3.3 Jacobian Matrix

Relationship between the force $\mathbf{F}^O$/moment $\mathbf{M}^O$ exerted on the output link and driving force/moment of active joints τ are written in base coordinate frame as

$$\begin{bmatrix} \mathbf{F}^O \\ \mathbf{M}^O \end{bmatrix} = -J^{\mathrm{T}}\tau, \quad J^{\mathrm{T}} = J_1 J_2^{-1}, \tag{4}$$

where

$$J_1 = \begin{bmatrix} \mathbf{t}_1^O & \mathbf{t}_2^O & \mathbf{t}_3^O & \mathbf{z}_1^O & \mathbf{z}_2^O & \mathbf{z}_3^O \\ \mathbf{r}_1^O \times \mathbf{t}_1^O & \mathbf{r}_2^O \times \mathbf{t}_2^O & \mathbf{r}_3^O \times \mathbf{t}_3^O & \mathbf{r}_1^O \times \mathbf{z}_1^O & \mathbf{r}_2^O \times \mathbf{z}_2^O & \mathbf{r}_3^O \times \mathbf{z}_3^O \end{bmatrix},$$

$$J_2 = \begin{bmatrix} \{\mathbf{a}_1^O \times \mathbf{t}_1^O, \mathbf{e}_1^O\} & 0 & 0 & \{\mathbf{a}_1^O \times \mathbf{z}_1^O, \mathbf{e}_1^O\} & 0 & 0 \\ 0 & \{\mathbf{a}_2^O \times \mathbf{t}_2^O, \mathbf{e}_1^O\} & 0 & 0 & \{\mathbf{a}_2^O \times \mathbf{z}_2^O, \mathbf{e}_1^O\} & 0 \\ 0 & 0 & \{\mathbf{a}_3^O \times \mathbf{t}_3^O, \mathbf{e}_1^O\} & 0 & 0 & \{\mathbf{a}_3^O \times \mathbf{z}_3^O, \mathbf{e}_1^O\} \\ \{\mathbf{t}_1^O, \mathbf{e}_{2,1}^O\} & 0 & 0 & \{\mathbf{z}_1^O, \mathbf{e}_{2,1}^O\} & 0 & 0 \\ 0 & \{\mathbf{t}_2^O, \mathbf{e}_{2,2}^O\} & 0 & 0 & \{\mathbf{z}_2^O, \mathbf{e}_{2,2}^O\} & 0 \\ 0 & 0 & \{\mathbf{t}_3^O, \mathbf{e}_{2,3}^O\} & 0 & 0 & \{\mathbf{z}_3^O, \mathbf{e}_{2,3}^O\} \end{bmatrix},$$

and $\tau = [\tau_{1,1}, \tau_{1,2}, \tau_{1,3}, \tau_{2,1}, \tau_{2,2}, \tau_{2,3}]^{\mathrm{T}}$. $\tau_{1,i}$ denotes the driving torque of the revolute joint on the base, and $\tau_{2,i}$ denotes the driving force of the prismatic joint of i-th connecting chain. J is the Jacobian matrix including closed-loop constraint.

3.4 Singular Configuration and Predominant Factors of Orientation Capability

When the rank of matrix J in Eq. (4) is deficient, the mechanism is located at a singular configuration. Singular configurations of the mechanism are classified as follows.

(a) *Uncertain configuration*: At this configuration, the matrix J_1 loses its full rank. Typical configurations are: (a-1) axes of revolute joints on the output link z_i and points A_i of two connecting chains are located in a plane, (a-2) points A_i of two connecting chains coincide. The case (a-2) is not physically available due to the collision of two arms. Example of (a-1) is shown in Fig. 8(a).

(b) *Stationary configuration*: At this configuration, the matrix J_2 loses its rank. This configuration occurs when the curve S_i is tangent to the circle of radius l in Fig. 7 and Eq. (3) in one connecting chain has a multiple root. Example of this singular configuration is shown in Fig. 8(b). Here, let us note that the singular configuration of this mechanism is theoretically equivalent to the well-known 6-3 fully parallel mechanism with 6-SPS structure.

As well as the singular configurations mentioned above, motion range of the spherical joint at A_i is one of the predominant factors of the orientation capability because motion range of spherical joint is quite limited. In the next section, orientation capability of the 3-RPSR mechanism is investigated taking into consideration these factors.

4 Analysis of Orientation Capability of 3-RPSR Mechanism

4.1 Evaluation Indices

Taking into consideration typically required movement of a movable-die drive mechanism in bending a spiral pipe, output motion for evaluation was given as

$$\left.\begin{array}{l} X_P = X''\cos\theta_z,\ Y_P = X''\sin\theta_z,\ Z_P = Z_O - S_Z + 2\frac{S_Z}{S_X}X'',\ \theta_y = \frac{\theta_{y,\max}}{S_X}X'', \\ \Psi = 0,\ \theta_z = (i-1)\pi/9\ (i = 1,\ldots,6),\ X'':\ [-S_X, S_X] \end{array}\right\} \quad (5)$$

Here, S_X and S_Z are maximum strokes of P_B in the XY plane and in Z direction, and $\theta_{y,\max}$ is the maximum inclination angle of the movable die.

As an example of result, changes of $\det J$ and the swing angle of spherical joint $\max(\theta_{A,i})_{i=1,2,3}$ (denoted as $\theta_{A,\max}$) with respect to orientation angle θ_y for a mechanism of $r = 45$ mm, $l = 150$ mm, $\beta_B = 0$ deg are shown in Fig. 9 for $S_X = S_Z = 16$ mm, $\theta_{y,\max} = 90$ deg. At a pose $X_P = Y_P = 0$, $Z_P = Z_O$, $\theta_y = \theta_z = \psi = 0$, the direction of swing center of spherical joint $\mathbf{s}_{A,i}$ was determined where Z_O was given such that orientation capability would be optimal. From the figure, we found the following results.

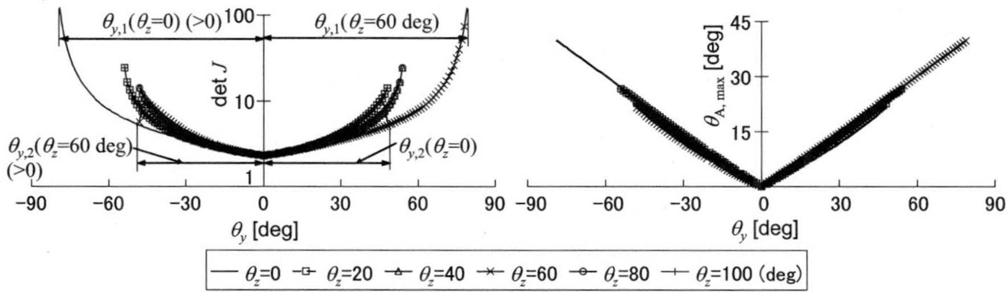

(a) det$J$ (b) $\theta_{A,\max}$

Fig. 9 Change of det$J$ and $\theta_{A,\max}$ with respect to θ_y ($r = 45$ mm, $l = 150$ mm, $\beta_B = 0$ deg).

(1) Maximum inclination angle, $\max(|\theta_y|)$, depends on the direction (θ_z).
(2) Swing angle of spherical joint $\theta_{A,\max}$ linearly changes according to θ_y.

Considering the application of the mechanism to pipe bender, both the maximum of the maximum inclination angles at a constant θ_z, denoted as $\max(\theta_{y,1}(\theta_z = \text{const.}))$, and the minimum of the maximum inclination angle at a constant θ_z, denoted as $\min(\theta_{y,2}(\theta_z = \text{const.}))$, are important, where $\theta_{y,1}(\theta_z = \text{const.})$ and $\theta_{y,2}(\theta_z = \text{const.})$ are shown in the figure. As for the requirement to spherical joint, smaller value of ratio of the maximum swing angle of spherical joint $\theta_{A,\max}$ at $\max(\theta_{y,1}(\theta_z = \text{const.}))$ to $\max(\theta_{y,1}(\theta_z = \text{const.}))$ is better. Then, we determined $\max(\theta_{y,1}(\theta_z = \text{const.}))$, $\min(\theta_{y,2}(\theta_z = \text{const.}))$ and $\theta_{A,\max}/\max(\theta_{y,1}(\theta_z = \text{const.}))$ as the evaluation indices.

4.2 Results

Relationship between design parameters and evaluation indices are shown in Fig. 10. Here, $\min(\theta_{y,2}(\theta_z = \text{const.}))$ vs. design parameters is not shown. We found that design parameters have less effect on $\min(\theta_{y,2}(\theta_z = \text{const.}))$ than $\max(\theta_{y,1}(\theta_z = \text{const.}))$. It is known from the results that β_B is the most important design parameters among the kinematic constants and that set of kinematic constants should be optimized in order to obtain a mechanism that can achieve a high orientation capability and reduce requirement to the swing angle of spherical joint.

Based on the analysis results, we optimized kinematic constants, then we designed a mechanism. Its 3D-CAD diagram is shown in Fig. 11. As shown in the figure, our designed mechanism can achieve a very high orientation capability while practical dimension of links and motion range of joints are considered as well as singularity.

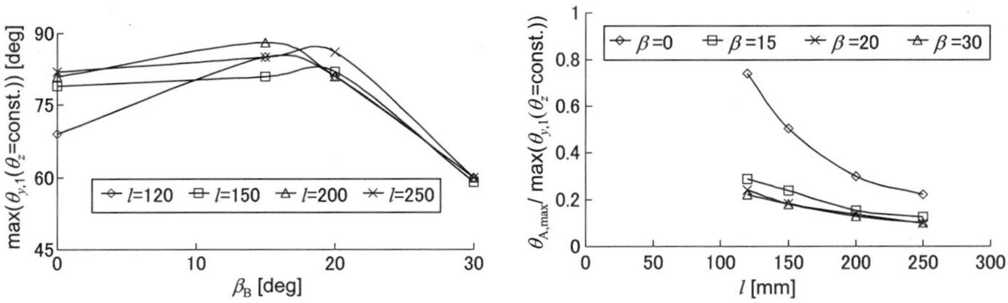

(a) $\max(\theta_{y,1}(\theta_z = \text{const.}))$ (b) $\theta_{A,\max}/\max(\theta_{y,1}(\theta_z = \text{const.}))$

Fig. 10 Relationships between design parameters and evaluation indices (kinematic constants that are not specified in the figure are $r = 45$ mm, $l = 150$ mm, $\beta_B = 15$ deg).

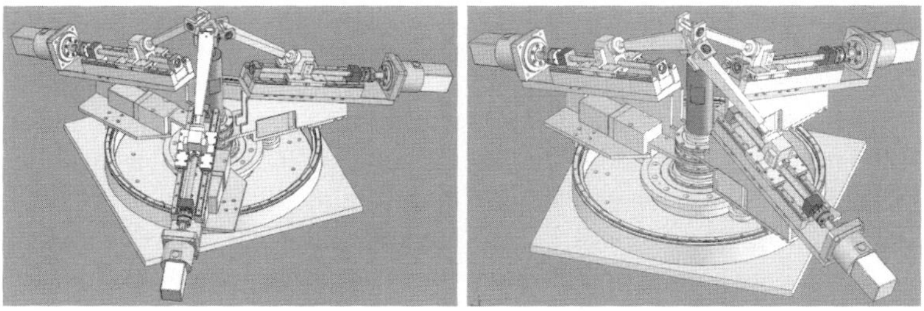

(a) At home configuration (b) At a very inclined configuration ($\theta_y = 90$ deg)

Fig. 11 3D-CAD diagram of designed mechanism ($r = 45$ mm, $l = 260$ mm, $\beta_B = 20$ deg).

5 Conclusions

Kinematic analysis of a 3-RPSR parallel mechanism with six DOF, which has been applied to a movable-die drive mechanism of pipe bender, has been done to clarify the relationship of its design parameters and orientation capability. Based on the results of analysis, a mechanism that can achieve a high orientation capability was designed.

References

1. Takeda, Y., Xiao, X., Higuchi, M., Inada, S., Hirose, K., Yoshida, Y., Ishikura, Y.: Development of a pipe bender using a parallel mechanism with 3-RPSR structure with six degrees of freedom. In: Proceedings of 13th World Congress in Mechanism and Machine Science, pp. 19–25. Guanajuato, Mexico (2011)
2. Takeda, Y., Xiao, X., Hirose, K., Yoshida, Y., Ichiryu, K.: Kinematic analysis and design of 3-RPSR parallel mechanism with triple revolute joints on the base. Int. J. Autom. Technol. **4**(4), 346–354 (2010)

Investigation of a Cable-Driven Parallel Mechanism for Interaction with a Variety of Surfaces, Applied to the Cleaning of Free-Form Buildings

K.H.J. Voss, V. van der Wijk and J.L. Herder

Abstract In this paper, the capability of a specific cable-driven parallel mechanism to interact with a variety of surfaces is investigated. This capability could be of use in for example the cleaning of large building surfaces. A method is presented to investigate the workspace for which the cables do not interfere and a surface interaction force can be generated. This method takes into account the influence of cable mass. As an example, this method is used for the design of a mechanism with a workspace conform to the dimensions of a typical building facade. The mechanism is concluded to be feasible as long as there is room to locate the pulleys at an adequate distance from the surface.

Key words: Cable-driven parallel mechanism, cable interference, workspace, cable mass influence, surface interaction

1 Introduction

The cleaning of building surfaces, especially large ones, is a challenging task because of the difficulty in reaching them. With the increasing number of buildings with free-form architecture, so called 'blobitecture', this task has become even more challenging, and sometimes even impossible to do with conventional equipment such as suspended platforms [13]. Another problem is that conventional cleaning is expensive, because of the considerable amount of human labor involved [4] and the necessary adaptations to the building to make it cleanable.

For the cleaning of glass surfaces, several automated devices have been proposed and developed, e.g. [3, 4]. However, they still require human labor or they are only applicable to a single surface or a limited number of simple surfaces. The current problem is therefore that a cost-effective automated device that can interact with a wide variety of large and small, straight and curved surfaces is still to be found.

K.H.J. Voss · V. van der Wijk · J.L. Herder
Department of Mechanical Automation and Mechatronics, Faculty of Engineering Technology, University of Twente, 7500 AE Enschede, The Netherlands, e-mail: k.h.j.voss@student.utwente.nl

J. Lenarčič, M. Husty (eds.), *Latest Advances in Robot Kinematics*,
DOI 10.1007/978-94-007-4620-6_33,

Cable Driven Parallel Mechanisms (CDPMs) are known to have several advantages: they have a large workspace to mass ratio, they are easy to reconfigure and they are relatively cheap [5, 9]. Therefore these mechanisms have a large potential for the cost-effective interaction with a wide variety of surfaces. However, to the authors best knowledge, the investigation of these mechanisms for this purpose is unknown.

The goal of this paper is therefore to propose a CDPM for the purpose of interaction with a variety of large surfaces.

The paper is structured as follows. Firstly, the device configuration is presented. Secondly, the method for the determination of the workspace in order to investigate its feasibility for interaction with large free-form surfaces is presented. In this method cable interferences, the influence of cable mass and the ability to provide a surface interaction force are taken into account. The method is applied to an example, yielding numerical results that are then discussed. The control of the device and the actual cleaning process are not treated in this paper.

2 Mechanism Configuration and Definitions

Figure 1 shows the configuration of an eight cable CDPM. It consists of actuated winches located at points A_i. For the application on a building, these winches can be mounted on the ground, the roof or on beams at a distance from the surface. The winches are connected by cables to a mobile platform at B_i. An end-effector (e.g. a cleaning head [2]) can be located at H. F is the fixed frame and M is the mobile frame attached to the platform. Vectors $\mathbf{A_{i,F}}$ describe the location of points A_i in F, vectors $\mathbf{B_{i,M}}$ describe the location of B_i in M.

The three translational degrees of freedom (DOFs) of the platform are described by the vector $\mathbf{r}$ directed from the origin of F to the origin of M. The three rotational DOFs of the platform are defined by a rotation about the y-, x- and z-axis of F respectively, contained in vector $\theta = [\theta_x, \theta_y, \theta_z]^{\mathrm{T}}$. This rotation from F to M is described by a rotation matrix $\mathbf{R}(\theta) = \mathbf{R_z}(\theta_z)\mathbf{R_x}(\theta_x)\mathbf{R_y}(\theta_y)$, where the latter three matrices are elementary rotation matrices. The pose is defined as $\mathbf{x} = [\mathbf{r}^{\mathrm{T}}, \theta^{\mathrm{T}}]^{\mathrm{T}}$, which thus includes all DOFs. The surface to be interacted with lies within quadrilateral $A_5A_6A_7A_8$ with possible features in y_F direction.

There is actuation redundancy because there are eight actuated cables actuating six DOFs; The seventh cable is necessary to keep tension in the cables, the eighth is added to achieve a proper workspace.

Surfaces conform those usually found on buildings have large dimensions in two directions, and also some smaller features in the third direction. However, the workspace of a CDPM usually has fairly equal dimensions in all directions (e.g. [1, 11]). In this paper it will be shown that the presented CDPM has the proper workspace with one small edge and two long edges, which could make it applicable on the set of mentioned surfaces. Within this workspace, the cables are always free of interference and an interaction force on the surface can be produced.

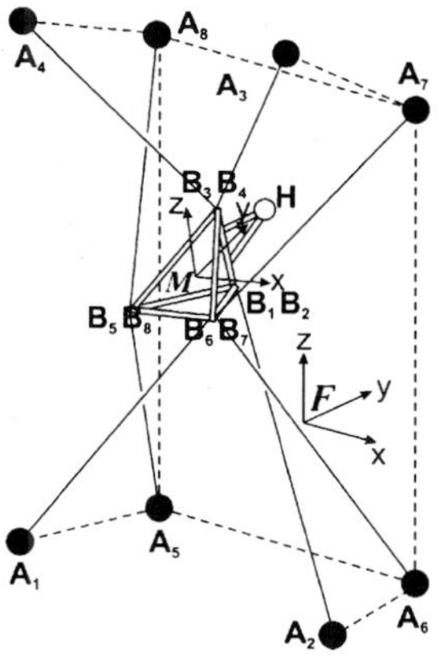

Fig. 1 CDPM configuration.

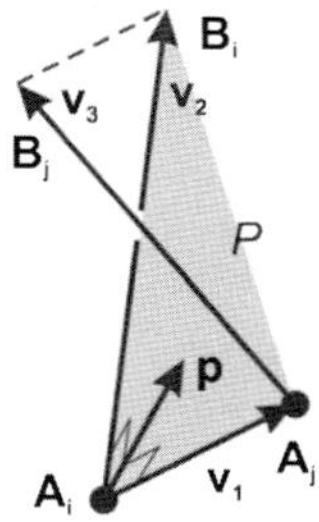

Fig. 2 Cable pair.

3 Method

To generate a workspace in which cables do not interfere, the interference needs to be detected. This needs to be done in such a way that the result is also valid for cables that sag due to their mass and that it is implementable in the method for calculating a wrench feasible workspace, which is explained later. This made existing methods to calculate cable interferences (e.g. [8, 10]) insufficient. Therefore a new method is developed specifically for this configuration.

Instead of investigating the interference between each possible pair of cables and also the mobile platform, a simplifying observation can be made. Namely, interference will always occur between cables in one of the pairs of cables 1-5, 2-6, 3-7 and 4-8, prior to any interference between cables belonging to different pairs or between a cable and the mobile platform. Therefore, interference is tested solely between two cables in one of these pairs. E.g. a rotation of the pose in Fig. 1 around the z_F axis will result in interference occurring first in pair 1-5 or 4-8.

Figure 2 shows a pair of cables i and j in a state of no interference. Plane P with normal vector $\mathbf{p}$ is spanned by $\mathbf{v_1}$ from pulley A_i to A_j and by $\mathbf{v_2}$ from A_i along cable i. The cables in this pair are defined to interfere if cable j ($\mathbf{v_3}$) lies in or on the other side of P, because it will then have crossed cable i. This is the case if $\angle(\mathbf{p}, \mathbf{v_3}) \geq 90°$, or in terms of the vectors $\mathbf{v_i}$:

$$\det([\mathbf{v_1}, \mathbf{v_2}, \mathbf{v_3}]) \leq 0. \tag{1}$$

Eq. (1) is still based on an assumption of massless, straight cables. To account for mass effects, i.e. sag of the cables, a distance margin d_{mar} is introduced. Should the minimal distance d_{min} between two straight cables fall below d_{mar}, the sagged cables are defined to interfere, see Eq. (2).

$$d_{\min} = \left| \mathbf{v_1} \cdot \frac{\mathbf{v_2} \times \mathbf{v_3}}{\|\mathbf{v_2} \times \mathbf{v_3}\|} \right| = \left| \frac{\det([\mathbf{v_1}, \mathbf{v_2}, \mathbf{v_3}])}{\|\mathbf{v_2} \times \mathbf{v_3}\|} \right| < d_{\text{mar}}. \tag{2}$$

Instead of testing for interference by evaluating Eqs. (1) and (2), they can be combined in Eq. (3), allowing for a single equation to test for interference.

$$\frac{\det([\mathbf{v_1}, \mathbf{v_2}, \mathbf{v_3}])}{\|\mathbf{v_2} \times \mathbf{v_3}\|} < d_{\text{mar}}. \tag{3}$$

To determine a proper d_{mar}, the sagged state of the cables will be compared to the state with straight cables. The method to calculate the sagged state is adapted from [7, 12], in which the location of one end of a cable in a local x-z plane with the other end clamped at $(0,0)$ is given by Eqs. (4) and (5).

$$x_{\text{end}} = \frac{F_x L_0}{E A_0} + \frac{|F_x|}{\rho_0 g}\left(\operatorname{asinh}\left(\frac{F_z}{F_x}\right) - \operatorname{asinh}\left(\frac{F_z - \rho_0 g L_0}{F_x}\right)\right), \tag{4}$$

$$z_{\text{end}} = \frac{F_z L_0}{E A_0} - \frac{\rho_0 g L_0^2}{2 E A_0} + \frac{1}{\rho_0 g}\left(\sqrt{F_x^2 + F_z^2} + \sqrt{F_x^2 + (F_z - \rho_0 g L_0)^2}\right). \tag{5}$$

For a given pose of the mobile platform, x_{end} and z_{end} can be calculated for each cable. Young's modulus E, cable cross-sectional area A_0, linear density ρ_0 and gravitational acceleration g are assumed to be known. The forces F_x and F_z at the end and the sagged cable length L_0 are to be calculated, which are three unknowns for two equations. Therefore there are $2m$ equations for $3m$ unknowns for an n-DOF, m-cable device. Then, with static equilibrium providing an additional n equations, the equations are solvable for an $n = m$ CDPM, as was done in [7, 12]. However, an $n = m - 2$ CDPM like the one investigated here has two extra free variables. These free variables are used in the adapted method presented here.

At first, suppose that the cables are massless and therefore straight. Then Eq. (6) [1] shows how 6×8 wrench matrix $\mathbf{W_M}$ relates the cable tension forces gathered in 8×1 vector $\mathbf{t}$ to the forces and torques (i.e. wrench), gathered in 6×1 wrench vector $\mathbf{f_M}$ in M.

$$\mathbf{W_M t} = \begin{bmatrix} \mathbf{u_{1,M}} & \cdots & \mathbf{u_{8,M}} \\ \mathbf{B_{1,M}} \times \mathbf{u_{1,M}} & \cdots & \mathbf{B_{8,M}} \times \mathbf{u_{8,M}} \end{bmatrix} \mathbf{t} = \mathbf{f_M}, \quad \mathbf{u_{i,M}} = \frac{\mathbf{l_{i,M}}}{\|\mathbf{l_{i,M}}\|} \tag{6}$$

The unknowns for a given pose $\mathbf{x}$ and a given external force vector $\mathbf{f_M}$ are calculated using a four step method. The first step is to solve the linear programming equation of Eq. (7) [6] with cable tensions $\mathbf{t}$ between specified bounds. $\mathbf{c} = -\mathbf{1}$ to find maximal valid cable tensions, which results in low sag.

$$\min_{\mathbf{t}} \mathbf{c}^{\mathrm{T}} \mathbf{t} \text{ such that } \mathbf{W_M t} = \mathbf{f_M}, \ \mathbf{t} \in [\mathbf{t}_{\text{low}}, \mathbf{t}_{\text{high}}]. \tag{7}$$

The second step is to solve the system of Eqs. (4) and (5) and equilibrium for the eight cables simultaneously. The initial values of F_x and F_z for each cable are calculated from the $\mathbf{t}$ found with Eq. (7). The initial values of the L_0 are $||\mathbf{L_{i,M}}|| = ||\mathbf{A_{i,M}} - \mathbf{B_{i,M}}||$.

The found solution for $\mathbf{t}$ is not unique because of the free variables, but starting from this solution, these variables can be used to affect the sag of all cables so that they stay below a pre-defined limit $\delta_{\max}$. This limit is defined by Eq. (8), and if this equation is not true, step three is performed.

$$\forall i \in [1,2,\ldots,8]: \delta_i \leq \delta_{\max}, \text{ where } \delta_i = \frac{L_{0,i} - ||\mathbf{l_i}||}{||\mathbf{l_i}||}. \tag{8}$$

Step three is to add Eq. (9) to the system of equations and solve it again, where i is the number of the cable with the largest δ_i, F_p is the solution for F_z for this cable and $\alpha > 1$. The effect is that the F_z for the cable with the most sag is increased, which reduces its δ_i. Step three is repeated until Eq. (8) is valid. $\delta_{\max}$ needs to be raised if this proves impossible.

$$F_{z,i} = \alpha F_p. \tag{9}$$

For step four, it is now possible to calculate the path of a sagged cable from $\mathrm{A_i}$ to $\mathrm{B_i}$ with [7, 12]. This allows minimal distances between the sagged cables to be calculated and the presence of interference to be investigated with a plot. By comparing these results at a number of poses for which Eq. (3) predicts interference, the optimal d_{mar} can be found.

Now that d_{mar} is known, the space that needs to be tested for interference is discretized into a finite number of poses. If for a specific grid point Eq. (3) is false for each pair of cables, it is considered that there is no interference and the corresponding pose is added to the interference free space (IFS).

In addition to being interference free, at a pose an adequate surface normal force also needs to be producible. This means that there should be equilibrium for an $\mathbf{f_M}$ in which a surface interaction is specified along the y_M direction and a gravity force is specified in the proper direction. This could be tested by checking whether the system of sag equations converges for this pose and this $\mathbf{f_M}$. However, a lack of convergence might also indicate a wrongly defined $\delta_{\max}$ or α and it is recommended to visually inspect the solution for a pose that does converge. This makes this approach infeasible and instead, a test for wrench feasibility with a method adapted from [6] to include the effects of cable mass is proposed.

A pose is wrench feasible if at this pose Eq. (6) can be satisfied with a $\mathbf{t} \in [\mathbf{t}_{\min}, \mathbf{t}_{\max}]$; this can be tested with [6]. Although this method is based on massless cables, the cable mass is accounted for in two ways. Firstly, the cable mass m_c given by $m_c = \rho_0 \sum_{i=1}^{8} ||\mathbf{l_i}||$ is added to the platform mass m_p, resulting in a larger gravity force $\mathbf{f_M}$. Not just a part, but the entire mass of all eight cables is added, because the pulleys at the top need to lift this entire mass.

Secondly, the $\mathbf{t}_{\min}$ is determined in the same way as d_{mar} is determined: for a number of representative poses with sagged cables the cable tensions are calculated. The cable tensions that are found to be minimally necessary to keep Eq. (8) valid are the input for $\mathbf{t}_{\min}$. In [6] wrench feasibility is tested for pose intervals. To reduce

Table 1 Numerical values, all coordinates are in meters

$\mathbf{A}_{1,\mathbf{F}}$	$[-30,-5,-50]^{\mathrm{T}}$	$\mathbf{A}_{5,\mathbf{F}}$	$[-30,5,-50]^{\mathrm{T}}$	$\mathbf{B}_{1,\mathbf{M}}$	$[0,3,-2]^{\mathrm{T}}$	$\mathbf{B}_{5,\mathbf{M}}$	$[-3,-3,0]^{\mathrm{T}}$
$\mathbf{A}_{2,\mathbf{F}}$	$[30,-5,-50]^{\mathrm{T}}$	$\mathbf{A}_{6,\mathbf{F}}$	$[30,5,-50]^{\mathrm{T}}$	$\mathbf{B}_{2,\mathbf{M}}$	$[0,3,-2]^{\mathrm{T}}$	$\mathbf{B}_{6,\mathbf{M}}$	$[3,-3,0]^{\mathrm{T}}$
$\mathbf{A}_{3,\mathbf{F}}$	$[30,-5,65]^{\mathrm{T}}$	$\mathbf{A}_{7,\mathbf{F}}$	$[30,5,60]^{\mathrm{T}}$	$\mathbf{B}_{3,\mathbf{M}}$	$[0,3,2]^{\mathrm{T}}$	$\mathbf{B}_{7,\mathbf{M}}$	$[3,-3,0]^{\mathrm{T}}$
$\mathbf{A}_{4,\mathbf{F}}$	$[-30,-5,65]^{\mathrm{T}}$	$\mathbf{A}_{8,\mathbf{F}}$	$[-30,5,60]^{\mathrm{T}}$	$\mathbf{B}_{4,\mathbf{M}}$	$[0,3,2]^{\mathrm{T}}$	$\mathbf{B}_{8,\mathbf{M}}$	$[-3,-3,0]^{\mathrm{T}}$
$\mathbf{H}_{\mathbf{M}}$	$[0,5,0]^{\mathrm{T}}$	ρ_0	0.24 kg/m	E	400 GPa	A_0	50 mm^2
m_p	250 kg	g	9.81 m/s2				

calculation time, here the IFS is used as an input grid of poses to test. The result is an interference free wrench feasible workspace (IFWFW).

4 Results

The method is applied to the configuration of Fig. 1 with the numerical values of Table 1. m_p is the mass of the platform; ρ_0, E and A_0 are consistent with commercially available fiber core wire rope. Using the discussed method, the table values were selected to make a good set for the cleaning of a building facade of 40 × 100 m with feature depths of 5 m.

$$\mathbf{f_M} = \left[[\mathbf{F_s}+\mathbf{F_g}]^{\mathrm{T}},0,0,0\right]^{\mathrm{T}}, \quad \mathbf{F_s} = [0,100,0]^{\mathrm{T}}, \quad \mathbf{F_g} = \mathbf{R}(\theta)^{\mathrm{T}}[0,0,mg]^{\mathrm{T}} \tag{10}$$

The wrench vector $\mathbf{f_M}$ used in this analysis is given by Eq. (10). Herein, a surface interaction force of 100 N is defined in $+y_M$ direction. A force mg in $+z_F$ direction counteracts the gravity on the platform. $\mathbf{R}(\theta)^{\mathrm{T}}$ rotates this force to the mobile frame. $m = m_p$ for the sag analysis and $m = m_p + m_c$ for the wrench feasibility analysis. The rest of the forces and the torques are zero for equilibrium.

Using this force vector, the influence of cable mass is investigated for a number of poses to find a suitable d_{mar} and $\mathbf{t}_{\text{min}}$. A $\delta_{\max} = 0.0005$ and $\alpha = 1.1$ were used for these investigations. Together with all entries of $\mathbf{t}_{\text{high}}$ equal to 6 kN in Eq. (7) this resulted in maximal cable tensions of around 9.5 kN for the final solution of all poses of the sagged state. This corresponds to a safety factor of 4 for the used wire rope and will also be used as the $\mathbf{t}_{\max}$ for the wrench feasibility analysis. A d_{mar} of 0.35 m was concluded to be adequate. This occurred e.g. at pose $\mathbf{x} = [20,0,40,19°,0°,20°]^{\mathrm{T}}$ where the cables 1 and 5 were observed to touch each other in sagged state, while Eq. (3) predicted a distance of 0.35 m. A cable tension of 1.6 kN insures that Eq. (8) is satisfied for each cable in each pose and therefore this is used as the $\mathbf{t}_{\min}$ for the wrench feasibility analysis.

Figure 3 shows the result of pulling the cables tight to make Eq. 8 valid for the pose $\mathbf{x} = [20,0,50,0°,0°,0°]^{\mathrm{T}}$. Cable 5 had too much sag at first, which is illustrated by the dashed line. After pulling the cable tight, the cable was located along the solid line. This increased the minimal distance between Cable 1 and 5 from 0.48 m to 1.26 m, which shows how this method can increase the interference free space. Another effect of pulling tight is that the stiffness added by Cable 5 to the

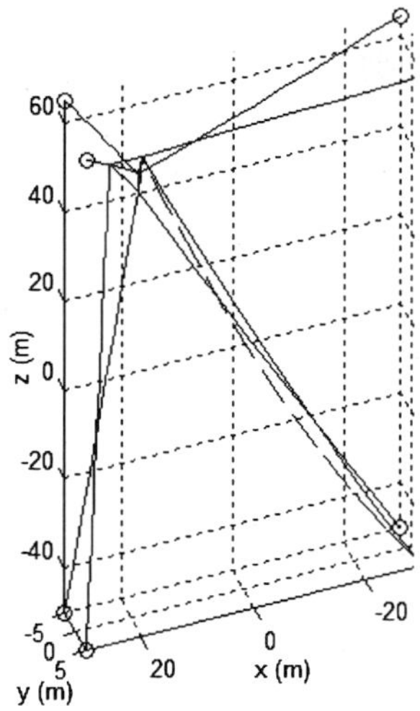

Fig. 3 Sagged cable states.

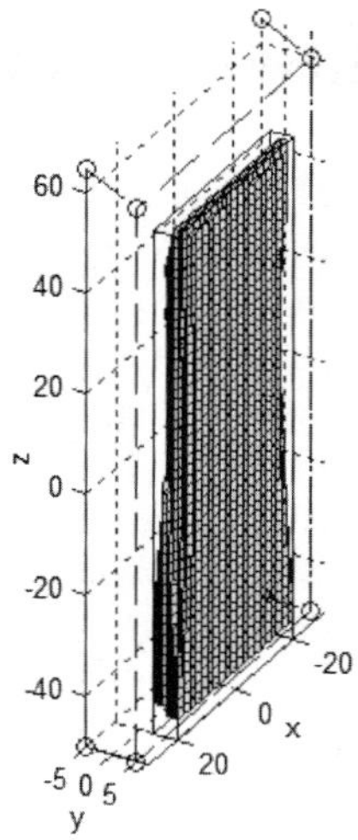

Fig. 4 IFWFW.

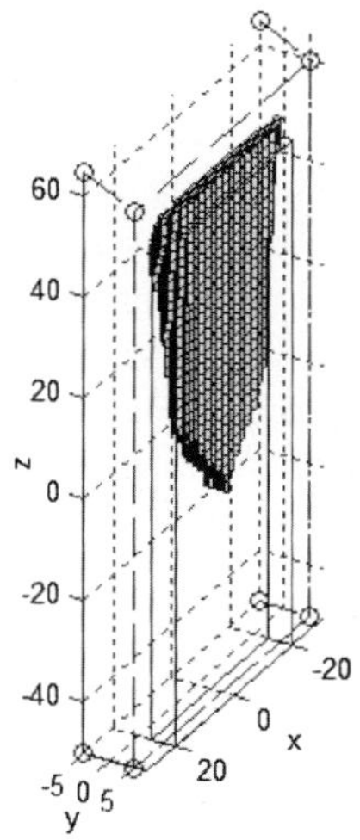

Fig. 5 IFWFW.

mobile platform was raised by more than a factor 5 in all directions. Although this had no real effect on the total stiffness of the mobile platform, since this is mostly determined by the very taut cables (e.g. Cable 3), it does mean that Cable 5 will be less influenced by outside factors like wind.

Using all the found data, the IFWFW can be calculated. In principle this is a 6 dimensional space, but to make the results presentable, three-dimensional IFWFWs are calculated for fixed sets of the vector θ. The grid for r is $40 \times 5 \times 100$ m in x, y and z direction respectively. Plotted are all the positions that point H can reach. Figure 4 shows the IFWFW for all rotations equal to zero. For 82% of the investigated grid points, the wrench was feasible and there was no interference. Figure 5 shows the IFWFW for a rotation of 45° around x_F. The cables interfere in the bottom part of the grid, but this rotation can be used to reach higher with point H. Now, 40% of the grid points were wrench feasible and interference free. For poses towards the core of the IFWFW, maximal surface interaction forces far higher than 100 N can be produced, e.g. 3.7 kN for $\mathbf{x} = [0,0,0,0°,0°,0°]^T$.

The found IFWFW conforms rather well to a flat box-like shape, which is a good shape for cleaning building facades. To achieve this shape, the pulleys had to be positioned at quite a distance away from its edges. As long as this is not a problem, the CDPM is a viable option for a facade cleaning device. Within the workspace the mechanism is independent of surface features, but interference of cables with the surface still needs to be checked.

5 Discussion and Conclusion

In this paper, the occurrence of interference was defined as a limit to the workspace.

However, even after interference has occurred between cables or with the surface, the mechanism might still be able to function. This could favorably increase the workspace of e.g. the situation in Fig. 5. The assumption of interference between cable pairs resulted in an interference calculation method fast enough to be used in real-time operation. This method should not only work for the flat box-like CDPM configuration presented here, but also for similar configurations with different box shapes and sizes. These could be used to clean solar panel arrays or cooling towers.

It can be concluded that a CDPM is suitable for interaction with a large surface. Within the workspace, this surface can have any shape. A method has been developed in which linear cable theory is used to analyze cable interference and wrench feasibility for an over-actuated CDPM while taking into account the effects of cable mass. This method can be used to design CDPMs for surface interaction, which was done for a facade cleaning application. The designed mechanism is viable for this application as long as the pulleys can be placed at sufficient distances from the surface.

References

1. Diao, X., Ma, O.: Workspace determination of general 6-d.o.f. cable manipulators. Adv. Robot. **22**(2), 261–278 (2008)
2. Dornier, I.: Brush cleaning head. U.S. Patent WO 98/461121998
3. Elkmann, N., Felsch, T., Sack, M., Saenz, J., Hortig, J.: Innovative service robot systems for facade cleaning of difficult-to-access areas. In: IEEE International Conference on Intelligent Robots and Systems, pp. 756–762 (2002)
4. Gambao, E., Hernando, M., Hernández, F., Pinilla, E.: Cost-effective robots for façade cleaning. In: Proceedings of the 21st ISARC (2004)
5. Gouttefarde, M., Gosselin, C.: Analysis of the wrench-closure workspace of planar parallel cable-driven mechanisms. IEEE Trans. Robot. **22**(3), 343–445 (2006)
6. Gouttefarde, M., Merlet, J.P., Daney, D.: Wrench-feasible workspace of parallel cable-driven mechanisms. In: IEEE International Conference on Robotics and Automation, pp. 1492–1497 (2007)
7. Kozak, K.: Static analysis of cable-driven manipulators with non-negligible cable mass. IEEE Trans. Robot. **22**(3), 425–433 (2006)
8. Merlet, J.P.: Analysis of the influence of wires interference on the workspace of wire robots. In: Lenarčič, J., Galletti, C. (eds.) Advances in Robot Kinematics, pp. 211–218. Kluwer (2004)
9. Merlet, J.P.: Parallel Robots, 2nd edn. Springer, Dordrecht (2006)
10. Otis, M.J.D., Perreault, S., Nguyen-Dang, T.-L., Lambert, P., Gouttefarde, M., Laurendeau, D., Gosselin, C.: Determination and management of cable interferences between two 6-dof foot platforms in a cable-driven locomotion interface. IEEE Trans. Syst. Man Cybern., Part A, Syst. Humans **39**(3), 528–544 (2009)
11. Riechel, A.T., Ebert-Uphoff, I.: Force-feasible workspace analysis for underconstrained, point-mass cable robots. In: IEEE International Conference on Robotics and Automation, pp. 4956–4962 (2004)
12. Riehl, N., Gouttefarde, M., Baradat, C., Pierrot, F.: On the determination of cable characteristics for large dimension cable-driven parallel mechanisms. In: IEEE International Conference on Robotics and Automation, pp. 4709–4714 (2010)
13. Schraft, R.D., Bräuning, U., Orlowski, T., Hornemann, M.: Automated cleaning of windows on standard facades. Autom. Constr. **9**(5–6), 489–501 (2000)

Auto Calibration Method for Cable-Driven Parallel Robots Using Force Sensors

Philipp Miermeister and Andreas Pott

Abstract This paper presents an auto calibration method for overconstrained cable-driven parallel robots using internal position and force sensors. The consideration of cable forces is necessary in order to regard the cable elasticity. A calibration workflow is proposed and implemented including pose selection, measurement, and parameter adjustment. The calibration procedure is not limited to the geometrical parameters, but also allows to identify force related parameters such as the cable stiffness and platform mass. The calibration results are shown for an unknown parameter set and the influence of sensor noise on the calibration results is presented.

Key words: Cable-driven parallel robot, auto calibration, least squares, force sensors

1 Introduction

A cable-driven parallel robot, here also simply called cable robot, is a parallel kinematic machine mainly consisting of a platform, cables and winches as shown in Figure 1. The cables connect the platform to the winches which control the platform pose by changing the cable length. In the recent years cable robots got lots of attention [1, 3] due to their advantages over serial kinematics and conventional parallel kinematics. On the one side, cable robots inherit the properties of all parallel kinematic robots as for example high structural stiffness, payload and good precision. On the other side, they outperform conventional parallel kinematics with regards to flexibility, workspace and speed, because the cables allow ultra light weight constructions in nearly arbitrary large areas. The winches can be easily attached to solid structures such as steel frames or walls and allow to change the robot configuration

Philipp Miermeister · Andreas Pott
Fraunhofer IPA, 70569 Stuttgart, Germany, e-mail: {philipp.miermeister, andreas.pott}@ipa.fhg.de

J. Lenarčič, M. Husty (eds.), *Latest Advances in Robot Kinematics*,
DOI 10.1007/978-94-007-4620-6_34,

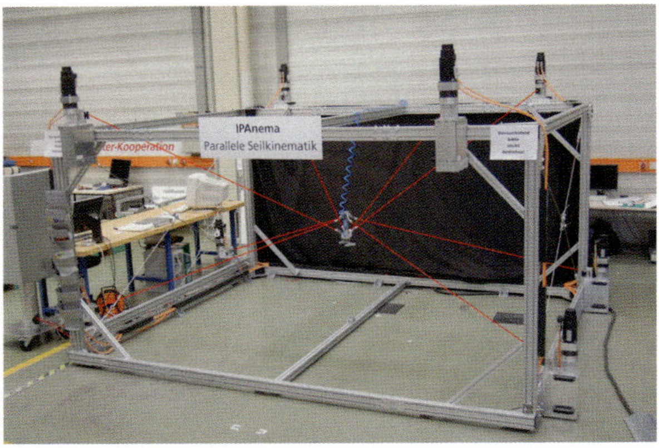

Fig. 1 Cable-driven parallel robot demonstrator at Fraunhofer IPA.

in a short time. Since each new configuration requires the identification of the actual geometric parameters, an out calibration procedure is highly desirable. Especially an auto calibration method is of interest where only internal sensors are used to determine the robot parameters avoiding the use of expensive and time consuming external metrology equipment such as laser trackers. The calibration task consists of multiple parts [2]. At first, a model of the robot has to be established including the parameters which have the largest influence on the robots behavior. This goes along with pose selection [6] in order to find a well conditioned identification matrix which reflects the sensitivity with respect to parametric errors. The well conditioned pose set is used to compute the actuator set points for the measurement procedure based on the nominal model. After measurement it is necessary to compare the predicted and measured quantities in order to minimize the error in the unknown parameters. Firstly we derive the equations for the kinematic model.

2 Robot Kinematics

The cable robot consists of m cables connecting the platform to the winches which in turn are attached to a rigid base. The platform pose is described by the generalized coordinates $\mathbf{x} = \begin{bmatrix} \mathbf{r}^{\mathrm{T}} & \mathbf{d}^{\mathrm{T}} \end{bmatrix}^{\mathrm{T}}$ where $\mathbf{r}$ and $\mathbf{d}$ refer to the platform position and orientation, respectively. Considering the base coordinates $\mathbf{a}_i$ and the cable attachment points at the platform described by $\mathbf{b}_i$, the kinematic loop for a single drive chain reads

$$\mathbf{l}_i = \mathbf{a}_i - \mathbf{r} - \mathbf{R}_{0\mathrm{P}}(\mathbf{d})\mathbf{b}_i, \tag{1}$$

where $\mathbf{l}_i$ refers to the cable vector and $\mathbf{R}_{0\mathrm{P}}$ is the rotation matrix between the platform frame $\mathscr{K}_{\mathrm{P}}$ and the inertial frame $\mathscr{K}_0$ shown in Figure 2. Considering the cables as massless elements in an ideal tightened state without elongation and sagging it is

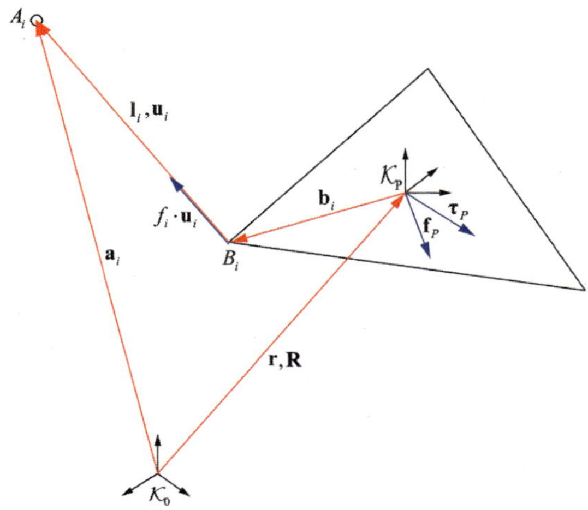

Fig. 2 Robot kinematics.

possible to compute the actuator variables $\mathbf{q}_\theta = \begin{bmatrix} \theta_1 & \cdots & \theta_m \end{bmatrix}^\mathrm{T}$ for a given platform pose $\mathbf{x}$ using the analytically and uniquely solvable inverse kinematics

$$\mathbf{q}_\theta = \mathbf{h}(\mathbf{p}, \mathbf{x}, \mathbf{x}_0) = \mathbf{q}_\mathrm{N}(\mathbf{x}) - \mathbf{q}(\mathbf{x}_0), \tag{2}$$

where $q_{\mathrm{N},i}(\mathbf{x}) = ||\mathbf{l}_{\mathrm{N},i}||_2$ denotes the nominal controlled cable length, $\mathbf{x}_0$ describes the initial configuration, and $\mathbf{p}$ relates to the system parametrization. The corresponding forward kinematics

$$\mathbf{x} = \mathbf{h}^{-1} = \mathbf{f}(\mathbf{p}, \mathbf{q}_\theta) \tag{3}$$

has to be solved numerically and provides multiple solutions. Both Eqs. (2) and (3) are fundamental to the solution of various problems such as the development of controllers. Therefore it is of highest importance to determine the correct geometrical parametrization $\mathbf{p}_\mathrm{G} = \begin{bmatrix} \mathbf{a}_1^\mathrm{T} & \mathbf{b}_1^\mathrm{T} & \cdots & \mathbf{a}_m^\mathrm{T} & \mathbf{b}_m^\mathrm{T} \end{bmatrix}^\mathrm{T}$ using a calibration strategy. Deriving the actuator variable $\mathbf{q}_\theta$ with respect to the platform pose $\mathbf{x}$ and using the kinetostatic principle one gets the following relation

$$\frac{\mathrm{d}\theta(\mathbf{x})}{\mathrm{d}\mathbf{x}} = \frac{\mathrm{d}\mathbf{q}(\mathbf{x})}{\mathrm{d}\mathbf{x}} = \mathbf{J}_\mathrm{qx} = -\mathbf{A}\mathbf{P} \tag{4}$$

therein

$$\mathbf{A}^\mathrm{T} = \begin{bmatrix} {}_{(0)}\mathbf{u}_1 & \cdots & {}_{(0)}\mathbf{u}_m \\ {}_{(0)}\mathbf{b}_1 \times {}_{(0)}\mathbf{u}_1 & \cdots & {}_{(0)}\mathbf{b}_m \times {}_{(0)}\mathbf{u}_m \end{bmatrix} \tag{5}$$

corresponds to the well known *structure matrix* $\mathbf{A}^\mathrm{T}$ [5]. Matrix $\mathbf{P}$ depends on the parametrization of the rotation and expresses the relation between the geometrical and analytical Jacobian matrix. For quaternions, matrix $\mathbf{P}$ can be computed as in [4].

3 Force Equations

While a calibration method, that is only based on robot kinematics, works well for robots with stiff links, this is not true for cable robots where the error introduced by the flexible cables is not negligible. Here a simplified linear spring model

$$\mathbf{f} = \mathbf{K}\mathbf{Q}^{-1}\Delta\mathbf{q}(\mathbf{x}, \mathbf{q}_\theta) \tag{6}$$

is used to describe the cable elasticity where $\mathbf{K} = \text{diag}(\mathbf{k})$ is the stiffness matrix and $\mathbf{Q} = \text{diag}(\mathbf{q})$ is a $m \times m$ diagonal matrix containing the cable lengths. Vector $\Delta\mathbf{q} = \mathbf{q}(\mathbf{x}) - \mathbf{q}_\mathrm{N}(\mathbf{x}_0, \mathbf{q}_\theta)$ represents the difference between the actual cable length $\mathbf{q}$ and the nominal cable length $\mathbf{q}_\mathrm{N}$. Damping effects are neglected, because all poses are measured in a state of rest after dynamic effects have decayed. For an exclusively kinematic model, forward kinematics is used to compute a valid pose which fulfills the kinematic constraints (1). Taking cable forces into account, the platform is described as a free floating body whose static equilibrium is described by constraint

$$\mathbf{g}(\mathbf{x}) = \mathbf{A}^\mathrm{T}(\mathbf{p}, \mathbf{x})\mathbf{f}(\mathbf{q}_\theta, \mathbf{x}) + \mathbf{w}_\mathrm{g}(\mathbf{x}) = \mathbf{0}. \tag{7}$$

During calibration, no process forces are applied to the platform so that the external wrench equates to the gravitational wrench $\mathbf{w}_\mathrm{g}$. Unknown external process forces would make calibration impossible. The same is true for the unknown platform dynamics wherefore all poses that are used for calibration must fulfill Eq. (7). Finding a static pose from a given initial pose $\hat{\mathbf{x}}$ with $\hat{\mathbf{g}} = \mathbf{g}(\hat{\mathbf{x}})$ is a nonlinear optimization problem. For small deviations $\Delta\mathbf{x}$ around the equilibrium point $\mathbf{x}_\mathrm{eq}$ it is possible to linearize Eq. (7) and solve the associated least squares problem

$$\min\left(\frac{1}{2}\mathbf{r}^\mathrm{T}\mathbf{r}\right) \tag{8}$$

with $\mathbf{r} = \Delta\mathbf{g} - \mathbf{J}_\mathrm{gx}\Delta\mathbf{x}$. The minimal solution can be computed by solving the well known normal equation

$$\mathbf{J}_\mathrm{gx}^\mathrm{T}\mathbf{J}_\mathrm{gx}\Delta\mathbf{x} = \mathbf{J}_\mathrm{gx}^\mathrm{T}\Delta\mathbf{g} \tag{9}$$

where the Jacobian equation reads

$$\mathbf{J}_\mathrm{gx} = \frac{\mathrm{d}\mathbf{g}(\mathbf{x})}{\mathrm{d}\mathbf{x}} = \frac{\mathrm{d}\mathbf{A}^\mathrm{T}(\mathbf{x})}{\mathrm{d}\mathbf{x}}\mathbf{f}(\mathbf{x}) + \mathbf{A}^\mathrm{T}(\mathbf{x})\frac{\mathrm{d}\mathbf{f}(\mathbf{x})}{\mathrm{d}\mathbf{x}} + \frac{\mathrm{d}\mathbf{w}_\mathrm{g}(\mathbf{x})}{\mathrm{d}\mathbf{x}}. \tag{10}$$

The derivative of the first expression in (10) yields

$$\mathbf{J}_\mathrm{Ax} = \frac{\mathrm{d}}{\mathrm{d}\mathbf{x}}\left(\mathbf{A}(\mathbf{x})^\mathrm{T}\right)\mathbf{f}(\mathbf{x}) = \sum_{i=1}^{m}\frac{\mathrm{d}}{\mathrm{d}\mathbf{x}}\left(\mathbf{v}_{\mathrm{A},i}(\mathbf{x})^\mathrm{T}\right)f_i(\mathbf{x}) \tag{11}$$

where $\mathbf{v}_{\mathrm{A},i}$ is the i-th column vector of the structure matrix $\mathbf{A}^\mathrm{T}$. The derivative of the force vector and the wrench vector yield

$$\mathbf{J}_{\mathrm{fx}} = \frac{\mathrm{d}\mathbf{f}(\mathbf{x})}{\mathrm{d}\mathbf{x}} = -\mathbf{K}\mathbf{Q}_{\mathrm{N}}\mathbf{Q}^{-2}\mathbf{A}\mathbf{P} \tag{12}$$

with $\mathbf{Q}_{\mathrm{N}} = \mathrm{diag}(\mathbf{q}_{\mathrm{N}})$ and

$$\mathbf{J}_{\mathrm{wx}} = \frac{\mathrm{d}}{\mathrm{d}\mathbf{x}}\mathbf{w}_{\mathrm{g}}(\mathbf{x}) = mg\begin{bmatrix}\mathbf{0} & \left(-\frac{\mathrm{d}}{\mathrm{d}\mathbf{x}}b_{g,y}\right)^{\mathrm{T}} & \left(\frac{\mathrm{d}}{\mathrm{d}\mathbf{x}}b_{g,x}\right)^{\mathrm{T}} & \mathbf{0}\end{bmatrix}^{\mathrm{T}} \tag{13}$$

where the local platform vector $\mathbf{b}_{\mathrm{g}}$ describes the point of action of the gravitational force.

4 Calibration Method

The aim of calibration is to find the parameter set $\mathbf{p}_{\mathrm{opt}}$ for a given model that minimizes the error between the measured cable forces $\mathbf{f}_{\mathrm{M}}$ and the predicted cable forces $\mathbf{f}_{\mathrm{T}}$. To make cable forces comparable it must be guaranteed that all influences are modeled or otherwise can be avoided. Therefore, during calibration, all process forces must be avoided and the platform must remain in a stable state as defined by (7) in order to avoid forces caused by system dynamics.

4.1 Objective Function

The cable forces computed by Eqs. (6) and (7) depend on the current platform pose $\mathbf{x}$, the reference platform pose $\mathbf{x_0}$ and the parameter set $\mathbf{p}$. Considering a set of u different poses with m cables and w parameters subject to calibration one gets the residual vector for all poses $\bar{\mathbf{r}}_{\mathrm{F}} = \left[\mathbf{r}_{\mathrm{F}}^{(1)\mathrm{T}} \; \cdots \; \mathbf{r}_{\mathrm{F}}^{(u)\mathrm{T}}\right]^{\mathrm{T}}$ where

$$\bar{\mathbf{r}}_{\mathrm{F}}(\mathbf{p}) = \bar{\mathbf{f}}_{\mathrm{M}} - \bar{\mathbf{f}}(\mathbf{p}, \mathbf{x}(\mathbf{p})) \tag{14}$$

expresses the difference between the measured $\bar{\mathbf{f}}_{\mathrm{M}} = \left[\mathbf{f}_{\mathrm{M}}^{(1)\mathrm{T}} \; \cdots \; \mathbf{f}_{\mathrm{M}}^{(u)\mathrm{T}}\right]^{\mathrm{T}}$ and predicted forces $\bar{\mathbf{f}} = \left[\mathbf{f}^{(1)\mathrm{T}} \; \cdots \; \mathbf{f}^{(u)\mathrm{T}}\right]^{\mathrm{T}}$. Using the linearized force equation as well as the difference between the measured and the nominal cable force $\Delta\bar{\mathbf{f}} = \bar{\mathbf{f}}_{\mathrm{M}} - \bar{\mathbf{f}}(\mathbf{p}_{\mathrm{N}})$ one gets $\bar{\mathbf{r}}_{\mathrm{F}}(\Delta\mathbf{p}) = \Delta\bar{\mathbf{f}} - \mathbf{H}\Delta\mathbf{p}$. The optimum of the corresponding objective function $\frac{1}{2}\bar{\mathbf{r}}_{\mathrm{F}}^{\mathrm{T}}\bar{\mathbf{r}}_{\mathrm{F}}$ can be computed by solving the weighted least squares problem

$$\mathbf{H}^{\mathrm{T}}\mathbf{W}_{\mathrm{T}}\mathbf{H}\Delta\mathbf{p} = \mathbf{H}^{\mathrm{T}}\mathbf{W}_{\mathrm{T}}\Delta\bar{\mathbf{f}}. \tag{15}$$

The diagonal weighting matrix $\mathbf{W}_{\mathrm{T}}$ is used to weight the different cables individually with respect to their force state. Cables with small forces are weighted less than cables with high forces since the linear elastic model only holds true when $f_i > f_{\mathrm{min}}$. Furthermore, calibration of parameters with vastly different magnitudes such as geometrical parameters in comparison with the spring parameters, make parameter scaling necessary in order to get a well conditioned identification matrix $\mathbf{H}$. This is achieved by right multiplication of $\mathbf{W}_{\mathrm{P}}$ so that $\Delta\bar{\mathbf{f}} = \mathbf{H}\mathbf{W}_{\mathrm{P}}^{-1}\mathbf{W}_{\mathrm{P}}\Delta\mathbf{p} = \tilde{\mathbf{H}}\Delta\tilde{\mathbf{p}}$ with $\tilde{\mathbf{H}} = \mathbf{H}\mathbf{W}_{\mathrm{P}}^{-1}$ and $\Delta\tilde{\mathbf{p}} = \mathbf{W}_{\mathrm{P}}\Delta\mathbf{p}$. The weighting matrix can be chosen as in [2]. Regarding the spring constants $\mathbf{k}$, the initial cable offsets $\mathbf{q}_{\mathrm{off}}$ as well as the platform mass m_{P}, one can combine the force related parameters in a single vector $\mathbf{p}_{\mathrm{F}} = \left[\mathbf{q}_{\mathrm{off}}^{\mathrm{T}}\ \mathbf{k}^{\mathrm{T}}\ \mathbf{b}_{\mathrm{g}}^{\mathrm{T}}\ m_{\mathrm{P}}\right]^{\mathrm{T}}$. Deriving force Eq. (6) with respect to the parameter set $\mathbf{p} = \left[\mathbf{p}_{\mathrm{G}}^{\mathrm{T}}\ \mathbf{p}_{\mathrm{F}}^{\mathrm{T}}\right]^{\mathrm{T}}$ one gets

$$\delta\mathbf{f} = \frac{\mathrm{d}\mathbf{f}(\mathbf{p},\mathbf{x}(\mathbf{p}))}{\mathrm{d}\mathbf{p}}\delta\mathbf{p} = \left(\frac{\partial\mathbf{f}}{\partial\mathbf{p}} + \frac{\partial\mathbf{f}}{\partial\mathbf{x}}\frac{\partial\mathbf{x}}{\partial\mathbf{p}}\right)\delta\mathbf{p} \tag{16}$$

Implicit differentiation of force constraint (7) yields

$$\frac{\mathrm{d}\mathbf{g}(\mathbf{p},\mathbf{x}(\mathbf{p}),\mathbf{x}_0(\mathbf{p}))}{\mathrm{d}\mathbf{p}} = \frac{\partial\mathbf{g}(\mathbf{p},\mathbf{x},\mathbf{x}_0)}{\partial\mathbf{p}} + \frac{\partial\mathbf{g}(\mathbf{p},\mathbf{x},\mathbf{x}_0)}{\partial\mathbf{x}}\frac{\partial\mathbf{x}}{\partial\mathbf{p}} = \mathbf{0} \tag{17}$$

and therefore the mapping of the parameters $\mathbf{p}$ on the platform pose $\mathbf{x}$ while staying on the constraint manifold $\mathbf{g}$ yields

$$\frac{\partial\mathbf{x}}{\partial\mathbf{p}} = -\frac{\partial\mathbf{g}(\mathbf{p},\mathbf{x})}{\partial\mathbf{x}}^{-1}\frac{\partial\mathbf{g}(\mathbf{p},\mathbf{x})}{\partial\mathbf{p}}. \tag{18}$$

Now the relation between the cable forces $\mathbf{f}$ and the robot parameters $\mathbf{p}$ can be described by substituting (18) and (16) resulting in

$$\delta\mathbf{f} = \left(\frac{\partial\mathbf{f}}{\partial\mathbf{p}} - \frac{\partial\mathbf{f}}{\partial\mathbf{x}}\frac{\partial\mathbf{g}(\mathbf{p},\mathbf{x})}{\partial\mathbf{x}}^{-1}\frac{\partial\mathbf{g}(\mathbf{p},\mathbf{x})}{\partial\mathbf{p}}\right)\delta\mathbf{p} \tag{19}$$

where the derivatives with respect to $\mathbf{x}$ are given by Eqs. (10) and (12). Deriving force Eqs. (6) and (7) with respect to the parameter set $\mathbf{p}$ while holding $\mathbf{x}$ constant, one gets

$$\frac{\partial\mathbf{f}}{\partial\mathbf{p}_{\mathrm{G}}} = -\mathbf{K}\mathbf{Q}^{-2}\frac{\partial\mathbf{Q}}{\partial\mathbf{p}_{\mathrm{G}}}\Delta\mathbf{q} + \mathbf{K}\mathbf{Q}^{-1}\frac{\partial\Delta\mathbf{q}}{\partial\mathbf{p}_{\mathrm{G}}}, \tag{20}$$

$$\frac{\partial\mathbf{f}}{\partial\mathbf{q}_{\mathrm{off}}} = -\mathbf{K}\mathbf{Q}^{-1}, \quad \frac{\partial\mathbf{f}}{\partial\mathbf{k}} = \mathbf{Q}^{-1}\Delta\mathbf{q}, \tag{21}$$

$$\frac{\partial\mathbf{g}(\mathbf{p},\mathbf{x}(\mathbf{p}))}{\partial\mathbf{p}} = \left(\frac{\partial}{\partial\mathbf{p}}\left(\mathbf{A}(\mathbf{p},\mathbf{x})^{\mathrm{T}}\right)\mathbf{f}(\mathbf{p},\mathbf{x}) + \mathbf{A}(\mathbf{p},\mathbf{x})^{\mathrm{T}}\frac{\partial\mathbf{f}(\mathbf{p},\mathbf{x})}{\partial\mathbf{p}} + \frac{\partial\mathbf{w}_{\mathrm{g}}(\mathbf{p},\mathbf{x})}{\partial\mathbf{p}}\right). \tag{22}$$

5 Results

The proposed robot model depends on $8m+4$ parameters. For a use case scenario, the parameter values of the robot model are chosen similar to the parameters of the actual cable robot prototype which consists of $m = 8$ cables. Therefore the robot model is defined by 68 parameters. The winch parameters are chosen according to the actual robot frame which has the size of $8\text{m} \times 6\text{m} \times 5\text{m}$. The cable stiffness is approximately $k_i = 50000\text{N}$ and the platform mass is around 20kg. It is not possible to regard all m parameters as unknowns at the same time because then the identification matrix $\mathbf{H}$ becomes singular, which means that some parameters are not identifiable. For instance, one has to fix 6 parameters of the base vectors $\mathbf{a}$ in order to define the position and orientation of the robot frame. In reality, the uncertainties in some parameters of the cable robot are larger than the uncertainties in some other parameters. After manually assembling the cable robot, the uncertainties in the winch positions $\mathbf{a}$, the initial cable lengths $\mathbf{q}_{\text{off}}$ and the spring constants $\mathbf{k}$ may be large, while the platform geometry $\mathbf{b}$ is already well known. Considering w unknown parameters $\mathbf{a}_i$, $\mathbf{q}_{\text{off}}$, $\mathbf{k}$, m_{P} and $\mathbf{b}_{\text{g}}$, it is necessary to define at least w poses such that the identification matrix becomes regular. A higher number of poses yields a better conditioned identification matrix. Here a pose set of 200 different poses is used, which are equally distributed in a 2m cube around the origin. The condition number with respect to inversion yields $\kappa = \text{cond}(\mathbf{H}) = 250$. The uncertainties in the model parameters are modeled by introducing normal distributed errors for all unknown parameters. The standard deviation for the different parameters are chosen as $\sigma(\mathbf{a}_i) = 0.3\text{m}$, $\sigma(\mathbf{q}_{\text{off}}) = 0.02\text{m}$, $\sigma(\mathbf{k}) = 500\text{N}$, $\sigma(m_{\text{P}}) = 5\text{kg}$ and $\sigma(\mathbf{b}_{\text{g}}) = 0.01\text{m}$. The result of the calibration procedure can be seen in Figure 3(a) using a Levenberg–Marquardt algorithm for optimization. The platform position error is determined by adding 50 equally distributed platform poses to the cube and computing the root mean sqaured error over all platform poses. While the calibration procedure works very well when assuming ideal force sensors without noise, in reality on has to take noise into account depending on the sensor quality. The influence of the sensor noise for 200 measured platform poses is shown in Figure 3(b) using random erroneous parameter sets. The curve shows the platform position error that remains after the calibration procedure has finished. For force sensors with a standard deviated noise up to 15N one can expect an average pose error smaller than 10mm. The effect of noise can be reduced using a larger number of pose measurements.

6 Conclusion and Outlook

In this paper we presented an auto calibration method for cable-driven parallel robots using force sensors at the platform. It was shown with help of a simulation model, that the here proposed robot model and calibration workflow can be used to determine the geometrical and force related parameters of a cable robot. The calibration method addresses the cable flexibility explicitly and therefore allows to

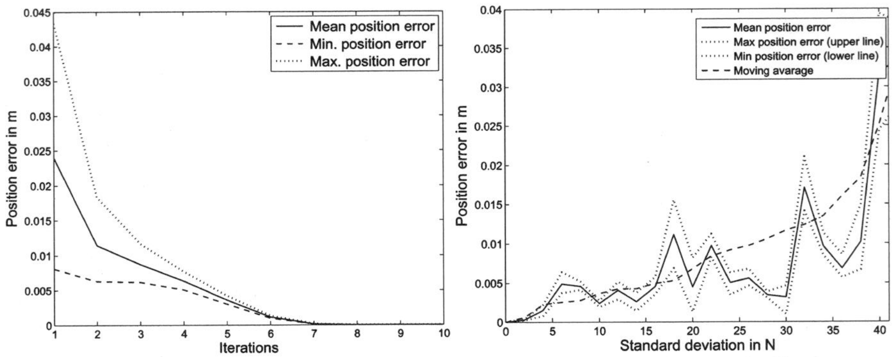

(a) Platform position error.

(b) Effect of sensor noise on the calibration error.

Fig. 3 Platform positioning accuracy and with respect to iteration steps and sensor noise.

determine the robots geometrical parameters more accurately than a solely geometrical calibration method. In the future, the calibration method will be used with the cable robot demonstrator IPAnema.

Acknowledgements This work was supported by the FhG Internal Programs under Grant No. WISA 823 244.

References

1. Hiller, M., Fang, S., Mielczarek, S., Verhoeven, R., Franitza, D.: Design, analysis and realization of tendon-based parallel manipulators. Mech. Mach. Theory **40**(4), 429–445 (2005)
2. Hollerbach, J.M., Wampler, C.W.: The calibration index and taxonomy for robot kinematic calibration methods. Int. J. Robot. Res. **15**, (1996)
3. Merlet, J.P., Daney, D.: A new design for wire-driven parallel robot. In: 2nd Int. Congress, Design and Modelling of Mechanical Systems. Monastir (2007)
4. Schiehlen, W., Eberhard, P.: Technische Dynamik, Modelle für Regelung und Simulation. Teubner (1986)
5. Verhoeven, R.: Analysis of the workspace of tendon-based Stewart platforms. Ph.D. thesis, University of Duisburg-Essen (2004)
6. Verl, A., Boye, T., Pott, A.: Measurement pose selection and calibration forecast for manipulators with complex kinematic structures. CIRP Ann. – Manuf. Technol. **57**(1), 425–428 (2008)

Protein Folding Pathways Implementing Dihedral Angle Variable Speed

Mikel Diez, Victor Petuya, Mónica Urizar and Alfonso Hernández

Abstract Protein folding remains as an impossible riddle biologist must solve. Its huge computational requirements make it difficult to obtain clues regarding protein folding nature. The procedure presented in this paper presents a fast algorithm capable of obtaining hundreds of intermediate positions between the unfolded and the folded structures of several proteins. Presented algorithms make use of the balls and rods approach for protein structure modelization. In this paper, structures are normalized in order to minimize differences between experimental data and the kinematic model used in this paper.

Key words: Protein, folding, biokinematics, serial robot

1 Introduction

Protein folding problem is one of the most challenging riddles that remains unsolved on biochemistry. Actually there is little option when trying to obtain a protein folded structure. Experimental methods, such as X-ray diffraction or Nuclear Magnetic resonance are still lords and masters in this field. Nevertheless these methods give little information about the folding process itself.

Currently, in order to solve or to obtain information about the folding process analytical methods are the only available choice. The mayor problem of these methods is their high computational requirements. Ab initio methods are focused on the search of a potential energy minima. This potential energy minima is supposed to correspond to the folded structure of the protein. The search is carried out by statistical minimization procedures such as Monte Carlo methods[6]. The mayor drawback of these methods is their high computational cost. Recent works [11] are capable of

Mikel Diez · Victor Petuya · Mónica Urizar · Alfonso Hernández
Department of Mechanical Engineering, Faculty of Engineering in Bilbao, Bilbao, Spain, e-mail: {mikel.diez, victor.petuya, monica.urizar, a.hernandez}@ehu.es

J. Lenarčič, M. Husty (eds.), *Latest Advances in Robot Kinematics*,
DOI 10.1007/978-94-007-4620-6_35,

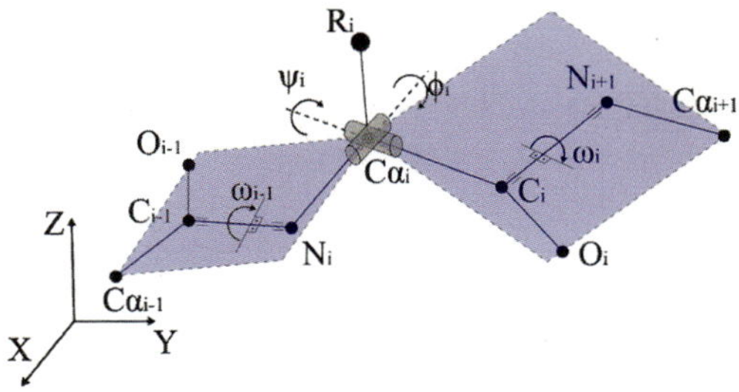

Fig. 1 Structural conformation. Degrees of freedom of the protein.

simulating 1.5 millisecond of movement using the Folding @Home distributed computing platform. Other important drawback is the high number of local minima that potential energy fields present, making possible that the minimization process finish on an incorrect and not fully folded structure. Molecular dynamics [8] approaches the problem without any simplification. Molecular dynamics tries to simulate the whole real movement of every atom considering every possible factor. This yields to enormous computational efforts that actually limits its use to small molecules or linear fibrous proteins. Lattice models [5] approach the same problem from the simplification side. On this models proteins are simplified so that each amino acid is modeled as a single "bead" and their movement is restricted to a lattice. These methods work similarly to interval analysis methods used to solve the forward kinematic problem in mechanism [7].

2 Kinematic Model of Protein Structure

For protein structure modeling ball and rods approach has been used. This model has already been used in other protein simulations works such as [9, 10]. In this paper apart from bond length and angles peptide planes *omega* angle is also considered constant. This approach allow us to only consider dihedral angles ϕ_i and ψ_i as the only protein degrees of freedom (see Fig. 1), thus reducing the computational cost of the model.

The procedure presented in this paper is an evolution of the one presented in [3]. The algorithm presented in [3] considered only one possible rotation direction on each protein degree of freedom resulting on a computationally very cheap method of simulating protein movement but obtaining poor results on complex protein motions such as protein folding. In this paper the proposed algorithm allows changes on the rotation direction under certain circumstances, resulting on a computationally more expensive method but greatly improving the results obtained with the previous algorithm. The procedure progresses during the simulation using proteins' natural degrees of freedom, i.e. dihedral angles, in order to better reflect the true nature of the movement and maintain the kinematic continuity of the motion. Folded structures of the proteins under study are taken from the Protein Data Bank (PDB) and used as input data for the procedure.

As the procedure calculates dihedral angles values for the simulation process and considering that available data comes from experimental procedures, data quality must be rated. On a protein, atom-atom interactions make them to vibrate around an equilibrium position. Since experimental methods take “photographs” of proteins the internal disposition of atoms in two “photographs” of the same protein is never the same. Thus bond length and angles values differ between initial and final positions of the same protein function. As a result, calculated dihedral angles increments will not make initial position to reach the final one. That being so, it has been chosen to normalize bond length and peptide planes applying the normalization algorithm presented in [3]. Normalization process is carried out ensuring both bond length values and peptide planes existence maintaining protein biological sense. This normalization process does allow us to avoid potential energy bonded terms evaluation during the simulation process. Since bond lengths and peptide ω angle are equal on both initial and final structures there is no change on their energies during the simulation process. Actual normalization algorithm does not normalize bond angles, nevertheless their contribution to the proteins’ potential energy can be neglected rarely representing more than 2% of proteins total energy.

3 Algorithm for Dihedral Angle Rotation Implementing Variable Speed

The simulations were carried out using a software developed by our research group called GIMPRO [4].

The inputs for the simulation process were the unfolded and folded conformations of the proteins. The motion of the protein is defined by a series of incremental steps, applied on the dihedral angle values, from the initial to the final conformations. In the analysis, the dihedral angles ϕ_i and ψ_i of each amino acid are calculated for the initial (0) and final (f) conformations.

Three indicators have been used in order to verify the validity of the results regarding both obtained precision and biological meaning of the simulation process. Firstly the root mean squared deviation (rmsd) is used as a global indicator assessing the geometric similarity between the obtained structure and the pattern folded structure. Secondly proteins’ potential energy is used as a local indicator. Proteins’ potential energy is highly dependent on interatomic distances and its value changes drastically when two or more atoms are about to collide. From the several available potential energy fields, AMBER force field, with the parameters proposed by Cornel [1] has been chosen. Finally Ramachandran plots are obtained to biologically validate folded structures.

Two algorithms have been developed for the simulation of the molecular mechanism. First algorithm, described in detail in [2], works by blocking degrees of freedom that, in each iteration, lead to greater increases in potential energy.

Algorithm 1 Potential energy algorithm for the *k*-th step of the procedure

foreach DoF i in the protein **do**
  Rotate the i-th DoF $(\Delta\psi_i||\Delta\phi_i)$
  $E_i^k \leftarrow$ Evaluate potential energy after rotation of the i-th DoF
  $\Delta E_i^k = E_i^k - E_{i-1}^k \leftarrow$ Energy increment associated with the i-th DoF
end foreach
$E^k = E_i^k \leftarrow$ Potential energy after step k
while $\left[\left(E^k - E^0\right)/E^0\right] \cdot 100 \geq \varepsilon^k$ **do**
  Unrotate DoF i $(-\Delta\psi_i|| - \Delta\phi_i)$ related to the highest ΔE_i^k
  Rotate DoF i with reduced angular increment $(\Delta\psi_i/n||\Delta\phi_i/n)$
  $E^k \leftarrow$ Evaluate new potential energy
  if $\left[\left(E^k - E^0\right)/E^0\right] \cdot 100 \geq \varepsilon^k$ **then**
    Unrotate DoF i $(-\Delta\psi_i/n|| - \Delta\phi_i/n)$
  end if
  $\Delta E_i^k = 0$
  if i-th DoF is slowed more than m times **then**
    $\Delta\psi_i = -\Delta\psi_i||\Delta\phi_i = -\Delta\phi_i$
  end if
  $E^k \leftarrow$ Evaluate new potential energy
end while

The second algorithm, presented in Algorithm 1, is an evolution of the first one. This algorithm prior to blocking the degree of freedom tries to make the rotation with a reduced increment value. First, the algorithm starts by calculating protein's initial conformation potential energies' non bonded term, E_0, value. Once calculated, simulation process starts rotating sequentially proteins' degrees of freedom, from the first amino acid to the last one. After each rotation, the variation in protein's potential energy, ΔE_i^k, is stored for later use and associated with the degree of freedom i that has generated it. Once all degrees of freedom are rotated the final conformations' potential energy value is obtained E^k. This value cannot exceed the allowed threshold for the current k iteration, $E_0 + E_0 * \varepsilon_k$, where ε_k is an energy tolerance intended to distribute energy changes uniformly across the simulation, and is calculated as follows:

$$\varepsilon_k = \frac{k \cdot \varepsilon}{p} \tag{1}$$

where ε is the total energy tolerance and p is the number of steps for the simulation process, both parameters are defined by the user. In case E_k exceeds the threshold, the algorithm begins by analyzing the degree of freedom that has generated greater potential energy increments. The rotation of this degree of freedom is rolled back and applied again, halving its angular increment value $(\Delta\psi_i/2, \Delta\phi_i/2)$. In the event that this new rotation does not achieve to reduce the value of the potential energy below the threshold, the degree of freedom is blocked. The process is repeated with the next degrees of freedom that have generated greater potential energy increases, until the energy requirements are fulfilled. Along the simulation, the algorithm keeps a record of the number of times a degree of freedom has been blocked or slowed down. If a degree of is blocked m times (user defined) the algorithm considers that it

Table 1 Simulation results for Algorithms 1 and 2

Protein	Algorithm 1			Algorithm 2		
	rmsd	energy %	RP (atoms in favored regions)	rmsd	energy %	RP (atoms in favored regions)
1k9p	–Å	–%	–%	10.63Å	7.2%	96%
1vii	9.05Å	9.8%	41%	8.66Å	9.1%	41%
1zac	20Å	6.4%	89%	23.47Å	8.7%	86%

cannot move in that direction and changes its rotation direction ($\Delta\psi_i = -\Delta\psi_i, \Delta\phi_i = -\Delta\phi_i$) for the next n iterations (user defined).

As can be seen the algorithm has total freedom to change dihedral angle values it may happen that when reached the p number of steps defined by the user the dihedral angles did not have enough time to reach their final value. Because of this fact p is incremented by 1 whenever on an iteration at least one degree of freedom has been either blocked or reversed. This also makes the energy tolerance to be adjusted dynamically during the simulation process (see (1)).

The simulation may end by three events. Firstly the dihedral angles have reached their final values. Second the algorithm consumes the energy tolerance defined by the user. Finally as the number of steps is incremented each time a degree of freedom is blocked or reversed the simulation may continue indefinitely. It has been introduced a maximum number of iterations of $10 * p$ in order to avoid too large simulations.

Both algorithms use information from the evaluation of the proteins' potential energy to guide the simulation process. The objective of feeding the model with data of the potential energy is to avoid peaks in potential energy. These increments reflect the imminent collision of two atoms. The algorithm is intended to favor rotations that cause moderate changes in potential energy, in order that the energy changes as smoothly as possible during the course of the motion.

4 Results

Both algorithms have been tested in several proteins with different order of final structure complexity. For the simulations, parameters have been set as follows: $p = 100$, $n = 2$, $m = 2$ and the maximum energy tolerance $\varepsilon = 10\%$. A video of the simulation of 1vii folding process can be viewed on www.ehu.es/compmech. On Table 1 results for both algorithms are represented.

From Table 1 several conclusions can be made. The first one, the algorithm presented in this paper is able to obtain a solution where the first algorithm cannot, such as the folding of 1k9p protein. On 1vii folding the algorithm presented in this paper slightly improves the rmsd value obtained with the first algorithm. It is to mention that although on this folding both algorithms obtains quite low Ramachandran plot values, folded structure used as data also presents low Ramachandran plot values,

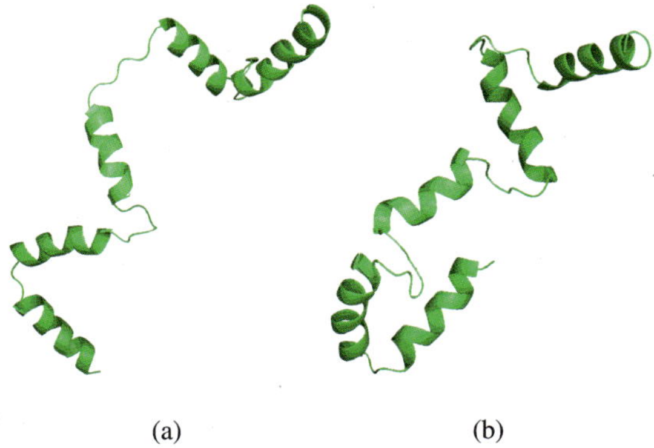

Fig. 2 Partially folded structure obtained with algorithm 1 (a) and algorithm 2 (b). See how both algorithms obtain 5 α-helixes. Represented with pymol.

being only 58% of the atoms inside the favored zones. On the other hand 1zac folding process results on high rmsd values for both the first and the second algorithm. This high values are due to the complexity of the structure that is composed of five α-helixes. In this case both algorithms achieve secondary structures folding but fail in the tertiary structure one (see Fig. 2) due to the interferences between main chain loops radicals.

Folding process is highly dependent on the secondary chains or R groups configuration. Currently the algorithm treats these chains as rigid bodies simply translating them through the simulation process. Those chains are supposed to rotate around their axes in order to favor the folding process. With the purpose of testing the influence of the secondary chains stiffness in the folding simulation, another simulation has been made with 1vii protein, without considering their energy contribution into the simulation process. This simulation does not represent any real protein movement since not considering secondary chains energy allow atoms of those chains to collide between each other. Ramachandran plots are neither calculated for these simulation since atomic collisions between radicals allowance makes impossible for those positions to fall into the allowed zones of the Ramachandran plots.

This simulation reduces dramatically the rmsd value obtaining a minimum value of 5.35Å for the 1vii folding. In conclusion, taking into account R groups energy contribution without considering their rotational degrees of freedom penalizes the simulation process. Blocking those degrees of freedom makes the energy landscape more restrictive. This leads to a noticeable reduction on the available paths for the folding process, thus, the algorithm has a limited available research space. This problem is greater in proteins with more complex structures such as 1k9p or 1zac proteins. The same simulation has been done with both proteins obtaining lower rmsd values. On the 1k9p rmsd is reduced to 6.26Å and on 1zac to 10.7Å.

New evolutions of the proposed procedure will be focused on the introduction of R groups' degrees of freedom into the simulation process. It is important to evaluate which of the approaches obtains the best compromise between solution precision and computational effort, aways maintaining the biological sense of the obtained structures. Furthermore, computational cost will play a determinant role on these new algorithms because of their models higher complexity.

5 Conclusion

Proteins and mechanisms similar characteristics allow us to use simplified models of protein structures that are suitable for kinematic simulations. The algorithm proposed in this paper is capable of simulating different proteins foldings obtaining promising results. Its implementation of dihedral angle variable speed and capacity of changing rotational direction allows it to search a wider configurational space. Future work is focused on developing a more complex model of protein structure that considers secondary chains rotation, so as to reduce the restriction imposed by the rigid secondary chains assumption resultant energy landscape.

Acknowledgements The authors wish to acknowledge the financial support received from the Spanish Government through the Ministerio de Economía y Competitividad (Project DPI2011-22955), the European Union (Project FP7-CIP-ICT-PSP-2009-3) and the Regional Government of the Basque Country through the Departamento de Educación, Universidades e Investigación (Project IT445-10).

References

1. Cornell, W., Cieplak, P., Bayly, C., Gould, I., Merz, K., Ferguson, D., Spellmeyer, D., Fox, T., Caldwell, J., Kollman, P.: A second generation force field for the simulation of proteins, nucleic acids, and organic molecules. J. Am. Chem. Soc. (1995)
2. Diez, M., Petuya, V., Macho, E., Hernandez, A.: Protein kinematic motion simulation including potential energy feedback. In: New Trends in Mechanism Science. Springer (2010)
3. Diez, M., Petuya, V., Martínez-Cruz, L.A., Hernández, A.: A biokinematic approach for the computational simulation of proteins molecular mechanism. Mech. Mach. Theory (2011)
4. Diez, M., Petuya, V., Urizar, M.: Hernandez: A biokinematic computational procedure for protein function simulation. In: Reconfigurable Mechanisms and Robots, 2009, ReMAR 2009, ASME/IFToMM International Conference on (2009)
5. Klimov, D.K., Thirumalai, D.: Cooperativity in protein folding: From lattice models with side chains to real proteins. arXiv.org (1998)
6. Leach, A.: Molecular Modelling: Principles and Applications, 2nd edn. Prentice Hall (2001)
7. Merlet, J.P.: Solving the forward kinematics of a Gough-type parallel manipulator with interval analysis. Int. J. Robot. Res. (2004)
8. Schlick, T.: Molecular Modeling and Simulation: An Interdisciplinary Guide (Interdisciplinary Applied Mathematics). Springer (2010)
9. Sharma, G., Badescu, M., Yarmush, M.L.: Kinematics and workspace analysis of protein based nano-actuators. J. Mech. Des. (Trans. ASME) (2005)

10. Subramanian, R., Kazerounian, K., Fellow, A.: Improved molecular model of a peptide unit for proteins. J. Mech. Des. (2007)
11. Voelz, V.A., Bowman, G.R., Beauchamp, K., Pande, V.S.: Molecular simulation of ab initio protein folding for a millisecond folder NTL9(139). J. Am. Chem. Soc. (2010)

Sufficient Conditions for the Mobility of Overconstrained Mechanisms

René Bartkowiak and Christoph Woernle

Abstract A single-loop mechanism with $n \leq 6$ helical joints which has a finite mobility with a degree of freedom $f \geq 1$ is overconstrained. Here the case $f = 1$ is considered. For the finite mobility of an overconstrained mechanism it is sufficient, that the first-order closure condition is fulfilled in an open neighborhood of the actual position of the joint axes. Since the explicit functional dependencies of the closure condition are unknown, a Taylor-series expansion yields necessary conditions for the mobility of the mechanism. According to [6] it is sufficient for the finite mobility of the mechanism if all higher-order mobility conditions up to a maximum, but unknown, order m are fulfilled. Special solutions for necessary mobility conditions are discussed which guarantee that all higher-order necessary conditions are fulfilled. By this these special solutions are sufficient for finite mobility. Some of this special mobility conditions are presented.

Key words: Overconstrained mechanisms, kinematical synthesis, loop closure condition, mobility condition

1 Mobility Conditions for Overconstrained Mechanisms

The screw axes of a single-loop mechanism with n helical joints (nH mechanism) according to Fig. 1 are described by 6-vectors

$$\underline{\hat{\boldsymbol{a}}}_i \equiv \begin{bmatrix} \boldsymbol{a}_i \\ \boldsymbol{a}_{\varepsilon i} \end{bmatrix}, \qquad i = 1, \ldots, n, \tag{1}$$

René Bartkowiak · Christoph Woernle
Department of Mechanical Engineering and Marine Technology, University of Rostock, Rostock, Germany, e-mail: {rene.bartkowiak, woernle}@uni-rostock.de

J. Lenarčič, M. Husty (eds.), *Latest Advances in Robot Kinematics*,
DOI 10.1007/978-94-007-4620-6_36,

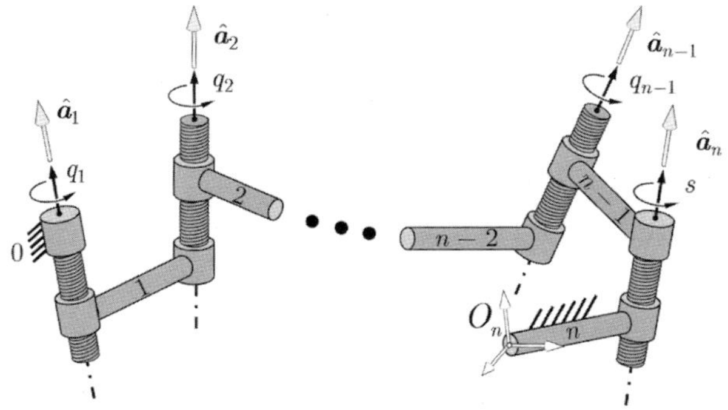

Fig. 1 Mechanism with n helical joints (nH).

containing the real part $\boldsymbol{a}_i$ and the dual part $\boldsymbol{a}_{\varepsilon i}$ of the corresponding dual vector $\hat{\boldsymbol{a}}_i = \boldsymbol{a}_i + \varepsilon \boldsymbol{a}_{\varepsilon i}$ with $\varepsilon^2 = 0$. Without loss of generality the first and nth axes $\hat{\boldsymbol{a}}_1$ and $\hat{\boldsymbol{a}}_n$ are fixed. The joint angles are q_i.

In the following, overconstrained nH mechanisms, $n \le 6$, with one degree of freedom (dof) are considered. If the motion of the mechanism is parametrized, without loss of generality, by the joint coordinate $s = q_n$ of the nth helical joint, the first-order closure condition is

$$\boldsymbol{A}(s)\,\boldsymbol{\lambda}(s) = \underline{\hat{\boldsymbol{a}}}_n \quad \text{with} \quad \boldsymbol{A} = \left[\,\underline{\hat{\boldsymbol{a}}}_1 \;\dots\; \underline{\hat{\boldsymbol{a}}}_{n-1}\right], \quad \boldsymbol{\lambda} = \begin{bmatrix} \lambda_1 \\ \vdots \\ \lambda_{n-1} \end{bmatrix}, \quad \lambda_i = -\frac{\dot{q}_i}{\dot{q}_n}. \tag{2}$$

For finite mobility of the mechanism it is sufficient that the nth joint axis $\hat{\boldsymbol{a}}_n$ is a linear combination of the other $n-1$ screw axes in an open neighborhood of an actual position $s_0 = q_{n0}$. Since the functional dependencies $\boldsymbol{A}(s)$ and $\boldsymbol{\lambda}(s)$ are a priori not known, the loop closure condition (2) is expanded to a TAYLOR series at a position $s_0 = q_{n0}$ with $s = q_{n0} + \delta s$ yielding

$$\underline{\hat{\boldsymbol{a}}}_{n0} = \overbrace{\left(\boldsymbol{A}_0 + \frac{\boldsymbol{A}_0'}{1!}\,\delta s + \frac{\boldsymbol{A}_0''}{2!}\,\delta s^2 + \dots\right)}^{\boldsymbol{A}(s)} \overbrace{\left(\boldsymbol{\lambda}_0 + \frac{\boldsymbol{\lambda}_0'}{1!}\,\delta s + \frac{\boldsymbol{\lambda}_0''}{2!}\,\delta s^2 + \dots\right)}^{\boldsymbol{\lambda}(s)} \tag{3}$$

with $\underline{\hat{\boldsymbol{a}}}_{n0} = \underline{\hat{\boldsymbol{a}}}_n$, $\boldsymbol{A}_0 = \boldsymbol{A}(s_0)$, $\boldsymbol{A}_0' = \frac{\mathrm{d}\boldsymbol{A}(s)}{\mathrm{d}s}|_{s_0}, \dots$ and $\boldsymbol{\lambda}_0 = \boldsymbol{\lambda}(s_0)$, $\boldsymbol{\lambda}_0' = \frac{\mathrm{d}\boldsymbol{\lambda}(s)}{\mathrm{d}s}|_{s_0}, \dots$. The derivative $\boldsymbol{A}_0' = [\,\underline{\hat{\boldsymbol{a}}}_{10}' \;\cdots\; \underline{\hat{\boldsymbol{a}}}_{n-1,0}'\,]$ of matrix $\boldsymbol{A}$ with respect to s is expressed by the derivatives of the screw axes using the dual vector product

$$\underline{\hat{\boldsymbol{a}}}_{k0}' \equiv \left.\frac{\mathrm{d}\underline{\hat{\boldsymbol{a}}}_k(s)}{\mathrm{d}s}\right|_{s_0} = -\sum_{i=1}^{k-1} \widetilde{\underline{\hat{\boldsymbol{a}}}}_{i0}\,\underline{\hat{\boldsymbol{a}}}_{k0}\,\lambda_{i0} \quad \text{with} \quad \widetilde{\underline{\hat{\boldsymbol{a}}}} \equiv \begin{bmatrix} \widetilde{\boldsymbol{a}} & \boldsymbol{0} \\ \widetilde{\boldsymbol{a}}_\varepsilon & \widetilde{\boldsymbol{a}} \end{bmatrix}. \tag{4}$$

The tilde operator $\widetilde{\ }$ transforms a vector $\boldsymbol{a}$ to the skew-symmetric tensor $\widetilde{\boldsymbol{a}}$, thus $\widetilde{\boldsymbol{a}}\boldsymbol{b} = \boldsymbol{a} \times \boldsymbol{b}$, see [3].

Since the closure condition (3) must hold for arbitrary values of δs, the coefficients of the powers of δs must independently vanish

$$
\begin{aligned}
\delta s^0: \quad & \boldsymbol{A}_0\,\boldsymbol{\lambda}_0 = \underline{\hat{\boldsymbol{a}}}_{n0}\,,\\
\delta s^1: \quad & \boldsymbol{A}_0\,\boldsymbol{\lambda}_0' = -\boldsymbol{A}_0'\,\boldsymbol{\lambda}_0\,,\\
\delta s^2: \quad & \boldsymbol{A}_0\,\boldsymbol{\lambda}_0'' = -\boldsymbol{A}_0''\,\boldsymbol{\lambda}_0 - 2\boldsymbol{A}_0'\,\boldsymbol{\lambda}_0'\,,\\
& \qquad\vdots
\end{aligned}
\tag{5}
$$

The mechanism is mobile, if solutions $\boldsymbol{\lambda}_0, \boldsymbol{\lambda}_0', \boldsymbol{\lambda}_0'', \ldots$ of the loop closure equations (5) exist. These solutions exist, if each right-hand side lies in the column space of matrix $\boldsymbol{A}_0$, thus

$$
\underline{\hat{\boldsymbol{a}}}_{n0}, \quad -\boldsymbol{A}_0'\,\boldsymbol{\lambda}_0, \quad -\boldsymbol{A}_0''\,\boldsymbol{\lambda}_0 - 2\boldsymbol{A}_0'\,\boldsymbol{\lambda}_0', \quad \ldots \quad \in \operatorname{span}\left\{\underline{\hat{\boldsymbol{a}}}_{10}, \ldots, \underline{\hat{\boldsymbol{a}}}_{n-1,0}\right\}. \tag{6}
$$

Then the solution of (5) can be written by means of a left inverse $\boldsymbol{A}_0^+$, for example the MOORE-PENROSE pseudo-inverse,

$$
\begin{aligned}
\boldsymbol{\lambda}_0 &= \boldsymbol{A}_0^+\,\underline{\hat{\boldsymbol{a}}}_{n0},\\
\boldsymbol{\lambda}_0' &= -\boldsymbol{A}_0^+\,\boldsymbol{A}_0'\,\boldsymbol{A}_0^+\,\underline{\hat{\boldsymbol{a}}}_{n0},\\
\boldsymbol{\lambda}_0'' &= \boldsymbol{A}_0^+\left(2\boldsymbol{A}_0'\,\boldsymbol{A}_0^+\,\boldsymbol{A}_0' - \boldsymbol{A}_0''\right)\boldsymbol{A}_0^+\,\underline{\hat{\boldsymbol{a}}}_{n0},\\
&\ \ \vdots
\end{aligned}
\tag{7}
$$

The condition (6) means that all screws on the right-hand side of (5) and the column screws of matrix $\boldsymbol{A}_0$ have the same reciprocal screws $\underline{\hat{\boldsymbol{k}}}_{j0}$, $j = 1, \ldots, (6 - \operatorname{rank}(\boldsymbol{A}_0))$,

$$
\underline{\hat{\boldsymbol{k}}}_{j0}^{\mathrm{T}}\,\Delta\,\underline{\hat{\boldsymbol{x}}} \stackrel{!}{=} 0 \quad \text{with} \quad \Delta = \begin{bmatrix} \boldsymbol{0} & \boldsymbol{I} \\ \boldsymbol{I} & \boldsymbol{0} \end{bmatrix}, \quad j = 1, \ldots, (6 - \operatorname{rank}(\boldsymbol{A}_0)). \tag{8}
$$

Here $\underline{\hat{\boldsymbol{x}}}$ stands for any right-hand side of (5). The reciprocal screws $\underline{\hat{\boldsymbol{k}}}_{j0}$ are obtained from the reciprocity condition

$$
\underline{\hat{\boldsymbol{k}}}_{j0}^{\mathrm{T}}\,\Delta\,\boldsymbol{A}_0 = \boldsymbol{0}\,, \quad j = 1, \ldots, (6 - \operatorname{rank}(\boldsymbol{A}_0))\,. \tag{9}
$$

Inserting (5) and (7) into the reciprocity condition (8) yields a system of m nonlinear equations for the unknown screws $\underline{\hat{\boldsymbol{a}}}_{i0}$, $i = 1, \ldots, n$,

$$
\begin{aligned}
0 &= \underline{\hat{\boldsymbol{k}}}_{j0}^{\mathrm{T}}\,\Delta\,\underline{\hat{\boldsymbol{a}}}_{n0} && \equiv g_1\left(\underline{\hat{\boldsymbol{a}}}_{n0}, \boldsymbol{A}_0\right),\\
0 &= \underline{\hat{\boldsymbol{k}}}_{j0}^{\mathrm{T}}\,\Delta\,\boldsymbol{A}_0'\,\boldsymbol{A}_0^+\,\underline{\hat{\boldsymbol{a}}}_{n0} && \equiv g_2\left(\underline{\hat{\boldsymbol{a}}}_{n0}, \boldsymbol{A}_0\right),\\
0 &= \underline{\hat{\boldsymbol{k}}}_{j0}^{\mathrm{T}}\,\Delta\left(-2\boldsymbol{A}_0'\,\boldsymbol{A}_0^+\,\boldsymbol{A}_0' + \boldsymbol{A}_0''\right)\boldsymbol{A}_0^+\,\underline{\hat{\boldsymbol{a}}}_{n0} && \equiv g_3\left(\underline{\hat{\boldsymbol{a}}}_{n0}, \boldsymbol{A}_0\right),\\
&\ \ \vdots\\
0 &= \ldots && \equiv g_m\left(\underline{\hat{\boldsymbol{a}}}_{n0}, \boldsymbol{A}_0\right),
\end{aligned}
\tag{10}
$$

where m is the highest order of the closure conditions from (5) that are taken into consideration. The equations (10) are necessary conditions for the finite mobility of an overconstrained nH mechanism with one dof [2]. The solution of (10) are those screw axes $\underline{\hat{\boldsymbol{a}}}_{i0}$, $i = 1,\dots,n$, in the actual configuration, which fulfill all closure conditions from (5) up to the order m. The nonlinear system of equations (10) can be numerically solved. In [6] it is shown that there exists a finite maximum number of closure conditions for the finite mobility of overconstrained mechanisms. However the order m of the sufficient closure conditions is still unknown and depends on the number of links and the types of the joints.

There exist special solutions of (10) which provide sufficient but not necessary conditions for the mobility of overconstrained mechanisms. Some examples of such solutions are presented in the following.

2 Sufficient but not Necessary Conditions for Finite Mobility

It is well known that for the mobility of a mechanism with $n = 4$ revolute joints (4R) the closure conditions from (5) up to the maximum order $m = 2$ are sufficient and necessary [1].

4R mechanism The closure conditions for the 4R mechanism are solved up to the second order. In order to fulfill the first-order closure condition from (5) the fourth screw axis $\underline{\hat{\boldsymbol{a}}}_{40}$ is expressed as a linear combination of the other three axes with the unknown ratios $\boldsymbol{\lambda}_0 = \left[\lambda_{10}\ \lambda_{20}\ \lambda_{30}\right]^{\mathrm{T}}$, thus

$$\underline{\hat{\boldsymbol{a}}}_{40} = \boldsymbol{A}_0\,\boldsymbol{\lambda}_0. \tag{11}$$

With the derivatives of the screw axes according to (4),

$$\underline{\hat{\boldsymbol{a}}}'_{10} = \boldsymbol{0}, \quad \underline{\hat{\boldsymbol{a}}}'_{20} = -\underline{\tilde{\hat{\boldsymbol{a}}}}_{10}\,\underline{\hat{\boldsymbol{a}}}_{20}\,\lambda_{10}, \quad \underline{\hat{\boldsymbol{a}}}'_{30} = -\underline{\tilde{\hat{\boldsymbol{a}}}}_{10}\,\underline{\hat{\boldsymbol{a}}}_{30}\,\lambda_{10} - \underline{\tilde{\hat{\boldsymbol{a}}}}_{20}\,\underline{\hat{\boldsymbol{a}}}_{30}\,\lambda_{20}, \tag{12}$$

the second-order closure condition $\boldsymbol{0} = \boldsymbol{A}_0\,\boldsymbol{\lambda}'_0 + \boldsymbol{A}'_0\,\boldsymbol{\lambda}_0$ from (5) becomes

$$\boldsymbol{0} = \underline{\hat{\boldsymbol{a}}}'_{10}\,\lambda_{10} + \underline{\hat{\boldsymbol{a}}}'_{20}\,\lambda_{20} + \underline{\hat{\boldsymbol{a}}}'_{30}\,\lambda_{30} + \underline{\hat{\boldsymbol{a}}}_{10}\,\lambda'_{10} + \underline{\hat{\boldsymbol{a}}}_{20}\,\lambda'_{20} + \underline{\hat{\boldsymbol{a}}}_{30}\,\lambda'_{30}\,. \tag{13}$$

By rearranging (13) the axis $\underline{\hat{\boldsymbol{a}}}_{30}$ is obtained by

$$\underline{\hat{\boldsymbol{a}}}_{30} = \left(\underline{\tilde{\hat{\boldsymbol{a}}}}_{10}\,\lambda_{10}\,\lambda_{30} + \underline{\tilde{\hat{\boldsymbol{a}}}}_{20}\,\lambda_{20}\,\lambda_{30} - \boldsymbol{I}\,\lambda'_{30}\right)^{-1}\left(\underline{\hat{\boldsymbol{a}}}'_{20}\,\lambda_{20} + \underline{\hat{\boldsymbol{a}}}_{10}\,\lambda'_{10} + \underline{\hat{\boldsymbol{a}}}_{20}\,\lambda'_{20}\right), \tag{14}$$

if $\det\left(\underline{\tilde{\hat{\boldsymbol{a}}}}_{10}\,\lambda_{10}\,\lambda_{30} + \underline{\tilde{\hat{\boldsymbol{a}}}}_{20}\,\lambda_{20}\,\lambda_{30} - \boldsymbol{I}\,\lambda'_{30}\right) \neq 0$. With (11) and (14) the axes $\underline{\hat{\boldsymbol{a}}}_{30}$ and $\underline{\hat{\boldsymbol{a}}}_{40}$ are calculated for arbitrary values of $\underline{\hat{\boldsymbol{a}}}_{10}, \underline{\hat{\boldsymbol{a}}}_{20}, \boldsymbol{\lambda}_0, \boldsymbol{\lambda}'_0$ in such a way that the first- and second-order closure conditions are fulfilled. In order to obtain revolute axes, all axes have to fulfill the Plücker conditions

$$\boldsymbol{a}_{i0}^{\mathrm{T}}\,\boldsymbol{a}_{\varepsilon i0} = 0 \quad \text{with} \quad |\boldsymbol{a}_{i0}| = 1, \quad i = 1,\dots,4. \tag{15}$$

It can be shown that these conditions for $\underline{\hat{a}}_{30}$ and $\underline{\hat{a}}_{40}$ are fulfilled with the special values

$$\boldsymbol{\lambda}_0 \equiv \begin{bmatrix} \lambda_{10} \\ \lambda_{20} \\ \lambda_{30} \end{bmatrix} = \begin{bmatrix} \lambda_{10} \\ \mu_1 \\ \mu_2 \lambda_{10} \end{bmatrix} \quad \text{and} \quad \boldsymbol{\lambda}'_0 \equiv \begin{bmatrix} \lambda'_{10} \\ \lambda'_{20} \\ \lambda'_{30} \end{bmatrix} = \begin{bmatrix} \lambda'_{10} \\ 0 \\ \mu_2 \lambda'_{10} \end{bmatrix} \tag{16}$$

with $\mu_1, \mu_2 = \pm 1$. The arbitrarily given axes $\underline{\hat{a}}_{10}, \underline{\hat{a}}_{20}$ and axes $\underline{\hat{a}}_{30}, \underline{\hat{a}}_{40}$ obtained from (11), (14), and (16) fulfill the implicit Bennett condition for the fixed angles and distances β_{12}, b_{12} between axes $\underline{\hat{a}}_1$, $\underline{\hat{a}}_2$ and β_{23}, b_{23} between axes $\underline{\hat{a}}_2$, $\underline{\hat{a}}_3$,

$$\frac{b_{23}}{\sin \beta_{23}} = \mu_1 \mu_2 \frac{b_{12}}{\sin \beta_{12}}. \tag{17}$$

Since the Bennett mechanism has finite mobility all necessary mobility conditions (10) are fulfilled. Thus the fulfillment of the closure conditions up to the maximum order $m = 2$ is a sufficient condition for the mobility of a 4R mechanism.

***n*H mechanism** For general nH mechanisms, $n \leq 6$, the maximum order m is unknown. However special sufficient conditions can be found. For example there exist special solutions of the second necessary mobility condition from (10)

$$0 = \underline{\hat{\boldsymbol{k}}}_{j0}^{\mathrm{T}} \Delta \boldsymbol{A}'_0 \boldsymbol{\lambda}_0 , \quad j = 1, \ldots, (6 - \mathrm{rank}(\boldsymbol{A}_0)) , \tag{18}$$

which are sufficient but not necessary for the mobility of overconstrained nH mechanisms, whereby it is required that the first-order condition from (5) is fulfilled, $\underline{\hat{a}}_{n0} = \boldsymbol{A}_0 \boldsymbol{\lambda}_0$.

First special solution For instance the special solution of the mobility condition (18)

$$\boldsymbol{A}'_0 \boldsymbol{\lambda}_0 = \mathbf{0} \tag{19}$$

is sufficient for finite mobility [4, 5]. This is shown by rearranging the higher-order closure conditions from (5) under consideration of (7) and (19),

$$\begin{aligned} \boldsymbol{A}''_0 \boldsymbol{\lambda}_0 &= 2 \boldsymbol{A}'_0 \boldsymbol{A}_0^+ \underbrace{\boldsymbol{A}'_0 \boldsymbol{\lambda}_0}_{\mathbf{0}}, \\ \boldsymbol{A}'''_0 \boldsymbol{\lambda}_0 &= 3 \boldsymbol{A}''_0 \boldsymbol{A}_0^+ \underbrace{\boldsymbol{A}'_0 \boldsymbol{\lambda}_0}_{\mathbf{0}} + 3 \boldsymbol{A}'_0 \boldsymbol{A}_0^+ \underbrace{\boldsymbol{A}''_0 \boldsymbol{\lambda}_0}_{\mathbf{0}} - 6 \boldsymbol{A}'_0 \boldsymbol{A}_0^+ \boldsymbol{A}'_0 \boldsymbol{A}_0^+ \underbrace{\boldsymbol{A}'_0 \boldsymbol{\lambda}_0}_{\mathbf{0}} , \\ &\vdots \end{aligned} \tag{20}$$

An example for the special solution (19) is the Bennett mechanism in a special position where all four joint axes intersect a common line orthogonally, see Fig. 2. In this position the coefficient matrix of the linear system of equations (14) is singular, $\det\big(\widetilde{\underline{\hat{a}}_{10} \lambda_{10} \lambda_{30} + \underline{\hat{a}}_{20} \lambda_{20} \lambda_{30}} - \boldsymbol{I} \lambda'_{30}\big) = 0$. The matrix $\boldsymbol{A}_0 = [\underline{\hat{a}}_{10}\ \underline{\hat{a}}_{20}\ \underline{\hat{a}}_{30}]$ is

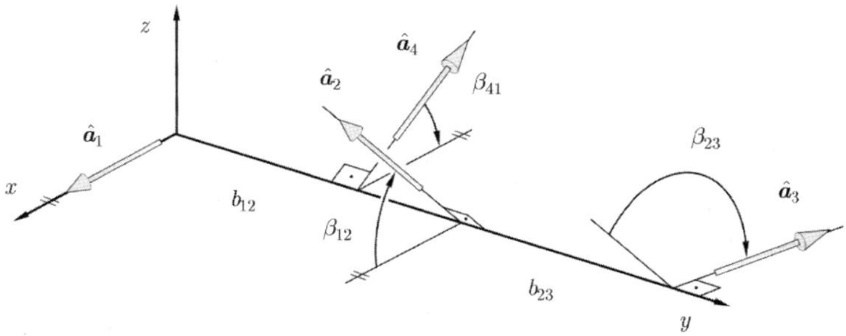

Fig. 2 Special configuration of the Bennett mechanism.

$$A_0 = \begin{bmatrix} 1 & \cos\beta_{12} & \cos\beta_{12}\cos\beta_{23}+\sin\beta_{12}\sin\beta_{23} \\ 0 & 0 & 0 \\ 0 & \sin\beta_{12} & \cos\beta_{23}\sin\beta_{12}-\cos\beta_{12}\sin\beta_{23} \\ 0 & b_{12}\sin\beta_{12} & -(b_{12}+b_{23})(\cos\beta_{12}\sin\beta_{23}-\cos\beta_{23}\sin\beta_{12}) \\ 0 & 0 & 0 \\ 0 & -b_{12}\cos\beta_{12} & -(b_{12}+b_{23})(\cos\beta_{12}\cos\beta_{23}+\sin\beta_{12}\sin\beta_{23}) \end{bmatrix}, \tag{21}$$

and the fourth screw axis is $\underline{\hat{a}}_{40} = \left[\cos\beta_{23}\ 0\ -\sin\beta_{23}\ -b_{23}\sin\beta_{23}\ 0\ -b_{23}\cos\beta_{23}\right]^{\mathrm{T}}$. The sufficient condition (19) here is

$$\underbrace{\begin{bmatrix} 0 & 0 & 0 \\ 0 & -\gamma\sin\beta_{12} & \sin\beta_{12} \\ 0 & 0 & 0 \\ 0 & 0 & 0 \\ 0 & \gamma b_{12}\cos\beta_{12} & -b_{12}\cos\beta_{12} \\ 0 & 0 & 0 \end{bmatrix}}_{A_0'} \underbrace{\begin{bmatrix} -\gamma \\ -1 \\ -\gamma \end{bmatrix}}_{\boldsymbol{\lambda}_0} = \underbrace{\begin{bmatrix} 0 \\ 0 \\ 0 \\ 0 \\ 0 \\ 0 \end{bmatrix}}_{\mathbf{0}} \tag{22}$$

with $\gamma = \frac{\sin\beta_{23}(b_{12}-b_{23})}{b_{23}(\cos\beta_{12}\sin\beta_{23}-\cos\beta_{23}\sin\beta_{12})}$.

Second special solution Another special solution of the second necessary mobility condition (18) is

$$\mathbf{0} = \underline{\hat{k}}_{j0}^{\mathrm{T}}\Delta A_0', \quad j = 1,\ldots,(6-\operatorname{rank}(A_0)). \tag{23}$$

The geometrical interpretation of (23) is, that the derivatives of the screw axes with respect to the independent parameter s lie in the same screw space as the screw axes of the helical joints. By rearranging the higher order mobility conditions from (10) under consideration of (7) and (23) it is shown that (23) is a sufficient condition

$$\underline{\hat{\boldsymbol{k}}}_{j0}^{\mathrm{T}} \Delta \boldsymbol{A}_0'' \boldsymbol{\lambda}_0 = 2 \underbrace{\underline{\hat{\boldsymbol{k}}}_{j0}^{\mathrm{T}} \Delta \boldsymbol{A}_0'}_{\boldsymbol{0}} \boldsymbol{A}_0^+ \boldsymbol{A}_0' \boldsymbol{\lambda}_0,$$

$$\underline{\hat{\boldsymbol{k}}}_{j0}^{\mathrm{T}} \Delta \boldsymbol{A}_0''' \boldsymbol{\lambda}_0 = 3 \underbrace{\underline{\hat{\boldsymbol{k}}}_{j0}^{\mathrm{T}} \Delta \boldsymbol{A}_0''}_{\boldsymbol{0}} \boldsymbol{A}_0^+ \boldsymbol{A}_0' \boldsymbol{\lambda}_0 + 3 \underbrace{\underline{\hat{\boldsymbol{k}}}_{j0}^{\mathrm{T}} \Delta \boldsymbol{A}_0'}_{\boldsymbol{0}} \boldsymbol{A}_0^+ \boldsymbol{A}_0'' \boldsymbol{\lambda}_0 - 6 \underbrace{\underline{\hat{\boldsymbol{k}}}_{j0}^{\mathrm{T}} \Delta \boldsymbol{A}_0'}_{\boldsymbol{0}} \boldsymbol{A}_0^+ \boldsymbol{A}_0' \boldsymbol{A}_0^+ \boldsymbol{A}_0' \boldsymbol{\lambda}_0$$

$$\vdots$$

An example for the special solution (23) is a Sarrus linkage with six helical joints. This special 6H mechanism has two groups of three adjacent parallel joint axes. Five screw axes are given by

$$\boldsymbol{A}_0 = \begin{bmatrix} \boldsymbol{a}_{10} & \boldsymbol{a}_{10} & \boldsymbol{a}_{10} & \boldsymbol{a}_{40} & \boldsymbol{a}_{40} \\ \boldsymbol{a}_{\varepsilon 10} & \boldsymbol{a}_{\varepsilon 20} & \boldsymbol{a}_{\varepsilon 30} & \boldsymbol{a}_{\varepsilon 40} & \boldsymbol{a}_{\varepsilon 50} \end{bmatrix} \tag{24}$$

and the sixth screw axis by $\underline{\hat{\boldsymbol{a}}}_{60} = \begin{bmatrix} \boldsymbol{a}_{40} \; \boldsymbol{a}_{\varepsilon 60} \end{bmatrix}^{\mathrm{T}}$. The reciprocal screw obtained from the reciprocity condition (9) is

$$\underline{\hat{\boldsymbol{k}}}_0 = \begin{bmatrix} \boldsymbol{0} \\ \widetilde{\boldsymbol{a}}_{10} \boldsymbol{a}_{40} \end{bmatrix}. \tag{25}$$

To proof the fulfillment of the second-order mobility condition from (10),

$$\underline{\hat{\boldsymbol{k}}}_0^{\mathrm{T}} \, \Delta \, \boldsymbol{A}_0' \boldsymbol{\lambda}_0 = 0, \tag{26}$$

the matrix $\boldsymbol{A}_0' = [\, \underline{\hat{\boldsymbol{a}}}_{10}' \cdots \underline{\hat{\boldsymbol{a}}}_{50}' \,]$ is calculated by means of (4). The dual vector product of the parallel screws $\underline{\hat{\boldsymbol{a}}}_{10}$, $\underline{\hat{\boldsymbol{a}}}_{20}$, $\underline{\hat{\boldsymbol{a}}}_{30}$ with $\boldsymbol{a}_{10} = \boldsymbol{a}_{20} = \boldsymbol{a}_{30}$ is given by

$$\widetilde{\underline{\hat{\boldsymbol{a}}}}_{10} \underline{\hat{\boldsymbol{a}}}_{i0} \equiv \widetilde{\begin{bmatrix} \boldsymbol{a}_{10} \\ \boldsymbol{a}_{\varepsilon 10} \end{bmatrix}} \begin{bmatrix} \boldsymbol{a}_{10} \\ \boldsymbol{a}_{\varepsilon i0} \end{bmatrix} = \begin{bmatrix} \boldsymbol{0} \\ \boldsymbol{b}_{1i} \end{bmatrix}, \quad i = 2, 3, \tag{27}$$

with the common normal vector $\boldsymbol{b}_{1i}$ from $\underline{\hat{\boldsymbol{a}}}_{10}$ to $\underline{\hat{\boldsymbol{a}}}_{i0}$. The dual vector product of the non-parallel screws $\underline{\hat{\boldsymbol{a}}}_{10}$, $\underline{\hat{\boldsymbol{a}}}_{40}$ and $\underline{\hat{\boldsymbol{a}}}_{10}$, $\underline{\hat{\boldsymbol{a}}}_{50}$ with $\boldsymbol{a}_{40} = \boldsymbol{a}_{50}$ is given by

$$\widetilde{\begin{bmatrix} \boldsymbol{a}_{10} \\ \boldsymbol{a}_{\varepsilon 10} \end{bmatrix}} \begin{bmatrix} \boldsymbol{a}_{40} \\ \boldsymbol{a}_{\varepsilon i0} \end{bmatrix} = \widetilde{\begin{bmatrix} \boldsymbol{a}_{10} \\ \boldsymbol{a}_{\varepsilon 10} \end{bmatrix}} \begin{bmatrix} \boldsymbol{a}_{40} \\ \boldsymbol{a}_{\varepsilon 40} \end{bmatrix} + \widetilde{\begin{bmatrix} \boldsymbol{a}_{10} \\ \boldsymbol{a}_{\varepsilon 10} \end{bmatrix}} \begin{bmatrix} \boldsymbol{0} \\ (h_i - h_4)\, \boldsymbol{a}_{40} + \widetilde{\boldsymbol{b}}_{4i} \boldsymbol{a}_{40} \end{bmatrix}, \; i = 4, 5, \tag{28}$$

whereby h_i are the screw pitches. The derivatives of the screw axes $\underline{\hat{\boldsymbol{a}}}_{i0}'$, $i = 1, \ldots, 5$, are put together to the matrix

$$\boldsymbol{A}_0' = \begin{bmatrix} \boldsymbol{0} & \boldsymbol{0} & \boldsymbol{0} & \widetilde{\hat{\boldsymbol{a}}}_{10} \hat{\boldsymbol{a}}_{40} (\lambda_{10} + \lambda_{20} + \lambda_{30}) & \widetilde{\hat{\boldsymbol{a}}}_{10} \hat{\boldsymbol{a}}_{40} (\lambda_{10} + \lambda_{20} + \lambda_{30}) \\ \boldsymbol{0} & \underline{\hat{\boldsymbol{a}}}_{\varepsilon 20}' & \underline{\hat{\boldsymbol{a}}}_{\varepsilon 30}' & \underline{\hat{\boldsymbol{a}}}_{\varepsilon 40}' & \underline{\hat{\boldsymbol{a}}}_{\varepsilon 50}' \end{bmatrix}. \tag{29}$$

The real part of the first-order closure condition from (5) yields

$$\boldsymbol{a}_{10} \, (\lambda_{10} + \lambda_{20} + \lambda_{30}) + \boldsymbol{a}_{40} \, (\lambda_{40} + \lambda_{50} - 1) = \boldsymbol{0}. \tag{30}$$

Without loss of generality the direction vector of the first screw axis is defined as $\boldsymbol{a}_{10} = \begin{bmatrix} 1 & 0 & 0 \end{bmatrix}^{\mathrm{T}}$. From (30) it then follows

$$\lambda_{10} + \lambda_{20} + \lambda_{30} = 0 \quad \text{and} \quad \lambda_{40} + \lambda_{50} = 1. \tag{31}$$

Introducing (31) into the second-order mobility condition (26) yields

$$\underbrace{\begin{bmatrix} \mathbf{0} \\ \widetilde{\boldsymbol{a}}_{10}\,\boldsymbol{a}_{40} \end{bmatrix}^{\mathrm{T}} \Delta \begin{bmatrix} \mathbf{0} & \mathbf{0} & \mathbf{0} & \mathbf{0} & \mathbf{0} \\ \mathbf{0} & \underline{\hat{\boldsymbol{a}}}'_{\varepsilon 20} & \underline{\hat{\boldsymbol{a}}}'_{\varepsilon 30} & \underline{\hat{\boldsymbol{a}}}'_{\varepsilon 40} & \underline{\hat{\boldsymbol{a}}}'_{\varepsilon 50} \end{bmatrix}}_{\mathbf{0}} \boldsymbol{\lambda}_0 = 0. \tag{32}$$

Thus it is shown that condition (23) is sufficient for finite mobility of an overconstrained nH mechanism.

Third special solution Another special solution of the mobility conditions (18) and, by this, a further sufficient but not necessary condition for the finite mobility is

$$\mathbf{0} = \boldsymbol{k}_{\varepsilon 0}^{\mathrm{T}} \begin{bmatrix} \boldsymbol{a}_1 & \ldots & \boldsymbol{a}_{n-1} \end{bmatrix} \quad \text{and} \quad \mathbf{0} = \boldsymbol{k}_0^{\mathrm{T}} \begin{bmatrix} \boldsymbol{a}_{\varepsilon 1} & \ldots & \boldsymbol{a}_{\varepsilon n-1} \end{bmatrix}. \tag{33}$$

The Sarrus mechanism with helical joints fulfills the condition (33), seen in (32).

3 Conclusions

For the synthesis of overconstrained 6H mechanisms sufficient but not necessary conditions for the position of the screw axes are found which guarantee finite mobility. These conditions are special solutions of the necessary conditions which are obtained by the requirement that the higher-order loop-closure conditions are fulfilled. Examples demonstrate some of these sufficient but not necessary conditions.

References

1. Bardavit, H., Duffy, J., Martinez, J.M.R.: Position analysis, via velocity and acceleration, in the Bennett linkage. In: Angeles, J., Zakhariev, E. (eds.) Computational Methods in Mechanisms. NATO Advanced Study Institute, Contributed Papers, vol. **II**, Varna, pp. 13–22 (1997)
2. Bartkowiak, R., Woernle, C.: Numerical synthesis of overconstrained mechanisms based on screw theory. In: Lenarčič, J., Stanišić, M.M. (eds.) Advances in Robot Kinematics, pp. 539–546. Springer, Berlin (2010)
3. Hiller, M., Woernle, C.: A unified representation of spatial displacements. Mech. Mach. Theory **19**, 477–486 (1984)
4. Sugimoto, K.: Existence criteria for over-constrained mechanisms: an extension of motor algebra. J. Mech. Des. **112**(3), 295–299 (1990)
5. Milenkovic, P.: Mobility of single-loop kinematic mechanisms under differential displacement. J. Mech. Des. **132**(4) (2010)
6. Alexandrov, V.: Sufficient condition for the extendibility of an nth order flex of polyhedra. Beiträge Algebra Geom. **39**(2), 367–378 (1998)

Some Rigid-Body Constraint Varieties Generated by Linkages

J.M. Selig

Abstract The set of rigid-body displacements allowed by three simple open-chain linkages are studied. These linkages consist of a cylindrical and spherical joint: the CS dyad, a revolute, a prismatic and a spherical joint: the RPS linkage, two revolutes and a spherical joint: the RRS linkage. Using the Study quadric to represent the group of all rigid-body displacements the constraint varieties for these examples are found. In the case of the CS and RPS linkages these are found to be quartic hypersurfaces while the constraint variety for the RRS linkage is a hypersurface of degree 8. Finally it is shown that all three constraint varieties are linear projections of a Segre variety in $\mathbb{P}^{15}$.

Key words: Rigid-body displacements, open-chain linkages, constraint varieties

1 Introduction

Previous work [6, 7] studied the set of all proper rigid-body displacements which transformed a point in such a way that it remained on a fixed plane or a fixed sphere. These sets of rigid displacements are important in kinematics because they can also be viewed as the possible displacements achievable by an ES or an SS dyad respectively. It was shown that if the space of rigid displacements is represented by the Study quadric, then the constraint manifolds for the point-plane and point-sphere constraints are the intersection of the Study quadric with another quadric hypersurface in $\mathbb{P}^7$.

The present work follows some work of McCarthy and co-workers [4] who investigated the set of displacements achievable by certain open-chain linkages. Some of this work is repeated here using the Study quadric as a model for the group of all rigid-body displacements. In a final section some general remarks are made con-

J.M. Selig
London South Bank University, London SE1 0AA, UK, e-mail: seligjm@lsbu.ac.uk

J. Lenarčič, M. Husty (eds.), *Latest Advances in Robot Kinematics*,
DOI 10.1007/978-94-007-4620-6_37,

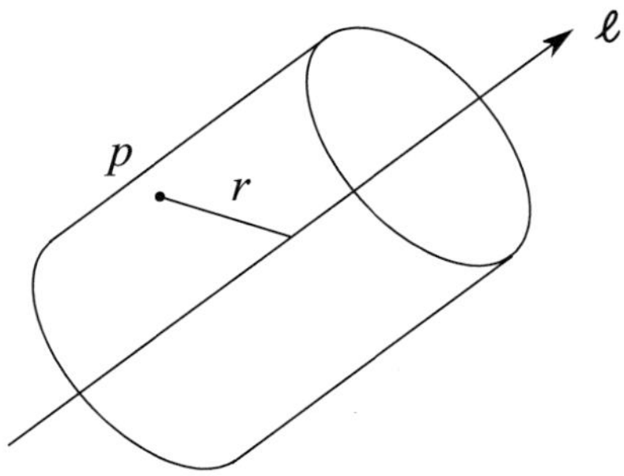

Fig. 1 A point on a cylinder.

cerning the geometry of the constraint varieties produced by the three examples considered below.

It is convenient to use the Clifford algebra $Cl(0,3,1)$ to perform some of the geometric computations. This algebra is particularly well suited to Euclidean geometry and contains the dual quaterions as a subalgebra. Unfortunately limitations on space preclude a review of this material, however a complete introduction to this algebra and its use for geometric computation may be found in [5, chap. 10].

2 The CS Dyad

The first linkage considered is a chain consisting of a cylindrical joint together with a spherical joint. Note that the same set of displacements could be achieved with a revolute joint, a prismatic joint and a spherical joint so long as the axes of the revolute and prismatic joints are parallel.

Another way to look at this set of possible displacements is as the set of group elements that preserve the incidence of a point with a cylinder. The point is the centre of the spherical joint and the axis of the cylinder is the axis of the cylindrical joint, see Fig. 1.

Using the Clifford algebra $Cl(0,3,1)$, it is now easy to find the equation satisfied by the group elements which preserve this incidence. In this algebra a point is given by $p = p_0e_1e_2e_3 + p_1e_2e_3e + p_2e_3e_1e + p_3e_1e_2e$ where the p_i are the projective coordinates of the point and The e_is are the basis elements of the algebra. A general line has the form $\ell = u_{01}e_2e_3 + u_{02}e_3e_1 + u_{03}e_1e_2 + u_{23}e_1e + u_{31}e_2e + u_{12}e_3e$, where the u_{ij} are the Plücker coordinates of the line. In general the square of the distance from a point p to a line ℓ is given by,

$$r^2 = (p \vee \ell)(p \vee \ell)^-/\ell\ell^-.$$

Here $\vee$ is the shuffle product, a derived product in the Clifford algebra and the superscript $()^-$, denotes the Clifford conjugate of an element of the algebra. If the cylinder has axis ℓ and radius r then the required rigid displacements g, will satisfy,

$$(gpg^{-}\vee\ell)(gpg^{-}\vee\ell)^{-}=r^{2}(\ell\ell^{-}).$$

If the point p lies on the cylinder to begin with, then this can be written,

$$(gpg^{-}\vee\ell)(gpg^{-}\vee\ell)^{-}-(p\vee\ell)(p\vee\ell)^{-}(gg^{-})^{2}=0. \tag{1}$$

For a fixed cylinder this equation is a homogeneous quartic in $\mathbb{P}^7$.

To make this a little clearer and to investigate some of the properties of these constraint varieties a particular example is introduced. Assume that the point p is initially located at the origin and the line ℓ is parallel to the z-axis but displaced a distance r in the x-direction. In the Clifford algebra these elements are given by

$$p=e_1e_2e_3 \qquad \text{and} \qquad \ell=e_1e_2-re_2e.$$

A general group element in this algebra has the form,

$$g=a_0+a_1e_2e_3+a_2e_3e_1+a_3e_1e_2+c_0ee_1e_2e_3+c_1e_1e+c_2e_2e+c_3e_3e, \tag{2}$$

where the coefficients a_i and c_i satisfy $a_0c_0+a_1c_1+a_2c_2+a_3c_3=0$, the quadratic equation defining the Study quadric. Using this to transform the point gives,

$$\begin{aligned} gpg^{-} &= (a_0^2+a_1^2+a_2^2+a_3^2)e_1e_2e_3+2(a_0c_1-a_1c_0+a_2c_3-a_3c_2)e_2e_3e+\\ &\quad 2(a_0c_2-a_1c_3-a_2c_0+a_3c_1)e_3e_1e+2(a_0c_3+a_1c_2-a_2c_1-a_3c_0)e_1e_2e \end{aligned}$$

and hence,

$$\begin{aligned} gpg^{-}\vee\ell = {} & -2r(a_0c_2-a_1c_3-a_2c_0+a_3c_1)e-\\ & 2(a_0c_2-a_1c_3-a_2c_0+a_3c_1)e_1+\\ & \big(2(a_0c_1-a_1c_0+a_2c_3-a_3c_2)-r(a_0^2+a_1^2+a_2^2+a_3^2)\big)e_2. \end{aligned}$$

Finally the quartic equation sought is,

$$\begin{aligned} &(gpg^{-}\vee\ell)(gpg^{-}\vee\ell)^{-}-(p\vee\ell)(p\vee\ell)^{-}(gg^{-})^{2}=\\ &\quad 4(a_0c_1-a_1c_0+a_2c_3-a_3c_2)^2+4(a_0c_2-a_1c_3-a_2c_0-a_3c_1)^2-\\ &\quad\quad 4r(a_0^2+a_1^2+a_2^2+a_3^2)(a_0c_1-a_1c_0+a_2c_3-a_3c_2)=0. \end{aligned}$$

Let us write,

$$\begin{aligned} F = {} & (a_0c_1-a_1c_0+a_2c_3-a_3c_2)^2+(a_0c_2-a_1c_3-a_2c_0-a_3c_1)^2-\\ & r(a_0^2+a_1^2+a_2^2+a_3^2)(a_0c_1-a_1c_0+a_2c_3-a_3c_2), \end{aligned}$$

so that the quartic constraint equation is $F=0$. Clearly F vanishes when $a_0=a_1=a_2=a_3=0$. This 3-plane also lies in the Study quadric $a_0c_0+a_1c_0+a_2c_2+a_3c_3=0$ but does not correspond to any rigid-body displacement. This 3-plane of ideal elements will be denoted A_∞. The above shows that A_∞ lies in this constraint

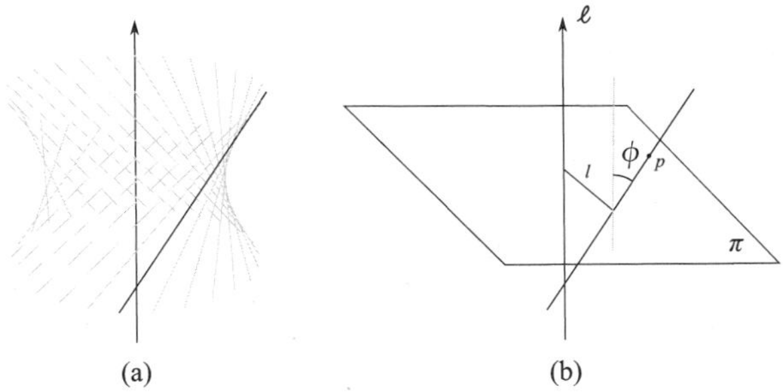

Fig. 2 A point on a hyperboloid.

variety. Moreover, it is easy to see that the partial derivatives $\partial F/\partial a_i$ and $\partial F/\partial c_i$ $(i = 0,1,2,3)$ all vanish on A_∞. So the 3-plane is in fact singular in the quartic.

Notice that this result doesn't depend on the value of r, the radius of the cylinder. Moreover, the corresponding equation for a CS linkage with a different axis can be found by an action of the group of rigid-body displacements. However, the 3-plane A_∞ is invariant under the action of the group and hence this 3-plane will be singular in the constraint variety for any CS linkage.

3 The RPS Linkage

Next we look at the RPS linkage. The group elements generated by such a linkage can be considered as the group elements that constrain the centre of the spherical joint to remain on a cylindrical hyperboloid. This cylindrical hyperboloid is the regulus generated by swinging the prismatic joint about the axis of the revolute, see Fig. 2.

The equation satisfied by the group elements satisfying this constraint can be found in much the same way as in the previous section. In this case we need to consider the distance of the point to the axis of the revolute joint, $(p\vee\ell)(p\vee\ell)^-$ as well as the distance from the point to the plane defined by the common perpendicular between the axes of the revolute and prismatic joints, see Fig. 2. If this plane is labelled π then the square of the perpendicular distance from the point p to the plane is given by the expression, $(p\vee\pi)(p\vee\pi)^-$. The equation of a general hyperbola is $x^2/\alpha^2 - y^2/\beta^2 = 1$, where α and β are constants. Hence the group elements g, which preserve the incidence of the point with a cylindrical hyperboloid will be,

$$\frac{1}{\alpha^2}(gpg^-\vee\ell)(gpg^-\vee\ell)^- - \frac{1}{\beta^2}(gpg^-\vee\pi)(gpg^-\vee\pi)^- = (gg^-)^2.$$

As in the previous section this constraint equation has degree 4 in the components of the group element g.

Suppose that the minimum distance from the axis of the revolute joint ℓ to the point p is l. This design parameter will be referred to as the link-length of the linkage. Clearly this minimum will occur when the point lies on the plane π and hence it is easy to see that $\alpha = l$ in the equation above. By considering a more general point on the hyperboloid, see Fig. 2(b), the constant β can be shown to satisfy $\beta^2 = l^2/\tan^2\phi$, where ϕ is the twist angle between the axes revolute joint and the prismatic joint. Hence in terms of the design parameters l and ϕ the equation for the constraint variety may be written as,

$$\cos^2\phi(gpg^- \vee \ell)(gpg^- \vee \ell)^- - \sin^2\phi(gpg^- \vee \pi)(gpg^- \vee \pi)^- = l^2(gg^-)^2. \quad (3)$$

As a concrete example consider a linkage with a similar design to the CS linkage studied above. The point p can initially be taken to be at the origin, $p = e_1e_2e_3$ and the axis of the revolute joint can be taken as $\ell = e_1e_2 - le_2e$, that is a line parallel to the z-axis but displaced l units in the x-direction. The plane π can be taken as the xy-plane, $\pi = e_3$. Notice that the location of the prismatic joint is not important, only its direction determined by the twist angle ϕ. With these choices many of the computations we need have already been done above, in fact the only new result needed is,

$$gpg^- \vee \pi = -2(a_0c_3 + a_1c_2 - a_2c_1 - a_3c_0).$$

Substituting the results into (3) above and rearranging a little gives,

$$\begin{aligned} 0 =& 4\cos^2\phi(a_0c_2 - a_1c_3 - a_2c_0 + a_3c_1)^2 + \\ & 4\cos^2\phi(a_0c_1 - a_1c_0 + a_2c_3 - a_3c_2)^2 - \\ & 4\sin^2\phi(a_0c_3 + a_1c_2 - a_2c_1 - a_3c_0)^2 - \\ & 4l\cos^2\phi(a_0c_1 - a_1c_0 + a_2c_3 - a_3c_2)(a_0^2 + a_1^2 + a_2^2 + a_3^2) - \\ & l^2\sin^2\phi(a_0^2 + a_1^2 + a_2^2 + a_3^2)^2. \end{aligned}$$

Notice that the quartic variety in $\mathbb{P}^7$ defined by this equation contains the 3-plane A_∞ and moreover is singular on this plane.

4 The RRS Linkage

Finally here consider the RRS linkage. It is well known that a point attached to a pair of revolute joints will trace out a general torus, see [1]. As in [1] the equation of the surface traced out by the point can be found by considering the radial components and z-components of the point p in Fig. 3 and then eliminating the second joint angle θ_2 to give,

$$\left((x^2 + y^2 + z^2) - (l^2 + d^2 + r^2)\right)^2 = 4l^2\left(r^2 - \left(\frac{z - d\cos\phi}{\sin\phi}\right)^2\right).$$

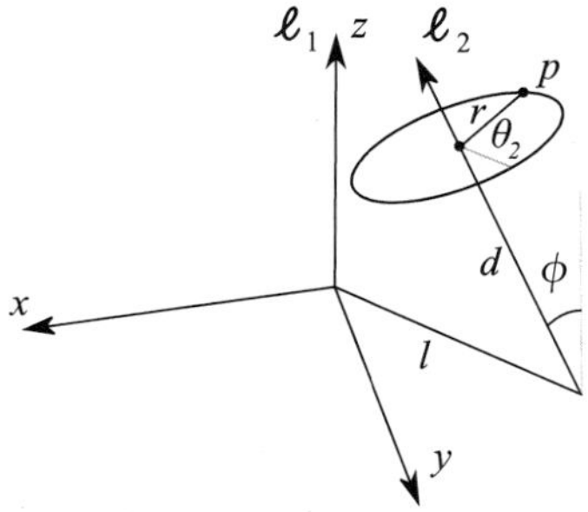

Fig. 3 The RRS linkage.

The design parameters l, d, r and ϕ are as shown in Fig. 3 and can be thought of as the link-length, offset, second radius and twist-angle respectively.

To find the equation in $\mathbb{P}^7$ satisfied by the group elements which preserve the incidence of the point with the torus we look again at the radial and z-components of p. The radial component of p is the distance from the first joint axis ℓ_1 to p. For a general point the square of this distance will be $(gpg^- \vee \ell_1)(gpg^- \vee \ell_1)^-$. For the coordinates of Fig. 3, this expression can be used to replace $x^2 + y^2$. The z-component of p is the distance to p from the plane containing the perpendicular common to ℓ_1 and ℓ_2. This distance can be expressed as $gpg^- \vee \pi$ which can be used to replace z in the equation above. Multiplying terms by the factor gg^- to produce a homogeneous equation gives,

$$\left((gpg^- \vee \ell_1)(gpg^- \vee \ell_1)^- + (gpg^- \vee \pi)(gpg^- \vee \pi)^- - (l^2 + d^2 + r^2)(gg^-)^2\right)^2 = 4(gg^-)^2 l^2 \left((gg^-)^2 r^2 - \left(\frac{(gpg^- \vee \pi) - d(gg^-)\cos\phi}{\sin\phi}\right)^2\right). \quad (4)$$

Clearly this equation has degree 8 in the components of the group elements g.

In [1] Fichter and Hunt look for circles in the general torus described above to find all possible mobile *RRSR* mechanisms. A revolute joint produces a one-parameter family of group elements which lie on a line in the Study quadric. So to find all mobile RRSR mechanisms we could look for lines in the degree 8 hypersurface given by equation (4). Actually, we only need to look for lines of the form $g(\lambda) = 1 + \lambda\ell$ where ℓ is a line—the axis of the final revolute joint. Substituting this form into the equation for the constraint variety will produce a degree 8 equation in the parameter λ. For the line to lie in the constraint variety the coefficients of each power of λ must vanish. Hence we can produce conditions for an RRSR mechanism to be mobile. In principle, this strategy for finding mobile mechanisms can be extended to other linkages, however the analysis of the conditions produced may not be straightforward.

5 The Segre Variety $\mathbb{P}^1 \times \mathbb{P}^1 \times \mathbb{P}^3$

The three examples studied above, together with a couple of other examples not considered here, are united by the fact that their constraint varieties are projections of the Segre variety $\mathbb{P}^1 \times \mathbb{P}^1 \times \mathbb{P}^3$.

To see this first consider the Segre variety $\mathbb{P}^1 \times \mathbb{P}^1 \times \mathbb{P}^3$. A point, with homogeneous coordinates, $(\alpha_0 : \alpha_1) \times (\beta_0 : \beta_2) \times (\gamma_0 : \gamma_1 : \gamma_2 : \gamma_3)$ can be mapped to a point $(X_{000} : X_{100} : X_{010} : \cdots : X_{113})$ in $\mathbb{P}^{15}$. This is the Segre map defined by $X_{ijk} = \alpha_i \beta_j \gamma_k$. That is $X_{000} = \alpha_0\beta_0\gamma_0$, $X_{100} = \alpha_1\beta_0\gamma_0$ and so forth. The image of this map is known as the Segre variety. Using the techniques described in [2] it is possible to show that this variety has degree 20. It is also easy to see that it lies on several quadric hypersurfaces in $\mathbb{P}^{15}$. There are 12 quadrics of the form $X_{ijk}X_{ilm} - X_{ijm}X_{ilk} = 0$ and another 12 of the form $X_{ijk}X_{ljm} - X_{ijm}X_{ljk} = 0$. Then there are 6 each of the forms $X_{00k}X_{11l} - X_{10l}X_{01k} = 0$, $X_{01k}X_{10l} - X_{00l}X_{11k} = 0$ and $X_{00k}X_{11l} - X_{01k}X_{10l} = 0$. Finally there are 4 of the form $X_{00k}X_{11k} - X_{10k}X_{01k} = 0$. This makes 46 linearly independent quadrics.

To see the connection with the constraint varieties discussed above we will look at just one example. Consider the CS dyad from Section 2, in particular the example given at the end of the section. The constraint variety can be parameterised as a product of Clifford algebra elements,

$$g = (\alpha_0 + \alpha_1(e_1e_2 - re_2e))(\beta_0 + \beta_1 e_3 e)(\gamma_0 + \gamma_1 e_2e_3 + \gamma_2 e_3e_1 + \gamma_3 e_1e_2).$$

Multiplying out the above product gives an element of the form given in (2) with components,

$$\begin{aligned}
a_0 &= \alpha_0\beta_0\gamma_0 - \alpha_1\beta_0\gamma_3, & c_0 &= -\alpha_0\beta_1\gamma_3 - \alpha_1\beta_1\gamma_0 + r\alpha_1\beta_0\gamma_2,\\
a_1 &= \alpha_0\beta_0\gamma_1 - \alpha_1\beta_0\gamma_2, & c_1 &= -\alpha_0\beta_1\gamma_2 - \alpha_1\beta_1\gamma_1 - r\alpha_1\beta_0\gamma_3,\\
a_2 &= \alpha_0\beta_0\gamma_2 + \alpha_1\beta_0\gamma_1, & c_2 &= \alpha_0\beta_1\gamma_1 - \alpha_1\beta_1\gamma_2 - r\alpha_1\beta_0\gamma_0,\\
a_3 &= \alpha_0\beta_0\gamma_3 + \alpha_1\beta_0\gamma_0, & c_3 &= \alpha_0\beta_1\gamma_0 - \alpha_1\beta_1\gamma_3 + r\alpha_1\beta_0\gamma_1.
\end{aligned}$$

Now suppose we choose new coordinates for $\mathbb{P}^{15}$, label these new coordinates $\bar{a}_0, \ldots, \bar{a}_3, \bar{b}_0, \ldots, \bar{b}_3, \bar{c}_0, \ldots, \bar{c}_3$ and set,

$$\bar{a}_0 = X_{000} - X_{103}, \qquad \bar{c}_0 = -X_{013} - X_{110} + rX_{102},$$

and so forth. The coordinates $\bar{b}_i$ can be chosen so that the coordinate transformation is non-singular. Now it is clear that mapping $\bar{a}_i \mapsto a_i$ and $\bar{c}_i \mapsto c_i$ maps the Segre variety to the constraint variety for the CS linkage. Geometrically this is a linear projection from $\mathbb{P}^{15}$ to $\mathbb{P}^7$, the centre of the projection is given by the 7-plane which is the intersection of the eight hyperplanes $\bar{a}_i = 0$, $\bar{c}_i = 0$. It is not difficult to see that similar constructions can be given for the other examples discussed above.

Note that is should be possible to compute the degree of the constraint variety from the degree of the Segre variety and a knowledge of how the Segre variety meets the centre of the projection, see [2].

6 Conclusions

In the space available it has only been possible to outline briefly the geometry of these constraint varieties and hint at some possible applications for this approach.

This work should be viewed as part of a general programme to investigate geometric constraint varieties. These are the possible rigid displacements allowed by some geometric problem. The rigid displacements allowed by simple linkages are of most relevance to Kinematics and Robotics. It is known that a revolute joint produces a line in the Study quadric, a general RR dyad generates the intersection of the Study quadric with a 3-plane. An RRR linkage gives a $\mathbb{P}^1 \times \mathbb{P}^1 \times \mathbb{P}^1$ Segre variety, [3]. Some quadratic constraint varieties were studied in [7].

The ubiquity of Segre manifolds is now clear and connected to the parameterisation of these varieties as products of subgroups. The constraint variety of a general *RR* dyad could be thought of as the Segre variety $\mathbb{P}^1 \times \mathbb{P}^1$. Moreover, it can be predicted with some confidence that the constraint variety formed by the displacements allowed by a general RRRR linkage will be the projection of a $\mathbb{P}^1 \times \mathbb{P}^1 \times \mathbb{P}^1 \times \mathbb{P}^1$ Segre variety.

By intersecting several of these constraint varieties we can study the properties of mechanisms formed by joining the corresponding linkages in parallel. Except in the case of point-plane constraints, this problem has not been investigated to any great extent.

McCarthy and co-workers use equations such as those found above to design mechanisms, see [4]. It would be useful to formalize this problem by embedding the space of all possible linkages of some type, in an algebraic variety. For example, the CS dyads could be thought of as (an open set in) $Q_K \times \mathbb{P}^3$, the product of the Klein quadric of lines in $\mathbb{P}^3$ with the space of points. This turns design synthesis into a problem in Algebraic geometry. Unfortunately these problems are still highly non-trivial.

References

1. Fichter, E.F., Hunt, K.H.: The fecund torus, its bitangent-circles and derived linkages. Mech. Mach. Theory **10**(2–3), 167–176 (1975). doi:10.1016/0094-114X(75)90017-8
2. Harris, J.: Algebraic Geometry. A First Course. Springer Verlag, New York (1992)
3. Husty, M.L., Pfurner, M., Schröcker, H.-P.: A new and efficient algorithm for the inverse kinematics of a general serial 6R manipulator. Mech. Mach. Theory **42**(1), 66–81 (2007). doi:10.1016/j.mechmachtheory.2006.02.001
4. McCarthy, J.M., Soh, G.S.: Geometric Design of Linkages, 2nd edn. Springer Verlag, New York (2011)
5. Selig, J.M.: Geometric Fundamentals of Robotics. Springer Verlag, New York (2005)
6. Selig, J.M.: On the geometry of point-plane constraints on rigid-body displacements. Acta Appl. Math. **116**(2), 133–155 (2011). doi:10.1007/s10440-011-9634-6
7. Selig, J.M.: Quadratic constraints on rigid-body displacements. ASME J. Mech. Robot. **2**(4), 041009 (2010). doi:10.1115/1.4002344

Constraint-Consistent Analysis of Muscle Force Contributions to Human Gait

Emel Demircan and Oussama Khatib

Abstract The goal of this study is to apply a task-space approach to characterize muscle force contributions to the body center of mass during human gait taking into account the contacts with the environment and the constraints in the musculoskeletal system. Motion capture, electromyography and force plate data were taken from a male subject walking at free speed. The obtained data were used together with a full-body musculoskeletal model to generate and to analyze the simulation of one complete gait cycle. The contribution of the muscles spanning the lower body joints to the body center of mass acceleration were calculated using a task-space approach which was successfully applied to analyze human dynamic motions in our previous studies. The results showed that gluteus medius, vasti, biceps femoris long head and short head, tibialis anterior, medial gastrocnemius, rectus femoris and soleus were the primary contributors to gait at free speed. The study provides an approach for in depth motion analysis including the effects of contact forces and joint mechanics as well as physiological constraints, muscle dynamics and actuation.

Key words: Human locomotion, operational space accelerations, muscle function, musculoskeletal modeling

1 Introduction

Understanding the control mechanisms involved in human movement presents a daunting challenge to the biomechanics and neuroscience communities. In the presence of this challenge there is a significant motivation to predict and emulate injury-free and high-performing human movement [2, 4, 8, 15]. In robotics research, similar challenges arise to improve the dynamic performance of multi-degree-of freedom

Emel Demircan · Oussama Khatib
Artificial Intelligence Laboratory, Stanford University, Stanford, CA 94305, USA,
e-mail: {emeld, khatib}@cs.stanford.edu

J. Lenarčič, M. Husty (eds.), *Latest Advances in Robot Kinematics*,
DOI 10.1007/978-94-007-4620-6_38,

manipulators without compromising for safety [16]. In both areas, optimal performance of a skill (or a task) is affected by the kinematic constraints of the system as well as by the torque generating capacities, or the strength of the actuators. Over the last decade, our efforts have concentrated on the synthesis and analysis of human motion using efficient techniques in robotics research [6, 7, 11].

Investigating human motion dynamics by determining how the muscles contribute to body center of mass accelerations in dynamic skills can help predict injury-causing motion patterns or to synthesize subject-specific optimal movement. Several studies have examined how muscles provide support and progression [12, 13] to overground and treadmill walking at several walking speeds. However, the characterization of human gait taking into account the effect of contacts (i.e. foot strike) and constraints (i.e. muscle force generating capacity) in subject's task space hasn't been studied. In this paper we address this problem and present a robotics method for the constraint-consistent analysis of human gait using simulated musculoskeletal model and data from motion capture, electromyography and force plates.

2 Constraint-Consistent Motion Analysis

2.1 Task-Space Formulation

In robotics research, the *Operational Space Formulation* [9] was introduced to address the dynamic interaction between a robot's task-space motion and force. To characterize the additional task redundancy, the operational space formulation defines a dynamically consistent task null space. Multiple operational tasks can be controlled if they are combined into a single task vector and additional criteria can be controlled within the task-consistent null-space. A task can be defined to be any formal description of desired activity that can be explicitly represented as a function of the joint coordinates, q, $\dot{q}$ and $\ddot{q}$. Multiple tasks can be combined into a single task definition in a higher dimensional space, as long as they are kinematically consistent with each other. The full task is represented as the $m \times 1$ vector, $x_t = x_t(q)$, formed by vertically concatenating the coordinates of the operational points. The Jacobian matrix associated with the task, x_t, is denoted by J_t. The joint space equations of motion can be expressed as,

$$A(q)\ddot{q} + b(q,\dot{q}) + g(q) + J_{ext}^T F_{ext} = \Gamma, \tag{1}$$

where q is the vector of n joint coordinates, $A(q)$ is the $n \times n$ kinetic energy matrix, $\mathrm{b}(q,\dot{q})$ is the $n \times 1$ vector of centrifugal and Coriolis joint forces, $g(q)$ is the $n \times 1$ vector of gravity, and Γ is the $n \times 1$ vector of generalized joint forces (torques). In the presence of external forces in the system, the associated Jacobian and reaction force vector are J_{ext} and F_{ext}, respectively. To simplify notation, we will often refrain from explicitly denoting the functional dependence of these quantities on q and $\dot{q}$.

The task dynamic behavior can be obtained by projecting the system dynamics (1) into the space associated with the task, using the generalized inverse of the Jacobian, J_t. This generalized inverse of the Jacobian has been showed to be unique and dynamically consistent [9, 10] and given by,

$$\bar{J}_t = A^{-1} J_t^T (J_t A^{-1} J_t^T)^{-1}, \quad (2)$$

The dynamic behavior associated with the task, x_t can be obtained by

$$\bar{J}_t^T (A\ddot{q} + b + g + \Gamma_{ext} = \Gamma) \Rightarrow \Lambda_t \ddot{x}_t + \mu_t + p_t + R_t = F_t. \quad (3)$$

In the operational space, Λ_t is the $m \times m$ kinetic energy matrix associated with the task, and μ_t, p_t, R_t and F_t are, respectively, the centrifugal and Coriolis force vector, gravity vector, reaction force vector and generalized force acting along the direction of the task, x_t. This process provides a description of the dynamics in task coordinates rather than joint space coordinates (while joint space coordinates are still present in (3), the inertial term involves task space accelerations rather than joint space accelerations). The control framework defined in terms of the relevant task coordinates, x_t can be represented using a relevant operational space force, F_t acting along the same direction. The forces acting along given task coordinates can be mapped to a joint torque, Γ_{task} by the relationship,

$$\Gamma_{task} = J_t^T F_t. \quad (4)$$

This model can be applied to analyze the acceleration characteristics in human dynamic skills shaped by the skeletal mechanics as well as the physiological parameters. For a human musculoskeletal system of n degrees of freedom and r muscles, a set of muscle forces, m, arises based on muscle activations, as well as the skeletal configuration, q and $\dot{q}$. These muscle forces are related to the joint torques, Γ, through the $r \times n$ muscle Jacobian matrix, L:

$$\Gamma = L^T m, \quad (5)$$

where m is the vector of net muscle forces (active and passive components) and Γ is the $n \times 1$ vector of muscle induced joint torques required to produce the desired motion. In presence of contacts with the environment (i.e. foot strike), the contact force vectors and the corresponding Jacobian matrices should be included in the equations of motion of the system. Thus, the muscle induced joint torque/operational space acceleration relationship can be given as,

$$\ddot{x} = J(q)A(q)^{-1}(\Gamma - g(q) - J_{c_1}^T F_{ext_1} - J_{c_2}^T F_{ext_2}). \quad (6)$$

where $J(q)$, $A(q)$ and $g(q)$ are respectively the Jacobian matrix, joint space kinetic energy matrix and the joint space gravity torque vector. F_{ext_1} and F_{ext_2} capture the external forces/moments in the system at two different contact points c_1 and c_2 with the corresponding Jacobian matrices $J_{c_1}^T$ and $J_{c_2}^T$.

Table 1 Functional groups of lower body muscles used in gait analysis.

Hip Adduction	Hip Abduction	Hip Flexion	Hip Extension
Biceps Femoris L. Head Gracilis	Gluteus Maximus Gluteus Medius Sartorius Tensor Fasciae Latae	Gluteus Medius Gracilis Iliacus Psoas Rectus Femoris Sartorius Tensor Fasciae Latae	Biceps Femoris L. Head Gluteus Maximus Gluteus Medius
Hip Intrarotation	**Hip Extrarotation**	**Knee Flexion**	**Knee Extension**
Gluteus Medius Iliacus Psoas Tensor Fasciae Latae	Gluteus Medius	Biceps Femoris L. Head Biceps Femoris S. Head Gracilis Medial Gastrocnemius Sartorius	Rectus Femoris Vastus Intermedius
Ankle Plantar Flexion	**Ankle Dorsi Flexion**		
Medial Gastrocnemius Soleus	Tibialis Anterior		

The feasible set of operational space accelerations can be determined using (6) given the bounds on the muscle induced joint torque capacities by,

$$0 < \Gamma < L^T m_{max}. \tag{7}$$

where m_{max} is the vector of muscle force generating capacities.

In order to evaluate the actual muscle contributions to gait, we can scale the force generating capacities of each individual muscle by its activation pattern obtained from the electromyography data and from the simulation results. The actual resulting torque values can be obtained through the relation,

$$\Gamma = L^T m_{max} a. \tag{8}$$

where a represents the activation level of the muscles of interest. In consequence, Equation 6 becomes,

$$\ddot{x} = J(q)A(q)^{-1}(L^T m_{max} a - g(q) - J_{c_1}^T F_{ext_1} - J_{c_2}^T F_{ext_2}). \tag{9}$$

To evaluate the contributions of muscle forces to the center of mass acceleration in operational space, we grouped the muscles according to their primary function (Table 1). These muscle groups were determined based on the their activity measured by the electromyography channels during gait. The torque generating capacities of 28 muscles spanning the right and left hip, knee and ankle joints were mapped into the operational space accelerations of the body center of mass.

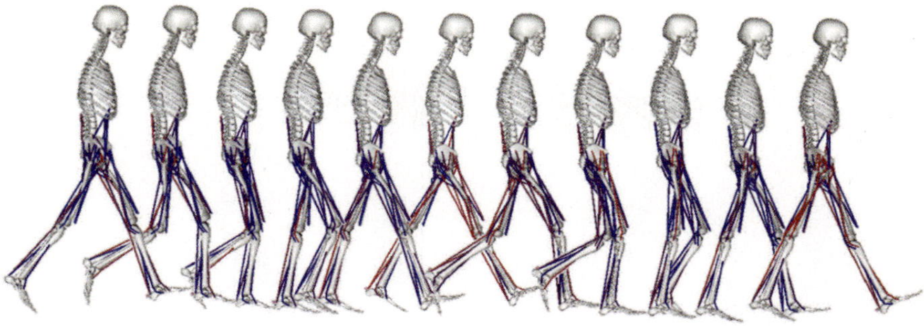

Fig. 1 Three-dimensional dynamic simulation of normal walking at free speed. The lower extremity and the back joint are actuated by 54 musculo-tendon actuators. Muscle color indicates simulated activation level from fully activated (red) to fully deactivated (blue).

2.2 Experiments and Musculoskeletal Simulations

Motion analysis data were collected for one male subject walking at 1.75 m/s. Three-dimensional marker trajectories, force plate, and electromyography data were recorded simultaneously using an 8-camera Vicon system (OMG plc, Oxford UK). A static trial was performed on the subject to assist scaling the musculoskeletal model (with markers attached to the medial and lateral femoral epicondyles and medial and lateral malleoli). Three-marker clusters were placed on the subject's feet, thigh, and shank for tracking purposes [3]. Four markers were placed on both the pelvis (anterior and posterior superior iliac spines) and torso (acromion processes, seventh cervical spine, and sternal notch). The ground reaction forces were sampled at 1200 Hz and low-pass filtered at 20 Hz. The muscle activity was recorded at 1200 Hz. The raw electromyography (EMG) data were processed by high-pass filtering at 30 Hz to remove motion artifacts, full-wave rectifying and then low-pass filtering at 6 Hz to generate a linear envelope. These data were then normalized to a maximum isometric muscle contraction measured from the subject.

One complete gait cycle was simulated for a subject walking at free speed (1.75 m/s) (Figure 1). A generic musculoskeletal model with 23 degrees of freedom, actuated by 54 muscle-tendon compartments [5], was scaled to match subject's anthropometry based on experimentally measured markers placed on anatomical landmarks. The hip was modeled as a ball-and-socket joint (3 dofs), the knee was modeled as a custom joint with 1 dof [14], and the foot and ankle were modeled as a custom joint with 2 dofs (dorsiflexion and plantar flexion at the ankle joint; eversion and inversion at the tarsal joint). Lumbar motion was modeled as a ball-and-socket joint (3 dofs) [1]. Muscle parameters and path geometries in the model were based on data reported by [5].

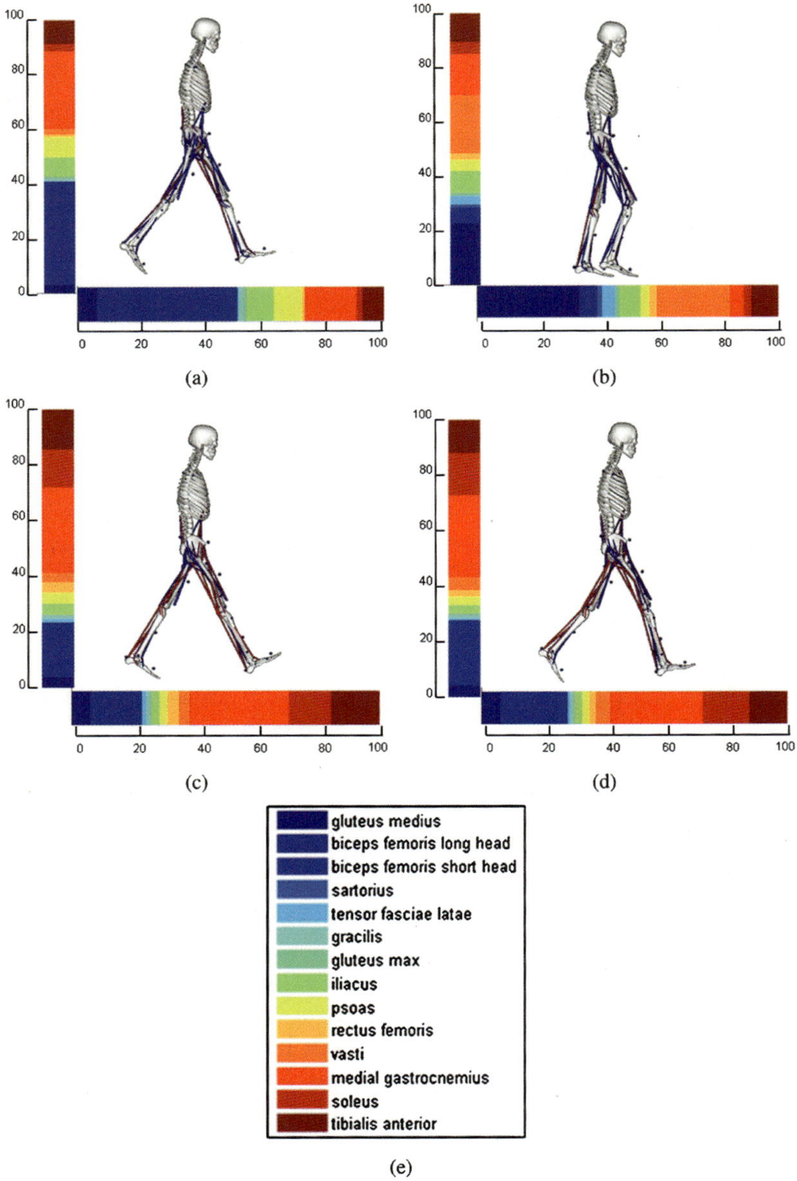

Fig. 2 Muscle force contributions to accelerations of body center of mass during gait at free speed. The vertical and horizontal color bars illustrate individual muscle contribution (%) to vertical and horizontal acceleration of the center of mass, respectively. Contributions of the muscles listed in (e) are given for four configurations: (a) heel strike, (b) early stance, (c) late stance, (d) toe-off. Gluteus medius, vasti, biceps femoris long head and short head, tibialis anterior, medial gastrocnemius, rectus femoris and soleus were the primary contributors to gait at free speed. During early stance, biceps femoris and vasti highly contributed to horizonal acceleration of center of mass while gluteus medius, biceps femoris, medial gastrocnemius, tibialis anterior an vasti contributed to support. During late stance, medial gastrocnemius and soleus were the main contributors to progression while tibialis anterior and soleus contributed to vertical acceleration of the center of mass.

3 Results

The three-dimensional dynamic simulation provided the joint kinematics, muscle dynamics and actuation. The operational space and contact-consistent task-space frameworks provided the effects of contacts and constraints to the resulting motion dynamics. The contributions of 28 muscles to the center of body accelerations during gait at free speed were calculated (Figure 2). The results showed that gluteus medius, vasti, biceps femoris (long head and short head), tibialis anterior, medial gastrocnemius, rectus femoris and soleus were the primary contributors to gait in both progression and support.

Muscles which provided vertical support also resisted progression in early stance (i.e. vasti) and assisted progression in late stance (i.e. soleus). During early stance, biceps femoris and vasti highly contributed to horizonal acceleration of center of mass while gluteus medius, biceps femoris, medial gastrocnemius, tibialis anterior an vasti contributed to support. During late stance, medial gastrocnemius and soleus were the main contributors to progression while tibialis anterior and soleus contributed to vertical acceleration of the center of mass.

4 Conclusions

We presented a robotics approach to analyze human dynamics in the context of the desired task and the contacts in the environment as well as the physiological constraints, muscle dynamics and actuation. Using this approach together with motion capture, force plate and EMG data we simulated a musculoskeletal model to characterize human gait. Three-dimensional muscle-actuated dynamic simulation of one complete gait cycle was created, the characteristics of the relationship governing the transmission of muscle forces to the body center of mass operational space accelerations were analyzed. The dynamic simulation provided the kinematics as well as muscle dynamics and actuation. The constraint-consistent task-space framework provided the effects of constraints and contacts in the environment. Individual muscle contributions to body center of mass acceleration during gait at free speed confirmed with the values reported in the literature [12]. The task-space analysis approach presented in this paper provides a comprehensive analysis of human motion dynamics including the effects of constraints and contacts.

Acknowledgements The financial support of Honda Company is gratefully acknowledged. The authors would like to thank Thor Besier and Jason Wheeler for their assistance with the motion capture experiments.

References

1. Anderson, F., Pandy, M.: A dynamic optimization solution for vertical jumping in three dimensions. Comput. Methods Biomech. Biomed. Eng. **2**(3), 201–231 (2007)
2. Besier, T., Loyd, D., Ackland, T., Cochrane, J.: Anticipatory effects on knee joint loading during running and cutting maneuvers. Med. Sci. Sports Exerc. **22**(7), 1176–1181 (2001)
3. Besier, T., Sturnieks, D., Alderson, J., Lloyd, D.: Repeatability of gait data using a functional hip joint centre and a mean helical knee axis. J. Biomech. **36**, 1159–1168 (2003)
4. Chaudhari, A., Hearn, B., Andriacchi, T.: Sport-dependent variations in arm position during single-limb landing influence knee loading - implications for anterior cruciate ligament injury. Am. J. Sports Med. **33**(6), 824–830 (2005)
5. Delp, S., Loan, P., Hoy, M., Zajac, F., Topp, E., Rosen, J.: An interactive graphics-based model of the lower extremity to study orthopaedic surgical procedures. IEEE Trans. Biomed. Eng. **37**(8), 757–767 (1990)
6. Demircan, E., Besier, T.F., Khatib, O.: Muscle force transmission to operational space accelerations during elite golf swings. In: Proc. of the IEEE International Conference on Robotics and Automation. St Paul, MN, USA (2012)
7. Demircan, E., Sentis, L., Sapio, V.D., Khatib, O.: Human motion reconstruction by direct control of marker trajectories. In: Proc. of the Eleventh International Symposium Advances in Robot Kinematics, pp. 263–272. Batz-sur-Mer, France (2008)
8. Fortenbaugh, D., Fleisig, G., Andrews, J.: Baseball pitching biomechanics in relation to injury risk and performance. Sports Health Multidiscip. Appr. **1**, 314–320 (2009)
9. Khatib, O.: A unified approach for motion and force control of robot manipulators: The operational space formulation. Int. J. Robot. Autom. **3**(1), 43–53 (1987)
10. Khatib, O.: Inertial properties in robotics manipulation: An object-level framework. Int. J. Robot. Res. **14**(1), 19–36 (1995)
11. Khatib, O., Demircan, E., DeSapio, V., Sentis, L., Besier, T., Delp, S.: Robotics-based synthesis of human motion. J. Physiol. (Paris) **103**, 211–219 (2009)
12. Liu, M.Q., Anderson, F.C., Schwartz, M.H., Delp, S.L.: Muscle contributions to support and progression over a range of walking speeds. J. Biomech. **41**, 3243–3252 (2008)
13. Neptune, R.R., Sasaki, K., Kautz, S.A.: The effect of walking speed on muscle function and mechanical energetics. Gait Posture **28**, 135–143 (2008)
14. Seth, A., Sherman, M., Eastman, P., Delp, S.: Minimal formulation of joint motion for biomechanisms. Nonlinear Dyn. (2010)
15. Steele, J.: Factors affecting performance in netball – Implications for improving performance and injury reduction. Sports Med. **10**(2), 88–102 (2001)
16. Zinn, M., Khatib, O., Roth, B., Salisbury, J.K.: A new actuation approach for human friendly robot design. Int. J. Robot. Res. **23**, 379–398 (2004)

Inherently Balanced 4R Four-Bar Based Linkages

V. van der Wijk and J.L. Herder

Abstract Synthesis of mechanisms with their center of mass (CoM) at an invariant point on one of the elements is useful for the design of statically balanced and shaking-force balanced mechanisms and manipulators. For this purpose, a kinematic architecture based on a general 4R four-bar linkage is found by applying the method of principal vectors as a linkage together with a similar four-bar linkage. The balance conditions are obtained for an arbitrary mass distribution of each of the elements and a balanced grasper mechanism and a balanced two-degree-of-freedom manipulator are derived as practical examples.

Key words: Center of mass, four-bar linkage, shaking-force balancing, static balancing

1 Introduction

When the center of mass (CoM) of a mechanism (i.e. manipulator, robot) is at a stationary point with respect to the base, the mechanism is shaking-force balanced. This means that for all motion of the mechanism the resultant dynamic forces on the base are zero [7]. Shaking-force balance therefore is important for high speed mechanisms with minimal vibrations of the base. A mechanism with a stationary CoM is also statically balanced with respect to gravity. Then a mechanism can be maintained in each posture with minimal effort [3].

The CoM of a mechanism is stationary if it is an invariant point on at least one of the elements with this point or element being (part of) the base. An elementary way to describe the CoM with respect to its elements is with the method of principal vectors [1]. This method has been applied to derive such inherently balanced linkages considering general mass distributions of all elements [5, 6].

V. van der Wijk · J.L. Herder
University of Twente, Enschede, The Netherlands, e-mail: {v.vanderwijk,j.l.herder}@utwente.nl

J. Lenarčič, M. Husty (eds.), *Latest Advances in Robot Kinematics*,
DOI 10.1007/978-94-007-4620-6_39,

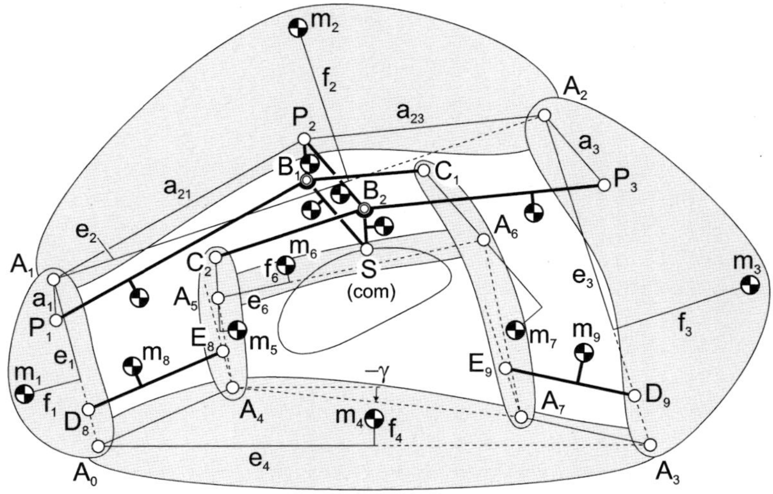

Fig. 1 Center of mass of four-bar linkage $A_0A_1A_2A_3$, similar linkage $A_4A_5A_6A_7$, the principal vector linkage $A_1P_1B_1SB_2P_3A_2P_2$, and links D_8E_8 and D_9E_9 is at invariant point S on links A_5A_6, B_1S, B_2S.

It was shown in [2, 8] that the CoM of a general 4R four-bar linkage also is an invariant point on the coupler link of a similar four-bar linkage moving synchronously. In addition, with the double contour method [4] similar linkages are found for the analysis of the CoM of more complex linkages. This method is based on principal vectors.

The goal of this paper is to combine and to apply the three mentioned approaches as linkages to obtain an inherently balanced 4R four-bar based kinematic architecture from which a wide variety of balanced mechanisms can be derived. First the kinematic architecture is found and subsequently its force balance conditions are derived.

2 Kinematic Architecture with CoM at Invariant Link Point

Figure 1 shows a kinematic architecture of which the CoM of all elements is at invariant link point S. The architecture is based on a general 4R four-bar linkage $A_0A_1A_2A_3$. S is a point on the coupler link A_5A_6 of a similar four-bar linkage $A_4A_5A_6A_7$ [2, 8] and is also a point on a linkage of parallelograms $A_1P_1B_1P_2$, $P_2B_1SB_2$, and $A_2P_2B_2P_3$ of which the (principal) dimensions a_1, a_{21}, a_{23}, and a_3 are defined by the principal points P_i [1, 6]. In addition to their coinciding joint at S, these two linkages can be linked by parallelograms $SA_6C_1B_1$ and $SB_2C_2A_5$ of which B_1C_1 and B_2C_2 are part of elements $P_1B_1C_1$ and $P_3B_2C_2$, respectively. These parallelograms are found with the double contour method of which one solution could be

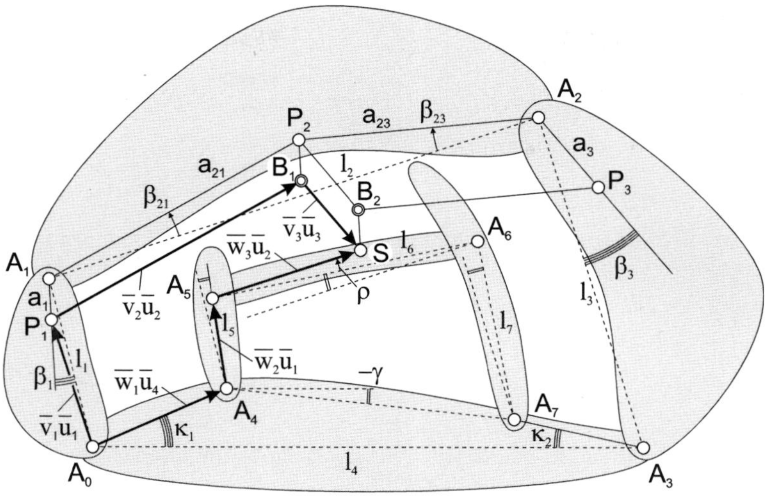

Fig. 2 By describing the CoM of all elements S along the principal vectors $\overline{v}_i$, the conditions $\overline{w}_i$ for which S is a coupler point of the similar linkage are found.

the linkage $P_1C_1A_6A_7A_4A_5$ [4]. Other possible links are D_8E_8 and D_9E_9 which are parallel to lines A_0A_4 and A_3A_7, respectively.

The conditions for which the CoM of all elements S is a coupler point of similar linkage $A_4A_5A_6A_7$ can be written as a function of the principal dimensions. To obtain the conditions, the position of S can be written with complex vectors as illustrated in Fig. 2 with

$$\overline{A_0S} = \overline{v}_1\overline{u}_1 + \overline{v}_2\overline{u}_2 + \overline{v}_3\overline{u}_3 = \overline{w}_1\overline{u}_4 + \overline{w}_2\overline{u}_1 + \overline{w}_3\overline{u}_2 \tag{1}$$

Vectors $\overline{u}_i$ are the time dependent vectors describing the relative positions of joints A_0, A_1, A_2, and A_3. Constant vectors $\overline{v}_i$ are the principal vectors describing the principal points P_i within each element. Vectors $\overline{w}_i$ are also constant and determine the size and pose of the similar linkage. These vectors can be written as

$$\overline{v}_1 = \frac{l_1 - a_1\cos\beta_1}{l_1} + \frac{a_1\sin\beta_1}{l_1}i \quad \overline{v}_2 = \frac{a_{21}\cos\beta_{21}}{l_2} + \frac{a_{21}\sin\beta_{21}}{l_2}i \quad \overline{v}_3 = \frac{a_3\cos\beta_3}{l_3} + \frac{a_3\sin\beta_3}{l_3}i$$
$$\overline{w}_1 = \kappa_1^R + \kappa_1^I i \qquad \overline{w}_2 = \gamma^R + \gamma^I i \qquad \overline{w}_3 = \overline{w}_2(\rho^R + \rho^I i)$$

with link lengths l_i and angles β_{ij} to describe the orientation of the principal dimensions a_{ij} with respect to the line connecting the joints. Vectors $\overline{w}_i$ are written with the real and imaginary parts of the orientation κ_1 of A_0A_4, the orientation γ of the similar linkage and the orientation ρ of A_5S with respect to A_5A_6. With the substitution of the loop equation $\overline{u}_1 + \overline{u}_2 + \overline{u}_3 = \overline{u}_4$ for $\overline{u}_4$, Eq. 1 can be rewritten as

$$(\overline{v}_1 - \overline{w}_1 - \overline{w}_2)\overline{u}_1 + (\overline{v}_2 - \overline{w}_1 - \overline{w}_3)\overline{u}_2 + (\overline{v}_3 - \overline{w}_1)\overline{u}_3 = 0 \tag{2}$$

and after substitution of the constant vectors it is written as

$$\begin{aligned}&\{(1-\frac{a_1\cos\beta_1}{l_1}-\kappa_1^R-\gamma^R)+(\frac{a_1\sin\beta_1}{l_1}-\kappa_1^I-\gamma^I)i\}\overline{u}_1+\\ &\{(\frac{a_{21}\cos\beta_{21}}{l_2}-\kappa_1^R-\gamma^R\rho^R+\gamma^I\rho^I)+(\frac{a_{21}\sin\beta_{21}}{l_2}-\kappa_1^I-\gamma^R\rho^I-\gamma^I\rho^R)i\}\overline{u}_2+\\ &\{(\frac{a_3\cos\beta_3}{l_3}-\kappa_1^R)+(\frac{a_3\sin\beta_3}{l_3}-\kappa_1^I)i\}\overline{u}_3=0\end{aligned} \tag{3}$$

Since generally this equation must hold for all motion, i.e. for all independent values of $\overline{u}_i$ not being restricted to the relative motions of the 4R four-bar linkage, each of the six terms needs to be zero. The terms for $\overline{u}_3$ are zero when

$$\kappa_1^R=\frac{a_3\cos\beta_3}{l_3},\quad \kappa_1^I=\frac{a_3\sin\beta_3}{l_3},\quad \kappa_1=\tan^{-1}\left(\frac{\kappa_1^I}{\kappa_1^R}\right)=\beta_3 \tag{4}$$

from which κ_1 is found to be equal to β_3 as was shown in another way also in [8]. Subsequently, γ is found from the terms for $\overline{u}_1$ being zero resulting in

$$\begin{aligned}&\gamma^R=1-\frac{a_1\cos\beta_1}{l_1}-\frac{a_3\cos\beta_3}{l_3},\gamma^I=\frac{a_1\sin\beta_1}{l_1}-\frac{a_3\sin\beta_3}{l_3}\\ &\gamma=\tan^{-1}\left(\frac{\gamma^I}{\gamma^R}\right)=\tan^{-1}\left(\frac{\frac{a_1}{l_1}\sin\beta_1-\frac{a_3}{l_3}\sin\beta_3}{1-\frac{a_1}{l_1}\cos\beta_1-\frac{a_3}{l_3}\cos\beta_3}\right)\\ &\eta=\sqrt{(\gamma^R)^2+(\gamma^I)^2}=\sqrt{(1-\frac{a_1}{l_1}\cos\beta_1-\frac{a_3}{l_3}\cos\beta_3)^2+(\frac{a_1}{l_1}\sin\beta_1-\frac{a_3}{l_3}\sin\beta_3)^2}\end{aligned} \tag{5}$$

η is the scaling factor of the similar linkage and equals $\eta=l_5/l_1=l_6/l_2=l_7/l_3=||\overline{A_4A_7}||/l_4$. ρ is found when the terms for $\overline{u}_2$ are zero, which is for

$$\begin{aligned}&\rho^R=\frac{(1-\frac{a_1\cos\beta_1}{l_1}-\frac{a_3\cos\beta_3}{l_3})(\frac{a_{21}\cos\beta_{21}}{l_2}-\frac{a_3\cos\beta_3}{l_3})-(\frac{a_1\sin\beta_1}{l_1}-\frac{a_3\sin\beta_3}{l_3})(\frac{a_{21}\sin\beta_{21}}{l_2}-\frac{a_3\sin\beta_3}{l_3})}{(\frac{a_1\sin\beta_1}{l_1}-\frac{a_3\sin\beta_3}{l_3})^2+(1-\frac{a_1\cos\beta_1}{l_1}-\frac{a_3\cos\beta_3}{l_3})^2}\\ &\rho^I=\frac{(1-\frac{a_1\cos\beta_1}{l_1}-\frac{a_3\cos\beta_3}{l_3})(\frac{a_{21}\sin\beta_{21}}{l_2}-\frac{a_3\sin\beta_3}{l_3})-(\frac{a_1\sin\beta_1}{l_1}-\frac{a_3\sin\beta_3}{l_3})(\frac{a_{21}\cos\beta_{21}}{l_2}-\frac{a_3\cos\beta_3}{l_3})}{(\frac{a_1\sin\beta_1}{l_1}-\frac{a_3\sin\beta_3}{l_3})^2+(1-\frac{a_1\cos\beta_1}{l_1}-\frac{a_3\cos\beta_3}{l_3})^2}\\ &\rho=\tan^{-1}\left(\frac{\rho^I}{\rho^R}\right),\quad \tau=\sqrt{(\rho^R)^2+(\rho^I)^2}\end{aligned} \tag{6}$$

From polygon $A_0A_3A_7A_4$ and with κ_1 and γ known, angle κ_2 can be derived as

$$\kappa_2^R=1-\kappa_1^R-\gamma^R=\frac{a_1\cos\beta_1}{l_1}\quad \kappa_2^I=\kappa_1^I+\gamma^I=\frac{a_1\sin\beta_1}{l_1}\quad \kappa_2=\tan^{-1}\left(\frac{\kappa_2^I}{\kappa_2^R}\right)=\beta_1 \tag{7}$$

Herewith the similar linkage has been fully defined with parameters based on the principal dimensions and the dimensions of the four-bar linkage $A_0A_1A_2A_3$ solely.

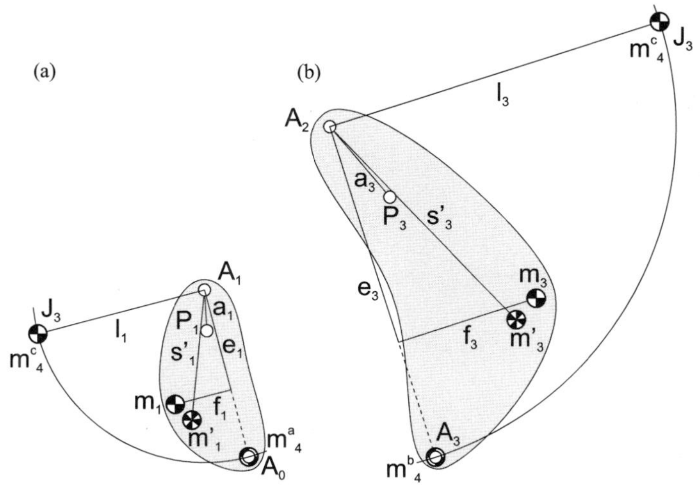

Fig. 3 The mass of link 4 is distributed on the other three links. Shown are (a) link 1 and (b) link 3 on which a mass m_4^a at A_0, a mass m_4^b at A_3, and a mass m_4^c at J_3 on each of the links are modeled.

3 Force Balance Conditions

To have S be the CoM of the complete kinematic architecture of Fig. 1, the principal dimensions a_1, a_{21}, a_{23}, and a_3 need to be calculated from the mass of each element and their positions. Since the principal dimensions are defined with respect to three elements of the four-bar linkage $A_0A_1A_2A_3$, the first step is to distribute the mass of the fourth element, m_4 of link 4, equivalently to the other elements. For links 1 and 3 this can be done by modeling a mass $m_4^a = m_4(1 - e_4/l_4)$ at A_0, a mass $m_4^b = m_4e_4/l_4$ at A_3 and a mass $m_4^c = m_4f_4/l_4$ at positions J_3 on both links 1 and 3 as indicated in Figs. 3a and b, respectively. m_4^a, m_4^b, and m_4^c also need to be modeled on link 2, which will be shown later on. For the analysis of link 1 it now has a total mass $m'_1 = m_1 + m_4^a + m_4^c$ centered at s'_1 from A_1 which is de CoM of the three masses. Similarly, for the analysis of link 3 it has a total mass $m'_3 = m_3 + m_4^b + m_4^c$ centered at s'_3 from A_2.

To include the masses of D_8E_8 and D_9E_9, also they can be distributed among the other elements in a similar way as with m_4. Unfortunately this paper leaves too little space to present this distribution in detail, for which they are not considered here.

With m_4 distributed, link 4 can be taken out resulting in the linkage of Fig. 4. This linkage is an extended composition of the linkage investigated in [5, 6] and the same method can be applied here to derive the principal dimensions. This means that P_i can be found independently from one another with linear momentum equations.

To find P_1, the linear momentum of the linkage for $\dot{\theta}_2 = \dot{\theta}_3 = 0$ (links 2 and 3 being immovable) can be written with respect to reference frame x_1y_1 aligned with a_1 as indicated in Fig. 4, to be equal to the total mass m_{tot} moving at S as

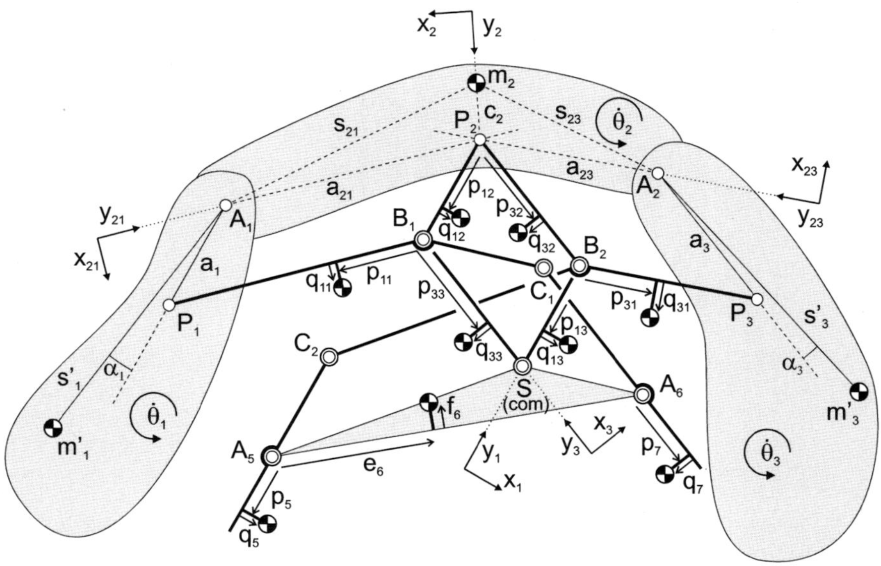

Fig. 4 Mass model of the kinematic architecture after distribution of the mass of link 4 to link 1 and link 2, and without considering links D_8E_8 and D_9E_9.

$$\frac{\overline{L}_1}{\dot{\theta}_1} = \begin{bmatrix} m_1' s_1' \cos\alpha_1 + (m_5+m_6+m_7+m_{11}+m_{33})a_1 + m_{12}p_{12} + m_{13}p_{13} + m_5 p_5 \\ m_1' s_1' \sin\alpha_1 - m_{12}q_{12} - m_{13}q_{13} - m_5 q_5 \end{bmatrix}$$
$$= \begin{bmatrix} m_{tot}a_1 \\ 0 \end{bmatrix} \tag{8}$$

with $m_{tot} = m_1+m_2+m_3+m_4+m_5+m_6+m_7+m_{11}+m_{12}+m_{13}+m_{31}+m_{32}+m_{33}$. From these equations a_1 and α_1 are obtained with a_1 resulting in

$$a_1 = \frac{\sqrt{m_1'^2 s_1'^2 - (m_{12}q_{12}+m_{13}q_{13}+m_5q_5)^2} + m_{12}p_{12} + m_{13}p_{13} + m_5 p_5}{m_{tot} - m_5 - m_6 - m_7 - m_{11} - m_{33}} \tag{9}$$

P_3 is found similarly by writing the linear momentum of the linkage for $\dot{\theta}_1 = \dot{\theta}_2 = 0$ with respect to frame x_3y_3 aligned with a_3 to be equal to m_{tot} moving at S as

$$\frac{\overline{L}_3}{\dot{\theta}_3} = \begin{bmatrix} m_3' s_3' \cos\alpha_3 + (m_5+m_6+m_7+m_{31}+m_{13})a_3 + m_{32}p_{32} + m_{33}p_{33} + m_7 p_7 \\ m_3' s_3' \sin\alpha_3 - m_{32}q_{32} - m_{33}q_{33} - m_7 q_7 \end{bmatrix}$$
$$= \begin{bmatrix} m_{tot}a_3 \\ 0 \end{bmatrix} \tag{10}$$

From these equations a_3 and α_3 are obtained with a_3 resulting in

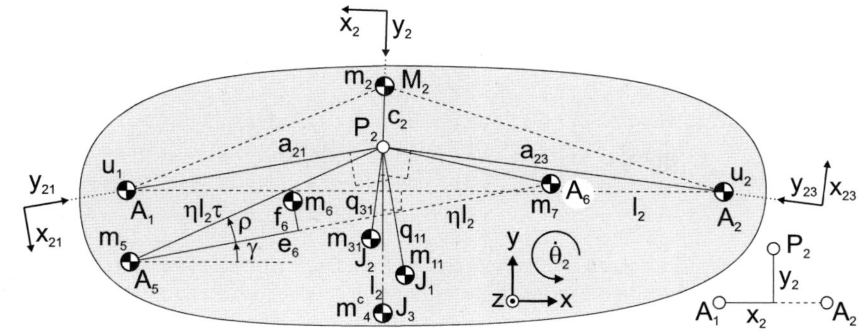

Fig. 5 Equivalent Linear Momentum System (ELMS) for $\dot{\theta}_2$ when $\dot{\theta}_1 = \dot{\theta}_3 = 0$ for which the masses of the moving elements are projected on link 2. P_2 is found as being the CoM of the ELMS.

$$a_3 = \frac{\sqrt{m_3'^2 s_3'^2 - (m_{32}q_{32} + m_{33}q_{33} + m_7 q_7)^2} + m_{32}p_{32} + m_{33}p_{33} + m_7 p_7}{m_{tot} - m_5 - m_6 - m_7 - m_{31} - m_{13}} \tag{11}$$

As in [5, 6], P_2 can be found by using an Equivalent Linear Momentum System (ELMS). This means that the mass of the moving elements for $\dot{\theta}_1 = \dot{\theta}_3 = 0$ (immovable parallelogram $P_2B_1SB_2$, link 2 rotating about P_2) are modeled on link 2 such that their linear momentum is equal to one of the reference frames $x_{21}y_{21}$, $x_{23}y_{23}$, and x_2y_2. Figure 5 shows the resulting ELMS with masses $u_1 = m_1 + m_4^a + m_{11}p_{11}/a_{21}$ and $u_2 = m_3 + m_4^b + m_{31}p_{31}/a_{23}$ at A_1 and A_2, respectively, masses m_5 and m_6 at A_5 and A_6, respectively, mass m_6 at distances e_6 and f_6 with respect to line A_5A_6, and masses m_{11} and m_{31} also placed at J_1 and J_3, respectively. u_1 and u_2 contain the distributed masses m_4^a and m_4^b of link 4 on link 2 and a mass m_4^c is placed at J_3. J_3 is located at a distance l_2 from P_2 normal to line A_1A_2 in indicated direction.

P_2 is found as being the CoM of the ELMS. With P_2 being located at a distance x_2 from A_1 along A_1A_2 and y_2 normal to A_1A_2 as indicated in Fig. 5, P_2 is found by solving the linear momentum equations of the ELMS

$$\begin{aligned}\frac{\overline{L}_2}{\dot{\theta}_2} &= u_1\begin{bmatrix} y_2 \\ -x_2 \end{bmatrix} + v_1\begin{bmatrix} x_2 \\ y_2 \end{bmatrix} + u_2\begin{bmatrix} y_2 \\ -(x_2 - l_2) \end{bmatrix} - v_2\begin{bmatrix} x_2 - l_2 \\ y_2 \end{bmatrix} + m_2\begin{bmatrix} y_2 - f_2 \\ -(x_2 - e_2) \end{bmatrix} + \\ &\quad m_5\eta l_2\tau\begin{bmatrix} \sin(\gamma+\rho) \\ -\cos(\gamma+\rho) \end{bmatrix} + m_7\eta l_2\begin{bmatrix} \tau\sin(\gamma+\rho) - \sin(\gamma) \\ -\tau\cos(\gamma+\rho) + \cos(\gamma) \end{bmatrix} + \\ &\quad m_6\begin{bmatrix} \eta l_2\tau\sin(\gamma+\rho) - e_6\sin(\gamma) - f_6\cos(\gamma) \\ -\eta l_2\tau\cos(\gamma+\rho) + e_6\cos(\gamma) - f_6\sin(\gamma) \end{bmatrix} + m_4^c\begin{bmatrix} -l_2 \\ 0 \end{bmatrix} = \overline{0}\end{aligned} \tag{12}$$

with $v_1 = m_{11}q_{11}/a_{21}$ and $v_2 = m_{31}q_{31}/a_{23}$. No algebraic solution for P_2 was found, for which the equations have to be solved numerically. The principal dimensions defining P_2 are calculated as $a_{21} = \sqrt{x_2^2 + y_2^2}$ and $a_{23} = \sqrt{(l_2 - x_2)^2 + y_2^2}$ with which all principal dimensions are obtained.

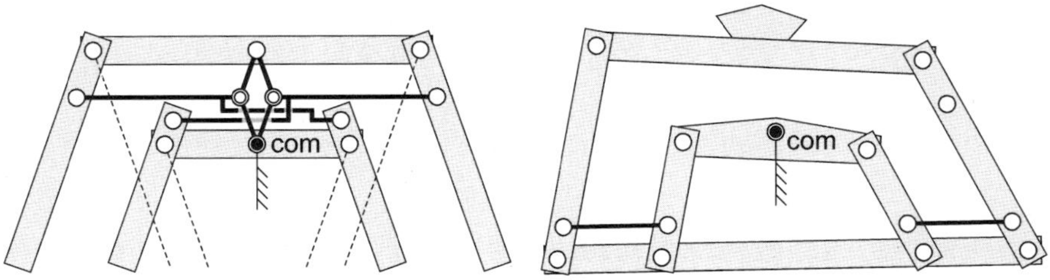

Fig. 6 Two examples of balanced mechanisms derived from Fig. 1 with the CoM being a joint with the base: (a) double grasper mechanism, (b) two-degree-of-freedom balanced manipulator.

4 Conclusion

An inherently balanced kinematic architecture of which the CoM is an invariant link point has been composed based on a general 4R four-bar linkage and by applying the method of principal vectors as a linkage together with a similar four-bar linkage. The conditions for the kinematic architecture were derived as a function of the principal dimensions. The principal dimensions were calculated from a generally defined mass and mass location of each element, resulting in the general force balance conditions of the kinematic architecture. Figure 6 shows examples of possible balanced devices that can be derived from the kinematic architecture such as a double grasper mechanism and a 2-DoF manipulator with end-effector considering an arbitrary mass distribution of each of the shown elements.

References

1. Fischer, O.: Theoretische Grundlagen für eine Mechanik der lebenden Körper. Teubner, Leipzig (1906)
2. Kreutzinger, R.: Über die Bewegung des Schwerpunktes beim Kurbelgetriebe. Getriebetechnik **10**(9), 397–398 (1942)
3. Lim, T.G., Cho, H.S., Chung, W.K.: Payload capacity of balanced robotic manipulators. Robotica **8**, 117–123 (1990)
4. Shchepetil'nikov, V.A.: The determination of the mass centers of mechanisms in connection with the problem of mechanism balancing. J. Mech. **3**, 367–389 (1968)
5. Van der Wijk, V., Herder, J.L.: On the development of low-mass force balanced manipulators. In: Lenarčič, J., Stanisic, M.M. (eds.) Advances in Robot Kinematics. Proceedings of the IFToMM 12th Int. Symposium on Advances in Robot Kinematics, pp. 411–420. Springer (2010). ISBN 978-90-481-9261-8
6. Van der Wijk, V., Herder, J.L.: Synthesis method for linkages with center of mass at invariant link point – Pantograph based mechanisms. Mech. Mach. Theory **48**, 15–28 (2012)
7. Van der Wijk, V., Herder, J.L., Demeulenaere, B.: Comparison of various dynamic balancing principles regarding additional mass and additional inertia. J. Mech. Robot. **1**(4), 04 1006 (2009)
8. Wunderlich, W.: Concerning the trajectory of the center of mass of the four-bar linkage and the slider-crank mechanism. J. Mech. **3**, 391–396 (1968)

Integrated Type and Dimensional Synthesis of Planar Four-Bar Mechanisms

Tim J. Luu and M. John D. Hayes

Abstract A novel approach to integrated type and approximate dimensional synthesis of planar four-bar mechanisms (i.e. linkages comprised of any two of *RR*, *PR*, *RP*, and *PP* dyads) for rigid-body guidance is proposed. The essence is to correlate coordinates of the coupler attachment points in two different coordinate frames, thereby reducing the number of independent variables defining a suitable dyad for the desired rigid-body motion from five to two. After applying these geometric constraints, numerical methods are used to size link lengths, locate joint axes, and decide between *RR*, *PR*, *RP* and *PP* dyads that, when combined, guide a rigid body through the best approximation, in a least-squares sense, of n specified positions and orientations, where $n \geq 5$. No initial guesses of type or dimension are required. An example is presented illustrating the effectiveness and robustness of this new approach.

Key words: Approximate type and dimensional synthesis, planar four-bar mechanisms, rigid body guidance, singular value decomposition

1 Introduction

Planar linkages contain either revolute (R-pairs), or prismatic (P-pairs). These kinematic pairs permit rotations about one axis, or translations parallel to one direction, respectively. In general, dimensional synthesis for rigid body guidance assumes a mechanism *type*, i.e., planar $4R$; slider-crank; crank-slider; trammel, etc. Our aim is to develop a completely general planar mechanism synthesis algorithm that in-

Tim J. Luu
Neptec Design Group Ltd., Ottawa, Ontario, Canada, e-mail: tluu@neptec.com

M. John D. Hayes
Department of Mechanical and Aerospace Engineering, Carleton University, Ottawa, Ontario, Canada, e-mail: jhayes@mae.carleton.ca

J. Lenarčič, M. Husty (eds.), *Latest Advances in Robot Kinematics*,
DOI 10.1007/978-94-007-4620-6_40,

tegrates both *type* and *dimensional* synthesis for n-position approximate synthesis for rigid body guidance. The pairing of the two types leads to four possible dyads: revolute-revolute (*RR*), prismatic-revolute (*PR*), revolute-prismatic (*RP*), and prismatic-prismatic (*PP*).

There is an extensive body of literature reporting research on approximate dimensional kinematic synthesis of planar four-bar mechanisms for rigid-body guidance, see for example [1, 4–6, 9, 12]. However, there are no methods reported in the substantial body of literature that successfully integrate both type and approximate dimensional synthesis of planar four-bar mechanisms for rigid body guidance, without a priori knowledge or initial guesses with the exception of two special cases reported in [2, 3]. In this paper a method for doing so is presented for the first time.

The minimization criteria of the algorithm presented in this paper is purely mathematical: the condition number of the synthesis matrix. The algorithm will be enhanced when the transmission angle is incorporated as an optimization objective. It would be additionally beneficial to examine the order and branch defect problems. It may be that advances made in [10] can be incorporated into the integrated type-dimensional synthesis algorithm to address these issues. These issues notwithstanding, the algorithm presented in this paper is a robust foundation upon which to build. The algorithm is being adapted for synthesis of spatial motion platforms.

2 Kinematic Constraints: Circular and Linear

The motion of the coupler link in a four-bar planar mechanism is determined by the relative displacements of all links in the kinematic chain. The relative displacement of two rigid bodies in the plane can be considered as the displacement of a Cartesian reference coordinate frame E attached to one of the bodies with respect to a Cartesian reference coordinate frame Σ attached to the other. Without loss of generality, Σ may be considered fixed with E free to move, see Figure 1. The homogeneous coordinates of points represented in E are given by the ratios $(x : y : z)$. Those of the same points represented in Σ are given by the ratios $(X : Y : Z)$. The mapping between the coordinates of points expressed in the two reference frames is given by the homogeneous coordinate transformation

$$\begin{bmatrix} X \\ Y \\ Z \end{bmatrix} = \begin{bmatrix} \cos\theta & -\sin\theta & a \\ \sin\theta & \cos\theta & b \\ 0 & 0 & 1 \end{bmatrix} \begin{bmatrix} x \\ y \\ z \end{bmatrix}, \tag{1}$$

where (a, b) are the $(\frac{X}{Z}, \frac{Y}{Z})$ Cartesian coordinates of the origin of E with respect to Σ, and θ is the orientation of E relative to Σ. Any point $(x : y : z)$ in E can be mapped to $(X : Y : Z)$ in Σ using this transformation. For rigid body guidance, each pose is defined by the position and orientation of E with respect to Σ, which is specified by the ordered triple (a, b, θ). Dyads are connected through the coupler link at the coupler attachment points M_1 and M_2, see Figure 1.

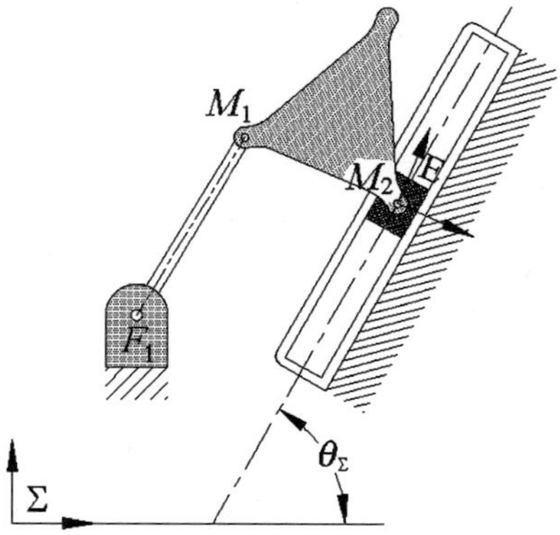

Fig. 1 Planar *RRRP* linkage.

There is a specific type of constrained motion corresponding to each one of the four types of planar lower-pair dyad. The ungrounded *R* pair in an *RR* dyad is constrained to move on a circle with a fixed centre. Because of this they are denoted *circular constraints*. *Linear constraints* result when *PR* and *RP* dyads are employed because the *R* pair attachment point is constrained to move on a line defined by the *P* pair translation direction. The *PP* dyad represents a planar constraint: the line of one *P* pair direction is constrained to translate on the direction line of the other.

It can be shown [2] that the model representing both circular and linear constraints for n Cartesian point coordinate pairs can be expressed in matrix form as

$$\mathbf{Ck} = \begin{bmatrix} X_j^2 + Y_j^2 & 2X_j & 2Y_j & 1 \end{bmatrix} \begin{bmatrix} K_0 \\ K_1 \\ K_2 \\ K_3 \end{bmatrix} = \mathbf{0}, \tag{2}$$

where $\mathbf{C}$ is an $n \times 4$ dimensional array with $j \in \{1, 2, \ldots, n\}$, with X and Y being the Cartesian coordinates of points on either a circle or line, and the K_i are constant shape parameters determined by the constraint imposed by the dyad [2].

For circular constraints the K_i are defined as

$$K_0 = 1, \quad K_1 = -X_c, \quad K_2 = -Y_c, \quad K_3 = K_1^2 + K_2^2 - r^2, \tag{3}$$

where (X_c, Y_c) are the Cartesian coordinates of the circle centre expressed in Σ and r is the circle radius.

Linear constraints require $K_0 = 0$ and the remaining K_i are proportional to line coordinates defined by

$$K_1 = -\frac{1}{2}F_{Z/\Sigma}\sin\theta_\Sigma, \quad K_2 = \frac{1}{2}F_{Z/\Sigma}\cos\theta_\Sigma, \quad K_3 = F_{X/\Sigma}\sin\theta_\Sigma - F_{Y/\Sigma}\cos\theta_\Sigma, \tag{4}$$

where $(F_{X/\Sigma} : F_{Y/\Sigma} : F_{Z/\Sigma})$ are homogeneous coordinates of a fixed point, expressed in Σ, on the line that makes an angle θ_Σ with the positive X-axis in Σ.

In the definitions of the K_i, the parameter K_0 acts as a binary switch between circular and linear constraints. When $K_0 = 1$ Equation (2) represents the implicit equation of points on a circle, and when $K_0 = 0$ the equation becomes that of a line. Nonetheless, K_0 is still an homogenizing parameter whose value is arbitrary. The K_i can be normalized by K_0, but only when K_0 is nonzero.

3 Integrating Type and Approximate Dimensional Synthesis

Equations (2), (3), and (4) are used to integrate type and approximate dimensional synthesis of planar for-bar mechanisms for rigid-body guidance. Constructing the required synthesis matrix **C** based on the prescribed poses is done by relating the position of the two rigid body attachment points M_1 and M_2 in both reference frames E and Σ, see Figure 1. Reference frames Σ and E are correlated in two ways:

1. Points M_1 and M_2 move on circles or lines in Σ;
2. Points M_1 and M_2 have constant coordinates in E.

Let (x, y) be the coordinates expressed in E of one of the coupler attachment points, M, and (X, Y) be the coordinates of the same point expressed in Σ. Carrying out the matrix multiplication in Equation (1) yields

$$\begin{aligned} X &= x\cos\theta - y\sin\theta + az, \\ Y &= x\sin\theta + y\cos\theta + bz, \\ Z &= z. \end{aligned} \tag{5}$$

Ignoring infinitely distant coupler attachment points, it is reasonable to set $z = 1$ in Equation (5) and substituting the result into Equation (2), with $j \in \{1, 2, \ldots, n\}$ yields

$$\mathbf{Ck} = \begin{bmatrix} \left(x\cos\theta_j - y\sin\theta_j + a_j\right)^2 + \left(x\sin\theta_j + y\cos\theta_j + b_j\right)^2 \\ 2\left(x\cos\theta_j - y\sin\theta_j + a_j\right) \\ 2\left(x\sin\theta_j + y\cos\theta_j + b_j\right) \\ 1 \end{bmatrix}^T \begin{bmatrix} K_0 \\ K_1 \\ K_2 \\ K_3 \end{bmatrix} = 0. \tag{6}$$

Prescribing $n > 5$ poses makes **C** an $n \times 4$ matrix. The parameters x and y possess constant values in E. The n-dimensional vector parameters **a**, **b**, and $\boldsymbol{\theta}$ in **C** are all known a priori because they are the specified poses of E with respect to Σ.

The only unknown parameters in **C** are x and y. Determining the x and y that best satisfy Equation (6) will solve the problem. Once values for x and y are obtained **C** is fully determined, which allows the vector **k** to be identified. The problem is now a two dimensional search for x and y. However, at least two dyads are required to form a planar mechanism. This implies that there must be at least two distinct values for x and y in order for a complete solution to exist. The x and y are found such that they satisfy Equation (6). For equations of the form $\mathbf{Ck} = 0$ the only real $\mathbf{k}$ that satisfies the equation is the zero vector if $\mathbf{C}$ is not singular. In other words, for non-trivial $\mathbf{k}$ to exist, $\mathbf{C}$ must be rank deficient [11]. The task becomes finding values for x and y that make $\mathbf{C}$ rank deficient, or failing that, the most ill-conditioned.

The conditioning of a matrix is measured by the ratio of the largest and smallest singular values of the matrix, which is called the condition number κ. It is computationally more convenient to use is the inverse of the condition number, γ

$$\gamma \equiv \frac{1}{\kappa} = \frac{\sigma_{MIN}}{\sigma_{MAX}}, \quad 0 \leq \gamma \leq 1, \tag{7}$$

because it is bounded both from above and below. A well conditioned matrix has $\gamma \approx 1$, while an ill-conditioned matrix has $\gamma \approx 0$. Therefore, selecting x and y that renders $\mathbf{C}$ the most rank deficient involves minimizing γ.

The Nelder–Mead *Downhill Simplex Method in Multidimensions* algorithm may be used for this minimization [7]. This method requires only function evaluations, not derivatives. It is not very efficient in terms of the evaluations it requires, but for the problem at hand the computational burden is relatively small. We will not discuss the convergence properties, because any optimization method may be employed.

Since the Nelder–Mead algorithm needs an initial guess, γ may be plotted in terms of x and y first, in the neighborhood of $(x, y) = (0, 0)$ up to a user-defined range of the maximum distance that the coupler attachment points can be from the moving frame E origin, denoted ϵ. As x and y represent the position of a coupler attachment point with respect to moving reference frame E. The x and y parameters may then be selected approximately corresponding to the smallest value of γ. These points represent the local minima of the entire γ plot, that is, with $\epsilon = \infty$. However, for practical reasons, with ϵ finite, these minima may be regarded as the global minima of the region of interest. At least two minima are required to obtain a planar four-bar mechanism, as each minimum corresponds to a single dyad. The Nelder-Mead algorithm is fed these approximate values as inputs, and converges to the values of x and y that minimize γ.

Once the values of x and y have been determined the matrix $\mathbf{C}$ in Equation (6) can be populated. The $\mathbf{k}$ parameters may then be estimated. We have elected to use singular value decomposition (SVD) because we are necessarily required to work with either singular, or numerically very-close-to-singular sets of equations. SVD decomposes any given $m \times n$ matrix $\mathbf{C}$ into the product of three matrix factors such that

$$\mathbf{C}_{m\times n} = \mathbf{U}_{m\times m}\mathbf{S}_{m\times n}\mathbf{V}^T_{n\times n}, \tag{8}$$

where $\mathbf{U}$ and $\mathbf{V}$ are orthogonal, and $\mathbf{S}$ is a rectangular matrix whose only non-zero elements are on the diagonal of the upper $n \times n$ sub-matrix. These diagonal elements are the singular values of $\mathbf{C}$ arranged in descending order, lower bounded by zero [8]. SVD constructs orthonormal bases spanning the range of $\mathbf{C}$ in $\mathbf{U}$ and the nullspace of $\mathbf{C}$ in $\mathbf{V}$. If $\mathbf{C}$ is rank deficient, then the last $n - \text{rank}(\mathbf{C})$ singular values of $\mathbf{C}$ are zero. Furthermore, the corresponding columns of $\mathbf{V}$ are unit basis vectors

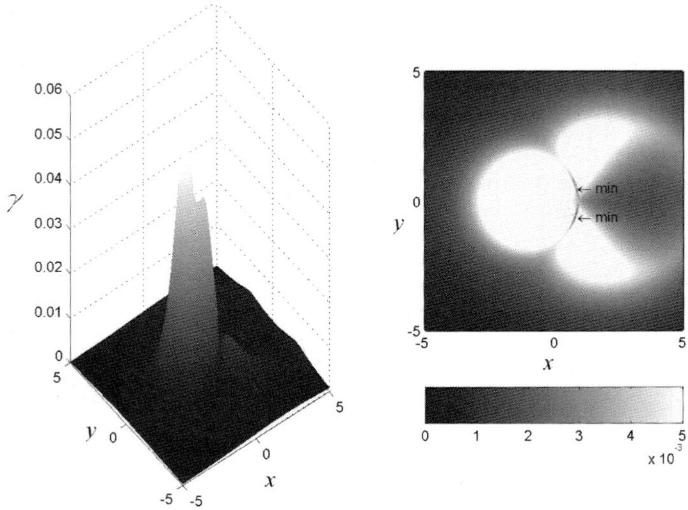

Fig. 2 The γ plot of the poses defining a square corner.

that span the nullspace of **C**. As such, any linear combination of these columns is a non-trivial solution that best satisfies the system $\mathbf{Ck} = 0$.

For overconstrained systems, where the $m \times n$ matrix **C** has $m > n$, in general no non-trivial exact solution exists, because in general an overconstrained synthesis matrix possesses full rank. In this case, the optimal approximate solution in a least-squares sense is last column of **V**, corresponding to the smallest singular value of **C**. Furthermore, the more ill-conditioned **C** is, the closer the optimal approximate solution is to being an exact solution. Because the K_i are homogeneous, the scaling posses no problem because **k** will be normalized by dividing through by K_0. In the case where $K_0 \approx 0$, or $K_0 = 0$ the linear definitions for K_1, K_2, and K_3 from Equation (4) are used. The switching threshold for K_0 to represent either an *RR* or *PR* (or *RP*) dyad must be user defined based on the geometry of the problem.

Note that *PP* dyads are a special case. Two serial *P* pairs restricts the distal link from changing its orientation. For type synthesis, given any set of poses with non constant orientation, the *PP* dyad is immediately ruled out.

3.1 Example

Consider an example that requires completely general integrated type and approximate dimensional synthesis by defining poses that are impossible to generate exactly by any four-bar planar mechanism. The poses define a square corner. A point on a rigid body moves linearly between the Cartesian coordinates from (0, 1) to (1, 0) via (1, 1). The orientation increases linearly from 0 to 90 degrees. The poses are listed in Table 1.

Table 1 Specified poses defining a square corner.

Pose	1	2	3	4	5	6	7	8	9	10	11	12	13	14	15	16	17	18
a	0.0	0.1	0.2	0.3	0.4	0.6	0.7	0.8	1.0	1.0	1.0	1.0	1.0	1.0	1.0	1.0	1.0	1.0
b	1.0	1.0	1.0	1.0	1.0	1.0	1.0	1.0	1.0	0.9	0.8	0.7	0.6	0.4	0.3	0.2	0.1	0.0
$\theta°$	0.0	4.5	9.0	13.5	18.0	27.0	31.5	36.0	45.0	49.5	54.0	58.5	63.0	72.0	76.5	81.0	85.5	90.0

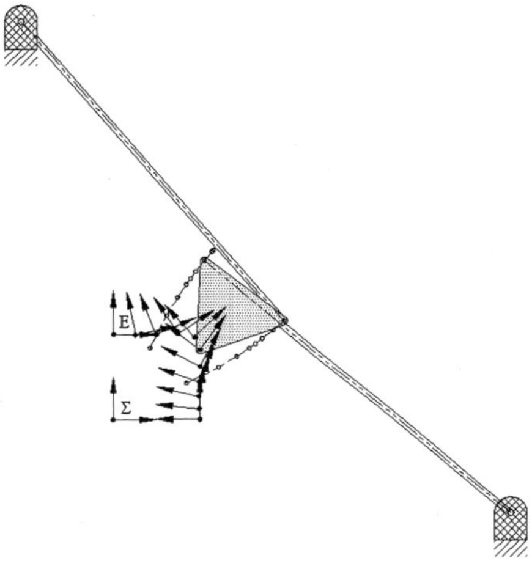

	Dyad 1	Dyad 2
x	0.8413	0.8413
y	0.5706	−0.5706
K_0	0.2010	0.2010
K_0/K_0	1	1
K_1/K_0	−4.5843	1.0549
K_2/K_0	1.0539	−4.5843
K_3/K_0	1.2704	1.2704

Fig. 3 Identified *RRRR* mechanism and corresponding dyads.

A planar four-bar mechanism cannot exactly replicate the motion defined above because points on the coupler generate either a 6th, 4th, or 2nd order curve. The curve

$$x^n + y^n = 1 \tag{9}$$

approaches a square corner as $n \to \infty$. With $n \leq 6$ for planar four-bar mechanisms, it is impossible to exactly replicate the desired motion. Although a *PPPP* mechanism may be able to generate the desired point translation, the change in orientation rules out this type of mechanism.

The pose data are substituted into Equation (6) to populate **C**. The γ of **C** are then plotted as a function of x and y and are shown in Figure 2. As can be seen in this figure, two distinct minima occur at approximately $(0.8, 0.6)$ and $(0.8, -0.6)$. Using the Nelder-Mead minimization and the pair of approximate x and y as initial guesses, the exact values of the two minima are found, and listed in Figure 3. These values are then substituted into Equation (6) to completely determine **C**. SVD is then applied to **C** to find **k** corresponding to each minimum. The values of **k** thus determined are also listed in Figure 3. The resulting synthesized mechanism, illustrated in Figure 3, is composed of two *RR* dyads centred on $(4.5843, -1.0539)$ and $(-1.0539, 4.5843)$, both with links having length 1.7307.

4 Conclusions

In this paper a novel method was presented that integrates type and approximate dimensional synthesis of planar four-bar mechanisms used for rigid-body guidance. Coupler attachment points are correlated between moving frame E and fixed frame Σ thereby reducing the number of independent variables defining a suitable dyad for the desired poses from five to two. Numerical methods are then used to determine both mechanism type and approximate dimensions. A numerical example was presented, illustrating the utility of the algorithm.

References

1. Akhras, R., Angeles, J.: Unconstrained nonlinear least-square optimization of planar linkages for rigid-body guidance. Mech. Mach. Theory **25**(1), 97–118 (1990)
2. Hayes, M.J.D., Luu, T., Chang, X.W.: Kinematic mapping application to approximate type and dimension synthesis of planar mechanisms. In: Lenarčič, J. and Galletti, C. (eds.) 9th Advances in Robotic Kinematics, pp. 41–48 Kluwer Academic Publishers, Dordrecht, the Netherlands (June 28–July 1, 2004)
3. Hayes, M.J.D., Zsombor-Murray, P.J.: Towards integrated type and dimensional synthesis of mechanisms for rigid body guidance. In: Proceedings of the CSME Forum 2004, pp. 53–61. University of Western Ontario, London, ON, Canada (June 1–4, 2004)
4. Kong, X.: Approximate kinematic synthesis of linkages using generalized inverse matrix and continuation. J. Mech. Sci. Technol. **18**(1), 38–40 (1999)
5. Larochelle, P.: Approximate motion synthesis of open and closed chains via parametric constraint manifold fitting: Preliminary results. Proceedings of the ASME Design Engineering Technical Conferences and Computers and Information in Engineering Conference **2**, 1049–1057 (2003)
6. Larochelle, P., McCarthy, J.M.: Designing planar mechanisms using a bi-invariant metric in the image space of SO. ASME, Design Engineering Division, 23rd Biennial Mechanisms Conference **70**(1), 221–228 (1994)
7. Nelder, J.A., Mead, R.: A simplex method for function minimization. Computer Journal **7**, 308–313 (1965)
8. Press, W., Teukolsky, S., Vetterling, W., Flannery, B.: Numerical Recipes in C, 2nd edn. Cambridge University Press, Cambridge, England (1992)
9. Ravani, B., Roth, B.: Motion synthesis using kinematic mappings. J. Mech. Transm. Autom. Des. **105**(3), 460–467 (1983)
10. Schröcker, H.P., Husty, M., McCarthy, J.M.: Kinematic mapping based evaluation of assembly modes for planar four-bar synthesis. ASME, J. Mech. Des. **129**(9), 924–929 (2007)
11. Strang, G.: Linear Algebra and its Applications. Academic Press (1980)
12. Vallejo, J., Aviles, R., Hernandez, A., Amezua, E.: Nonlinear optimization of planar linkages for kinematic syntheses. Mech. Mach. Theory **30**(4), 501–518 (1995)

Positional Workspace Boundary for Serial Manipulators with Revolute Joints

Ciprian S. Borcea and Ileana Streinu

Abstract We have recently shown that the boundary of the positional workspace for an nR serial manipulator with any pair of consecutive axes coplanar is made of circular arcs with an underlying combinatorial structure of interconnectivity. In the orthogonal case, the boundary is linear in size and can be computed efficiently with a polynomial time algorithm. In this paper we show that the general (non-orthogonal) case may be much more complex: in the worst case, the boundary may consist of 2^n such arcs.

Key words: Revolute-jointed serial manipulator, hyper-redundant, reach problem, positional workspace, boundary, descriptive complexity

1 Introduction and Overview

The positional workspace (in short, the *workspace*) $\mathscr{W}$ of a serial manipulator is defined as the set of points in R^3 which can be reached by a marked reference point T on the last link. The first link is considered fixed. The marked point may be imagined as the hand or end-effector reduced to a point (see Fig.1(a)). Exact description and computation of the workspace is of fundamental importance for manipulator design, placement in the environment and performance evaluation (see e.g. [1, 11]).

Historical Background. Workspace determination has been approached with a variety of methods: analysis of Jacobian singularities, recursive algebraic descriptions [9], tracing of extreme reaches, numerical or probabilistic calculations. Many papers addressed the workspace problem for general or specific robots with very few joints, e.g. [10, 12]. Despite the increased relevance of the prob-

Ciprian S. Borcea
Department of Mathematics, Rider University, Lawrenceville, NJ 08648, USA,
e-mail: borcea@rider.edu

Ileana Streinu
Computer Science Department, Smith College, Northampton MA 01063, USA,
e-mail: istreinu@smith.du

J. Lenarčič, M. Husty (eds.), *Latest Advances in Robot Kinematics*,
DOI 10.1007/978-94-007-4620-6_41,

lem for understanding nano-robots and protein structures, the case of an arbitrary number of joints has received little attention. The recursive procedure of [9], although general in principle, has been explicitly used only for instances with a small number of revolute axes. Numerical methods are fragile and lack any guarantee of sampling more than a small region of the workspace boundary.

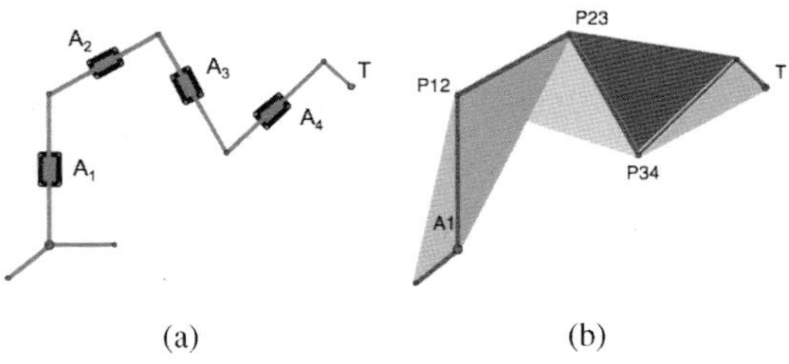

Fig. 1 A general 4R serial manipulator with intersecting consecutive joint axes, in (a) a schematic, standard representation and (b) viewed as a panel-and-hinge chain with labeled intersections of hinges $p_{i,i+1} = A_i \cap A_{i+1}$. The panel $\pi_{i,i+1}$ is the plane spanned by A_i and A_{i+1}.

Recent Results on Workspace and Extremal Reaches. In [5], we present a complete mathematical solution of the positional workspace boundary determination problem, for all spatial revolute-jointed serial manipulators with *any pair of consecutive joints coplanar*. Inspired by the terminology in Rigidity theory, and to emphasize the role of the planes spanned by the two joint axes carried by all interior links, we call them *panel-and-hinge chains*. The revolute joints, now called *hinges* or *hinge axes*, have no angular limitation in rotating around them. We denote the sequence of hinge axes by $A_1, \cdots, A_n$, and refer to the intersection of axes A_i and A_{i+1} by $p_{i,i+1}$. This notation is used in Figs. 1 and 2. The *natural order* of the hinges is $1, 2, 3, \cdots$ as they appear on the chain.

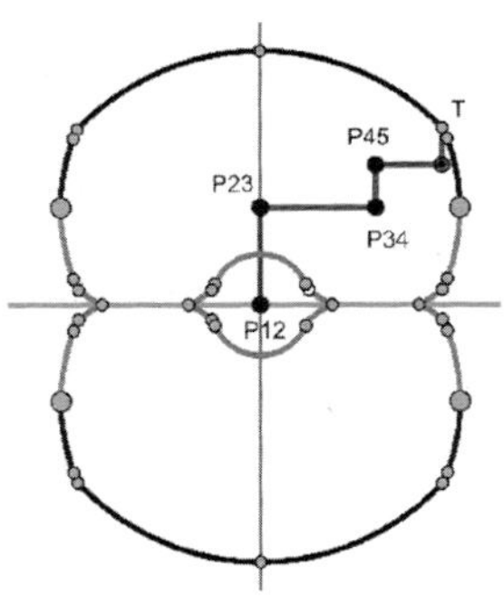

Fig. 2 Planar section $\partial\mathscr{W}_2$ through the workspace boundary $\partial\mathscr{W}$ of an orthogonal chain with $n = 5$ hinges. Shown are also the circular arc endpoints, and the max-boundary (black) and min-boundary (gray).

Our solution has a geometric and a combinatorial part. On the **geometric** side, we showed that, for panel-and-hinge chains, the intersection $\partial\mathscr{W}_2$ of the workspace boundary $\partial\mathscr{W}$ with a plane passing through the first hinge axis[1] is always made of *circular arcs* (Fig. 2). This boundary may have several connected components (the example in Fig. 2 has two, an outer layer and an inner "hole"). Each circular segment has a very clear relationship with an *extremal reach* problem (definitions are given in Sec. 2). We have shown that the entire 2D section of the workspace boundary $\partial\mathscr{W}_2$ is obtained by *tracing extreme reaches* relative to a base-point S which *sweeps the first hinge axis*. The workspace boundary $\partial\mathscr{W}$ is obtained as the surface of revolution of $\partial\mathscr{W}_2$ around the first axis. On the **combinatorial** side, we showed that the *articulation of the circular arcs* in $\partial\mathscr{W}_2$, the *positions of the corresponding centers* and their angular sectors are controlled through a discrete structure based on the notion of *fold points*. Fold points are key characteristics of extremal configurations of panel-and-hinge chains and a

[1] The subscript in $\partial\mathscr{W}_2$ serves as a reminder that we are referring to a planar region.

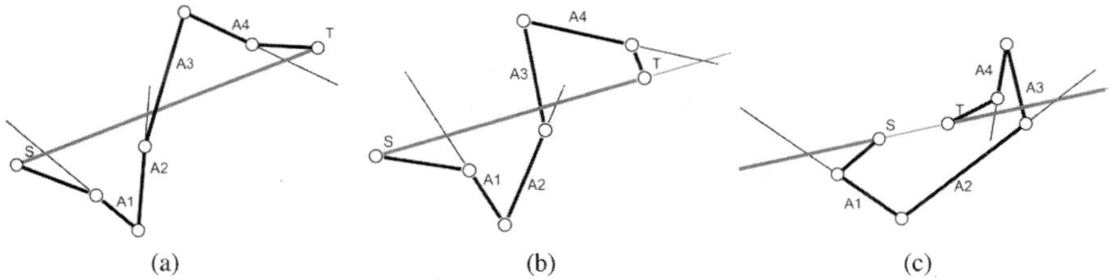

Fig. 3 The *natural order* criterion for maximum, resp. minimum reach of polygonal chains. Extensions of the thick hinge segments are shown with thinner lines. (a) This flat chain is in its global maximum position, since the oriented segment ST (thick gray) crosses the hinges in the natural order. (b) This flat configuration, although a critical one for the endpoint distance function, is not a maximum: the hinges are not crossed in natural order. (c) A flat non-zero minimum: the oriented projective complement of ST (thick gray) is crossed by the hinges in natural order.

distinctive feature of this restricted class of serial manipulators. Therefore, *computing the workspace boundary* amounts to giving a description of the connectivity of these circular arcs, along with their centers, radii and angular sectors. The number of such arcs gives the overall *complexity* of the boundary $\partial\mathscr{W}_2$.

These results make use of our recent complete solution [3] to the *Extreme Reaches Problem* for *nR* manipulators. Let the *endpoint axis* of the chain be the line through a base-point S on the first link and the end-point T. The necessary and sufficient condition given in our paper is that the global maximum of the distance between S and T is attained when *the oriented line segment between the base-point and the end-point meets all joint axes in their natural sequential order* $1,2,\cdots,n$. For an *nR* manipulator with *coplanar consecutive hinges* (the kind considered in this paper), non-zero minima have a similar characterization: *the complement of the segment between S and T, oriented as a projective arc from the base-point S to the end-point T and passing through the point at infinity, meets all joint axes in their natural sequential order.*

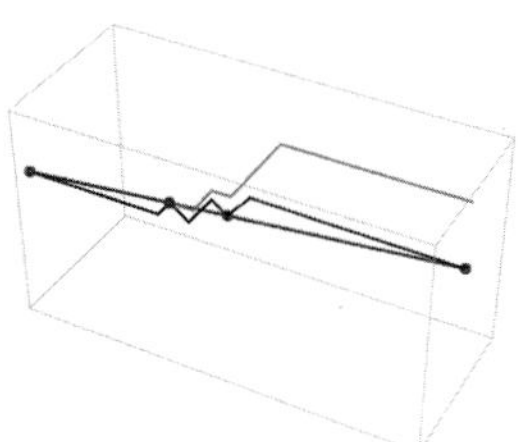

Fig. 4 A polygon folded to a 3D maximum configuration. The black dots are the endpoints and the two foldpoints; notice the endpoint axis, going through all of them. In light gray, for visual reference, is a flat configuration of the chain.

Fig. 3 illustrates these criteria for *flat* configurations, i.e. those where *all the panels* (and thus all hinges, plus S and T) lie in a common plane. Fig. 4 illustrates the general case, where the maximum is attained in a 3D configuration, and where *flat pieces* in maximum reach position are connected at *fold points* (see Sec. 2); the end-to-end axis, joining the endpoints of the chain, passes through the fold points, while globally crossing all the hinge axes in natural order. Note from Fig. 3 that in general our criterion is applied to hinge axes conceived as *entire lines*, not just line segments.

The two Max and Min Reach criteria are essential for tracing the *max-boundary* and the *min-boundary* of the workspace, that is, those parts of the workspace boundary that can be obtained as the maximum reach, respectively non-zero minimum reach of the end-point distance function between a base-point S on the first joint axis and the end-point T. Together, the max-boundary and the min-boundary yield the whole workspace boundary; see Fig. 2 for an example.

Algorithm for Computing the Workspace Boundary. In [5], we present an efficient algorithm for computing the *description* of $\partial \mathscr{W}_2$ in the case of *orthogonal chains*, i.e. when the angles between two consecutive hinge axes are $\frac{\pi}{2}$. For a panel-and-hinge chain with n links, the boundary $\partial \mathscr{W}_2$ consists of $O(n)$ circular arcs. We show how to efficiently compute each one, and their connectivity, in linear $O(n)$ time. This case makes use of special properties of orthogonal chains and generalizes to a larger class of manipulators identified in [4, 5]). It is natural to ask: *Is it possible to extend these algorithmic ideas to all panel-and-hinge chains?* Our previous theoretical results suggest a positive answer. Here, however, we uncover an intrinsic difficulty.

Our Result. A *general* algorithm for positional workspace boundary calculation would take at least as much time as it takes to describe the result. In this paper, we give a family of examples showing that the size of the workspace boundary can sometimes be $O(2^n)$, i.e. exponentially large.

2 Workspace Boundary

The *endpoint distance function* assigns a real non-negative value (the distance between the endpoints S and T) to each spatial configuration of the chain. The endpoint distance varies between two extreme values, the *global* minimum and maximum. As a function, it may have local minima, maxima and other critical values, attained in *critical configurations*. It is known that in all critical configurations, the endpoint *axis* meets all the hinge *axes*.

Fold Points and Flat Pieces. Two or more consecutive hinges cut by the endpoint axis away from their intersection point must be coplanar: the panels between them are folded over in a flat configuration. This leads to a *structural decomposition* of a critical configuration into *flat pieces* and identifies certain *fold points*. The flat pieces arise from contiguous segments of the chain (i.e. within an interval i to j of panel indices), in which several coplanar consecutive hinges are cut (simultaneously, in their common plane) by the endpoint axis (as in Fig. 3(a,c)). The flat pieces are connected at fold points, which are those intersections of hinge axes which *meet the endpoint axis* (see Fig. 4). The two hinges incident at each fold point determine, in addition, a simpler "triangular" *folding panel*, which is met by the endpoint axis only at the fold point.

Algorithms for Extremal Reaches. A *dual characterization* of the global maximum as a constrained shortest-path is given in [3]: *the global maximum of the endpoint distance function coincides with the length of the* shortest path *from S to T which meets all hinges in their natural order.* This result allowed us to recast the reach calculation as a constrained shortest path problem, and to give a polynomial time algorithm in [6] for extremal reaches of *arbitrary* panel-and-hinge chains.

Workspace Boundary. The *workspace* is the locus of the end-point T in R^3, as the chain assumes all possible configurations. Since the abstract configuration space can be parametrized by n dihedral angles and thus is the n-dimensional torus $(S^1)^n$, we obtain the workspace as the image $\mathscr{W}$ of the *end-point map* $e : (S^1)^n \to R^3$, which takes a configuration $\theta = (\theta_1, \ldots, \theta_n) \in (S^1)^n$ to the corresponding position of the end-point $T(\theta) \in R^3$. The symmetry under the circle group S^1 given by rotations of the whole part from the second body on, around the fixed first hinge, implies that the workspace $\mathscr{W}$ is determined by any planar section $\mathscr{W}_2$ through the first hinge.

Singularities of the Endpoint Map. By definition, a *singularity* of the end-point map e is a configuration θ where the differential $De(\theta)$ has rank strictly less than three. A geometric characterization for the singular configurations is known from [2, 7, 8]: a configuration is singular if and only if *there is a line through* $T = e(\theta)$ *which intersects projectively all hinges.* A configuration θ where T reaches a point in the boundary $\partial\mathscr{W}$ of the workspace must be singular, for otherwise the image of e would cover a neighborhood of $T = e(\theta)$ in R^3. Therefore, *the boundary of the workspace is contained in the image of the singularity locus* $e(\Sigma)$: $\partial\mathscr{W} \subset e(\Sigma)$.

The Workspace Boundary $\partial\mathscr{W}_2$ for Panel-and-Hinge Chains. We combine the above characterization of the singularity locus with our characterization of extremal configurations. We choose an arbitrary base-point or *start point* S on the fixed first panel. The configurations achieving the maximum distance between S and T must have T on the workspace boundary, and the oriented endpoint segment ST intersects all hinges in natural order. Obviously, we can move S along this segment until it reaches the first hinge A_1, with the same T as maximum reach. This means that *all maximum reach positions of* T *relative to a base-point* S *can be obtained with* S *on the first hinge* A_1. As we vary the position of S on A_1, the end-point T will trace a portion of the workspace boundary which we call the *max-boundary*. Since the sphere centered at S and passing through T obviously contains the whole workspace, the max-boundary is contained in (actually, coincides with, cf. [5]) the intersection of the workspace with the workspace convex hull boundary.

Non-zero minima are treated similarly. For a given position of a base-point S with a non-zero value for the minimum reach, the end-point T must be on the workspace boundary and on the projective complement of the endpoint segment. There are two possibilities, illustrated schematically in Fig. 5: (a) the intersection with A_1 is on the ray from S to the point at infinity; in this case we can move S until it reaches A_1 at a point denoted by S_-, with the same T as non-zero minimum reach, or (b) the intersection with A_1 (and thereby with all hinges) is on the ray from the point at infinity to T, when repositioning S on A_1 at a point denoted by S_+ produces a maximum reach configuration instead of a minimum one. This means that the corresponding position of T belongs to the max-boundary. *By restricting the choice of the base-point* S *to the first hinge* A_1, *the corresponding non-zero minimum reach positions of* T *trace all possible non-zero minimum reaches which are*

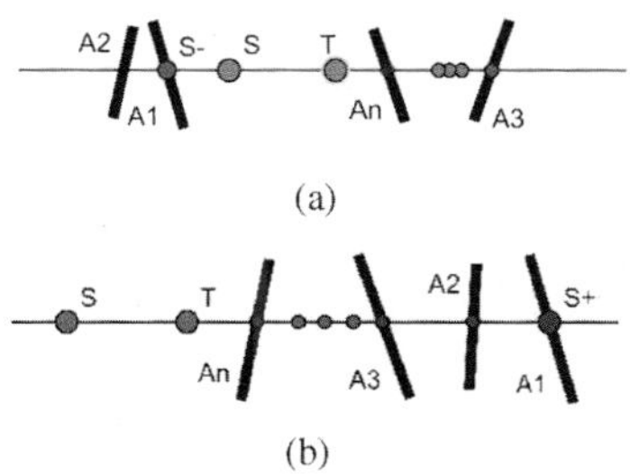

Fig. 5 (a) A minimum reach configuration with the same T for both S and S_-. (b) A minimum reach configuration for S is a maximum reach configuration for S_+.

not already maximum reaches (relative to a repositioned S). The closure of the locus obtained by varying S on A_1 and marking these reaches of T gives our *min-boundary*.

Pivoting at Fold Points. A maximum or non-zero minimum reach configuration has the endpoint axis projectively incident with all hinges. We may assume that T, and thus the endpoint line ST as well, are in the reference plane, intersecting all hinges (possibly at infinity). If the endpoint line ST avoids all points $p_{i,i+1} = A_i \cap A_{i+1}$, then all panels, one after another, must be in the same plane, namely the reference plane, and thus we have a flat configuration. Non-flat extreme reaches must therefore have one or more points $p_{j,j+1}$ on the end-point line ST. These are the *fold points* of the extremal configuration. If $p_{j,j+1}$ is a fold point, the corresponding panel $\pi_{j,j+1}$ may have a different normal direction from its two neighbors $\pi_{j-1,j}$ and $\pi_{j+1,j+2}$, but all panels $\pi_{j+1,j+2}, \dots, \pi_{k-1,k}$ between one fold point $p_{j,j+1}$ and the next fold point $p_{k,k+1}$, with $j+1 < k$, must be coplanar. All panels up to the first fold point must be in the reference plane and all panels after the last fold point are coplanar. One may imagine now all hinges as 'locked', except for the two meeting at $p_{j,j+1}$. This creates a short chain of 2-hinges: when T is near the given initial position, it is tracing a circular arc with center at $p_{j,j+1}$ in the reference plane. For some time, depending on the position of S on the first axis, the line through $p_{j,j+1}$ and T will intersect the first two axes in the same order as before. This scenario will be called **pivoting** (at the fold point $p_{j,j+1}$) and will be used repeatedly for tracing the max- and min-boundaries.

Algorithm for Computing the Workspace Boundary (Sketch). We lay our chain flat in the reference plane and conceptually sweep a point S on the first axis A_1, starting from the intersection with the second hinge axis A_2, stopping at certain events $S_1, S_2, \cdots$. Using our general algorithm [6], we compute the maximum reach from S to T at each corresponding event S_i: the first fold point on the line from S_i to T gives the center of a circular arc on the workspace boundary $\partial\mathscr{W}_2$, with the radius given by the actual length of the maximum reach value from that fold point on. The circular arc is traced during a pivoting step for as long as the combinatorics of the maximum reach (imagined for S moving on the first axis) still goes through this first fold point. The max-boundary ends with either S going to infinity, or, in the other direction, when a maximum reach morphs into a min reach. In [5] we describe how to do the sweep efficiently in the orthogonal case (when consecutive hinges meet at an angle of $\pi/2$). This method can be generalized to a larger family of chains, but - as we now show - it will not always work in polynomial time. The main source of difficulty arises from the sheer complexity of the boundary, in terms of number of circular arcs making it.

3 An Exponentially Large Workspace Boundary

If for a choice of a starting point S on the first axis, a *flat* chain is in a maximum reach configuration with respect to the terminus T, then the line segment ST ("endpoint axis") meets all the hinges in the natural order. Imagine the ST line segment drawn

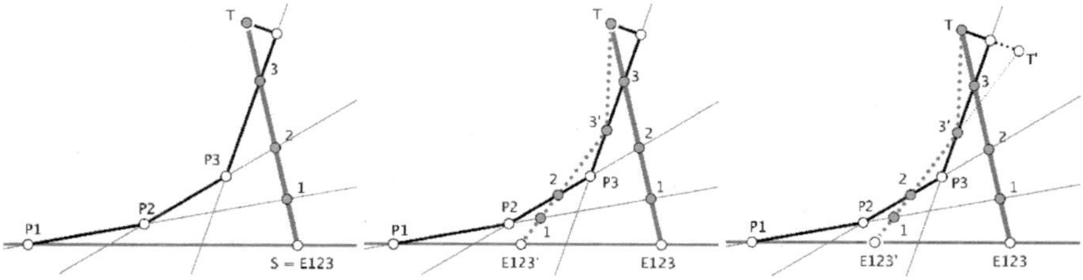

(a) (b) (c)

Fig. 6 Two endpoint axes, in gray and dotted gray, for two positions of S on the first hinge. (a) A straight (solid, from T to $S = E123$) endpoint axis, crossing the hinges in the natural order 123. (b) The second (dotted, from T to $S = E123'$) is reflected in the third axis, and corresponds to a flip of the last panel, bringing the terminus T to the new position T', as shown in (c).

in "red" on the plane containing the flat chain. Then, when the chain is folded into a different flat configuration (with rotations by π of some panels along incident hinges), the "red" line bends at the folds and appears as if *reflected* in those hinge axes which were folded. Walking along such a *folded endpoint axis* from S to T, *reflection at a hinge line* corresponds to some flipping of the next panel along that hinge, as to achieve another flat configuration where the reflected endpoint axis aligns to become straight. Fig. 6 illustrates this observation. We use it to represent, on a single flat configuration, multiple endpoint axes (called from now on *paths*) corresponding to maximum reach configurations originating from several starting points S chosen on the first hinge line.

To simplify the description of our construction, we relabel the hinges so that the first one is now A_0. We describe now the worst-case scenario of 2^n possible paths (going through or reflecting into each of n hinges, in natural order) and refer to Fig. 7. The first hinge axis A_0 is taken to be horizontal, and above it we construct a convex chain with n hinges $A_1, \cdots, A_n$. We take A_{i+1} such that it makes a left turn angle of $2^{n-i}\alpha$ with A_i, for some *very* small angle α. The terminus T is placed slightly above the last hinge A_n, so that a vertical line through T will meet the horizontal axis A_1 in a point $S = S_{12\cdots n}$. When taken as the starting point S, this leads to chain (with $n-1$ hinges) which is, in this flat position, in a global maximum reach position; indeed, its endpoint segment ST is straight and meets all the hinges in natural order. For appropriate choices of the edge lengths of the chain (which also grow exponentially), there are 2^n ways of reflecting this ST path, in all the hinges, so that these reflected paths will meet the base A_0 hinge in 2^n points; these are denoted by $S_{1^*2^*\cdots n^*}$, with i^* being either i (to denote that hinge i is crossed) or i' (reflection at hinge i). In Fig. 7 the symbol S is dropped, and just the sequence of signed indices is shown. Because of the very specific choice of slopes for the hinges, a precise straight-line drawing is impossible to draw (due to exponential explosion in its size) or to decipher (if scaled too small). We have therefore, for clarity, illustrated this intersection and reflection pattern with curved lines instead, as in Fig. 7(b). The choice of slopes (exponentially increasing from A_n to A_1) and edge lengths (inductively, proceeding from A_n to A_1, so that the crossing pattern

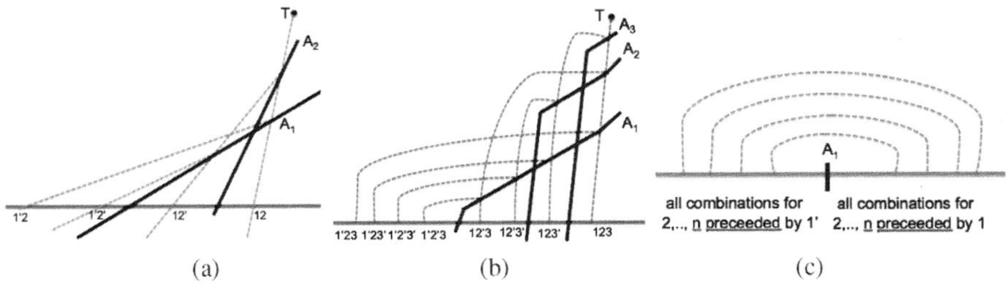

Fig. 7 (a) The base case, of two hinges A_1,A_2 and 4 reflected paths. The reflection angles in this picture are not exact. (b) The reflection pattern for 3 hinges, illustrated with curved lines: black solid, the hinges; gray dashed, the reflection paths. Labeled on the horizontal A_0 axis are the $2^3 = 8$ reflection patterns. (c) The general pattern in which the paths intersect the first axis, generalized from (b). The arcs illustrate a matching of shorter patterns (on $n-1$ symbols), before being preceded by a 1 (to indicate crossing straight through line 1) or 1' (reflected through line 1) for the larger pattern on n symbols.

illustrated in Fig. 7(b) is realized) allow for this very special chain, with 2^n circle arcs on its max boundary.

This family of examples precludes a general, worst case polynomial time algorithm for computing the entire workspace boundary. We leave for future work the problem of finding an output-sensitive method, as well as devising algorithms for simpler tasks, such as the determination of topological parameters of the workspace boundary (connected components).

References

1. Angeles, J.: Fundamentals of Robotic Mechanical Systems: Theory, Methods and Algorithms. Mechanical Engineering Series. Springer (2007)
2. Borcea, C.S., Streinu, I.: Singularities of hinge structures. arXiv:0812.1373
3. Borcea, C.S., Streinu, I.: Extremal configurations of manipulators with revolute joints. In: Reconfig. Mech. and Robots (ReMAR), Proc. ASME/IFToMM Int. Conf., pp. 279–284. KC Edizioni (2009). arXiv:0812.1375
4. Borcea, C.S., Streinu, I.: How far can you reach. In: Proc. ACM-SIAM Symposium on Discrete Algorithms, pp. 928–937 (2010)
5. Borcea, C.S., Streinu, I.: Exact workspace boundary by extremal reaches. In: Proc. 27th Symp. Comp. Geometry, pp. 481–490. ACM Press (2011)
6. Borcea, C.S., Streinu, I.: Extremal reaches in polynomial time. In: Proc. 27th Symp. Comp. Geometry, pp. 472–480. ACM Press (2011)
7. Burdick, J.W.: A classification of 3R regional manipulator singularities and geometries. Mech. Mach. Theory **30**(1), 71–89 (1995)
8. Burdick, J.W.: A recursive method for finding revolute-jointed manipulator singularities. Trans. ASME, J. Mech. Des. **117**, 55–63 (1995)
9. Ceccarelli, M.: A formulation for the workspace boundary of general n-revolute manipulators. Mech. Mach. Theory **31**(5), 637–646 (1996)
10. Ottaviano, E., Husty, M.L., Ceccarelli, M.: Identification of the workspace boundary of a general 3R manipulator. J. Mech. Des. **128**, 236–242 (2006)
11. Siciliano, B., Khatib, O. (eds.): Handbook of Robotics. Springer Verlag (2008)
12. Zein, M., Wenger, P., Chablat, D.: An exhaustive study of the workspace topologies of all 3R orthogonal manipulators with geometric simplifications. Mech. Mach. Theory **41**(8), 971–986 (2006)

Kinematics Analysis of a Parallel Surgical Robot

A. Szilaghyi, A. Stoica, D. Pisla, C. Vaida and N. Plitea

Abstract The geometric and kinematic analysis of a surgical parallel robot, used for camera and active instruments positioning is presented in this paper. Its workspace is also illustrated. The robot structure consists of two modules: the PARAMIS robot and the new parallel positioning module. The use of this new parallel structure in surgery presents the following advantages: it releases some of the pressure on the abdominal wall in certain positions of the robot and the new robot may be used either as laparoscope holder or as manipulator of active instruments, used for cutting, suturing, grasping etc. The new robot provides the necessary control of motion to fulfil the imposed tasks in the case of surgical applications. Numerical results and conclusions from the performed simulations are described.

Key words: Kinematics, parallel robot, surgical applications, numerical results

1 Introduction

In the last 25 years, minimally invasive surgery (MIS) has evolved and developed continuously. MIS refers to any procedure which is less invasive than open surgery for the same purpose [1]. This procedure typically involves the use of several small sized incisions to introduce the surgical instruments together with an endoscope which provides an indirect view of the surgical field. In a MIS procedure, some of the demonstrated patient benefits are: less trauma, reduced blood loss, shorter recovery time, less emotional discomfort, less morbidity. As a result of the latest developments in this field, robots have been used in complex surgical procedures. However, these robots are not autonomous machines that can carry out simple, pre-programmed instructions by themselves, but they are able to supplement the sur-

A. Szilaghyi · A. Stoica · D. Pisla · C. Vaida · N. Plitea
Technical University of Cluj Napoca, 400114 Cluj-Napoca, Romania, e-mail: {andras.szilaghyi, alin.stoica,doina.pisla,calin.vaida,nicolae.plitea}@mep.utcluj.ro

J. Lenarčič, M. Husty (eds.), *Latest Advances in Robot Kinematics*,
DOI 10.1007/978-94-007-4620-6_42,

geon's abilities, translating human movements into more steady and accurate robotic movements, which, in turn, manipulate instruments to aid delicate operations [2].

Initial use of robots in surgery began in 1985, with PUMA 560 serial robot which has been used to hold and to guide tools for stereotactic biopsy in neurosurgery [3]. The first surgical robotic application for positioning the laparoscopic camera within the surgical field was performed, in 1993 with the help of AESOP 1000 robotic system [4]. Prosurgics provides FreeHand [5], a robotic system that positions the laparoscopic camera based on the head motions of the surgeon. Another surgical robot is ViKY developed by EndoControl [6, 7], which is a compact motorized endoscope holder for a wide range of laparoscopic and thoracic surgeries. One of the most complex robot used in MIS is the da Vinci surgical system [8] developed in 1997 by Intuitive Surgical Inc. and currently at the second generation, approved for medical use, namely in surgery and urology. Another system, the laparoscope manipulator Naviot was developed in Japan. It is based on a five-bar linkage mechanism that has two independent motors on the bottom. In addition, the zoom-up mechanism of the laparoscope was applied to this manipulation system [9, 10]. Titan Medical Corporation developed another important system, called Amadeus® [11]. It uses cutting edge hardware and software from medical, defence and aerospace industries [11]. Another robot for surgery is MiroSurge, a configuration of the Miro platform (MRSP), that allows bimanual endoscopic telesurgery with force feedback [12]. Medrobotics [13] developed a flexible robot based on a highly articulated multi-link structure, for minimally invasive surgery.

The Technical University of Cluj-Napoca achieved the first steps in this field of research starting with 2005. The first experimental model of the laparoscopic holder PARAMIS was made in 2008 [14–16]. Taking into account, earlier results of the PARAMIS robot as laparoscope holder, encouraged further researches to expand the applicability domain of the robot, by transforming the passive joint from the tip of the arm, in active joint. This paper proposes the kinematic study of a parallel robot, consisting of PARAMIS robot and a new parallel module, which is described in detail in Section 2. Sections 3 and 4 deal with the kinematical analysis of the parallel robot, followed by some numerical simulation results in Section 5. Conclusions are given in Section 6.

2 Description of the New Robot

The main requirements of surgeons imposed on a surgical robot used for minimally invasive surgery are the following: the robot control has to be accurate, it has to be stable and rigid in the operating room, it should occupy a minimum space in the patient proximity; in addition to these technical characteristics the safety features must prevent the patient or surgeon harm in case of a robot malfunction [17].

The existing PARAMIS robot had actually all of these characteristics, but from practice it has been proved that there are situations when the abdominal wall is under a lot of pressure in certain positions from the laparoscope, requiring an additional

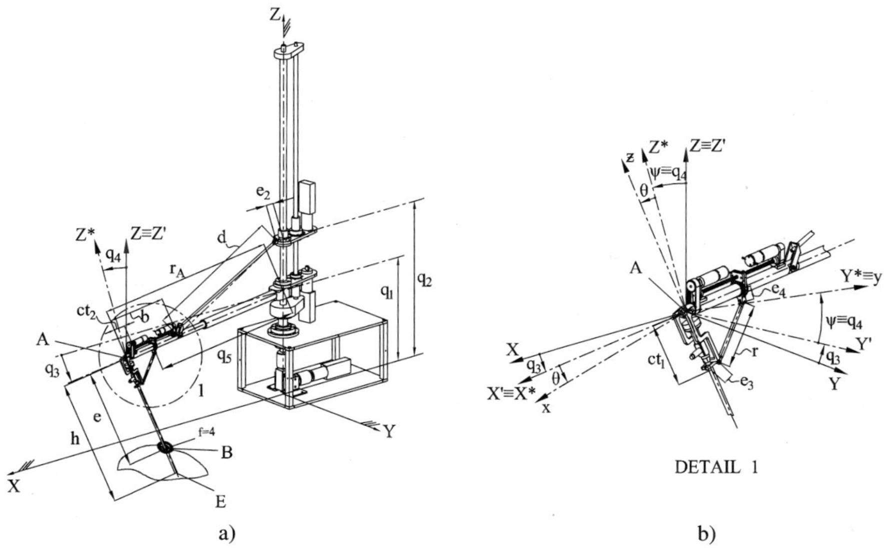

Fig. 1 The structure of the robot: (a) kinematic scheme; (b) detail of the parallel module.

mechanism that could release some of that pressure. The new robot structure has also the advantage that it may be used either as laparoscope holder or as a positioning arm for the active instruments used for cutting, suturing, grasping etc.

The new parallel robot with five motors is based on an already existing robot and an attachable parallel orientation module, assuring high rigidity due to its closed chains. The motivation for choosing a parallel configuration is given by its architectural advantages [18]. There is a restriction the mechanism must accomplish: the surgical instrument must pass through a fixed point in space. Such as example is the point $B(X_B, Y_B, Z_B)$ in Figure 1a which represent the point of incision of the laparoscope in the abdomen. The robotic arm consists of a positioning parallel module with 3 motors (PARAMIS), and an orientation parallel module with 2 motors, presented in Figure 1a. From the first three active joints, actuated from the robot basis, two of them are prismatic and one is rotational. From the other two active joints on the tip of the arm r_A, one of them is rotational and the other one is a prismatic joint.

In this case the robot uses its five active joints and several passive joints (four rotational joints and one prismatic joint) to fulfil its task. The last two active joints are q_4 and q_5, where q_4 achieves a rotational motion around the axis X', which is an axis situated in a parallel plane with the fixed plane XOY and q_5 achieves a rotational motion around the axis $Y*$, which is an axis situated in a plane obtained after the last motion. These two active joints, have the role to maintain a virtual point, fixed, to eliminate to leading the instrument, using the abdominal wall. Angles ψ and θ are also presented in Figure 1b. The laparoscope can be positioned in any point of the surgical field and the advantage of using this orientation parallel module is a better guidance of the camera and the possibility to position an active instrument.

3 The Geometric Model and the Robot Workspace

To obtain the geometric model, a mobile frame *AX'Y'Z'*, attached to the endoscope in the point *A* is used. As one can see in Fig. 1, r_A has the following form:

$$r_A = b + e_2 + \sqrt{d^2 - (q_2 - q_1)^2} \tag{1}$$

where q_1, q_2, q_3, q_4 and q_5 are the active coordinates of the robot and b, h, e_2 and d are geometrical parameters. A relation between some of these coordinates and the coordinates of point $A(X_A, Y_A, Z_A)$ may be defined:

$$X_A = r_A \cdot \cos(q_3), \quad Y_A = r_A \cdot \sin(q_3), \quad Z_A = q_1, \tag{2}$$

Direct geometric model. The displacements in motor coordinates q_1, q_2 and q_3 are given. One must take into consideration that the laparoscope must pass through the fixed point *B*, and for that to operate the robot without pressing on the abdominal wall, the coordinates q_4 and q_5 have to be determined. The end-effector coordinates are namely the point $E(X_E, Y_E, Z_E)$ – the tip of the laparoscope, and the two rotation angles ψ and θ. The angles ψ and θ are defined by equations:

$$\psi = \arctan 2(Y_B - Y_A, Z_A - Z_B), \quad \theta = \arctan 2(X_A - X_B, \quad \sqrt{e^2 - (X_A - X_B)^2}), \tag{3}$$

Where:

$$e = \sqrt{(X_B - X_A)^2 + (Y_B - Y_A)^2 + (Z_B - Z_A)^2} \tag{4}$$

Coordinates q_4 and q_5 are determined by the expressions:

$$q_4 = \psi \tag{5}$$

$$q_5 = ct_1 \cdot \sin(\theta) + e_3 \cdot \cos(\theta) + r \cdot \cos(\alpha) - ct_2 \tag{6}$$

where

$$\alpha = \frac{\arcsin(ct_1 \cdot \cos(\theta) - e_3 \cdot \sin(\theta) - e_4)}{r} \tag{7}$$

The notations ct_1, ct_2, e_3, e_4 and r were used for some geometrical parameters 1. The coordinates of point *E* are determined by the following relations:

$$\begin{cases} X_E = X_A - h \cdot \sin(\theta), \\ Y_E = Y_A + h \cdot \sin(\psi) \cdot \sin(\theta), \\ Z_E = Z_A - h \cdot \cos(\psi) \cdot \cos(\theta), \end{cases} \tag{8}$$

Inverse geometric model. In this case, the generalized coordinates q_1, q_2, q_3, q_4 and q_5 of the robot and the rotational angles ψ and θ, have to be determined using the generalized coordinates of the end effector X_E, Y_E, Z_E.

To solve the problem, the expressions of the coordinates of point $A(X_A, Y_A, Z_A)$, are determined using the coordinates X_E, Y_E, Z_E of point E.

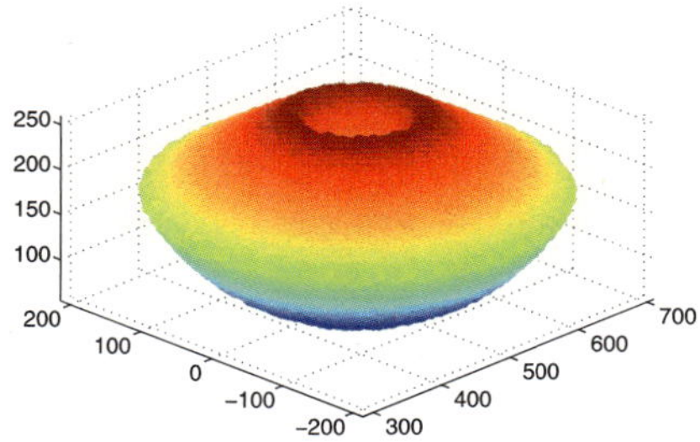

Fig. 2 The robot workspace.

$$\begin{cases} X_A = X_E + h \cdot \sin(\theta), \\ Y_A = Y_E - h \cdot \sin(\psi) \cdot \cos(\theta), \\ Z_A = Z_E + h \cdot \cos(\psi) \cdot \cos(\theta), \end{cases} \tag{9}$$

where the rotational angles ψ and θ are defined by equations:

$$\psi = \arctan 2(Y_E - Y_B, Z_B - Z_E) \tag{10}$$

$$\theta = \arctan 2(X_B - X_E, \sqrt{e_1^2 - (X_B - X_E)^2}) \tag{11}$$

where:

$$e_1 = \sqrt{(X_B - X_E)^2 + (Y_B - Y_E)^2 + (Z_B - Z_E)^2} \tag{12}$$

$$\begin{aligned} q_1 &= Z_E + h \cdot \cos(\theta) \cdot \cos(\psi), \\ q_2 &= Z_E + h \cdot \cos(\psi) \cdot \cos(\theta) + \sqrt{d^2 - (r_A - b - e_2)^2}, \\ q_3 &= \arctan 2(Y_E - h \cdot \sin(\psi) \cdot \cos(\theta), X_E + h \cdot \sin(\theta)), \\ q_4 &= \psi, \\ q_5 &= ct_1 \cdot \sin(\theta) + e3 \cdot \cos(\theta) + r \cdot \cos(\alpha) - ct_2, \end{aligned} \tag{13}$$

The active coordinates are obtained with the equations (13).

Robot workspace. Figure 2 shows the workspace of the robot, pointing out the constraints imposed. In order to analytically generate the workspace inside the abdominal cavity, the inverse geometric model has been used where the insertion point into the abdomen: $B(X_B, Y_B, Z_B)$ and the coordinates of point $E(X_E, Y_E, Z_E)$ are considered known. There are several constraints that must be taken into consideration: the angle between a normal axis on the insertion surface and the instrument must not exceed 60 degrees; denoting with A the outer extremity and with E the tip of the instrument the following restrictions apply: length $AB \leq 220$ mm and length $BE \geq 50$ mm, restrictions imposed by the dimensions of the trocar and the total laparoscope length.

4 Kinematics

The relations for the kinematics, including velocities and accelerations uses the equations from the geometric model.

$$F_i(q_i, X_E, Y_E, Z_E, \psi, \theta) = 0; \quad i = 1 \dots 5 \tag{14}$$

The system of equations in implicit form is obtained using the equations of the previous section:

$$\begin{cases} F_1 = q_1 - Z_E - h \cdot \cos(\theta) \cdot \cos(\psi), \\ F_2 = (q_2 - Z_E - h \cdot \cos(\psi) \cdot \cos(\theta))^2 - d^2 + RAD, \\ F_3 = (X_E + h \cdot \sin(\theta)) \cdot \sin(q_3) - (Y_E - h \cdot \sin(\psi) \cdot \cos(\theta)) \cdot \cos(q_3), \\ F_4 = q_4 - \psi, \\ F_5 = q_5 - ct_1 \cdot \sin(\theta) - e_3 \cdot \cos(\theta) - r \cdot \cos(\alpha) - ct_2, \end{cases} \tag{15}$$

where:

$$RAD = (\sqrt{(X_E + h \cdot \sin(\theta))^2 + (Y_E - h \cdot \sin(\psi) \cdot \cos(\theta))^2} - b - e_2)^2 \tag{16}$$

The relations for velocities result by differentiating the system of equations (15) with respect to time. Using the matrix representation, the kinematic model for velocities is:

$$A \cdot \dot{X} + B \cdot \dot{q} = 0, \tag{17}$$

Differentiating two times the system of equations (15), with respect to time, results the relations for accelerations.

$$\dot{A} \cdot \dot{X} + A \cdot \ddot{X} + \dot{B} \cdot \dot{q} + B \cdot \ddot{q} = 0, \tag{18}$$

where $\dot{q} = [\dot{q}_1\ \dot{q}_2\ \dot{q}_3\ \dot{q}_4\ \dot{q}_5]^T$ are the driving velocities, $\dot{X} = [\dot{X}_E\ \dot{Y}_E\ \dot{Z}_E\ \dot{\psi}\ \dot{\theta}]^T$ are the end-effector and the angular velocities, $\ddot{q} = [\ddot{q}_1\ \ddot{q}_2\ \ddot{q}_3\ \ddot{q}_4\ \ddot{q}_5]^T$ are the driving accelerations and $\ddot{X} = [\ddot{X}_E\ \ddot{Y}_E\ \ddot{Z}_E\ \ddot{\psi}\ \ddot{\theta}]^T$ are the end-effector and the angular accelerations. Both the *direct kinematic model (DKM)* and the *inverse kinematic model (IKM)* for velocities and accelerations were determined from relations (17) and (18). The solving of the both kinematic models leads to analytical solutions, which represents an advantage in the achievement of the control system.

5 Numerical Results

In the case of the parallel robot used in minimally invasive surgery, the geometric and kinematic models have been developed, in order to perform the imposed movements (given by the surgeon). It is supposed that the end effector axis always passes through a fixed point defined in space, which is the intersection point between the abdomen and the end effector. For validation, the developed models were

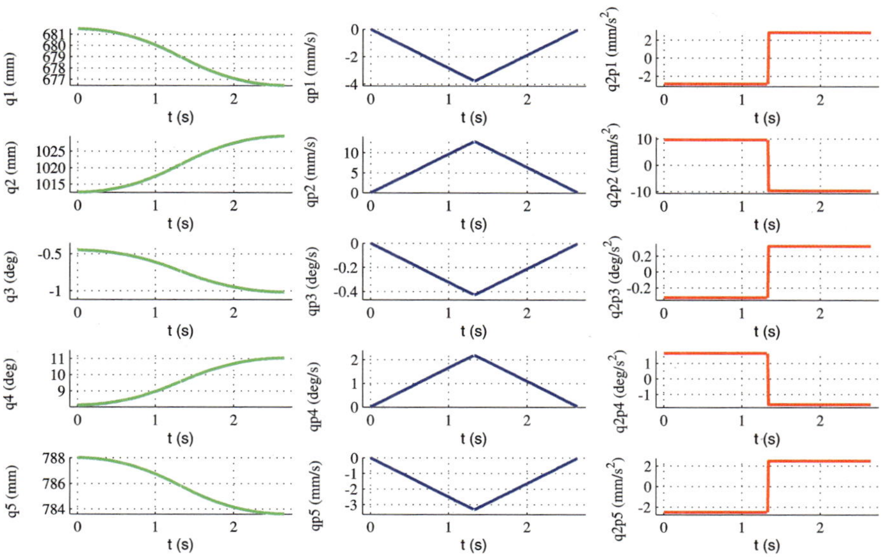

Fig. 3 Time history diagrams for driving coordinates, velocities and accelerations.

implemented as functions in the MATLAB simulation software, finally obtaining numerical results and graphics for an imposed movement. As input data were used the following constructive values of components involved directly in equations: $b = 326$ mm, $d = 654$ mm, $h = 270$ mm, $e_2 = 44$ mm, $ct_1 = 125$ mm, $ct_2 = 53$ mm, $e_3 = 28$, $e_4 = 27$ mm, $r = 144$ mm. For simulation it was chosen the command *MOVE UP* [19] which means a motion of the end-effector from the initial position and orientation with 5 degrees on the upper direction, from the surgeon's view. In kinematic terms, the motion represents a rotation around an axis with the origin in B, perpendicular on the longitudinal axis of the laparoscope, this axis being contained in a plane parallel with the horizontal one (XOY). Using a set of initial coordinates ($X_{Ei} = 990$ mm, $Y_{Ei} = 30$ mm, $Z_{Ei} = 420$ mm) for the end-effector and following the *MOVE UP* command the coordinates of the final position were determined ($X_{Ef} = 995$ mm, $Y_{Ef} = 33.3$ mm, $Z_{Ef} = 421.8$ mm), respecting the imposed geometrical restriction represented by a fixed point B with coordinates ($X_B = 975$ mm, $Y_B = 20$ mm, $Z_B = 490$ mm), where the longitudinal axis of the end effector must pass through. Figure 3 represents the obtained simulation results of the kinematics.

6 Conclusions

A new orientation parallel module was described in this paper. Starting from an existing robot with three DOF and adding a new parallel module, a structure was obtained which no longer induce any pressure on the abdominal wall which can handle both a laparoscope or an active instrument for different operations like cutting, suturing, grasping. The direct and inverse kinematics were solved using an analytical method. The obtained numerical results have shown that the kinematic models could be successfully implemented in the control algorithms of the experimental model.

Acknowledgements This paper was supported by the projects "Doctoral studies in engineering sciences for developing the knowledge based society-SIDOC" contract no. POSDRU/88/1.5/S/60078, "Development and support of multidisciplinary postdoctoral programmes in major technical areas of national strategy of RDI" 4D-POSTDOC, POSDRU/89/1.5/S/52603, both co-funded from European Social Fund through Sectorial Operational Program Human Resources 2007-2013 and Scopes International Grant IZ74Z0-137361 entitled "Creative Alliance in Research and Education focused on Medical and Service Robotics (CARE-Robotics)".

References

1. Gomes, P.: Surgical robotics: reviewing the past, analysing the present, imagining the future. Robot. Comput.-Integr. Manuf. **27**, 261–268 (2011)
2. Lobontiu, A., Loisance, D.: Robotic surgery and tele-surgery: Basic principles and description of a novel concept. J. Chir., Iasi **3**(3), 208–214 (2007).
3. Kragic, D., Miller, A.T., Allen, P.K.: Real-time tracking meets online grasp planning. In: 2001 ICRA IEEE International Conference on Robotics and Automation, vol. 3, pp. 2460–2465 (2001). doi:10.1109/ROBOT.2001.932992
4. Brown University, Division of Biology and Medicine http://biomed.brown.edu (2012)
5. Finlay, P.A.: A new miniature manipulator for laparoscopic camera control. In: WC on Medical Physics and Biomedical Eng., Germany, (2009). doi:10.1007/978-3-642-03906-5-34
6. Gumbs, A., et al.: Totally transumbilical laparoscopic cholecystectomy. J. Gastrointest. Surg. **13**, 533–534 (2009). doi:10.1007/s1105-008-0614-8
7. Voros, S., et al.: ViKY robotic scope holder: Initial clinical experience and preliminary results using instrument tracking. IEEE/ASME Trans. Mechatron. **15**(6), 879–886 (2010)
8. http://www.riverviewregional.com/Services/da-Vinci-Robot/da-Vinci-Background-information/Default.aspx
9. Hashizume, M., Tsugawa, K.: Robotic surgery and cancer: The present state, problems and future vision. Jpn. J. Clin. Oncol. **34**(5), 227–237 (2004)
10. Yasunaga, T., et al.: Remote-controlled laparoscope manipulator system, Naviot® for endoscopic surgery. Int. Congr. Ser. **1256**, 678–683 (June 2003)
11. Titan Medical Homepage, http://www.titanmedicalinc.com/ (2012)
12. Hagn, U., et al.: DLR MiroSurge: A versatile system for research in endoscopic telesurgery. Int. J. Comput. Assisted Radiol. Surg. **5**(2), 183–193 (2009)
13. Medrobotics, http://www.medrobotics.com/index.html (2012)
14. Plitea, N., et al.: Innovative development of parallel robots. Acta Electrotehnica, Meditech 2007, Cluj-Napoca, Romania, 201–206, (2007). ISSN 1841-3323
15. Pisla, D., Plitea, N., Vaida, C.: Kinematic modeling and workspace generation for a new parallel robot used in minimally invasive surgery. In: Advances in Robot Kinematics (2008)
16. Vaida, C., et al.: Development of a control system for a parallel robot used in minimally invasive surgery. In: Meditech 2009. IFMBE Proceedings Series (2009). ISSN 1680-0737
17. Taylor, R.H., Kazanzides, P.: Medical robotics and computer-integrated interventional medicine. Adv. Comput. 73, 219–260 (2008)
18. Merlet, J.-P.: Parallel Robots. Kluwer Academic Publisher (2006)
19. Vaida, C., et al.: Development of a voice controlled surgical robot. In: New Trends in Mechanism Science. Mechanisms and Machine Science, vol. 5, Part 10, pp. 567–574 (2010). doi:10.1007/978-90-481-9689-0-65

Constraint Singularity-Free Design of the IRSBot-2

Coralie Germain, Sébastien Briot, Stéphane Caro and Philippe Wenger

Abstract This paper deals with the constraint analysis of a novel two-degree-of-freedom (DOF) spatial translational parallel robot for high-speed applications named the IRSBot-2 (acronym for IRCCyN Spatial Robot with 2 DOF). Unlike most two-DOF robots dedicated to planar translational motions this robot has two spatial kinematic chains that provide a very good intrinsic stiffness. First, the robot architecture is presented and its constraint singularity conditions are given. Then, its constraint singularities are analyzed in its parameter space based on a cylindrical algebraic decomposition. Finally, a deep analysis is carried out in order to determine the sets of design parameters of the IRSBot-2 that prevent it from reaching any constraint singularity. To the best of our knowledge, such an analysis is performed for the first time.

Key words: Parallel manipulator, constraint singularity, cylindrical algebraic decomposition, design

1 Introduction

Several robot architectures with two translational degrees of freedom (DOF) for high-speed operations have been proposed in the past decades. Brogårdh proposed in [2] an architecture made of a parallelogram joint (also called Π joint) located between the linear actuators and the platform. Another two-DOF translational robot was presented in [5], where the authors use two Π joints to link the platform with two vertical prismatic actuators. Its equivalent architecture actuated by revolute joints is presented in [4].

Coralie Germain · Sébastien Briot · Stéphane Caro · Philippe Wenger
Institut de Recherche en Communications et Cybernétique de Nantes, Nantes, France, e-mail: {germain,briot,caro,wenger}@irccyn.ec-nantes.fr

J. Lenarčič, M. Husty (eds.), *Latest Advances in Robot Kinematics*,
DOI 10.1007/978-94-007-4620-6_43,

The foregoing architectures are all planar, i.e., their elements are constrained to move in the plane of motion. As a result, their elements are all subject to bending effects in the direction normal to the plane of motion. In order to guarantee a minimum stiffness in this direction, the elements have to be bulky, leading to high inertia and low acceleration capacities. In order to overcome these problems, a new Delta-like robot, named the Par2, was proposed in [7]. However, even if its acceleration capacities are impressive, its accuracy is poor.

A two-DOF spatial translational robot, named IRSBot-2, was introduced in [3] to overcome its counterparts in terms of mass in motion, stiffness and workspace size. The IRSBot-2 has a spatial architecture and the distal parts of its legs are subject only to traction/compression/torsion. As a result, its stiffness is increased and its total mass can be reduced. Nevertheless, the IRSBot-2 may reach some constraint singularities [1, 8]. In this paper, a deep analysis is carried out in order to determine the sets of design parameters of the IRSBot-2 that prevent it from reaching any constraint singularity.

This paper is organized as follows. First, the robot architecture is described and its constraint singularity conditions are given. Then, its constraint singularities are analyzed in its parameter space based on a cylindrical algebraic decomposition. Finally, the set of design parameters for the robot to be free of constraint singularity are determined.

2 Robot Architecture and Constraint Singularity Conditions

The IRSBot-2 is shown in Fig. 1 and is composed of two identical legs linking the fixed base to the moving platform. Each leg contains a proximal module and a distal module, which are illustrated in Fig. 2.

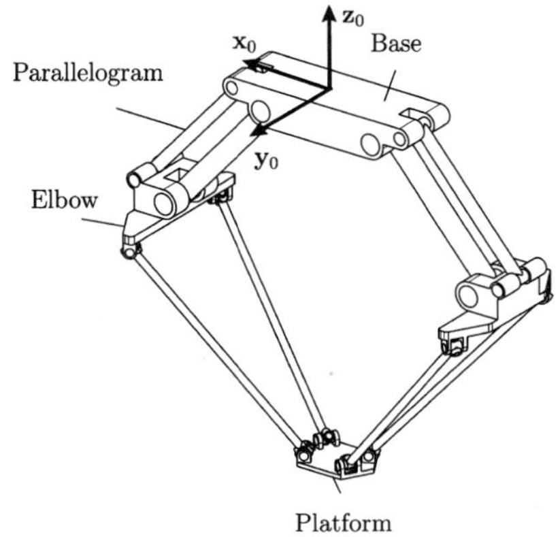

Fig. 1 CAD Modeling of the IRSBot-2

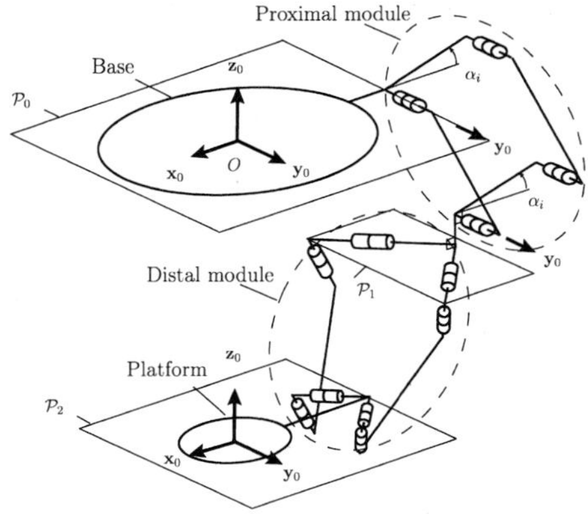

Fig. 2 Kinematic chain of the ith leg ($i = 1, 2$)

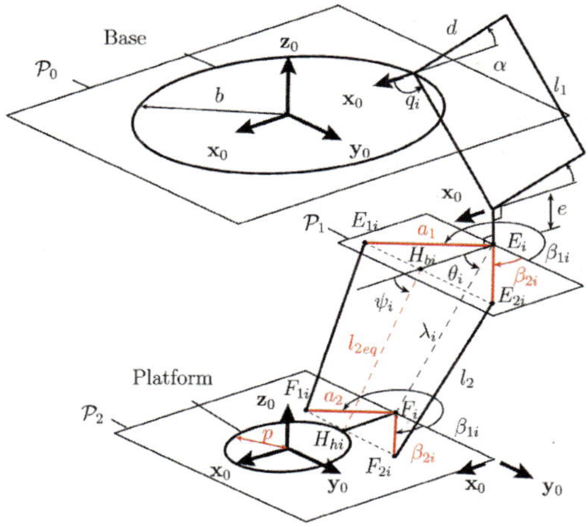

Fig. 3 Parameterization of the ith leg ($i = 1,2$)

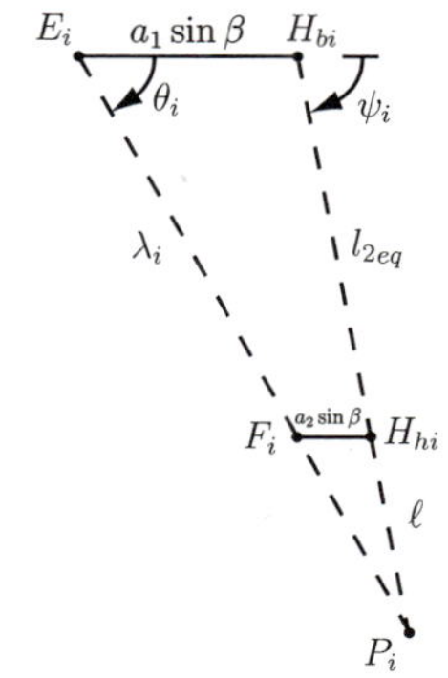

Fig. 4 Closed-loop E_i–H_{bi}–H_{hi}–F_i: projection of the distal module on the plane $(\mathbf{x}_0\, O\, \mathbf{z}_0)$

The parameters of the IRSBot-2 used throughout this paper are depicted in Figs. 3 and 4. From [3], the IRSBot-2 reaches a constraint singularity iff:[1]

$$\theta_1 = \theta_2 + k\pi, \quad k = 0,1 \tag{1}$$

and

$$(x_{P_2} - x_{P_1})\cos^2\beta\cos\theta_2 - (z_{P_2} - z_{P_1})\sin\theta_2 = 0 \tag{2}$$

It is noteworthy that Eqs. (1) and (2) depend only on the design parameters associated with the distal module. Therefore, the proximal modules of the IRSBot-2 do not affect its constraint singularities and we focus only on the constraint singularities associated with the distal modules.

3 Constraint Singularity Analysis of the IRSBot-2 in Its Parameter Space

This section aims to find the sets of design parameters $(a_1, a_2, \beta, p, l_{2eq})$ that allow the IRSBot-2 to reach some constraint singularities. Note that the foregoing five design parameters are shown in Fig. 3. a_1, a_2 and l_{2eq} are the lengths of segments E_iE_{1i}, F_iF_{1i} and $H_{bi}H_{hi}$, respectively. p is the moving-platform radius. The coordinates of vector $\overrightarrow{P_1P_2}$ can be expressed as:

[1] Let β denote β_{22}, then $\beta_{11} = \pi + \beta$, $\beta_{21} = -\beta$ and $\beta_{12} = \pi - \beta$

$$x_{P_2} - x_{P_1} = 2p + \ell(\cos\psi_2 - \cos\psi_1) \quad (3)$$
$$z_{P_2} - z_{P_1} = \ell(\sin\psi_2 - \sin\psi_1) \quad (4)$$
$$\ell = \frac{a_2 l_{2eq}}{a_1 - a_2} \quad (5)$$

Angles ψ_1 and ψ_2 are depicted in Figs. 3 and 4. From the closed-loop E_i–H_{bi}–H_{hi}–F_i $(i = 1,2)$ and Fig. 4, the following relations between λ_i, θ_i and ψ_i are obtained:

$$l_{2eq}\cos\psi_i = \lambda_i\cos\theta_i - (a_1 - a_2)\sin\beta \quad (6)$$
$$-l_{2eq}\sin\psi_i = -\lambda_i\sin\theta_i \quad (7)$$

λ_i is depicted in Fig. 4 and is derived from Eqs. (3) to (7):

$$\lambda_i = \sqrt{l_{2eq}^2 + (a_1 - a_2)^2\sin^2\beta + 2(-1)^{i+1} l_{2eq}\cos\psi_i(a_1 - a_2)\sin\beta} \quad (8)$$

The following three cases, obtained from Eqs. (1) and (8), allow us to simplify Eqs. (3) to (7) to end up with a univariate polynomial form of constraint singularity condition (2):

Case I: $\theta_1 = \theta_2 + \pi$ and $\lambda_1 = \lambda_2 \neq 0$
Case II: $\theta_1 = \theta_2 + \pi$ and $\lambda_1 \neq \lambda_2$
Case III: $\theta_1 = \theta_2$

For Case I, Eq. (2) takes the form:

$$P_I(X) = A_1 X^2 + B_1 X + C_1 = 0 \quad (9)$$

with

$$\begin{cases} A_1 = -\, l_{2eq}^2 \sin^2\beta\, a_2/(a_1 - a_2) \\ B_1 = l_{2eq}\,(1 - \sin^2\beta)\,(p - a_2\,\sin\beta) \\ C_1 = -\,p\,(a_1 - a_2)\,(1 - \sin^2\beta)\,\sin\beta + l_{2eq}^2\, a_2/(a_1 - a_2) \\ \quad X = \cos\psi,\ \psi = \psi_2,\ X \in [-1,\, 1],\ [a_1, a_2, \beta, p] \in \mathscr{D}, \\ \quad l_{2eq} \in]0,\, +\infty[\end{cases}$$

$\mathscr{D} =]0, +\infty[\times]0, a_1[\times [0, \pi/2] \times]0, +\infty[$.

For Case II, Eq. (2) takes the form:

$$P_{II}(X) = A_2 X^2 + C_2 = 0 \quad (10)$$

with

$$\begin{cases} A_2 = a_2 \sin^3\beta \\ C_2 = p(1 - \sin^2\beta) - a_2\sin^3\beta \\ \quad X = \cos\theta,\ \theta = \theta_2,\ X \in [-1, 0],\ [a_1, a_2, \beta, p] \in \mathscr{D}, \\ \quad l_{2eq} \in](a_1 - a_2)\sin\beta|\sin\theta|,\ (a_1 - a_2)\sin\beta[\end{cases}$$

For Case III, Eq. (2) takes the form:

$$P_{III}(X) = A_3 X^2 + C_3 = 0 \quad (11)$$

Table 1 Formulae describing the boundaries of the cells in Tables 2, 3 and 4.

$a_{11} = 0$	$p_1 = 0$
$a_{12} = +\infty$	$p_2(a_1, a_2, \beta) = \frac{1-\sin\beta}{1+\sin\beta} a_2 \sin\beta$
$a_{21} = a_1$	$p_3(a_1, a_2, \beta) = \frac{1-\sin^2\beta}{1+\sin^2\beta} a_2 \sin\beta$
$a_{22} = +\infty$	$p_4(a_1, a_2, \beta) = a_2 \sin\beta$
$\beta_1 = 0$	$p_5(a_1, a_2, \beta) = \frac{1+\sin^2\beta}{1-\sin^2\beta} a_2 \sin\beta$
$\beta_2 = \arcsin(1/\sqrt{3})$	$p_6(a_1, a_2, \beta) = \frac{1+\sin\beta}{1-\sin\beta} a_2 \sin\beta$
$\beta_3 = \pi/4$	$p_7 = +\infty$
$\beta_4 = \pi/2$	$p_8(a_1, a_2, \beta) = a_2 \sin\beta \tan^2\beta$
$l_{2eq_1}(a_1, a_2, \beta, p) = \frac{a_1 - a_2}{a_2} p$	
$l_{2eq_2}(a_1, a_2, \beta, p) = (a_1 - a_2) \sin\beta$	
$l_{2eq_3}(a_1, a_2, \beta, p) = \frac{a_1 - a_2}{2 a_2 \sin\beta} \sqrt{(\sin^2\beta - 1)\left[(\sin^2\beta - 1)(p - a_2 \sin\beta)^2 + 4 p a_2 \sin^3\beta\right]}$	
$l_{2eq_4}(a_1, a_2, \beta, p) = (a_1 - a_2) \sin\beta \lvert \sin\theta \rvert$	
$l_{2eq_4}(a_1, a_2, \beta, p) = +\infty$	

with

$$\left\{ \begin{array}{ll} A_3 & = a_2 \sin^3\beta \\ C_3 & = p(1 - \sin^2\beta) - a_2 \sin^3\beta \\ & X = \cos\theta,\ \theta = \theta_2,\ X \in [-1, 1],\ [a_1, a_2, \beta, p] \in \mathscr{D}, \\ & l_{2eq} \in](a_1 - a_2)\sin\beta,\ +\infty[\end{array} \right.$$

As a matter of fact, the IRSBot-2 reaches a constraint singularity as long as one of the univariate polynomials (9), (10), (11) admits one solution at least. The set of design parameters (a_1, a_2, β, p, l_{2eq}) for which the constraint singularities associated with Cases I, II and III can be reached are obtained with a method based on the notion of Discriminant Varieties and Cylindrical Algebraic Decomposition. This method resorts to Gröbner bases for the solutions of systems of equations and is described in [6]. Besides, the tools used to perform the computations are implemented in a Maple library called Siropa.[2]

Table 1 provides the different formulae bounding the five-dimensional cells associated with Cases I, II and III. a_1 and β can be chosen independently. Then, the boundaries for a_2, p are l_{2eq} are determined successively. Table 2 characterizes all the cells where the IRSBot-2 can reach a constraint singularity, namely, where P_I, P_{II} or P_{III} has at least one real root. It is noteworthy that a real root of one the three foregoing polynomials amounts to two symmetrical singular configurations of the distal module. It is apparent that six cells arise where P_I has a single real root, two cells arise where P_I has two real roots. P_{II} and P_{III} can get two real roots in one cell only. Some constraint singularities of the IRSBot-2 are shown on http://irccyn.ec-nantes.fr/Robotique/IRSBot2.

[2] http://www.irccyn.ec-nantes.fr/~chablat/SIROPA/files/siropa-mpl.html

Table 2 Cells of $\mathbb{R}^5$ where the IRSBot-2 can reach constraint singularities.

Case I			
$(]a_{11}, a_{12}[,]a_{21}, a_{22}[,]\beta_1, \beta_4[)$	$]p_1, p_2[$	$(]l_{2eq_1}, l_{2eq_2}[)$	Two singular configs.
	$]p_2, p_3[$	$(]l_{2eq_1}, l_{2eq_2}[)$	
	$]p_3, p_4[$	$(]l_{2eq_1}, l_{2eq_2}[)$	
	$]p_4, p_5[$	$(]l_{2eq_2}, l_{2eq_1}[)$	
	$]p_5, p_6[$	$(]l_{2eq_2}, l_{2eq_1}[)$	
	$]p_6, p_7[$	$(]l_{2eq_2}, l_{2eq_1}[)$	
	$]p_3, p_4[$	$(]l_{2eq_3}, l_{2eq_1}[)$	Four singular configs.
	$]p_4, p_5[$	$(]l_{2eq_3}, l_{2eq_2}[)$	
Case II			
$(]a_{11}, a_{12}[,]a_{21}, a_{22}[,]\beta_1, \beta_4[)$	$]p_1, p_8[$	$(]l_{2eq_4}, l_{2eq_2}[)$	Four singular configs.
Case III			
$(]a_{11}, a_{12}[,]a_{21}, a_{22}[,]\beta_1, \beta_4[)$	$]p_1, p_8[$	$(]l_{2eq_1}, l_{2eq_2}[)$	Four singular configs.

4 Design Parameters for the IRSBot-2 to Be Free of Constraint Singularity

This section aims to find the sets of design parameters (a_1, a_2, β, p, l_{2eq}) that prevent the IRSBot-2 from reaching any constraint singularity. It amounts to find the intersection of cells where P_I, P_{II} and P_{III} do not have any real root over their mutual domain.

It turns out to be quite difficult to obtain the intersection of cells contrary to their union. As a consequence, we will search for the cells where the product of P_I, P_{II} and P_{III} does not have any real root. From (10) and (11), it is apparent that the expressions of P_{II} and P_{III} are the same, but their domains are disjointed and complementary because of the bounds of l_{2eq}. Therefore, the sets of design parameters (a_1, a_2, β, p, l_{2eq}) that prevent the IRSBot-2 from reaching any constraint singularity correspond to the union of cells that do not provide any real root for the following two univariate polynomials:

$$P_{IV}(X) = P_I P_{II}(X) = (A_1 X^2 + B_1 X + C_1)(A_2((X-1)/2)^2 + C_2) = 0 \qquad (12)$$

with

$$\begin{cases} X \in [-1, 1], [a_1, a_2, \beta, p] \in \mathscr{D}, \\ l_{2eq} \in]|\sin\theta|(a_1 - a_2)\sin\beta, (a_1 - a_2)\sin\beta[\end{cases}$$

and

$$P_V(X) = P_I P_{III}(X) = (A_1 X^2 + B_1 X + C_1)(A_3 X^2 + C_3) = 0 \qquad (13)$$

with

$$\begin{cases} X \in [-1, 1], [a_1, a_2, \beta, p] \in \mathscr{D}, \\ l_{2eq} \in](a_1 - a_2)\sin\beta, +\infty[\end{cases}$$

$A_1, B_1, C_1, A_2, C_2, A_3, C_3$ and $\mathscr{D}$ being defined in Eqs. (9) to (11).

Table 3 Cells where Eq. (12) does not have any real root with $a_1 \in]a_{11}, a_{12}[$ and $a_2 \in]a_{21}, a_{22}[$.

$[\beta_1, \beta_2[$	$(]p_8, p_3[,]l_{2eq_4}, l_{2eq_1}[), (]p_3, p_4[,]l_{2eq_4}, l_{2eq_3}[), (]p_4, p_5[,]l_{2eq_4}, l_{2eq_3}[), (]p_5, p_7[,]l_{2eq_4}, l_{2eq_2}[)$
$[\beta_2, \beta_3[$	$(]p_8, p_4[,]l_{2eq_4}, l_{2eq_3}[), (]p_4, p_5[,]l_{2eq_4}, l_{2eq_3}[), (]p_5, p_7[,]l_{2eq_4}, l_{2eq_2}[)$
$[\beta_3, \beta_4]$	$(]p_8, p_5[,]l_{2eq_4}, l_{2eq_3}[), (]p_5, p_7[,]l_{2eq_4}, l_{2eq_2}[)$

Table 4 Cells where Eq. (13) does not have any real root with $a_1 \in]a_{11}, a_{12}[$ and $a_2 \in]a_{21}, a_{22}[$.

$[\beta_1, \beta_2[$	$(]p_8, p_3[,]l_{2eq_2}, l_{2eq_5}[), (]p_3, p_4[,]l_{2eq_2}, l_{2eq_5}[), (]p_4, p_5[,]l_{2eq_1}, l_{2eq_5}[), (]p_5, p_7[,]l_{2eq_1}, l_{2eq_5}[)$
$[\beta_2, \beta_3[$	$(]p_8, p_4[,]l_{2eq_2}, l_{2eq_5}[), (]p_4, p_5[,]l_{2eq_1}, l_{2eq_5}[), (]p_5, p_7[,]l_{2eq_1}, l_{2eq_5}[)$
$[\beta_3, \beta_4]$	$(]p_8, p_5[,]l_{2eq_1}, l_{2eq_5}[), (]p_5, p_7[,]l_{2eq_1}, l_{2eq_5}[)$

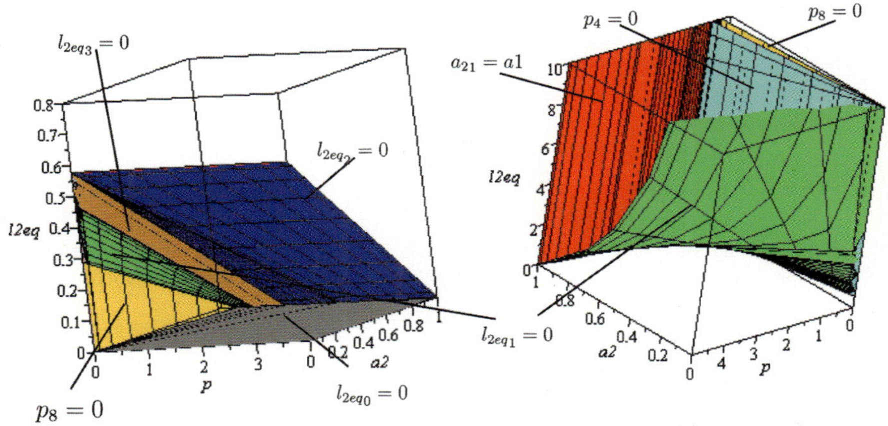

(a) (b)

Fig. 5 Cells where the IRSBot-2 cannot reach any constraint singularity for: (a) $a_1 = 1, \beta = \arcsin(1/\sqrt{3})$ and $l_{2eq} < (a_1 - a_2)\sin\beta$; (b) $a_1 = 1, \beta = \arcsin(1/\sqrt{3})$ and $l_{2eq} > (a_1 - a_2)\sin\beta$.

Eq. (12) amounts to the product of P_I and P_{II} with a change a variable for P_{II} and the most restrictive domain for l_{2eq} defined in (10), whereas Eq. (13) amounts to the product of P_I and P_{III} with the most restrictive domain for l_{2eq} defined in (11). Table 1 gives the different formulae bounding the five-dimensional cells associated with (12) and (13). The cells where P_{IV} and P_V do not have any real root, i.e., the sets of design parameters $(a_1, a_2, \beta, p, l_{2eq})$ that prevent the IRSBot-2 from reaching any constraint singularity, are expressed in Tables 3 and 4, respectively.

Figure 5(a) (Fig. 5(b), resp.) illustrates the cells where Eq. (12) (Eq. (13), resp.) does not have any real root, namely, the sets of design parameters that prevent the IRSBot-2 from reaching any constraint singularity for $a_1 = 1, \beta = \arcsin(1/\sqrt{3})$ and $l_{2eq} < (a_1 - a_2)\sin\beta$ ($l_{2eq} > (a_1 - a_2)\sin\beta$, resp.). We can notice that the amount of constraint singularity-free designs is higher with $l_{2eq} > (a_1 - a_2)\sin\beta$ than with $l_{2eq} < (a_1 - a_2)\sin\beta$.

5 Conclusions

This paper dealt with the constraint analysis of the IRSBot-2 throughout its parameter space. Its constraint singularities were analyzed in its parameter space with a method based on the notion of Discriminant Varieties and Cylindrical Algebraic Decomposition. This method allowed us to convert a kinematic problem into an algebraic one. Then, a deep analysis was carried out in order to determine the sets of design parameters of the distal modules that prevent the IRSBot-2 from reaching any constraint singularity. To the best of our knowledge, such an analysis had never been performed before. The design parameters associated with the proximal modules for the IRSBot-2 to be assembled will be determined in a future work.

Acknowledgements This work was conducted with the support of the French National Research Agency (Project ANR-2011-BS3-006-01-ARROW). The authors also thank Damien Chablat for his great help with the Siropa Maple library.

References

1. Amine, S., Tale-Masouleh, M., Caro, S., Wenger, P., Gosselin, C.: Singularity conditions of 3T1R parallel manipulators with identical imb structures. ASME J. Mech. Robot. (2012). 10.1115/1.4005336
2. Brogardh, T.: Device for relative movement of two elements. Patent US 6301988 B1 (2001)
3. Germain, C., Briot, S., Glazunov, V., Caro, S., Wenger, P.: Irsbot-2: A novel two-dof parallel robot for high-speed operations. In: Proceedings of the ASME 2011 International Design Engineering Technical Conferences & Computers and Information in Engineering Conference. Washington, DC, USA, August 29–31 2011
4. Huang, T., Li, M., Li, Z., Chetwynd, D., Whitehouse, D.: Planar parallel robot mechanism with two translational degrees of freedom. Patent WO 03055653 A1 (2003)
5. Liu, X., Kim, J.: Two novel parallel mechanisms with less than six degrees of freedom and the applications. In: Proc. Workshop on Fundamental Issues and Future Research Directions for Parallel Mechanisms and Manipulators, pp. 172–177. Quebec City, Quebec, Canada (2002)
6. Moroz, G., Chablat, D., Wenger, P., Rouiller, F.: Cusp points in the parameter space of RPR-2PRR parallel manipulator. In: 3rd European Conference on Mechanism Science, Cluj-Napoca, Romania, pp. 29–37. Springer (2010)
7. Pierrot, F., Krut, S., Company, O., Nabat, V., Baradat, C., Fernandez, A.S.: Two degree-of-freedom parallel manipulator. Patent WO 2009/089916 A1 (2009)
8. Zlatanov, D., Bonev, I., Gosselin, C.: Constraint singularities of parallel mechanisms. In: Proc. 2002 IEEE International Conference on Robotics and Automation, pp. 496–502 (2002)

Human Muscle Fatigue Model in Dynamic Motions

Ruina Ma, Damien Chablat, Fouad Bennis and Liang Ma

Abstract Human muscle fatigue is considered to be one of the main reasons for Musculoskeletal Disorder (MSD). Recent models have been introduced to define muscle fatigue for static postures. However, the main drawbacks of these models are that the dynamic effect of the human and the external load are not taken into account. In this paper, each human joint is assumed to be controlled by two muscle groups to generate motions such as push/pull. The joint torques are computed using Lagrange's formulation to evaluate the dynamic factors of the muscle fatigue model. An experiment is defined to validate this assumption and the result for one person confirms its feasibility. The evaluation of this model can predict the fatigue and MSD risk in industry production quickly.

Key words: Muscle fatigue model, dynamic motions, human simulation

1 Introduction

Muscle fatigue is defined as "any reduction in the ability to exert force in response to voluntary effort" [2] and is one of the main reasons leading to MSD [7]. From Hill's muscle model [4] to today's muscle fatigue models, this topic has been researched from different scientific field with special point of views. In general, mainly two approaches have been adopted to evaluate muscle fatigue [10], either in theoretical methods or in empirical methods. In [11], Wexleret et al. proposed a new muscle fatigue model based on Ca^{2+} cross-bridge mechanism and verified the model with simulation experiments. Although this model can be used to predict the muscle force fatigue under different simulation frequencies, the large number of variables

Ruina Ma · Damien Chablat · Fouad Bennis
Institut de Recherche en Communications et Cybernétique de Nantes, UMR CNRS 6597, Nantes, France, e-mail: {ruina.ma, damien.chablat, fouad.bennis}@irccyn.ec-nantes.fr

Liang Ma
Department of Industrial Engineering, Tsinghua University, Beijing, China, e-mail: liangma@tsinghua.edu.cn

J. Lenarčič, M. Husty (eds.), *Latest Advances in Robot Kinematics*,
DOI 10.1007/978-94-007-4620-6_44,

make it difficult to use compare with other models. In [8], Liu et al. proposed a fatigue and recovery models based on motor units pattern. They demonstrated the relationship among muscle activation, fatigue and recovery. This model is available under maximum voluntary contraction situation; this condition is rare in the manual working situation. Another muscle fatigue model was developed by Giat [3] based on force-pH relationship. This fatigue model was obtained by curve fitting of the pH level with time in the course of stimulation and recovery, but it cannot used in evaluating the muscle fatigue in the whole working process. In [9], Ma et al. proposed a muscle fatigue model from the macroscopic point of view. External physical factors and personal factors were taken into consideration to construct the model. This model can predict the muscle fatigue trend in static working posture ($\theta_{\text{elbow}} = 90^\circ$, $\theta_{\text{shoulder}} = 30^\circ$), but in dynamic working situation this model was limited.

The purpose of this work is to extend muscle fatigue model to dynamic working situations. The difference force generation between static working posture and dynamic working motions is dependent on the activation of different muscle types. There are three types of fibers of muscle: slow-twitch fibers, fast-twitch A fibers and fast-twitch B fibers [5]. In every postures and motions all of the three muscle fibers are used, but the percentage of every fibers in static and dynamic situation is different. In a static working posture fast-twitch fibers is mostly used and this type of muscle fibers have a low resistance to fatigue. In a low speed dynamic working motions slow-twitch fibers are mainly used and this type of muscle fibers have a high resistance to fatigue. Meanwhile, the blood circulation during dynamic motions is better than in a static working posture. For these reasons, the behavior of the muscle and its fatigue rate are different in the two types of situations. In this paper, a muscle fatigue model in dynamic situation is proposed. A new approach to identify the fatigue rate parameter k is used. An experimental setup is defined to validate this assumption.

Firstly, some assumptions are given and a new dynamic muscle fatigue model is proposed. Secondly, an experiment is designed to verify this model. Thirdly, a case-study for one person is illustrated and the fatigue parameter k is evaluated. Finally, some perspectives are presented.

2 Proposal of a New Muscle Fatigue Model

Dynamic muscle fatigue model: Muscles in the human body have one most important function, such as force generating devices. They can work only in a single direction. Hence, for each single joint, at least two groups of muscles (agonistic muscle and antagonistic muscle) are necessary to control the motion. The co-contraction of the two groups of muscles provide stability to joint and balance to the posture. From the articulation point of view, it is assumed that a joint is controlled by two groups of muscles (one for flexion and one for extension). These muscle groups create a torque on the joint. This torque drives the human movement and whether it is pos-

Table 1 Parameters in dynamic fatigue model.

Item	Unit	Description
Γ_{MVC}	N.m	Maximum voluntary contraction of joint torque, i.e. $\Gamma_{\max}$
$\Gamma_{\mathrm{cem}}(t)$	N.m	Current exertable maximum joint torque
Γ_{joint}	N.m	Joint torque, i.e. the torque which the joint needs to generate
k	min^{-1}	Fatigue rate

itive or negative depends on the angle and direction of joint rotation. Based on the previous model of Ma et al., we propose that:

1. the fatigue of muscle is proportional to the joint torque, i.e. in the same period of time, the larger the torque of joint exerted, the more fatigue people feel;
2. the fatigue of muscle is inversely proportional to the muscle torque capacity, i.e. the smaller the capacity is, the quicker the muscle becomes tired.

This can be mathematically described by the following equation.

$$\frac{d\Gamma_{\mathrm{cem}}(t)}{dt} = -k \cdot \frac{\Gamma_{\mathrm{cem}}(t)}{\Gamma_{\mathrm{MVC}}} \cdot \Gamma_{\mathrm{joint}}(t) \tag{1}$$

where the set of parameters are listed in Table 1. If we assume that $\Gamma_{\mathrm{cem}}(0) = \Gamma_{\mathrm{MVC}}$ and k is a constant, the integration result of the previous equation is given by

$$\Gamma_{\mathrm{cem}}(t) = \Gamma_{\mathrm{MVC}} \cdot e^{-\frac{k}{\Gamma_{\mathrm{MVC}}}\int_0^t \Gamma_{\mathrm{joint}}(u)du} \tag{2}$$

The value of Γ_{MVC} is a fixed value determined by individual person. In the first approximation, we assume that Γ_{MVC} is a constant of a joint torque during a limited period of time. According to robotic dynamic model [6], $\Gamma_{\mathrm{joint}}(u)$ can be modeled by a variable depending on the angle, the velocity, the acceleration and the internal/external load.

$$\Gamma_{\mathrm{joint}}(u) \stackrel{\mathrm{def}}{=} \Gamma(u, \theta, \dot{\theta}, \ddot{\theta}) \tag{3}$$

This way, Equation (2) can be further simplified in the form.

$$\Gamma_{\mathrm{cem}}(t) = \Gamma_{\mathrm{MVC}} \cdot e^{-\frac{k}{\Gamma_{\mathrm{MVC}}}\int_0^t \Gamma(u,\theta,\dot{\theta},\ddot{\theta})du} \tag{4}$$

Equation (4) defines our new dynamic muscle fatigue model. The model takes consideration of the motion by the variations of the torque Γ_{joint} from joint level. This torque which is computed using robotic method is integrated to obtain the current exertable maximum joint torque. At first stage, we do not take into account the muscle co-contraction factor [1]. This work enlarges the muscle fatigue model usefulness range.

The new dynamic fatigue model is in joint level. As mentioned in the assumption, the motion of joint is driven by a pair of muscles. Obviously, this model can be easily applicated in muscle level. For one cycle, the Γ_{joint} is negative or positive related with elbow rotation range. If Γ_{joint} is positive, we suppose it is the effect of agonistic muscle. Inversely, if Γ_{joint} is negative, we suppose it is the effect of antagonistic muscle. Based on this consideration, there will be two fatigue rate parameters k (k_{agonist} and $k_{\text{antagonist}}$) for a dynamic operation in the muscle level. Our new dynamic fatigue model can applicate fatigue evaluation in muscle level.

3 Experiment Design for Validation

The aim of the experiment design is to evaluate the muscle fatigue model. We suppose that in a push operation the agonistic muscle is mainly used whereas in a pull operation the antagonistic muscle is mainly used. Based on this assumptions, we concentrate the study on the elbow joint and use a push/pull operation to simulate dynamic motions. This evaluation consists in measuring the maximal push and pull strength after a continuous movement of the lower arm.

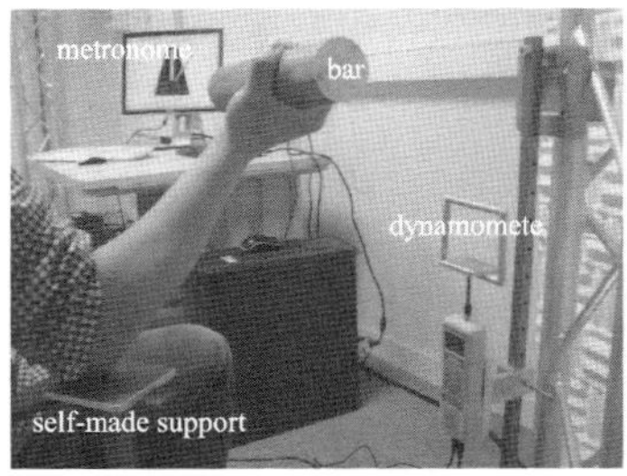

Fig. 1 Measurement device.

Experiment materials (Figure 1):
1. *A dynamometer*. This device is used to measure the maximum push/pull force after lengths of time's movement of the lower arm.
2. *A bar*. This weight is grabbed by the hand of the participant and is used to simulate the weight of an operation tool in industrial environment. For our experiment the bar weight is 3 Kg.
3. *A metronome*. This tool is used to define the sample times of the motion. For our experiment the frequency is 1Hz.
4. *A self-made support*. This support is used to maintain the elbow posture during the motion and measure the torque after the operation.

Experiment procedure: The participant seats in a chair and puts his elbow on the support. The procedure is to repeat a rotation of the elbow joint from 0 to 75 degrees and then from 75 degrees to 0 during t_i unit of time. This movement is done with the bar in hand. The experiment procedure is as follows:
1. Measure Γ_{MVC} before starting the operation ($\Gamma_{\text{MVC}}(0) = \Gamma_{\text{cem}}(0)$);
2. Perform the dynamic operation during t_i unit of time;
3. Measure the remained maximum torque of elbow joint $\Gamma_{\text{cem}}(t_i)$;
4. Take a rest about 1-2 hours until complete recovery;
5. Repeat steps 2, 3 and 4 for different values of t_i, $t_i \in \{0,1,2,3,4,5\}$ minutes.

4 Case Study of Muscle Fatigue for the Elbow Joint

Kinematics and dynamics of the arm: In this section we will illustrate the dynamic calculation in our dynamic muscle fatigue model using robotic method. An example of the lower arm cyclic periodic movement during 2 seconds is demonstrated in details.

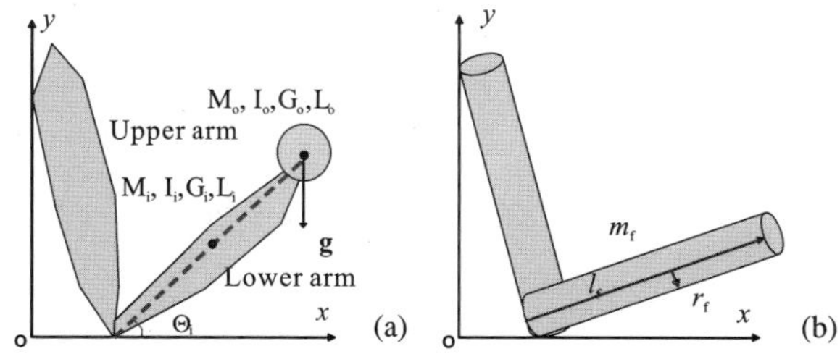

Fig. 2 (a) Geometric and inertia parameters of one arm, (b) simplified arm structure.

Geometric modeling of arm: As presented in Figure 2, the arm model is composed of an upper arm, a lower arm and a hand. Figure 2(b) gives the simplified model of the arm used for the calculation. In order to determine the torque Γ_{joint}, several parameters of the arm need to be obtained. These parameters are the length of the lower arm (ℓ_{f}), the length of the hand (ℓ_{h}), the radius of the lower arm (r_{f}) and the mass of the lower arm (m_{f}). If the human has a height of H and a weight of M, according to the anthropometry database [2], the related geometric human parameters are: $\ell_{\text{f}} = 0.146H$, $r_{\text{f}} = 0.125\ell_{\text{f}}$, $\ell_{\text{h}} = 0.108H$, $m_{\text{f}} = 0.023M$.

Trajectory generation: We suppose that for one motion of lower arm up and down movement both the initial and the final velocity and acceleration are null. We use a polynomial function to describe this movement. According to the hypothesis, the minimum degree of the polynomial satisfying the constraints is at least five and has the following form:

$$P = a_0 + a_1 t + a_2 t^2 + a_3 t^3 + a_4 t^4 + a_5 t^5 \tag{5}$$

where the coefficients a_i are determined from the boundary conditions:

$$\begin{aligned} \theta(0) &= \theta^{initial}, \ \dot{\theta}(0) = 0, \ \ddot{\theta}(0) = 0 \\ \theta(t_f) &= \theta^{end}, \quad \dot{\theta}(t_f) = 0, \ \ddot{\theta}(t_f) = 0 \end{aligned} \tag{6}$$

The trajectory between θ^{initial} and θ^{end} is determined by

$$\theta(t) = \theta^{initial} + r(t) \cdot \left(\theta^{\text{end}} - \theta^{\text{initial}}\right), \quad 0 \le t \le t_f \tag{7}$$

Solving Equation (7) with the above mentioned condition we can get the following interpolation function

$$r(t) = 10(t/t_f)^3 - 15(t/t_f)^4 + 6(t/t_f)^5 \tag{8}$$

Based on this interpolation function we can get the velocity and acceleration of every moment in the joint trajectory. Figure 3 represents the evolution of θ, $\dot{\theta}$ and $\ddot{\theta}$ for the considering experiment with the angle change between 0 to $5\pi/12$.

Dynamic model and joint torque evaluation: The Lagrange method is applied to compute the dynamic model [6]. Firstly, we calculate the joint kinetic energy and

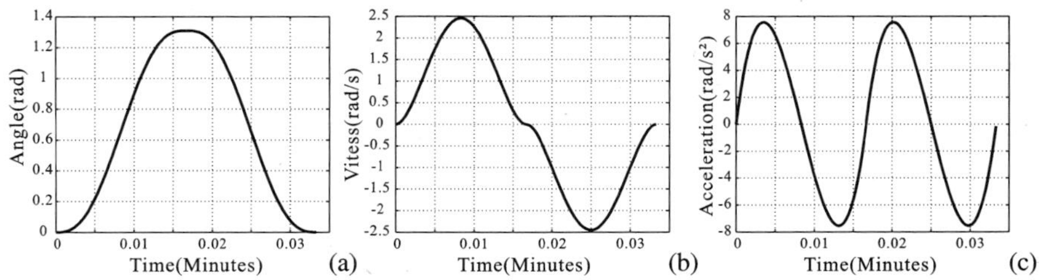

Fig. 3 (a) Joint angle, (b) joint angular velocity, (c) joint angular acceleration evolution during one cycle motion.

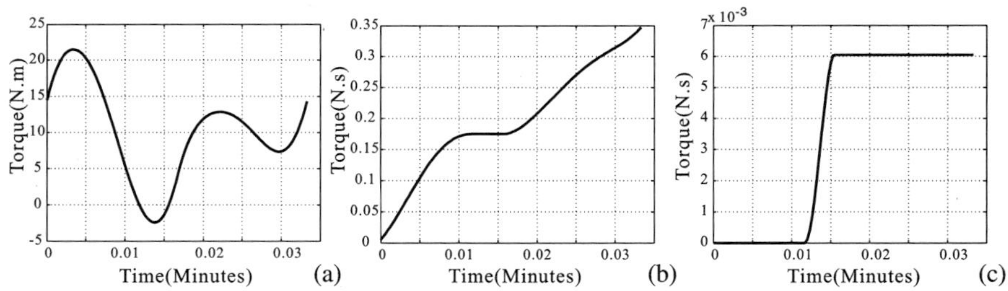

Fig. 4 (a) Elbow joint torque, (b) momentum of agonistic muscle, (c) momentum of antagonistic muscle.

the joint potential energy

$$E = E_{\text{joint}} + E_{\text{object}}, \quad U = U_{\text{joint}} + U_{\text{object}} \tag{9}$$

Then, the joint torque is given by

$$\Gamma = \frac{d}{dt}\left(\frac{\partial L}{\partial \dot{\theta}}\right) - \left(\frac{\partial L}{\partial \theta}\right) \tag{10}$$

where $L = E - U$.

In our dynamic muscle fatigue model, $\int_0^t \Gamma(u, \theta, \dot{\theta}, \ddot{\theta})du$ is the joint momentum. This is the most important difference between dynamic muscle fatigue model and static muscle fatigue model. In static situation the joint torque is a constant, and with time goes by the joint momentum is a linear function. In dynamic muscle fatigue model the joint torque is changing with joint angle and time. The joint momentum is a non-linear function. Figure 4(a) is the torque of elbow joint in a cyclic motion of 2 seconds and there are a part of positive torque and a part of negative torque. We consider the positive torque as a result of the effect of angonistic muscle and the negative torque as a result of antagonistic muscle. Figure 4 presented joint torque and momentum evolution of two groups of muscles during 2 seconds.

Experimental results: The experiment part is an implementation and verification of above mentioned experiment design. At the first stage we just measure one person to test its operability and feasibility. A large number of tests will be carried out in the future stage.

Table 2 Current exertable maximum joint torque for push and pull action.

Γ_{cem} [N · m]	0 min	1 min	2 min	3 min	4 min	5 min
Push	31.46	30.08	28.07	29.33	26.32	26.82
Pull	31.71	28.33	22.94	22.31	20.05	18.67

Experiment result for one person: A male subject ($H = 188$ cm, $M = 80$ Kg) took part in the presented experiment. Push and Pull torque of the lower arm are measured for the different operation times of t in $\{0, 1, 2, 3, 4, 5\}$ minutes. The experiment results are presented in Table 2.

Fatigue rate parameter k evaluation: The parameter k represents fatigue rate and it depends on individual person itself. To evaluate the parameter k of our model, we suppose k is constant. The following Eq. (11) which is deduced from Eq. (4) is used to calculate k_i with the help of using the experiment measurement of Γ_{cem_i} for each operation.

$$k_i = -\ln\left(\frac{\Gamma_{\text{cem}}(t)}{\Gamma_{\text{MVC}}}\right) / \int_0^{t_i} \Gamma(u, \theta, \dot{\theta}, \ddot{\theta})du \qquad (11)$$

For $t = 1, 2, 3, 4, 5$ minutes, the agonistic and antagonistic muscle group fatigue rate were evaluated as follows:

$$k_{agonist} = [0.13, 0.17, 0.07, 0.13, 0.09][\text{min}^{-1}]$$

$$k_{antagonist} = [19.56, 28.07, 20.32, 19.86, 18.36][\text{min}^{-1}]$$

In Ma [9], the values of k obtained are around 0.87, so $k_{agonist}$ is a realistic value due to that the fact the blood circulation is better during dynamic motions. Conversely, $k_{antagonist}$ seems to be too high. In fact, due to the co-contraction activities influence, the torque of the antagonistic muscle group is higher than the results computed by the dynamic model. To characterize $k_{antagonist}$ more precisely, another experimental measurement is necessary to make the same motion with a pulley based system that inverse the gravity force. Because of the measurement errors of forces, the calculated k is not exactly the same for each time t. To evaluate the confidence of the fatigue rate parameter k, with the minimum, average and maximum values of $k_{agonist}$, Γ_{cem} is evaluated separately and compared with the experimental measurements in Fig. 5. It seems that the first two experimental measurements overestimate $k_{agonist}$. This means that we have to wait three minutes to have a good evaluation of the muscle fatigue properties. In fact, we can consider the force capacity of one muscle group can increase in the beginning of the activity as a warming-up

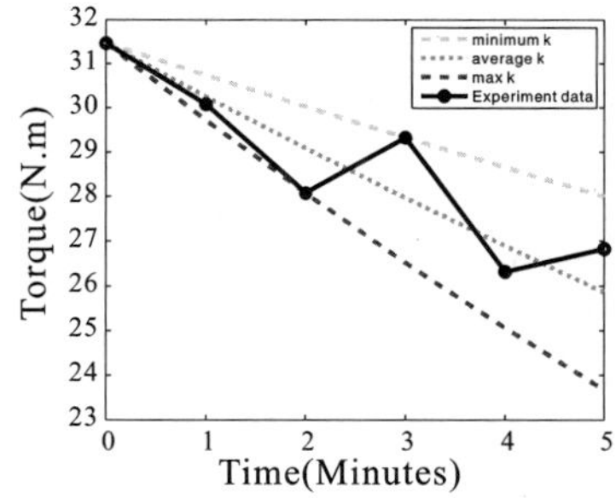

Fig. 5 Theoretical evolution of Γ_{cem} and experiment data using different values of $k_{agonist}$.

period of the muscle. As only one person participated the experiment, the conclusion cannot be generalized but we have obtained interesting informations. The test will be done for a representative number of participants in the future works.

5 Conclusion and Perspectives

In this paper, a new muscle fatigue model for dynamic motions is presented. Thanks to the robotic method, dynamic factors have been introduced to characterize a new dynamic muscle fatigue model from the joint level. This model can be explained theoretically. Meanwhile, an experiment has been designed to validate it. This model could demonstrate the potential for predicting muscle fatigue in dynamic motions. The limit of this work is that it still lacks experimental validation for more participants. In the future, validations of experiments for a number of participants will be carried out.

References

1. Ait-Haddou, R., Binding, P., Herzog, W.: Theoretical considerations on cocontraction of sets of agonistic and antagonistic muscles. J. Biomech. **33**(9), 1105–1111 (2000)
2. Chaffin, D.B., Andersson, G.B.J., Martin, B.J.: Occupational Biomechanics, 3rd edn. Wiley-Interscience (1999)
3. Giat, Y., Mizrahi, J., Levy, M.: A musculotendon model of the fatigue profiles of paralyzed quadriceps muscle under fes. IEEE Trans. Biomed. Eng. **40**(7), 664–674 (1993)
4. Hill, A.V.: The heat of shortening and dynamic constants of muscle. Biol. Sci. **126**, 136–195 (1983)
5. Karp, J.R.: Muscle fiber types and training. Strength Cond. J. **23**(5), 21 (2001)
6. Khalil, W.: Modeling and control of manipulators. Ecole Central de Nantes (2009–2010)
7. Kumar, R., Kumar, S.: Musculoskeletal risk factors in cleaning occupation: A literature review. Int. J. Ind. Ergon. **38**, 158–170 (2008)
8. Liu, J., Brown, R., Yue, G.: A dynamical model of muscle activation, fatigue, and recovery. Biophys. J. **82**(5), 2344–2359 (2002)
9. Ma, L.: Contributions pour l'analyse ergonomique de mannequins virtuels. Ph.D. thesis, Ecole Central de Nantes (2009)
10. Westgaard, R.H.: Work-related musculoskeletal complaints: some ergonomics challenges upon the start of a new century. Appl. Ergon. **31**, 569–580 (2000)
11. Wexler, A.S., Ding, J., Binder-Macleod, S.A.: A mathematical model that predicts skeletal muscle force. IEEE Trans. Biomed. Eng. **44**(5), 337–348 (1997)

Solution Regions in the Parameter Space of a 3-RRR Decoupled Robot for a Prescribed Workspace

D. Chablat, G. Moroz, V. Arakelian, S. Briot and P. Wenger

Abstract This paper proposes a new design method to determine the feasible set of parameters of translational or position/orientation decoupled parallel robots for a prescribed singularity-free workspace of regular shape. The suggested method uses Groebner bases to define the singularities and the cylindrical algebraic decomposition to characterize the set of parameters. It makes it possible to generate all the robot designs. A 3-RRR decoupled robot is used to validate the proposed design method.

Key words: Parallel robot, design, singularities, Groebner basis, discriminant varieties, cylindrical algebraic decomposition

1 Introduction

Parallel robots are attractive for various reasons but one has to cope with their singularities. There exists three main ways of coping with singularities, which have their own merits. A first approach consists in eliminating the singularities at the design stage by properly determining the kinematic architecture, the geometric parameters and the joint limits [1, 8]. This approach is difficult to apply in general and restricts the design possibilities but it is safe. A second approach is the determination of the singularity-free regions in the workspace [2, 3]. This solution does not involve a priori design restrictions but it may be difficult to determine safe regions that are sufficiently large. Finally, a third way consists in planning singularity-free trajectories in the manipulator workspace [4]. In this paper, the first approach is used. Designing a parallel robot that will operate in a singularity-free workspace is a first requirement but the designer often needs to optimize the robot as function

D. Chablat
Institut de Recherche en Communications et Cybernétique de Nantes, Nantes, France,
e-mail: damien.chablat@irccyn.ec-nantes.fr

J. Lenarčič, M. Husty (eds.), *Latest Advances in Robot Kinematics*,
DOI 10.1007/978-94-007-4620-6_45,

of various criteria [5]. Our goal is to generate the set of geometric parameters for a given singularity-free workspace. The resulting solution regions in the parameter space are of primary interest for the designer. Accordingly, this paper proposes a new design method to determine these solution regions. This method holds for parallel translational robots and for parallel robots with position/orientation decoupled architecture. Groebner bases are used to define the singularities and Cylindrical algebraic decomposition is applied to characterize the set of design parameters. The paper is organized as follows. Section 2 introduces the design method to generate the solution regions in the parameter space for a prescribed workspace of regular shape. Then, Section 3 applies this method to a 3-RRR planar parallel robot with position/orientation decoupled architecture.

2 Design Method

2.1 Definition of the Prescribed Regular Workspace

A robot should have sufficiently large, regular workspace with no singularity inside [9]. For planar (resp. spatial) translational robots, a regular workspace can be defined by a circle, a square or a rectangle (resp. a cylinder, a cube or parallelepiped). A circle, a cylinder or a sphere can be modeled with one single algebraic equation. A rectangle or a parallelepiped can be defined with a set of linear equations. It can be approximated using a Lamé curve (resp. surface). This approximation makes it possible to handle only one equation, thus simplifying the problem resolution as will be shown further. In the rest of the paper, the problem is formulated in the plane for practical reasons. A Lamé curve based workspace $\mathscr{W}_L$ can be defined by the following boundary algebraic equation:

$$\mathscr{W}_L : \left(\frac{x-x_c}{l_x/2}\right)^n + \left(\frac{y-y_c}{l_y/2}\right)^n = 1 \tag{1}$$

l_x and l_y being the edge lengths of the desired rectangle, n being a strictly positive integer. For the purpose of this paper, $n = 4$ and $l_x = l_y = 4$. A rectangle based workspace can be modeled by four parametric lines, noted $\overline{\mathscr{W}}_{Ci}$

$$\overline{\mathscr{W}}_{Ci} : \begin{cases} x = P_{(i)x}t + P_{(i+1)x}(1-t) \\ y = P_{(i)y}t + P_{(i+1)y}(1-t) \end{cases} \text{with } t \in [0\ 1], i = 1,2,3,4 \tag{2}$$

$P_{(i)x} = x_c \pm l_x/2 \quad P_{(i)y} = y_c \pm l_y/2$ where P_i denote the rectangle vertices. For position/orientation decoupled robot architectures, the regular workspace is defined using the same approach for the translational module and the orientation module is considered separately, as it will be shown in the next section.

2.2 Method to Generate the Solution Regions in the Parameter Space

The problem can be stated as follows: find the regions in the parameter space where the boundaries $\overline{\mathscr{W}}$ of the workspace $\mathscr{W}$ have no intersection with the serial and parallel singularities loci δ_i, namely:

$$\mathscr{P} : [a_1 \dots a_n] / \delta_i \cap \overline{\mathscr{W}} = \emptyset, a_j > 0, j = 1, \dots, n \tag{3}$$

where $[a_1 \dots a_n]$ are the set of design parameters. This approach stands if and only the singularity curves or points are never fully included in the prescribed region. In order to find the design parameters for which the intersection is empty, the design parameters will be sorted according to the number of intersections between the singularities and $\overline{\mathscr{W}}$. It is then necessary to decompose the design parameter space into cells $C_1, \dots, C_k$, such that: (a) C_i is an open connected subset of the design parameter space; (b) for all design parameter values in C_i, the design parameter space has a constant number of solutions. This analysis is done in 3 steps [6]:

(a) computation of a subset of the joint space (workspace, resp.) where the number of solutions changes: the *Discriminant Variety*;
(b) description of the complementary of the discriminant variety in connected cells: the *Generic Cylindrical Algebraic Decomposition*;
(c) connecting the cells that belong to the same connected component of the complementary of the discriminant variety: *interval comparisons*.

The results are sets of regions with the same number of intersections between δ_i and $\overline{\mathscr{W}}$. These three steps were integrated in a single function in the Siropa Library implemented in Maple (Moroz, 2010). For the purpose of this study, only the solutions with zero intersections are considered. When a decoupled robot is analyzed, problem $\mathscr{P}$ is first treated for a prescribed workspace and a slightly modified problem $\mathscr{P}'$ is then treated, in which the set of design parameters include the orientation parameters. This approach is illustrated in the next section.

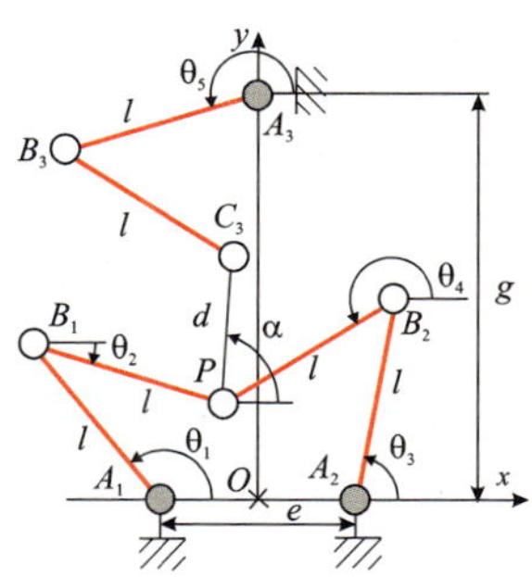

Fig. 1 The 3-RRR decoupled parallel robot under study.

3 Application to a 3-RRR Decoupled Parallel Robot

The robot under study is a planar 3-RRR robot with a modified mobile platform design [10] (Fig. 1), thus decoupling the position and the orientation of the platform [11]. It is assumed to have three identical legs. The loop (A_1, B_1, P, B_2, A_2) corresponds to a five-bar robot that defines the position of point P and the leg (A_3, B_3, C_3) adjusts the orientation according to the position. If position of point

P is given, this third leg is equivalent to a four-bar linkage. For this 3-RRR robot, thus, the problem can be split into two parts: (i) design of the five-bar robot (the translational module) so that the end-effector can move in a prescribed singularity-free workspace and (ii) design of the third leg (the four-bar linkages (A_3, B_3, C_3, P)) so that the platform can be oriented within desired bounds throughout the prescribed workspace.

3.1 Translational Module: Five-Bar Robot

The constraint equations of the five-bar robot are defined as:

$$\mathscr{C}_i : \begin{cases} x - l\cos(\theta_1) - l\cos(\theta_2) + e/2 = 0 & y - l\sin(\theta_1) - l\sin(\theta_2) = 0 \\ x - l\cos(\theta_3) - l\cos(\theta_4) - e/2 = 0 & y - l\sin(\theta_3) - l\sin(\theta_4) = 0 \end{cases} \tag{4}$$

where $||\mathbf{A}_1\mathbf{B}_1|| = ||\mathbf{A}_2\mathbf{B}_2|| = ||\mathbf{B}_1\mathbf{P}|| = ||\mathbf{B}_2\mathbf{P}|| =$ and $||\mathbf{A}_1\mathbf{A}_2|| = e$. The differentiation of the relation between the input variables $\mathbf{q}$ and the output variables $\mathbf{X}$ with respect to time leads to the velocity model $\mathbf{A}\mathbf{t} + \mathbf{B}\dot{\mathbf{q}} = 0$ where $\mathbf{A}$ and $\mathbf{B}$ are $n \times n$ Jacobian matrices, t is the platform twist and $\dot{\mathbf{q}}$ is the vector of joint rates. The roots of the determinant of $\mathbf{A}$ and $\mathbf{B}$ define the parallel and serial singularities, respectively. The first ones are directly characterized in the workspace and the second ones have to be projected from the joint space onto the workspace. The singularities are calculated using Groebner bases [6] as in [7].

The parallel singularities can be factored into a sextic, denoted δ_{p1}, and two quadratic polynomial equations, denoted δ_{p2} and δ_{p3}

$$\delta_{p1} : 16(y^6 + x^6) + 8(e^2y^4 - e^2x^4) + 48(y^4x^2 + y^2x^4) + e^4y^2 + e^4x^2 - 16l^2e^2y^2 = 0$$

$$\delta_{p2} : x^2 + \left(y - \frac{1}{2}\sqrt{4l^2 - e^2}\right)^2 - l^2 = 0 \quad \delta_{p3} : x^2 + \left(y + \frac{1}{2}\sqrt{4l^2 - e^2}\right)^2 - l^2 = 0$$

The serial singularities are two quadratic equations

$$\delta_{s1} : (2x + e)^2 + 4y^2 - 16l^2 = 0 \quad \delta_{s2} : (2x - e)^2 + 4y^2 - 16l^2 = 0$$

Due to the symmetry of the robot with respect to y-axis, the design parameters are restricted to $(l\ f)$ i.e. the size of the legs and the distance from axis x to the geometric center of the robot's workspace $\mathscr{W}$, respectively. Parameter e is set to 1 to have a two dimensional representation of the solution regions. For robots with two degrees of freedom, the intersection of the boundaries of $\mathscr{W}$ and the singularities is generically a finite set of points. Thus, as mentioned in Section 2.2, the singularity curves or points are never fully included in the prescribed region.

Lamé curve based workspace: The problem to be solved is:

$$\mathscr{P}_L : [f\ l]/\mathscr{S}_{p1} \cap \mathscr{S}_{p2} \cap \mathscr{S}_{p3} \cap \mathscr{S}_{s1} \cap \mathscr{S}_{s2} \cap \overline{\mathscr{W}} = \emptyset, f > 0, l > 0$$

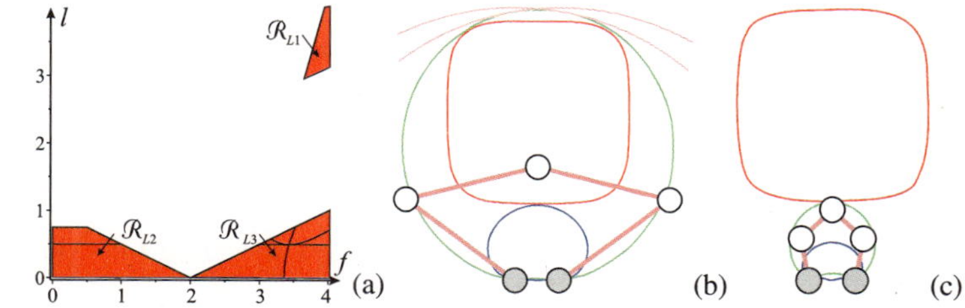

Fig. 2 (a) Solution regions $\mathscr{R}_{L1}$, $\mathscr{R}_{L2}$ and $\mathscr{R}_{L3}$ of problem $\mathscr{P}_L$ and five-bar robot design when (b) $f=3.7, l=3$ (c) $f=3.7, l=0.9$.

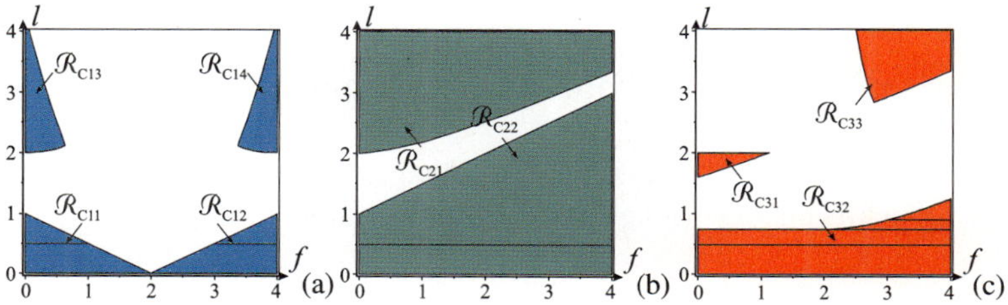

Fig. 3 Solution regions for problems (a) $\mathscr{P}_{C1}$, (b) $\mathscr{P}_{C2}$, (c) $\mathscr{P}_{C3}$ and (d) intersection regions $\mathscr{R}_{Cf1}$, $\mathscr{R}_{Cf2}$ and $\mathscr{R}_{Cf3}$.

Only the solutions with zero intersections are kept. Fig. 2 depicts the three solution regions obtained $\mathscr{R}_{L1}$, $\mathscr{R}_{L2}$ and $\mathscr{R}_{L3}$, i.e. the parameter sets for which the prescribed workspace is singularity-free.

It turns out that in $\mathscr{R}_{L1}$, $\mathscr{W}_L$ is inside the workspace (Fig. 2b). Conversely, in $\mathscr{R}_{L2}$ and $\mathscr{R}_{L3}$, $\mathscr{W}_L$ is outside the workspace (Fig. 2c). Thus the only feasible region is $\mathscr{R}_{L1}$. A feasible solution should not be taken on the boundary of $\mathscr{R}_{L1}$ since any solution on the boundary could touch a singularity curve. Fig. 2b shows a solution near the boundary of $\mathscr{R}_{L1}$.

Square based workspace: In this case, four separate problems need to be solved:

$$\mathscr{P}_{Ci} : [f\ l] / \mathscr{S}_{p1} \cap \mathscr{S}_{p2} \cap \mathscr{S}_{p3} \cap \mathscr{S}_{s1} \cap \mathscr{S}_{s2} \cap \overline{\mathscr{W}}_{Ci} = \emptyset, f>0, l>0, t \in [0,1], i=1,\ldots,4$$

where are the parametric equations defining the boundaries of the square. Only the solutions with zero intersections are kept. Due to the symmetry of the square with respect to the y-axis, $\mathscr{P}_{C3}$ and $\mathscr{P}_{C4}$ yield the same regions in the design parameters space. Fig. 3 depicts (a) four connected solution regions for problem $\mathscr{P}_{C1}$, (b) two solution regions for $\mathscr{P}_{C2}$ and (c) three solution regions for $\mathscr{P}_{C3}$. As compared to the Lamé curve based workspace, there is an additional step here: the final regions must be obtained by intersecting all these regions, thus yielding the three regions $\mathscr{R}_{Cf13}$, $\mathscr{R}_{Cf2}$ and $\mathscr{R}_{Cf3}$ as shown in Fig. 4. As expected, the solution regions obtained are similar to those associated with a Lamé curve (Fig.2) and only $\mathscr{R}_{Cf1}$ is solution to the problem for the same reasons. Fig. 5a shows a solution near the boundary of $\mathscr{R}_{Cf3}$.

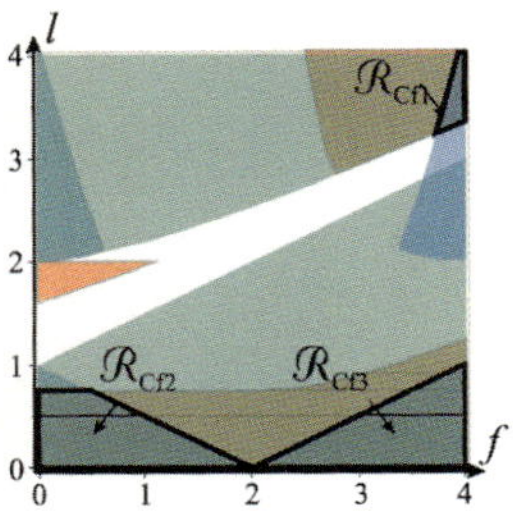

Fig. 4 Intersection regions $\mathscr{R}_{Cf1}$, $\mathscr{R}_{Cf2}$ and $\mathscr{R}_{Cf3}$.

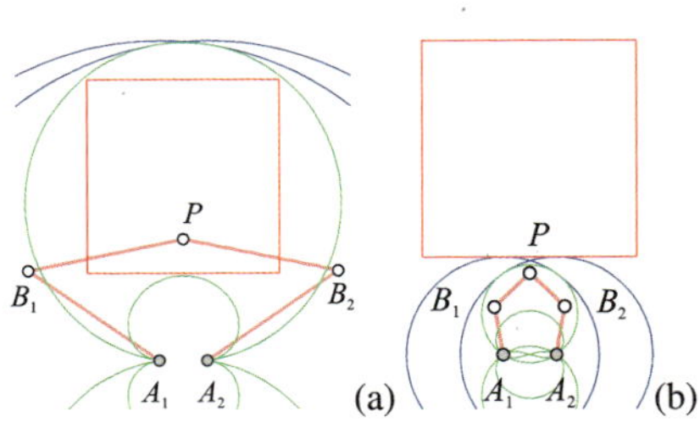

Fig. 5 A five-bar solution robot when $f = 3.8$, $l = 3.3$ from $\mathscr{R}_{Cf1}$ (a) and when $f = 3.8$, $l = 0.9$ from $\mathscr{R}_{Cf2}$(b).

3.2 Orientation Module: Four-Bar Linkages

One of the two base points of the four-bar linkages is the reference point $P(x,y)$ of the moving platform. Accordingly, the constraint equation of the four-bar linkage is:

$$\mathscr{C}_2 : (x + d\cos(\alpha) - l\cos(\theta_5))^2 + (y + d\sin(\alpha) - g - l\sin(\theta_5))^2 = l^2 \quad (5)$$

where θ_5 and α are the input and output angles, respectively, $\|\mathbf{A}_3\mathbf{B}_3\| = \|\mathbf{B}_3\mathbf{P}\| = l$, $\|\mathbf{C}_3\mathbf{P}\| = d$ and $\|\mathbf{A}_3\mathbf{O}\| = g$. A serial (resp. parallel) singularity is reached whenever $(\mathbf{A}_3\mathbf{B}_3)$ is aligned with $(\mathbf{B}_3\mathbf{C}_3)$ (resp. when $(\mathbf{B}_3\mathbf{C}_3)$ is aligned with $(\mathbf{C}_3\mathbf{P})$). These singularities are defined as follows:

$$\delta_{s3} : (2g\sin(\alpha) - 2x\cos(\alpha) - 2y\sin(\alpha))d - d^2 - x^2 - g^2 - y^2 + 4l^2 + 2yg = 0$$

$$\delta_{p4} : \begin{array}{l} g^2 + 2(l\sin(\alpha) - d\sin(\alpha) - y)g + x^2 \\ (d\cos(\alpha)) - 2l\cos(\alpha))x + y^2 + (2d\sin(\alpha) - 2l\sin(\alpha))y + d^2 - 2ld = 0 \end{array}$$

$$\delta_{p5} : \begin{array}{l} g^2 - 2(l\sin(\alpha) + d\sin(\alpha) + y)g + x^2 + \\ (d\cos(\alpha)) + 2l\cos(\alpha))x + y^2 + (2d\sin(\alpha) + 2l\sin(\alpha))y + d^2 + 2ld = 0 \end{array}$$

It is proposed to find those designs for which the platform can be oriented within desired bounds throughout the prescribed workspace. Accordingly, the parameters considered here are the orientation angle α of the moving platform plus only one geometric parameter to handle a two-dimensional parameter space. For the purpose of this study, we choose the distance between the fixed base point C_3 and the geometric center of the prescribed workspace: $h = g - f$ and parameter d is set to 1 to have a two dimensional representation of the solution regions.

Lamé curve prescribed workspace: From Fig. 2, the smallest value of parameter l is equal to 3. This value is chosen for the four-bar linkage design. The following problem has then to be solved:

$$\mathscr{P}_{L'} : [h\ \alpha]/\delta_{p4} \cap \delta_{p5} \cap \delta_{s3} \cap \overline{\mathscr{W}}_L = \emptyset, h > 0 \quad (6)$$

There exist two solution regions, $\mathscr{R}_{L'1}$ and $\mathscr{R}_{L'2}$ (Fig. 6), each one being associated with a single working mode and a single assembly mode. These regions describe the orientation ranges as function of parameter h, for which the robot can reach the full prescribed workspace without crossing singularities. It is then possible to choose h such that the range of the angular displacement α is greater than a prescribed value.

Square prescribed workspace: From Fig. 3, the smallest value of parameter l is equal to 3.3. This value is chosen for the 4-bar linkage design. The following problems have to be solved:

$$\mathscr{P}_{C'i} : [h\ \alpha]/\delta_{p4} \cap \delta_{p5} \cap \delta_{s3} \cap \overline{\mathscr{W}}_i = \emptyset, h > 0, t \in [0\ 1], i = 1, \dots, 4 \qquad (7)$$

The solutions regions of these problems and the intersection regions are shown in Figs. 7 and 8, respectively. Figure 9 depicts two 3-RRR parallel robots obtained for a square regular workspace. The solution obtained in Fig. 9b is more compact than in Fig. 9a and its angular range interval is greater but the design should take into account the self collisions.

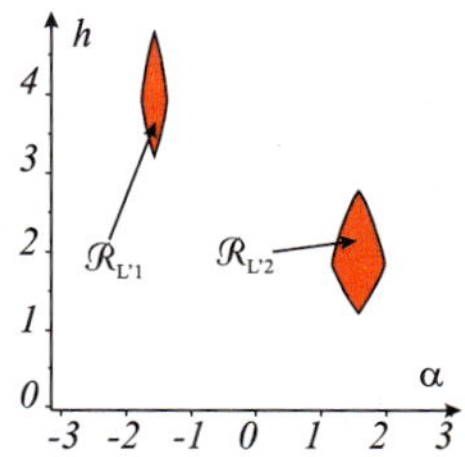

Fig. 6 Solution regions of problem $\mathscr{P}_{L'}$ for a four-bar linkage when $l = 3$.

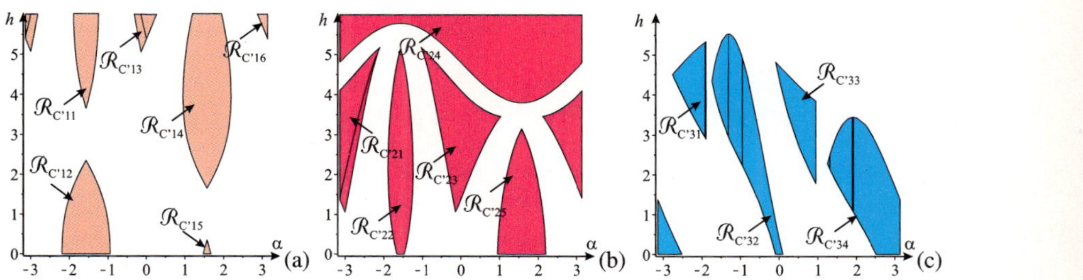

Fig. 7 Solution regions for problems (a) $\mathscr{P}_{C'1}$, (b) $\mathscr{P}_{C'2}$, (c) $\mathscr{P}_{C'3}$

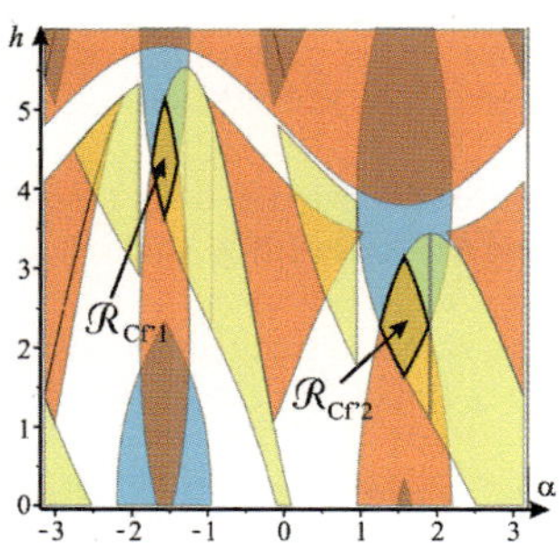

Fig. 8 Intersection regions when $l = 3.3$.

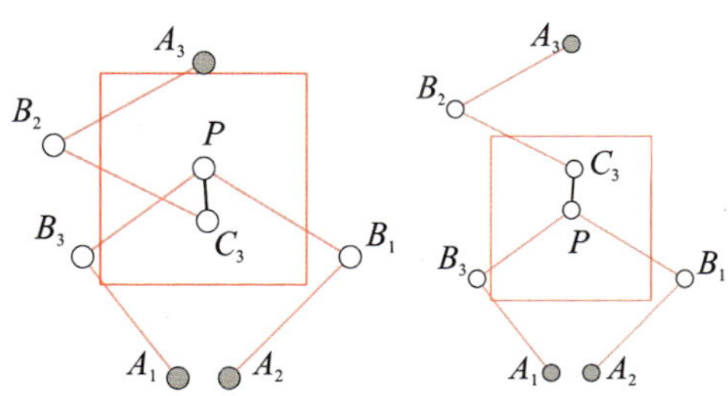

Fig. 9 Two examples from the regions $\mathscr{P}_{c5}$ and $\mathscr{P}_{c6}$ for (a) $h = 4.25$, $\alpha = [-1.717\ -1.424]$ and (b) $h = 2.2$, $\alpha = [1.306\ 1.835]$.

4 Conclusions

This paper presented a new design method to determine the feasible set of parameters of parallel manipulators for a prescribed singularity-free regular workspace. Rather than giving a single feasible or optimal solution, this method provides the solution regions in the parameter space. Groebner bases, discriminant varieties and cylindrical algebraic decomposition were used to generate the solution regions. As a result, their boundaries have an exact formulation. Solutions close to the boundaries of these regions correspond to robots for which the prescribed workspace is close to a singularity curve. The prescribed workspace can be defined in a more restrictive way to ensure that the robot will remain far enough from singularities. A solution would be to add a condition relying on some kinetostatic index [9]. The method was applied to a 3-RRR parallel planar robot with position/orientation decoupled architecture. It can handle other types of translational or decoupled robots but there are some limits that are due to the algebraic tools used. In particular, the number of parameters involved in the elimination process should not be too high.

References

1. Arsenault, M., Boudreau, R.: The synthesis of three-degree-of-freedom planar parallel robots with revolute joints (3-RRR) for an optimal singularity-free workspace. J. Robot. Syst. **21**(5), 259–274 (2004)
2. Merlet, J.P.: Determination of the presence of singularities in 6D workspace of a Gough parallel manipulator. In: Proc. Int. Symposium on Advances in Robot Kinematics, pp. 39–48 (1998)
3. Li, H., Gosselin, C.M.: Determination of maximal singularity-free zones in the workspace of planar three-degree-of-freedom parallel mechanisms. Mech. Mach. Theory **41**(10), 1157–1167 (2006)
4. Sen, S., Dasgupta, B., Mallik, A.K.: Variational approach for singularity-free-path-planning of parallel manipulators. Mech. Mach. Theory **38**(11), 1165–1183 (2003)
5. McCarthy, J.M., Soh, G.S.: Geometric Design of Linkages, 2nd edn. Springer (2010)
6. Lazard, D., Rouillier, F.: Solving parametric polynomial systems. J. Symb. Comput. **42**(6), 636–667 (2007)
7. Moroz, G., Chablat, D., Wenger, P., Rouillier, F.: Cusp points in the parameter space of RPR-2PRR parallel manipulator. In: Proc. 3rd European Conference on Robot Science, Cluj-Napoca, Romania, pp. 29–38 (2010)
8. Wenger, P., Gosselin, C., Maillé, B.: A comparative study of serial and parallel robot topologies for machine tools. In: Int. Workshop on Parallel Kinematic Machines 1999, Milan, Italy, pp. 23–35 (1999)
9. Chablat, D., Wenger, P., Majou, F., Merlet, J.P.: An interval analysis based study for the design and the comparison of 3-DOF parallel kinematic machines. Int. J. Robot. Res. **23**(6), 615–624 (2004)
10. Arakelian, V., Briot, S., Yatsun, S., Yatsun, A.: A new 3-DoF planar parallel manipulator with unlimited rotation capability. In: 13th World Congress in Robot and Machine Science, Guanajuato, Mexico (2011)
11. Gogu, G.: Structural Synthesis of Parallel Robots. Springer, the Netherlands (2009)

Inverse Geometrico-Static Analysis of Under-Constrained Cable-Driven Parallel Robots with Four Cables

Marco Carricato, Ghasem Abbasnejad and Dominic Walter

Abstract This paper presents the inverse geometrico-static analysis of under-constrained cable-driven parallel robots with 4 cables. The problem consists in finding all equilibrium configurations of the end-effector when either its orientation or the center-of-mass's position is assigned. In both cases, a further point of the end-effector is constrained to lie on a given plane. A major challenge is posed by the intrinsic coupling between kinematics and statics, which must be tackled simultaneously. The problems at hand are solved by analytical elimination procedures, thus leading to univariate polynomials free of spurious factors. All potential solutions may be real.

Key words: Cable-driven parallel robots, under-constrained robots, kinematic analysis, static analysis

1 Introduction

Cable-driven parallel robots (CDPRs) employ cables in place of rigid-body extensible legs in order to control the end-effector (EE) pose, thus strengthening classic advantages characterizing closed-chain architectures versus serial ones. A CDPR is *under-constrained* if the EE preserves some freedoms once actuators are locked. Typically, this occurs when the EE is controlled by a number of cables n smaller than the number of degrees of freedom (dofs) that the EE possesses with respect to the base [2].

Marco Carricato · Ghasem Abbasnejad
Department of Mechanical Engineering, University of Bologna, Bologna, Italy, e-mail: marco.carricato@unibo.it

Dominic Walter
Institute for Basic Sciences in Engineering, University of Innsbruck, Innsbruck, Austria

J. Lenarčič, M. Husty (eds.), *Latest Advances in Robot Kinematics*,
DOI 10.1007/978-94-007-4620-6_46,

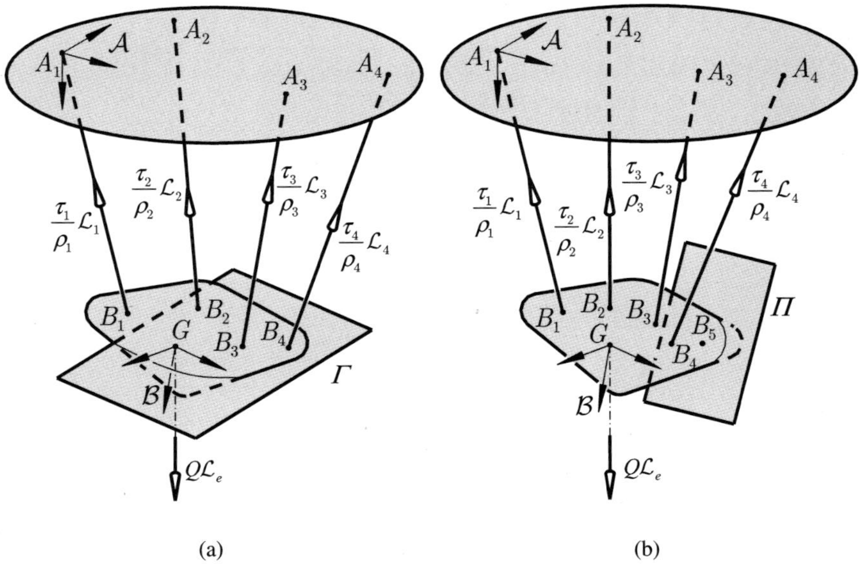

Fig. 1 A cable-driven parallel robot with 4 cables: (a) model for the IGP with assigned orientation, (b) model for the IGP with assigned position.

A major challenge in the kinematic study of under-constrained CDPRs emerges from the fact that the EE configuration depends on both the cable lengths and the applied forces (e.g. gravity). Accordingly, *loop-closure* and *mechanical-equilibrium* equations must be solved simultaneously and displacement-analysis problems become significantly more complex than analogous tasks concerning rigid-link fully-constrained manipulators [7, 9, 10]. Recently, Carricato and Merlet [2–4] proposed a general methodology for the kinematic, static and stability analysis of under-constrained CDPRs equipped with $n \leq 5$ cables. By properly formulating the mathematical model, the method allows one to find the entire set of equilibrium configurations when either n EE-pose coordinates (*inverse problem*) or n cable lengths (*direct problem*) are assigned.

In this paper, the inverse geometrico-static problem of CDPRs with 4 cables is solved. Two instances are considered, depending on whether the EE orientation or the EE's center-of-mass position is assigned. Since 4 dofs of the EE are to be controlled, an additional constraint must be set on the EE-pose coordinates: when the EE orientation is assigned, the EE's center-of-mass is constrained to lie on a given plane (this condition may be useful, for instance, to set the center-of-mass at a given height); when the EE orientation is assigned, an additional point of the EE is set to lie on a given plane (this condition may be useful for obstacle avoidance, in order to guide the EE tilt). In both cases, the overall robot configuration, the cable lengths and the cable tensions are to be computed.

2 Geometrico-Static Model

The EE is connected to the fixed base by 4 cables (Fig. 1). The ith cable exits from the base at point A_i and it is connected to the EE at point B_i. The geometric parameters of both the EE and the base are *arbitrary*. $\mathcal{A}$ is a fixed Cartesian frame attached at A_1, whereas $\mathcal{B}$ is a Cartesian frame attached to the EE at the platform's center of mass G. The platform pose is described by $\mathbf{X}^T = [\mathbf{g}^T; \mathbf{\Phi}^T]$, where $\mathbf{g}^T = [x, y, z]^T$ is the position of G in $\mathcal{A}$ and $\mathbf{\Phi}^T = [e_1, e_2, e_3]^T$ is the array grouping the Rodrigues variables parameterizing the EE orientation with respect to $\mathcal{A}$. The EE is acted upon by a constant force of magnitude Q, applied at G. This force is described as a 0-pitch wrench $Q\mathcal{L}_e$, where $\mathcal{L}_e$ is the normalized Plücker vector of the line of action. The normalized Plücker vector of the line associated with the ith cable is $\mathcal{L}_i/\rho_i$, where ρ_i is the cable length, $\mathbf{p}_i$ is any vector from the reduction pole of moments O to the cable line and $\mathcal{L}_i^T = [(A_i - B_i)^T; \{\mathbf{p}_i \times (A_i - B_i)\}^T]$. Accordingly, the wrench exerted by the ith cable on the EE is $(\tau_i/\rho_i)\mathcal{L}_i$, with τ_i being a positive scalar representing the intensity of the cable tensile force. Without loss of generality, one may assume $O \equiv A_1$.

When all cables of the robot are in tension, the set of *geometrical constraints* imposed on the EE is

$$||A_i - B_i||^2 = \rho_i^2, \quad i = 1 \ldots 4, \tag{1}$$

with the overall pose being determined by the EE *static equilibrium*, i.e.

$$\sum_{i=1}^{4} \frac{\tau_i}{\rho_i} \mathcal{L}_i + Q\mathcal{L}_e = \underbrace{\left[\mathcal{L}_1\ \mathcal{L}_2\ \mathcal{L}_3\ \mathcal{L}_4\ \mathcal{L}_e\right]}_{\mathbf{M}(O)} \begin{bmatrix} (\tau_1/\rho_1) \\ (\tau_2/\rho_2) \\ (\tau_3/\rho_3) \\ (\tau_4/\rho_4) \\ Q \end{bmatrix} = \mathbf{0}, \tag{2}$$

with $\tau_i \geq 0$, $i = 1 \ldots 4$.

Equations (1)–(2) amount to 10 scalar relations in 14 variables, namely $\mathbf{g}$, $\mathbf{\Phi}$, ρ_i and τ_i, $i = 1 \ldots 4$.

Generally, a finite set of system configurations may be determined if 4 additional constraints are assigned on the variables. When these constraints concern the EE-pose coordinates, an inverse geometrico-static problem (IGP) must be solved. Two relevant cases may be considered, depending on whether: (i) the orientation $\mathbf{\Phi}$ is assigned and G is constrained to lie on a given plane (*IGP with assigned orientation*); or (ii) the position of G is known and a further point B_5 of the EE is required to lie on a given plane (*IGP with assigned position*). In both cases, the pose coordinates are subject to 4 linear constraints, i.e. $q_i(\mathbf{X}) = 0$, $i = 1 \ldots 4$.

By following the method presented in [2], cable tensions may be eliminated from the set of unknowns by observing that Eq. (2) holds only if

$$\text{rank}\,[\mathbf{M}(O)] \leq 4, \tag{3}$$

which is a purely geometrical condition, since $\mathbf{M}(O)$ is a 6×5 matrix only depending on $\mathbf{X}$. By setting all 5×5 minors of $\mathbf{M}(O)$ equal to zero, 6 polynomial relations that do not contain cable tensions may be obtained, 5 of which are linearly independent, namely $p_j(\mathbf{X}) = 0$, $j = 1 \dots 5$. The 0-dimensional variety V of the ideal generated by $\{q_1, \dots, q_4, p_1, \dots, p_5\}$ provides the equilibrium configurations of the EE that satisfy the imposed constraints.

Once $\mathbf{X}$ is known, cable lengths may be computed by Eq. (1) and cable tensions may be obtained by 4 linearly-independent relations chosen within Eq. (2). Stability may be assessed as in [2]. Once a configuration is found, it proves *feasible* only if it is *stable* and therein cable tensions are *positive*.

3 IGP with Assigned Orientation

In this case, $\boldsymbol{\Phi}$ is known and G is constrained to lie on a plane Γ (Fig. 1(a)). The constraints $q_i(\mathbf{X}) = 0$, $i = 1 \dots 4$, are

$$e_1 = \bar{e}_1,\ e_2 = \bar{e}_2,\ e_3 = \bar{e}_3, \quad \mathbf{g} \cdot \mathbf{n} - d_\Gamma = x n_1 + y n_2 + z n_3 - d_\Gamma = 0, \qquad (4)$$

where $\bar{e}_1$, $\bar{e}_2$ and $\bar{e}_3$ are known scalars, $\mathbf{n} = [n_1, n_2, n_3]^T$ is a unit vector perpendicular to Γ and $|d_\Gamma|$ is the distance of Γ from A_1. Two subcases may be identified, depending on whether $n_3 \neq 0$ or $n_3 = 0$.

3.1 $n_3 \neq 0$

When $n_3 \neq 0$, Γ is a *nonvertical* plane and z may be expressed, from the last relationship in Eq. (4), as $z = -(n_1/n_3)x - (n_2/n_3)y + d_\Gamma/n_3$. By taking advantage of this expression and by imposing the first three constraints in Eq. (4), the 5 relations $p_j(\mathbf{X}) = 0$, $j = 1 \dots 5$, emerging from Eq. (3) become cubic relations in x and y comprising 10 monomials, i.e. $[y^3, y^2x, yx^2, x^3, y^2, yx, x^2, y, x, 1]$.

The problem may be efficiently solved by implementing a Sylvester dialytic method, namely by rewriting the relations $p_j = 0$ as linear equations in all monomials involving the original unknowns except one, which is 'hidden' in the equation coefficients. If these monomials are treated as linear unknowns, a square homogeneous system is obtained and the determinant of the coefficient matrix provides a resultant in the hidden variable. In the case at hand, by hiding y, 4 monomials in x emerge and, thus, *four* relations $p_j = 0$, $j = 1 \dots 4$, may be used to build up a square Sylvester matrix. However, the corresponding resultant exhibits a spurious solution. In order to get rid of the extraneous factor, all *five* relations $p_j = 0$, $j = 1 \dots 5$, may be linearized in the 5 monomials contained in the array $\boldsymbol{\kappa}_1 = [y^3, x^3, x^2, x, 1]^T$, namely

Table 1 A 4-4 CDPR whose IGP with assigned orientation admits 5 *real* potential solutions ($d_\Gamma = 3/\sqrt{6}$).

$[A_2]_\mathcal{A}$	$[A_3]_\mathcal{A}$	$[A_4]_\mathcal{A}$	$[B_1 - G]_\mathcal{A}$	$[B_2 - G]_\mathcal{A}$	$[B_3 - G]_\mathcal{A}$	$[B_4 - G]_\mathcal{A}$	$\sqrt{6}\mathbf{n}$
8	9	1	−3	3	2	3	−2
0	7	8	4	−2	1	1	1
5	6	4	3	1	0	2	1

$$\mathbf{S}_1(y)\boldsymbol{\kappa}_1 = \mathbf{0}, \tag{5}$$

where $\mathbf{S}_1(y)$ is a 5×5 matrix whose entries are known polynomial functions of y. Letting the determinant of $\mathbf{S}_1(y)$ vanish yields a 5th-degree univariate equation in y. This was obtained in symbolic form and it is devoid of spurious roots. For each root, a unique value for x may be obtained by solving the linear system (5).

Solutions may be either *complex* or *real*, with only the latter ones having physical interest. By varying the robot's geometry, the count of real roots may change. Table 1 reports an example for which the IGP with assigned orientation admits 5 *real* solutions (not all of them necessarily feasible). Examples of this kind may be obtained by using the algorithms developed in [1], namely a continuation procedure adapted from a routine originally proposed by Dietmaier [6] and two evolutionary techniques based on a genetic algorithm and particle swarm optimization.

3.2 $n_3 = 0$

When $n_3 = 0$ (without loss of generality, $n_2 \neq 0$), Γ is *vertical* and y may be expressed, from the last relation in Eq. (4), as $y = -(n_1/n_2)x + (d_\Gamma/n_2)$.

By substituting this expression in the relations $p_j(\mathbf{X}) = 0$, $j = 1 \ldots 5$, one obtains 5 cubics in the monomials $[z^2x, zx^2, x^3, z^2, zx, x^2, z, x, 1]$. By a procedure similar to that described in Sec. 3.1, a least-degree univariate equation free from extraneous polynomial factors may be obtained by linearizing all *five* relations $p_j = 0$, $j = 1 \ldots 5$, in the 5 monomials contained in the array $\boldsymbol{\kappa}_2 = [z^2, x^3, x^2, x, 1]^T$, and by writing them in the form

$$\mathbf{S}_2(z)\boldsymbol{\kappa}_2 = \mathbf{0}, \tag{6}$$

where $\mathbf{S}_2(z)$ is a 5×5 matrix whose entries are known polynomial functions of z. Letting the determinant of $\mathbf{S}_2(z)$ vanish yields a 4th-degree univariate polynomial in z, which is available in symbolic form and devoid of spurious roots. For each root, a unique value for x may be obtained by solving the linear system (6). Even in this case, all roots may possibly be *real*.

4 IGP with Assigned Position

In this case, the position of G is known and a further point B_5 of the EE is constrained to lie on an assigned plane Π (Fig. 1(b)). The constraints $q_i(\mathbf{X}) = 0$, $i = 1 \ldots 4$, become

$$x = \bar{x}, \; y = \bar{y}, \; z = \bar{z}, \quad \mathbf{r}_5 \cdot \mathbf{n} - d_\Pi = 0, \tag{7}$$

where $\bar{x}$, $\bar{y}$ and $\bar{z}$ are known scalars, $\mathbf{r}_5$ is the position vector of B_5 in $\mathcal{A}$, $\mathbf{n}$ is a unit vector perpendicular to Π and $|d_\Pi|$ is the distance of Π from A_1.

If $\mathbf{R}(\boldsymbol{\Phi})$ is the rotation matrix between $\mathcal{B}$ and $\mathcal{A}$, the position vector of B_i, $i = 1 \ldots 5$, in $\mathcal{A}$ may be expressed as $\mathbf{r}_i = [B_i]_\mathcal{A} = \mathbf{g} + \mathbf{R}(\boldsymbol{\Phi})[B_i]_\mathcal{B}$, where the position of B_i in $\mathcal{B}$ is known. By substituting these expressions in the 5 relations $p_j(\mathbf{X}) = 0$, $j = 1 \ldots 5$, and by imposing the first three constraints in Eq. (7), one obtains 5 sextics in e_1, e_2 and e_3. By a similar expansion, the fourth relationship in Eq. (7), i.e. $q_4 = 0$, becomes a quadratic equation in the Rodrigues parameters.

By denoting the ideal generated by the set $J = \{p_1, \ldots, p_5, q_4\}$ as $\langle J \rangle$, the solutions of the IGP with assigned position form the variety V of $\langle J \rangle$. The high order of the polynomials in J suggests applying elimination procedures based on Groebner bases in order to solve the problem. Even though the lexicographic monomial order is, in general, particularly suitable to solve polynomial systems, for it provides equation sets whose variables may be eliminated successively, it is highly inefficient in terms of computation time and memory requirements. A Groebner basis $G[J]$ of $\langle J \rangle$ with respect to a *graded reverse lexicographic order*, instead, i.e. grevlex(e_1, e_2, e_3), may be computed in a very expedited way (tenths of seconds, for the case at hand, on a PC with a 2.67GHz Intel Xeon processor and 4GB of RAM). Once $G[J]$ is known, the FGLM algorithm [8], converting a Groebner basis from one monomial order to another, may be called upon to compute a univariate polynomial in $\langle J \rangle$. However, a more efficient method is provided by the Groebner-Sylvester hybrid approach proposed in [5]. The method is based on the observation that $G[J]$ comprises 12 polynomials and these contain 12 monomials in e_1 and e_2, i.e. $\boldsymbol{\kappa}_3 = [e_1 e_2^4, e_2^5, e_1 e_2^3, e_2^4, e_1 e_2^2, e_2^3, e_1^2, e_1 e_2, e_2^2, e_1, e_2, 1]^T$. Accordingly, $G[J]$ may be set up as a square system of homogeneous linear equations in the form

$$\mathbf{S}_3(e_3)\boldsymbol{\kappa}_3 = \mathbf{0}, \tag{8}$$

where $\mathbf{S}_3(e_3)$ is a 12×12 matrix polynomial in e_3. Letting the determinant of $\mathbf{S}_3(e_3)$ vanish yields a spurious-root-free univariate polynomial of degree 32 in e_3. For each root, unique values for e_1 and e_2 may be obtained by solving the linear system (8).

An alternative procedure is based on the properties of the *normal set* $\mathbf{N}[J]$, which is the array grouping all monomials which are *not* multiples of any leading monomial in $G[J]$ [11]. $\mathbf{N}[J]$ contains 32 monomials in $\boldsymbol{\Phi}$, i.e. $\mathbf{N}[J] = [\eta_1, \ldots, \eta_{32}]^T$. If r_h is the remainder on division of $e_3\eta_h$ by $G[J]$, r_h is a linear combination of the monomials of $\mathbf{N}[J]$, i.e. $r_h = \sum_{k=1}^{32} a_{hk}\eta_k$, with a_{hk} being a *constant coefficient*. Since $r_h - e_3\eta_h$ belongs to $\langle J \rangle$, it must vanish on the variety

Table 2 A 4-4 CDPR whose IGP with assigned position admits 32 *real* potential solutions ($d_\Pi = -0.0543588$).

$[A_2]_\mathcal{A}$	$[A_3]_A$	$[A_4]_\mathcal{A}$	$[B_1]_\mathcal{B}$	$[B_2]_\mathcal{B}$	$[B_3]_\mathcal{B}$	$[B_4]_\mathcal{B}$	$[B_5]_\mathcal{B}$	$\mathbf{n}$	$(\bar{x}, \bar{y}, \bar{z})$
0.0096715	0.1602038	0.3227272	0.8585338	0.3187879	0.6598471	0.6273182	0.4579794	0.8894538	0.1339193
0	0.0649423	0.7151215	0	0.4888859	0.6661870	0.9610494	0.5558744	0.4392272	0.2021438
0.5484151	0.6597958	0.5378416	0	0	0.8744523	0.0797477	0.8222827	−.1262981	0.1180386

V, for any h. Thus, one may assemble all equations of this kind in the form

$$(\mathbf{A}[J, e_3] - e_3\mathbf{I}_{32})\,\mathbf{N}[J] = \mathbf{0}, \tag{9}$$

where $\mathbf{A}[J, e_3] = [a_{hk}]$ is a 32×32 numeric matrix (called *multiplication matrix* for e_3) and $\mathbf{I}_{32}$ is the 32×32 identity matrix. Equation (9) is a linear eigenvalue problem, whose 32nd-degree characteristic polynomial is the desired resultant in e_3. Equation (9) provides an efficient way to numerically compute all solutions of the problem at hand as the eigenvalues of $\mathbf{A}[J, e_3]$.

By taking advantage of the algorithms developed in [1], several sets of robot parameters were found proving that all 32 solutions may be *real*. Table 2 reports an example.

5 Conclusions

This study solved the inverse geometrico-static problem (IGP) of under-constrained cable-driven parallel robots with 4 cables. The problem consists in finding *all* equilibrium configurations of the robot when a subset of the end-effector pose coordinates is assigned. Two relevant cases were considered.

In the former, the orientation of the end-effector is assigned and the center of mass is constrained to lie on a plane (*IGP with assigned orientation*). The problem was solved by an elimination procedure based on a Sylvester dialytic method. A least-degree univariate polynomial was obtained in symbolic form. This has degree 4 or 5, depending on the orientation of the plane constraining the center of mass being vertical or not, and all roots may be *real*.

In the latter case, the position of the center of mass is assigned and a further point of the end-effector is required to lie on a known plane (*IGP with assigned position*). The problem was solved by exact-arithmetic procedures based on Groebner-basis computation. In this case, at the most 32 solutions (and a corresponding univariate polynomial) were obtained. By the algorithms developed in [1], a numerical example was found that proves that all potential solutions may be *real*.

It is worth observing that all solution counts reported above refer to *potential* solutions of the problems at hand, since they do not take into account the constraints imposed by the sign of cable tensions and the stability of equilibrium. Once such constraints are imposed and solutions are sifted, the number of *feasible* configurations reduces. When multiple feasible configurations exist, the EE may switch across them, due to inertia forces or external disturbances. Accordingly, the computation of the entire set of equilibrium configurations is essential for robust trajectory planning. This motivates the relevance of the presented algorithms, even when they are not applicable to real-time computation (as for the IGP with assigned position). For real-time applications, these authors are exploring interval-analysis-based approaches, in collaboration with Dr. Merlet's team at INRIA.

References

1. Abbasnejad, G., Carricato, M.: Real solutions of the direct geometrico-static analysis of under-constrained cable-driven parallel robots with three cables. In: Proc. of AIMETA 2011, the 20th Congress of the Italian Association of Theoretical and Applied Mechanics, pp. 1–15. Bologna, Italy, (2011)
2. Carricato, M., Merlet, J.P.: Geometrico-static analysis of under-constrained cable-driven parallel robots. In: Lenarčič J., Stanišič M.M. (eds.) Advances in Robot Kinematics: Motion in Man and Machine, pp. 309–319. Springer, Dordrecht (2010)
3. Carricato, M., Merlet, J.P.: Direct geometrico-static problem of under-constrained cable-driven parallel robots with three cables. In: Proc. of the 2011 IEEE Int. Conf. on Robotics and Automation, pp. 3011–3017. Shanghai, China, (2011)
4. Carricato, M., Merlet, J.P.: Inverse geometrico-static problem of under-constrained cable-driven parallel robots with three cables. In: Proc. of the 13th World Congress in Mechanism and Machine Science, pp. 1–10, Paper No. A7_283. Guanajuato, Mexico (2011)
5. Dhingra, A.K., Almadi, A.N., Kohli, D.: A Gröbner–Sylvester hybrid method for closed-form displacement analysis of mechanisms. ASME J. Mech. Des. **122**(4), 431–438 (2000)
6. Dietmaier, P.: The Stewart–Gough platform of general geometry can have 40 real postures. In: Lenarčič J., Husty M.L. (eds.) Advances in Robot Kinematics: Analysis and Control, pp. 7–16. Kluwer Academic Publishers, Dordrecht (1998)
7. Fattah, A., Agrawal, S.K.: On the design of cable-suspended planar parallel robots. ASME J. Mech. Des. **127**(5), 1021–1028 (2006)
8. Faugère, J.C., Gianni, P., Lazard, D., Mora, T.: Efficient computation of zero-dimensional Gröbner bases by change of ordering. J. Symb. Comput. **16**(4), 329–344 (1993)
9. Jiang, Q., Kumar, V.: The direct kinematics of objects suspended from cables. In: Proc. of the ASME 2010 Int. Design Engineering Technical Conferences, pp. 1–10, Paper no. DETC2010–28,036. Montreal, Canada, (2010)
10. McCarthy, J.M.: 21st century kinematics: Synthesis, compliance, and tensegrity. ASME J. Mech. Robot. **3**(2), 020201 (2011)
11. Stetter, H.J.: Matrix eigenproblems are at the heart of polynomial system solving. ACM SIGSAM Bull. **30**(4), 22–25 (1996)

The Kinematotropic 3-CPU Parallel Robot: Analysis of Mobility and Reconfigurability Aspects

Luca Carbonari and Massimo Callegari

Abstract This paper investigates the mobility analysis of a 3-CPU parallel machine, aiming at checking the possibility to perform both pure rotational and pure translational motions. Machine kinematics is formalized by taking advantage of algebraic geometry principles that allow extracting the necessary constraint equations in form of polynomial ideals. The analysis of the sub-ideals deriving from the decomposition of the starting constraint equations yields the conclusion that several kinds of motion are actually achievable by the same 3-CPU architecture. Among them, pure rotational and pure translational mobilities are present. Finally, the existence of machine configurations allowing the transition between such modes is proved and the related poses are explicitly worked out.

Key words: Parallel kinematics machines, kinematotropic mechanisms, reconfigurable mechanisms, algebraic geometry

1 Introduction

During past years, two different minor mobility parallel robots have been developed and prototyped at the Laboratory of Robotics of the Polytechnic University of Marche in Ancona (Italy): a pure translational machine, called I.Ca.Ro. [2], and a spherical wrist, called Sphe.I.Ro. [1]. Notwithstanding the quite different kinematic performances of the machines, they are both based on the same 3-CPU architecture, with of course a different setting of the joints. The present work was aimed at investigating whether a common mechanical architecture might be able to provide both motions by a simple reconfiguration or even if the same machine yields the two different kinds of motions by meeting some "switching configuration", i.e. it shows a kinematotropic behaviour [4, 8].

Luca Carbonari · Massimo Callegari
Polytechnic University of Marche, Ancona, Italy, e-mail: {l.carbonari,m.callegari}@univpm.it

J. Lenarčič, M. Husty (eds.), *Latest Advances in Robot Kinematics*,
DOI 10.1007/978-94-007-4620-6_47,

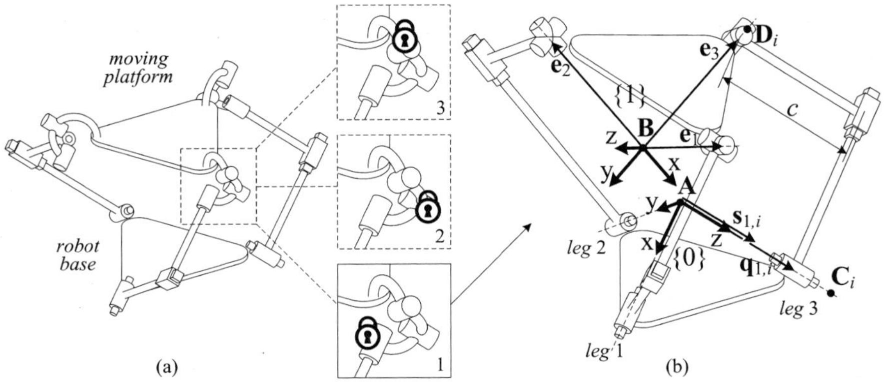

Fig. 1 Possible arrangement of the 3 revolute joints equivalent to a reconfigurable spherical joint (a) and 3-CPU architecture able to yield different motions (b).

The key idea was to realize a reconfigurable universal joint that allows to vary robot kinematics without changes in legs structure. A simple solution to this problem is using a spherical joint made of three consecutive revolute pairs (see Figure 1(a)): in this way, three different universal joints are obtained by locking, one at a time, the three rotations of the spherical joint. It is worth to remark that both I.Ca.Ro. and Sphe.I.Ro. are comprehended within these joints configurations. An extended analysis [3] pointed out that the only chance to yield different kinds of motions with the 3-CPU configuration is to lock the first rotation of spherical joint, thus obtaining the kinematics scheme of Sphe.I.Ro. for which the universal joint is made of two revolute pairs, the first one perpendicular to the plane of the leg and the second lying on leg plane intersecting the center of platform reference frame.

2 Kinematics Equations

In order to use a parametrization free of representation singularities, the transformation matrix ${}^0\mathbf{T}_1$ between the two frames $\{0\}$ and $\{1\}$ (see Figure 1(b)) has been expressed as a function of Study's parameters:

$$ {}^0\mathbf{T}_1 = \begin{pmatrix} {}^0\mathbf{R}_1 & {}^0\mathbf{q}_1 \\ \mathbf{0} & x_0^2 + x_1^2 + x_2^2 + x_3^2 \end{pmatrix} \tag{1} $$

where rotation matrix ${}^0\mathbf{R}_1$ and translation vector ${}^0\mathbf{p}_1$ are:

$$ {}^0\mathbf{R}_1 = \begin{pmatrix} x_0^2 + x_1^2 - x_2^2 - x_3^2 & 2\,(x_1x_2 - x_0x_3) & 2\,(x_1x_3 - x_0x_2) \\ 2\,(x_1x_2 - x_0x_3) & x_0^2 - x_1^2 + x_2^2 - x_3^2 & 2\,(x_2x_3 - x_0x_1) \\ 2\,(x_1x_3 - x_0x_2) & 2\,(x_2x_3 - x_0x_1) & x_0^2 - x_1^2 - x_2^2 + x_3^2 \end{pmatrix} \tag{2} $$

$$
{}^{0}\mathbf{p}_1 = \begin{pmatrix} 2\left(-x_0y_1+x_1y_0-x_2y_3+x_3y_2\right) \\ 2\left(-x_0y_2+x_1y_3-x_2y_0-x_3y_1\right) \\ 2\left(-x_0y_3-x_1y_2-x_2y_1+x_3y_0\right) \end{pmatrix} \tag{3}
$$

It is noted that Study's notation is based on 8 parameters x_0, x_1, x_2, x_3, y_0, y_1, y_2, y_3 which define a point in the 6-dimensional quadric $S \in \mathbb{P}^7$, a semi-algebraic set defined by [6]:

$$
\begin{aligned} x_0y_0+x_1y_1+x_2y_2+x_3y_3 &= 0 \\ x_0^2+x_1^2+x_2^2+x_3^2 &\neq 0 \end{aligned} \tag{4}
$$

If the inequality in (4) is assigned an arbitrary constant value, the expression becomes a normalizing equation ensuring that matrix (1) is effectively a non singular transformation matrix; typically it is assumed $x_0^2+x_1^2+x_2^2+x_3^2=1$. In the following of the paper it is called: $\langle \gamma_1 : x_0y_0+x_1y_1+x_2y_2+x_3y_3, \gamma_2 : x_0^2+x_1^2+x_2^2+x_3^2-1 \rangle$.

In order to obtain a complete algebraic description of robot kinematics, Study quadric equations (4), that are intrinsic of the used parametrization, must be juxtaposed to relations characteristic of the specific legs architecture. Among these constraint equations, a distinction can be made based on their dependence on robot actuation. Indeed, each leg is made of a serial kinematic chain whose joints constrain the manipulator mobility regardless of the actuation parameters. On the other hand some equations are needed to describe the influence of actuation displacements on end effector pose.

To this aim, it is observed that in each leg the universal joint constrains the end effector to maintain the axis of the last revolute joint in the plane of the leg itself. Such constraint can be geometrically visualized as the co-planarity of four points, i.e. origin of frame $\{0\}$, origin of frame $\{1\}$, any point of cylindrical pair axis and any point of revolute joint axis. Such coplanarity can be mathematically expressed by letting a matrix determinant vanish, i.e. the 4×4 matrix built by aligning the homogeneous vectors denoting the positions of these four points. Referring to Figure 1 the four points are respectively $\mathbf{A}$, $\mathbf{B}$, $\mathbf{C}_i$ and $\mathbf{D}_i$, thus:

$$
\alpha_i : \det\left(\mathbf{A}\ \mathbf{B}\ \mathbf{C}_i\ \mathbf{D}_i\right), \quad \alpha_i = 0 \tag{5}
$$

Points $\mathbf{A}$ and $\mathbf{C}_i$ are solid with mobile platform, therefore their coordinates do not vary when they are expressed in the local frame. On the contrary, $\mathbf{B}$ and $\mathbf{D}_i$ must be expanded through transformation (1) as $\mathbf{B} = {}^{0}\mathbf{T}_1 \left[0\ 0\ 0\ 1\right]^T$ and $\mathbf{D}_i = {}^{0}\mathbf{T}_1 \left[\mathbf{e}_i^T\ 1\right]^T$ where $\mathbf{e}_i$ denotes the direction of last revolute joint of the ith leg with respect to moving frame $\{1\}$. In order to simplify the three polynomial equations α_i, Study quadric equations (4) can be exploited for elimination of null parts and simplification of non vanishing factors, yielding:

$$
\begin{aligned} \alpha_1 &: x_0y_2+x_1y_3+x_2y_0+x_3y_1, \quad \alpha_1 = 0 \\ \alpha_2 &: x_0y_3+x_1y_2+x_2y_1+x_3y_0, \quad \alpha_2 = 0 \\ \alpha_3 &: x_0y_1+x_1y_0+x_2y_3+x_3y_2, \quad \alpha_3 = 0 \end{aligned} \tag{6}
$$

Equations (6) together with the Study-quadric (4) are not sufficient yet to uniquely characterize robot kinematics since no description of actuated joints has been provided so far.

Considering the ith leg, its architecture allows the connection point with the manipulator move only within leg's plane. In particular, its translation along the respective cylindrical joint axis is directly related to the displacement of the cylindrical joint:

$$\mathbf{D}_i^T \begin{bmatrix} \mathbf{s}_{1,i} \\ 0 \end{bmatrix} = \begin{bmatrix} \mathbf{e}_i^T & 1 \end{bmatrix} {}^0\mathbf{T}_1^T \begin{bmatrix} \mathbf{s}_{1,i} \\ 0 \end{bmatrix} = q_{1,i} \tag{7}$$

where $\mathbf{s}_{1,i}$ is the direction of joint 1 (the cylindrical joint) of ith leg and $q_{1,i}$ is its displacement. It is worth to remark that $q_{1,i}$ also represents the actuation parameter since it is the controlled distance between the center of reference frame $\{0\}$ and the center of the cylindrical pair. Expansion of (7) for each leg directly yields:

$$\begin{array}{ll} \beta_1 : x_0y_1 + x_1y_0 - x_2y_3 + x_3y_2 - e\,(x_0x_2 + x_1x_3) - \delta_1, & \beta_1 = 0 \\ \beta_2 : x_0y_2 + x_1y_3 - x_2y_0 + x_3y_1 - e\,(x_0x_3 + x_1x_2) - \delta_2, & \beta_2 = 0 \\ \beta_3 : x_0y_3 + x_1y_2 - x_2y_1 + x_3y_0 - e\,(x_0x_1 + x_2x_3) - \delta_3, & \beta_3 = 0 \end{array} \tag{8}$$

where $\delta_i = (q_{1,i} - c)/2$.

3 Equations Analysis: Primary Decomposition

At this point of the work, the robotic system is fully algebraically described by a set of polynomial constraint equations, that can be collected in the polynomial ideal $\mathscr{F} = \langle \alpha_1, \alpha_2, \alpha_3, \beta_1, \beta_2, \beta_3, \gamma_1, \gamma_2 \rangle$. As suggested by Walter et al. [6], a simpler formulation of the constraint equations $\mathscr{F}$ is fundamental for a deeper understanding of robot kinematic behaviour. With this aim, the authors developed a specialized study on that portion of the polynomial ideal collecting those constraints equations purely dependent on kinematic architecture, thus polynomials α_1, α_2 and α_3. This sub-ideal, that is hereby called $\mathscr{H}$, should also comprehend polynomial γ_1, without which the Study's representation is meaningless. Therefore, the sub-ideal $\mathscr{H} = \langle \alpha_1, \alpha_2, \alpha_3, \gamma_1 \rangle$ is analyzed through the computation of its primary decomposition; this method consists of splitting the ideal into several sub-ideals such that the union of their vanishing sets correspond to the vanishing set of the starting ideal. We remark that the *zero set*, or *vanishing set*, $\mathscr{V}(\mathscr{J})$ of a polynomial ideal $\mathscr{J}$ is the set of all points that simultaneously satisfy the homogeneous equations composing $\mathscr{J}$.

The primary decomposition of ideal $\mathscr{H}$ yields:

$$\mathscr{V}(\mathscr{H}) = \bigcup_{i=1}^{16} \mathscr{V}(\mathscr{H}_i) \tag{9}$$

with:

$$\mathscr{H}_1 = \langle y_0, y_1, y_2, y_3 \rangle,$$
$$\mathscr{H}_2 = \langle y_0 - y_3, y_1 - y_3, y_2 - y_3, x_0 + x_1 + x_2 + x_3 \rangle, \mathscr{H}_3 = \langle y_0 + y_3, y_1 - y_3, y_3 + y_2, x_0 - x_1 + x_2 - x_3 \rangle,$$
$$\mathscr{H}_4 = \langle y_0 + y_3, y_1 + y_3, y_2 - y_3, x_0 + x_1 - x_2 - x_3 \rangle, \mathscr{H}_5 = \langle y_0 - y_3, y_1 + y_3, y_3 + y_2, x_0 - x_1 - x_2 + x_3 \rangle,$$
$$\mathscr{H}_6 = \langle x_0 + x_3, x_1 + x_2, y_0 - y_3, y_1 - y_2 \rangle, \mathscr{H}_7 = \langle x_0 - x_3, x_1 - x_2, y_0 + y_3, y_1 + y_2 \rangle,$$
$$\mathscr{H}_8 = \langle x_0 + x_1, x_2 + x_3, y_0 - y_1, y_2 - y_3 \rangle, \mathscr{H}_9 = \langle x_0 - x_1, x_2 - x_3, y_0 + y_1, y_3 + y_2 \rangle,$$
$$\mathscr{H}_{10} = \langle x_0 + x_2, x_1 + x_3, y_0 - y_2, y_1 - y_3 \rangle, \mathscr{H}_{11} = \langle x_0 - x_2, x_1 - x_3, y_0 + y_2, y_1 + y_3 \rangle,$$
$$\mathscr{H}_{12} = \langle x_0 - x_3, x_1 - x_3, x_2 - x_3, y_0 + y_1 + y_2 + y_3 \rangle, \mathscr{H}_{13} = \langle x_0 + x_3, x_1 - x_3, x_2 + x_3, y_0 - y_1 + y_2 - y_3 \rangle,$$
$$\mathscr{H}_{14} = \langle x_0 - x_3, x_1 + x_3, x_2 + x_3, y_0 - y_1 - y_2 + y_3 \rangle, \mathscr{H}_{15} = \langle x_0 + x_3, x_1 + x_3, x_2 - x_3, y_0 + y_1 - y_2 - y_3 \rangle,$$
$$\mathscr{H}_{16} = \langle x_0, x_1, x_2, x_3 \rangle \tag{10}$$

The 16 sub-ideals (10) must be interpreted as different coexisting systems of equations able to describe the whole kinematic behaviour of the 3-CPU platform and to which correspond different types of solutions. Sub-ideal $\mathscr{H}_{16}$ represents a degenerate case since its vanishing set can not be satisfied together with $\gamma_2 = 0$: thus it represents an exception, since its solution is not compatible with $\mathscr{V}(\mathscr{F})$.

4 Robot Mobility

The sub-ideals $\mathscr{H}_i$ denote the different types of mobility that the robotic platform is able to perform. Such conclusion turns evident if homogeneous equations provided by $\mathscr{V}(\mathscr{H}_i)$ are substituted into end-effector transformation (1), which assumes a different shape for each one of the sub-ideals. Just to make an example, the vanishing set of $\mathscr{H}_1$ can be considered: in this case the translation vector ${}^0\mathbf{q}_1$ vanishes while both the rotation matrix and the scaling factor $x_0^2 + x_1^2 + x_2^2 + x_3^2$ do not change. Thus, the resulting transformation yields a pure rotational behaviour of the platform, namely the mobility expected in [1].

A deeper investigation on every sub-ideal revealed that several types of motions are allowed by the 3-CPU architecture. The transformations yielded by each substitution are not analytically reported but a brief description of the different kinds of mobility is provided in the following:

- $\mathscr{H}_1$: as previously mentioned, the vanishing set of this ideal refers to a pure rotational behaviour, already widely studied by several past works.
- $\mathscr{H}_2$, $\mathscr{H}_3$, $\mathscr{H}_4$, $\mathscr{H}_5$: all these ideals correspond to the same type of mobility, characterized by 3 spurious DOFs: the end-effector can change its orientation rotating about two distinct axes and it can translate along a direction which rotates solidly with the moving platform.
- $\mathscr{H}_6$, $\mathscr{H}_7$, $\mathscr{H}_8$, $\mathscr{H}_9$, $\mathscr{H}_{10}$, $\mathscr{H}_{11}$: the mobilities deriving from these ideals can not be exploited by the actuation chosen for the 3-CPU manipulator, since for each one of them the end-effector is only allowed to move on a plane perpendicular to one of the cylindrical joints passing through the origin of the absolute reference frame $\{0\}$. In these configurations, at least one of the cylindrical pairs is prevented from translating: the mechanical system gets stuck in an under-actuated configuration and one of platform DOFs becomes uncontrollable.

- $\mathscr{H}_{12}$, $\mathscr{H}_{13}$, $\mathscr{H}_{14}$, $\mathscr{H}_{15}$: here the manipulator is capable of pure translational motions, differing for the orientations of the manipulator, which are fixed in these cases. It is worth showing the homogeneous transformation is for at least one of these ideals. Thus, substitution of $\mathscr{V}(\mathscr{H}_{12})$ into (1) yields:

$$
{}^{0}\mathbf{T}_{1} = \begin{pmatrix} 0 & 0 & 4x_3^2 & -4x_3\,(y_1+y_3) \\ 4x_3^2 & 0 & 0 & -4x_3\,(y_2+y_1) \\ 0 & 4x_3^2 & 0 & -4x_3\,(y_3+y_2) \\ 0 & 0 & 0 & 4x_3^2 \end{pmatrix} \tag{11}
$$

Each mobility previously defined is fully determined only by one ideal that contemporary comprehends equations (8), the normalizing polynomial γ_2 and the respective ideal $\mathscr{H}_i$. For each mode it is defined $\mathscr{G}_i = \langle \beta_1, \beta_2, \beta_3, \gamma_2, \mathscr{H}_i \rangle$.

5 Transition Conditions

Previous section highlighted the possibility to obtain a multifunctional robot by using the same 3-CPU architecture, since it proved capable of both pure rotational and pure translational motions: therefore it is relevant to investigate the possibility to switch from a type of mobility to the other, which is done in the present section.

The configurations that may act as switching poses between two mobilities $\mathscr{G}_i$ and $\mathscr{G}_j$ must belong to both the characteristic vanishing sets $\mathscr{V}(\mathscr{G}_i)$ and $\mathscr{V}(\mathscr{G}_j)$. As a matter of fact, solutions that are common to a couple of vanishing sets are solutions of the intersection of the sets: $\mathscr{V}(\mathscr{G}_i) \cap \mathscr{V}(\mathscr{G}_j)$. As known, a solution is common to a couple of ideals if it satisfies all the polynomial equations collected in their zero sets: $\mathscr{V}(\mathscr{G}_i) \cap \mathscr{V}(\mathscr{G}_j) = \mathscr{V}(\mathscr{G}_i \cup \mathscr{G}_j)$.

It is not difficult to figure out what is the number of feasible solutions of $\mathscr{V}(\mathscr{G}_i) \cap \mathscr{V}(\mathscr{G}_j)$. To do that, Gröbner bases $\mathscr{G}_{i,j}$ of ideals $\mathscr{G}_i \cup \mathscr{G}_j$ are computed. It should be remarked that for this task both Study's parameters defining robot configuration and actuation displacements δ_1, δ_2 and δ_3 are unknown. For this reason the computation of ideals bases is performed on the polynomial ring defined on the complex field $\mathbb{C}$ by the lexicographic variables ordering $x_0 \succ x_1 \succ x_2 \succ x_3 \succ y_0 \succ y_1 \succ y_2 \succ y_3 \succ \delta_1 \succ \delta_2 \succ \delta_3$ [7]. The number of solutions of each ideal intersection is equal to the *dimension* of the vanishing set of the respective basis; thus, computation of $dim(\mathscr{G}_{i,j})$ [5], whose results are reported in table 1, directly provides the dimension of the space of solutions for each ideal intersection. Observation of such results gives information about the conditions that must be respected for the transition between two mobilities to take place:

- The intersection of vanishing sets having no common solutions has dimension -1; it means that no configuration is suitable for transition between mobilities.
- Dimension equal to 0 indicates that a finite number of solutions is available; therefore, the switch is possible in a restricted number (explicitly computable) of configurations.

Table 1 Dimensions of intersections of the ideals characteristic of the different mobilities.

$\cup$	$\mathscr{G}_{15}$	$\mathscr{G}_{14}$	$\mathscr{G}_{13}$	$\mathscr{G}_{12}$	$\mathscr{G}_{11}$	$\mathscr{G}_{10}$	$\mathscr{G}_{9}$	$\mathscr{G}_{8}$	$\mathscr{G}_{7}$	$\mathscr{G}_{6}$	$\mathscr{G}_{5}$	$\mathscr{G}_{4}$	$\mathscr{G}_{3}$	$\mathscr{G}_{2}$	$\mathscr{G}_{1}$
$\mathscr{G}_{1}$	0	0	0	0	1	1	1	1	1	1	2	2	2	2	3
$\mathscr{G}_{2}$	1	1	−1	1	0	2	0	2	0	2	1	1	1	3	
$\mathscr{G}_{3}$	1	1	1	−1	0	2	2	0	2	0	1	1	3		
$\mathscr{G}_{4}$	−1	1	1	1	2	0	0	2	2	0	1	3			
$\mathscr{G}_{5}$	1	−1	1	1	2	0	2	0	0	2	3				
$\mathscr{G}_{6}$	2	−1	−1	2	1	1	1	1	−1	3					
$\mathscr{G}_{7}$	−1	2	2	−1	1	1	1	1	3						
$\mathscr{G}_{8}$	−1	2	−1	2	1	1	−1	3							
$\mathscr{G}_{9}$	2	−1	2	−1	1	1	3								
$\mathscr{G}_{10}$	2	2	−1	−1	−1	3									
$\mathscr{G}_{11}$	−1	−1	2	2	3										
$\mathscr{G}_{12}$	−1	−1	−1	3											
$\mathscr{G}_{13}$	−1	−1	3												
$\mathscr{G}_{14}$	−1	3													
$\mathscr{G}_{15}$	3														

- A dimension greater than 0 denotes the existence of an infinite number of solutions; it is in general possible to extract a relation between actuation parameters that must be satisfied to make the transition possible.

The first row of table 1 arises a particular interest because its elements represent the transitions involving the pure rotational mobility: the zero set $\mathscr{V}(\mathscr{G}_1)$ shares solutions with all other ideals. Therefore, an intermediate passage through this behaviour makes possible, although not directly, all the transitions within table 1. Therefore, a deeper investigation on these specific cases looks reasonable.

- $\mathscr{V}(\mathscr{G}_1 \cup \mathscr{G}_2, \mathscr{G}_3, \mathscr{G}_4, \mathscr{G}_5)$: such spaces of solutions have dimension 2, thus the transition conditions are expressed by surfaces in the space of actuation parameters. The respective Gröbner bases contain, within others, also polynomials:

$$\begin{array}{ll} \mathscr{G}_{1,2} \subset \pi_1 : \delta_1 + \delta_2 + \delta_3 - 2e & \mathscr{G}_{1,3} \subset \pi_2 : \delta_1 - \delta_2 - \delta_3 - 2e \\ \mathscr{G}_{1,4} \subset \pi_3 : \delta_1 + \delta_2 - \delta_3 + 2e & \mathscr{G}_{1,5} \subset \pi_4 : \delta_1 - \delta_2 + \delta_3 + 2e \end{array} \tag{12}$$

  whose vanishing sets define 4 planes in the space of actuation displacements.
- $\mathscr{V}(\mathscr{G}_1 \cup \mathscr{G}_6, \mathscr{G}_7, \mathscr{G}_8, \mathscr{G}_9, \mathscr{G}_{10}, \mathscr{G}_{11})$: since the dimension 1, it is expected that the respective spaces are curves in δ_1, δ_2, δ_3. Indeed, the following equations are found:

$$\begin{array}{lll} \mathscr{G}_{1,6} \subset \rho_1 : \begin{matrix} \delta_2 - e \\ \delta_1 + \delta_3 \end{matrix} & \mathscr{G}_{1,7} \subset \rho_2 : \begin{matrix} \delta_2 + e \\ \delta_1 - \delta_3 \end{matrix} & \mathscr{G}_{1,8} \subset \rho_3 : \begin{matrix} \delta_3 - e \\ \delta_1 + \delta_2 \end{matrix} \\ \mathscr{G}_{1,9} \subset \rho_4 : \begin{matrix} \delta_3 + e \\ \delta_1 - \delta_2 \end{matrix} & \mathscr{G}_{1,10} \subset \rho_5 : \begin{matrix} \delta_1 - e \\ \delta_2 + \delta_3 \end{matrix} & \mathscr{G}_{1,11} \subset \rho_6 : \begin{matrix} \delta_1 + e \\ \delta_2 - \delta_3 \end{matrix} \end{array} \tag{13}$$

  that correspond to 6 different lines in the space of δ_1, δ_2 and δ_3.
- $\mathscr{V}(\mathscr{G}_1 \cup \mathscr{G}_{12}, \mathscr{G}_{13}, \mathscr{G}_{14}, \mathscr{G}_{15})$: as expected due to the null dimension of vanishing set intersection, a finite number of solutions is available here, represented by points identified by zero sets of bases generators:

$$\mathscr{G}_{1,12} \subset \varepsilon_1 : \begin{matrix} \delta_1 + e \\ \delta_2 + e \\ \delta_3 + e \end{matrix} \quad \mathscr{G}_{1,13} \subset \varepsilon_2 : \begin{matrix} \delta_1 + e \\ \delta_2 - e \\ \delta_3 - e \end{matrix} \quad \mathscr{G}_{1,14} \subset \varepsilon_3 : \begin{matrix} \delta_1 - e \\ \delta_2 + e \\ \delta_3 - e \end{matrix} \quad \mathscr{G}_{1,15} \subset \varepsilon_4 : \begin{matrix} \delta_1 - e \\ \delta_2 - e \\ \delta_3 + e \end{matrix} \tag{14}$$

6 Conclusions

The kinematic analysis of the kinematotropic 3-CPU manipulator has been shown by means of Study's quadric representation, thus allowing to use of an algebraic approach to the constraints analysis. The polynomial ideal collecting the actuation independent equations has been decomposed in several sub-ideals, each one characteristic of a particular mobility. The spaces of common solutions of each pair of sub-ideals have been studied to detect the transition conditions between different mobilities: the respective vanishing sets gave their dimensions, if achievable, in the unknown actuation variables. This allowed to conclude that a transition path between various behaviours is always possible due to the fact that the pure rotational mobility shares solutions with every other mode. Moreover, observation of bases generators provided a precise formulation of the spaces of common solutions for transitions between pure rotational behaviour and other mobilities.

References

1. Callegari, M., Marzetti, P., Olivieri, B.: Kinematics of a parallel mechanism for the generation of spherical motions. In: Lenarčič J., Galletti C. (eds.) On Advances in Robot Kinematics, pp. 449–458. Kluwer (2004)
2. Callegari, M., Palpacelli, M.C.: Prototype design of a translating parallel robot. Meccanica **43**, 133–151 (2008)
3. Carbonari, L.: Extended analysis of the 3-cpu reconfigurable class of parallel robotic manipulators. Ph.D. thesis, Polyt. Univ. of Marche, Ancona (2012)
4. Galletti, C., Fanghella, P.: Single-loop kinematotropic mechanisms. Mech. Mach. Theory **36**(6), 743–761 (2001)
5. Kredel, H., Weispfenning, V.: Computing dimension and independent sets for polynomial ideals. J. Symb. Comput. **6**(2–3), 231–247 (1988)
6. Walter, D.R., Husty, M.L., Pfurner, M.: A complete kinematic analysis of the SNU 3-UPU parallel robot. In: Interactions of classical and numerical algebraic geometry, vol. 496, pp. 331–344. American Mathematical Society (2008)
7. Weispfenning, V.: Comprehensive Gröbner bases and regular rings. J. Symb. Comput. **41**(3–4), 285–296 (2006)
8. Wohlhart, K.: Kinematotropic linkages. In: Lenarčič J., Parenti-Castelli V. (eds.) Recent Advances in Robot Kinematics, pp. 449–458. Kluwer (1996)

A Planar Compliant Mechanism with RRP Mobilities Based on the Singularity Analysis of a 3-US Parallel Mechanism

Lennart Rubbert, Stéphane Caro, Pierre Renaud and Jacques Gangloff

Abstract A new design method for parallel compliant mechanisms based on the singularity analysis of parallel mechanisms is presented in this paper. Here a 3-US parallel mechanism is introduced and its singular configurations are analyzed with Grassmann–Cayley algebra for the design of a compliant mechanism with RRP mobilities. A novel architecture of compliant mechanism, based on a 3-UU parallel mechanism, is presented and finally its stiffness properties are analyzed with a finite element method.

Key words: Singularity, parallel mechanism, mechanism analysis, Grassmann–Cayley algebra, compliant mechanism, mechanism design

1 Introduction

In a compliant mechanism, mobilities are obtained by material deformation. Compliant mechanisms are often obtained from a single part and they exhibit a high compactness without any backlash. They are thus considered for the design of MEMS or surgical tools, where small displacements are needed with a high accuracy.

Two open problems remain in the design of such mechanisms. First, their manufacturing remains difficult. Many manufacturing processes are only adapted to planar structures, which is a strong design constraint. Second, the synthesis of compliant mechanisms is more complex than rigid-link mechanisms. Kinematics and statics of the mechanism cannot be analyzed independently [7], and stresses in the structure need to be considered. One way to design a compliant mechanism is to syn-

Lennart Rubbert · Pierre Renaud · Jacques Gangloff
University of Strasbourg, Strasbourg, France, e-mail: {pierre.renaud,jacques.gangloff,rubbert}@unistra.fr

Stéphane Caro
IRCCyN/CNRS, Nantes, France, e-mail: stephane.caro@irccyn.ec-nantes.fr

J. Lenarčič, M. Husty (eds.), *Latest Advances in Robot Kinematics*,
DOI 10.1007/978-94-007-4620-6_48,

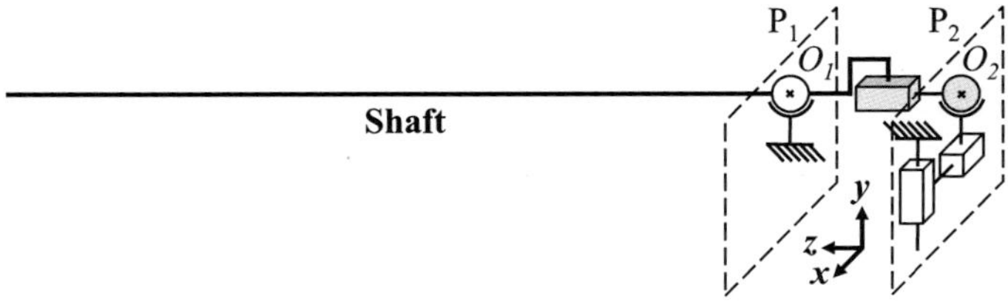

Fig. 1 Kinematic scheme of the compensation mechanism.

thesize a rigid-link mechanism and then transform the joints into compliant joints, also called flexure joints [5]. Revolute flexure joints can be easily manufactured contrary to spherical joints. The main issue during the design is therefore to identify rigid-link mechanisms that exhibit adequate kinematics and that can be transformed into compliant mechanisms.

In this paper, we propose to take advantage of singularity analysis of parallel mechanisms in order to select the architecture of a compliant mechanism. Parallel mechanisms are known for their high stiffness. We can design a compliant mechanism by considering a mechanism that exhibits a parallel singularity in a planar configuration and suppressing its actuated joints. Passive joints are then designed as flexure joints and are easy to manufacture. The obtained compliant mechanism exhibits small displacements in the directions determined from its singularity analysis as well as interesting stiffness properties. Such a design approach is, to the authors' knowledge, original and is illustrated in this paper by focusing on the design of a RRP compliant mechanism from a 3-US parallel mechanism, in the context of the design of a surgical tool.

The need for a RRP compliant mechanism in the design of a surgical tool is briefly described in Section 2. The singularity of the 3-US mechanism are analyzed with Grassmann–Cayley algebra in Section 3. The design and numerical simulation of the obtained compliant mechanism are shown in Section 4. Finally, conclusions and perspectives are written in Section 5.

2 Need for a RRP Compliant Mechanism

A cardiac stabilizer [2] is a surgical device that must actively immobilize the surface of a beating heart during surgery. It is mainly composed of a shaft that enters the body to reach the surface of the heart at one end. At the other end, the shaft is actuated with a mechanism which takes advantage of the oblong geometry of the shaft in order to maximize the compactness [3]. The kinematic scheme represented in Fig. 1 is considered. The shaft is controlled using a XY-stage located in a plane P_2 perpendicular to the shaft. The shaft is linked to the XY-stage with a spherical joint in O_2 and oriented with a spherical joint in O_1. A prismatic joint is however needed to take into account the distance variation between O_1 and O_2. As the rota-

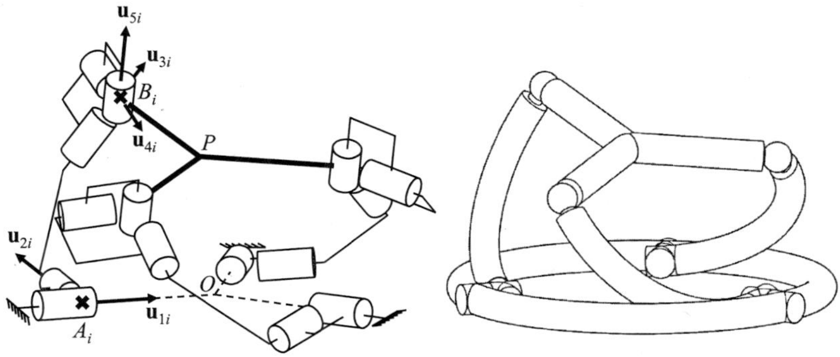

Fig. 2 Equivalent kinematic chains and architecture of the 3-US.

tion around the shaft axis is not necessary, a RRP mechanism, in grey in Fig. 1, is therefore considered to connect the shaft to the base.

3 Singularity Analysis with Grassmann–Cayley Algebra

The 3-US parallel mechanism has three identical legs where each leg is composed of a universal joint and a spherical joint. The first revolute joint axes of the universal joints are in the plane of the base and intersect in point O (Fig. 2). The 3-US mechanism has three degrees of freedom and its platform velocity can be described by three translational velocities and three angular velocities. In fact, the platform motions are coupled except when the platform is in the plane of its base. In this configuration the motion capabilities are decoupled in two angular velocities around the orthogonal in-plane axes and in one translational velocity along the axis normal to the base. In this configuration we get exactly the desired RRP motion capabilities described in Section 2.

The universal joint is similar to two revolute joints with perpendicular and intersecting axes. The spherical joint can be seen as three revolute joints with perpendicular and intersecting axes as shown in Fig. 2. This section aims to analyze the singularities of the 3-US for its five symmetrical actuation modes and the instantaneous gained motions in the vicinity of the singular configurations. The ith joint of each leg is actuated for the ith actuation mode of the mechanism, $i = 1,\ldots,5$. It is noteworthy that the architecture of the compliant mechanism will be directly related to the selected actuation scheme. The singularities of the 3-US are analyzed with the Grassmann–Cayley Algebra (GCA) [1, 4]. In the following, the singularity analysis for the first actuation mode is detailed as an illustrative example and the instantaneous gained motions are described for all actuation modes.

3.1 Analysis of the First Actuation Mode

Twist System of the 3-US: The twist system $\mathtt{T}^i$ associated with the ith leg of the 3-US is spanned by five zero-pitch twists defined as:

$$\hat{\varepsilon}_{01}^{i} = \begin{bmatrix} \mathbf{u}_{1i} \\ \mathbf{a}_i \times \mathbf{u}_{1i} \end{bmatrix}, \quad \hat{\varepsilon}_{02}^{i} = \begin{bmatrix} \mathbf{u}_{2i} \\ \mathbf{a}_i \times \mathbf{u}_{2i} \end{bmatrix}, \quad \hat{\varepsilon}_{03}^{i} = \begin{bmatrix} \mathbf{u}_{3i} \\ \mathbf{b}_i \times \mathbf{u}_{3i} \end{bmatrix}$$
$$\hat{\varepsilon}_{04}^{i} = \begin{bmatrix} \mathbf{u}_{4i} \\ \mathbf{b}_i \times \mathbf{u}_{4i} \end{bmatrix}, \quad \hat{\varepsilon}_{05}^{i} = \begin{bmatrix} \mathbf{u}_{5i} \\ \mathbf{b}_i \times \mathbf{u}_{5i} \end{bmatrix}, \quad i = 1,2,3 \tag{1}$$

$\mathbf{u}_{1i}$ and $\mathbf{u}_{2i}$ are the unit vectors of the first and second revolute joint axes of the universal joint of the ith leg. $\mathbf{u}_{3i}$, $\mathbf{u}_{4i}$ and $\mathbf{u}_{5i}$ are the unit vectors of the revolute joints associated with the spherical joint of the ith leg. $\mathbf{a}_i$ and $\mathbf{b}_i$ are the Cartesian coordinate vectors of points A_i and B_i shown in Fig. 2. The twist system $\mathtt{T}$ of the 3-US is the intersection of $\mathtt{T}^1$, $\mathtt{T}^2$ and $\mathtt{T}^3$.

Wrench System of the 3-US: In a non-singular configuration, the constraint wrench system $\mathtt{W}^c$ of the 3-US is a three-system spanned by the following three pure forces:

$$\hat{\mathscr{F}}_{i}^{c} = \begin{bmatrix} \mathbf{n}_i \\ \mathbf{b}_i \times \mathbf{n}_i \end{bmatrix}, \quad i = 1,2,3 \tag{2}$$

$\mathbf{n}_i$ being the unit vector of $\overrightarrow{A_iB_i}$. In turn, the actuation wrench system $\mathtt{W}^a$ of the 3-US depends on its actuation scheme. In case the first revolute joint of each leg is actuated, $\mathtt{W}^a$ is spanned by the following three pure forces:

$$\hat{\mathscr{F}}_{1i}^{a} = \begin{bmatrix} \mathbf{u}_{2i} \\ \mathbf{b}_i \times \mathbf{u}_{2i} \end{bmatrix}, \quad i = 1,2,3 \tag{3}$$

In case the second revolute joint of each leg is actuated, $\mathtt{W}^a$ is spanned by the following three pure forces:

$$\hat{\mathscr{F}}_{2i}^{a} = \begin{bmatrix} \mathbf{u}_{1i} \\ \mathbf{b}_i \times \mathbf{u}_{1i} \end{bmatrix}, \quad i = 1,2,3 \tag{4}$$

In case the jth revolute joint of each leg is actuated, $j = 3,4,5$, $\mathtt{W}^a$ is spanned by the following three pure forces:

$$\hat{\mathscr{F}}_{ji}^{a} = \begin{bmatrix} \mathbf{v}_{ji} \\ \mathbf{c}_{ji} \times \mathbf{v}_{ji} \end{bmatrix}, \quad i = 1,2,3 \tag{5}$$

$\mathbf{v}_{ji}$ being the unit vector of the intersection line $\mathscr{L}_{ji}$ of planes $\mathscr{P}_{1i}$ and $\mathscr{P}_{ji}$. $\mathbf{c}_{ji}$ is the Cartesian coordinate vector of any point C_{ji} on line $\mathscr{L}_{ji}$. $\mathscr{P}_{1i}$ is spanned by vectors $\mathbf{u}_{1i}$ and $\mathbf{u}_{2i}$ and passes through point A_i. $\mathscr{P}_{3i}$ is spanned by vectors $\mathbf{u}_{4i}$ and $\mathbf{u}_{5i}$ and passes through point B_i. $\mathscr{P}_{4i}$ is spanned by vectors $\mathbf{u}_{3i}$ and $\mathbf{u}_{5i}$ and passes through point B_i. $\mathscr{P}_{5i}$ is spanned by vectors $\mathbf{u}_{3i}$ and $\mathbf{u}_{4i}$ and passes through point B_i.

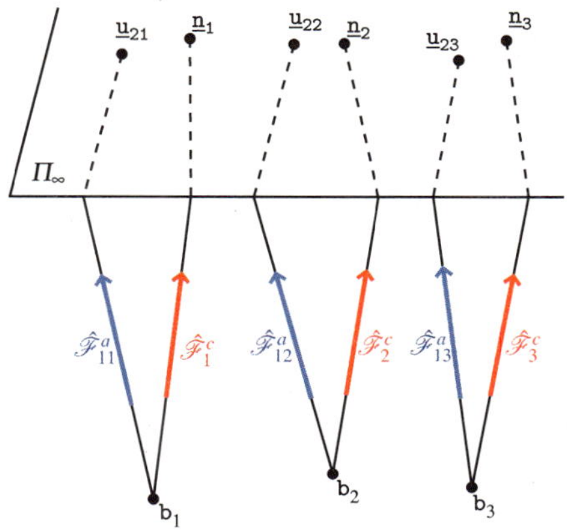

Fig. 3 Wrench graph of the 3-US associated with its first actuation mode.

As a result, the global wrench system W^k_{3US} of the 3-US associated with its kth actuation scheme is spanned by W^a and W^c, namely,

$$W^k_{3US} = span(\hat{\mathscr{F}}^c_1, \hat{\mathscr{F}}^c_2, \hat{\mathscr{F}}^c_3, \hat{\mathscr{F}}^a_{k1}, \hat{\mathscr{F}}^a_{k2}, \hat{\mathscr{F}}^a_{k3}), \quad k = 1, \dots, 5 \tag{6}$$

Wrench Graph of the 3-US: The six forces $\hat{\mathscr{F}}^c_1$, $\hat{\mathscr{F}}^c_2$, $\hat{\mathscr{F}}^c_3$, $\hat{\mathscr{F}}^a_{11}$, $\hat{\mathscr{F}}^a_{12}$ and $\hat{\mathscr{F}}^a_{13}$ form a basis of the global wrench system W^1_{3US}. Those wrenches are represented by six finite lines in P_3. To obtain the six extensors of the superbracket, we have to select twelve projective points on the six projective lines, *i.e.*, two points on each line. Let b_1, b_2, b_3 be the intersection points of $\hat{\mathscr{F}}^a_{11}$ and $\hat{\mathscr{F}}^c_1$, $\hat{\mathscr{F}}^a_{12}$ and $\hat{\mathscr{F}}^c_2$, $\hat{\mathscr{F}}^a_{13}$ and $\hat{\mathscr{F}}^c_3$, respectively. $\hat{\mathscr{F}}^a_{11}$, $\hat{\mathscr{F}}^c_1$, $\hat{\mathscr{F}}^a_{12}$, $\hat{\mathscr{F}}^c_2$, $\hat{\mathscr{F}}^a_{13}$ and $\hat{\mathscr{F}}^c_3$ intersect the infinite plane Π_∞ at points $\underline{u}_{21} = (\mathbf{u}_{21}, 0)^T$, $\underline{n}_1 = (\mathbf{n}_1, 0)^T$, $\underline{u}_{22} = (\mathbf{u}_{22}, 0)^T$, $\underline{n}_2 = (\mathbf{n}_2, 0)^T$, $\underline{u}_{23} = (\mathbf{u}_{23}, 0)^T$ and $\underline{n}_3 = (\mathbf{n}_3, 0)^T$, respectively. The wrench graph of the 3-US corresponding to its first actuation mode is illustrated in Fig. 3.

Superbracket of the 3-US: The expression of the superbracket of the 3-US associated with its first actuation mode is derived from the nine projective points depicted in Fig. 3 and takes the form: $[b_1\underline{u}_{21}\, b_1\underline{n}_1\, b_2\underline{u}_{22}\, b_2\underline{n}_2\, b_3\underline{u}_{23}\, b_3\underline{n}_3]$. This expression can be developed into a linear combination of 24 bracket monomials [4], each one being the product of three brackets of four projective points. The superbracket was simplified by using a novel graphical user interface:[1]

$$[b_1\underline{u}_{21}\, b_1\underline{n}_1\, b_2\underline{u}_{22}\, b_2\underline{n}_2\, b_3\underline{u}_{23}\, b_3\underline{n}_3] = [b_1\, \underline{u}_{21}\, \underline{n}_1\, \dot{b}_2][b_1\, b_2\, \dot{\underline{n}}_2\, b_3][\dot{\underline{u}}_{22}\, \underline{u}_{23}\, b_3\, \underline{n}_3] \tag{7}$$

$$= (b_1\, \underline{u}_{21}\, \underline{n}_1) \wedge (b_2\, \underline{u}_{22}\, \underline{n}_2) \wedge (b_3\, \underline{u}_{23}\, \underline{n}_3) \wedge (b_1\, b_2\, b_3) \tag{8}$$

[1] http://www.irccyn.ec-nantes.fr/~caro/SIROPA/GUIGCASiropa.jar

where dotted letters stand for permuted elements and $\wedge$ denotes the *meet* operator [4].

Geometric Parallel Singularity Conditions: Let Π_l be the plane passing through point B_l and spanned by $\mathbf{u}_{2l}$ and $\mathbf{n}_l$, $l = 1,2,3$. Let Π_4 be the plane passing through points B_1, B_2 and B_3. Equation (8) vanishes, namely, the 3-US reaches a parallel singularity for its first actuation mode, if and only if planes Π_1, Π_2, Π_3 and Π_4 intersect at least at one point.

3.2 Singularity Conditions and Instantaneous Gained Motions

Singularity conditions have been determined for the first actuation mode. A similar procedure was used to analyze the singularities associated with the second actuation mode of the 3-US. It is noteworthy that the simplification of the superbracket expression is not straightforward for the third, fourth and fifth actuation modes. Therefore, we can use GCA and Grassmann geometry as complementary approaches in order to derive the singularity conditions related to those five actuation modes as explained in [1]. In case of the third, fourth and fifth actuation modes, the Grassmann–Caley algebra is directly applied to the planar configuration when the plane of the platform normal to the $\mathbf{u}_{5i}$ is in the plane of the base defined by $\mathbf{u}_{1i}$ for the singularity analysis (Fig. 2).

For the five actuation modes, the 3-US turns to be singular in the planar configuration, *i.e.*, when the base and moving-platform are coplanar. As a matter of fact, from Eq. (7), the global wrench system of the 3-US becomes a 3-system composed of six coplanar pure forces in this configuration. Its twist system $\mathtt{T}_{3US}$ is reciprocal to $\mathtt{W}^k_{3US}$, $k = 1,\ldots,5$ and becomes also a 3-system in this configuration even if the actuated joints are locked. $\mathtt{T}_{3US}$ is spanned by two zero-pitch twists with non parallel axes belonging to Π_4 and one infinite-pitch twist of axis normal to Π_4. As a consequence, the instantaneous gained motions of the moving-platform correspond to the desired RRP motion, namely, one translation along the direction normal to Π_4 and two rotations about non parallel axes belonging to Π_4 for all actuation modes in the planar configuration.

4 Design of the Compliant Mechanism

For each actuation mode, the 3-US gains the desired RRP motion described in Section 2. Since the mechanism will be machined in a plane, it is more convenient to machine flexure joints with revolute axes in the plane. Therefore, among the five revolute joints per leg that can be suppressed, we choose to eliminate the fifth revolute joint of each leg ($\mathbf{u}_{5i}$, Fig. 2) which is the only one to have an axis normal "to the base and moving platform" in the planar configuration of the 3-US. Therefore,

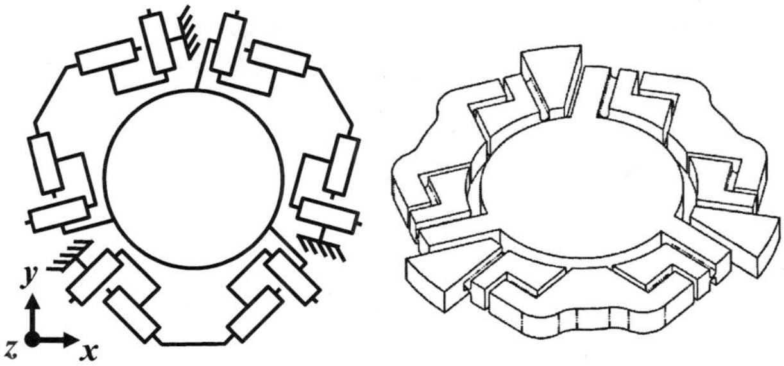

Fig. 4 Kinematic scheme and CAD view of the 3-UU compliant mechanism.

the 3-US becomes a 3-UU as illustrated in Fig. 4. The 3-UU compliant mechanism is then designed by replacing each revolute joint by a circular flexure hinge. The computer-aided design of the compliant 3-UU is shown in Fig. 4.

The compliant mechanism is assessed by computing the 6×6 compliance matrix, which relates the displacement $\mathbf{u} = [x; y; z; \theta_x; \theta_y; \theta_z]$ to the load $\mathbf{L} = [F_x; F_y; F_z; M_x; M_y; M_z]$ applied at the center of the platform [6]:

$$\mathbf{u} = \mathbf{C}.\mathbf{L} \tag{9}$$

The 6×6 compliance matrix $\mathbf{C}$ is evaluated using a finite element analysis (PTC ProMechanica):

$$\mathbf{C} = \begin{bmatrix} \mathbf{7.3e\text{-}5} & -2.5e\text{-}8 & -4.2e\text{-}8 & -3.4e\text{-}5 & 5.3e\text{-}8 & 1.4e\text{-}6 \\ 3.8e\text{-}8 & \mathbf{7.3e\text{-}5} & -2.7e\text{-}8 & -8.0e\text{-}6 & 9.0e\text{-}6 & 1.4e\text{-}6 \\ -1.1e\text{-}7 & -5.9e\text{-}8 & \mathbf{1.7e\text{-}2} & 4.1e\text{-}5 & 1.1e\text{-}4 & 1.2e\text{-}6 \\ -1.9e\text{-}6 & 0.0 & -8.7e\text{-}5 & \mathbf{1.4} & -6.0e\text{-}3 & 4.1e\text{-}5 \\ 0.0 & 1.5e\text{-}6 & -1.2e\text{-}5 & -8.4e\text{-}4 & \mathbf{1.4} & 3.0e\text{-}5 \\ 0.0 & -6.0e\text{-}8 & -9.6e\text{-}7 & 4.0e\text{-}6 & 0.0 & \mathbf{6.9e\text{-}3} \end{bmatrix} \tag{10}$$

The translational stiffness along the vertical axis is approximately 240 times lower than the in-plane translational stiffnesses. The rotational stiffness along the vertical axis is 210 times greater than the in-plane rotational stiffnesses. This is verified when three forces are applied at the center of the mechanism: there is only a displacement along the vertical axis (Fig. 5 left). When three moments are applied at the center of the mechanism, there are two rotations about in-plane axes (Fig. 5 right). Therefore, the compliant mechanism behaves as a RRP mechanism.

5 Conclusion

A new design method for parallel compliant mechanism based on singular configurations has been presented in this paper. A 3-US parallel mechanism has been

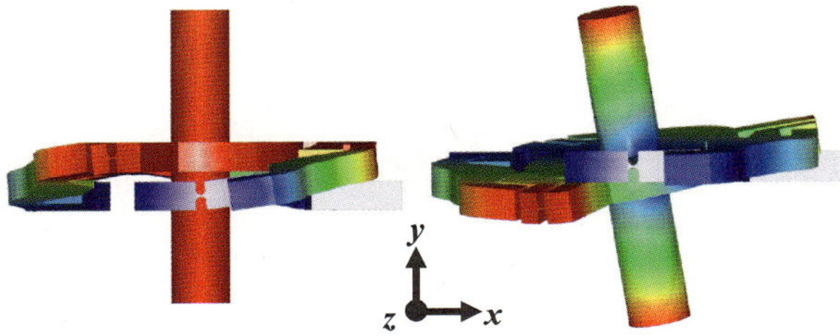

Fig. 5 Finite element analysis of the 3-UU compliant mechanism under a combined load (left) and moment (right).

introduced and based on the Grassmann–Cayley algebra it has been possible to analyze the singularities in this mechanism for the design of a compliant mechanism with RRP mobilities. The stiffnesses of the compliant mechanism have been evaluated with a finite element analysis. The high ratio between the stiffnesses in the different motion directions shows that the compliant mechanism behaves as a RRP mechanism. Hence, a new compliant planar parallel architecture for a RRP compliant mechanism has been proposed. The next step will consist in optimizing the mechanism to match the expected performances in terms of relative stiffnesses and stresses in the material.

References

1. Amine, S., Tale-Masouleh, M., Caro, S., Wenger, P., Gosselin, C.: Singularity analysis of 3T2R parallel mechanisms using Grassmann–Cayley algebra and Grassmann line geometry. Mech. Mach. Theory, (2012). doi:10.1016/j.mechmachtheory.2011.11.015
2. Bachta, W., Renaud, P., Laroche, E., Forgione, A., Gangloff, J.: Cardiolock: An active cardiac stabilizer, first in vivo experiments using a new robotized device. Comput. Aided Surg. **13**(5), 243–254 (2008)
3. Bachta, W., Renaud, P., Laroche, E., Gangloff, J.: Cardiolock2: Parallel singularities for the design of an active heart stabilizer. In: ICRA, pp. 3839–3844 (2009)
4. Ben-Horin, P., Shoham, M.: Singularity analysis of a class of parallel robots based on Grassmann–Cayley algebra. Mech. Mach. Theory **41**(8), 958–970 (2006)
5. Howell, L.: Compliant Mechanisms. Wiley-IEEE (2001)
6. Lobontiu, N.: Compliant Mechanisms – Design of Flexure Hinges. CRC Press (2003)
7. Rubbert, L., Renaud, P., Bachta, W., Gangloff, J.: Compliant mechanisms for an active cardiac stabilizer: Lessons and new requirements in the design of a novel surgical tool. Mech. Sci. **2**, 119–127 (2011)

Velocity Level Kinematic Analysis of Serial nA-Chains

James D. Robinson and M. John D. Hayes

Abstract The algebraic screw pair, or A-pair, represents a new class of kinematic constraint that exploits the self-motions inherent to a specific configuration of Griffis–Duffy platform. The A-pair causes a sinusoidal coupling of rotation and translation between adjacent links in the kinematic chain. The resulting linkage is termed an A-chain. This paper presents a derivation of the manipulator Jacobian of nA-chains in general, and a specific 4 degree-of-freedom hybrid serial-parallel 4A-chain.

Key words: Algebraic screw pair, Griffis–Duffy platform, nA Jacobian

1 Introduction

The algebraic screw pair [4], or A-pair, is a novel kinematic pair based on a specific configuration of parallel manipulator called the Griffis–Duffy platform (GDP) [1]. The GDP is a special configuration of the six legged, six degree-of-freedom (DOF) Stewart-Gough platform (SGP) that, in most configurations, is subject to self-motions regardless of the lengths of the actuated legs [2]. Kinematic chains composed of rigid links serially connected by A-pairs are denoted A-chains. The A-pair induces a sinusoidal coupling of rotation and translation between adjacent links. For this paper the derivation of the manipulator Jacobian of a 4A-chain, illustrated in Figure 1a, is used to demonstrate the method. While the method does not fail for $n > 4$, the terms become inconveniently large to express explicitly.

The A-pairs used in this paper are the midline-to-vertex GDP configuration, see Figure 1b. They are constrained by: the fixed base and moving platform anchor point triangles are congruent equilateral triangles with each side of the triangles being of length a and the six legs are all of a fixed length, l, equal to the height, h, of the triangles as illustrated in Figure 2a.

James D. Robinson · M. John D. Hayes
Department of Mechanical and Aerospace Engineering, Carleton University, Ottawa, Canada,
e-mail: jrobins7@connect.carleton.ca, jhayes@mae.carleton.ca

J. Lenarčič, M. Husty (eds.), *Latest Advances in Robot Kinematics*,
DOI 10.1007/978-94-007-4620-6_49,

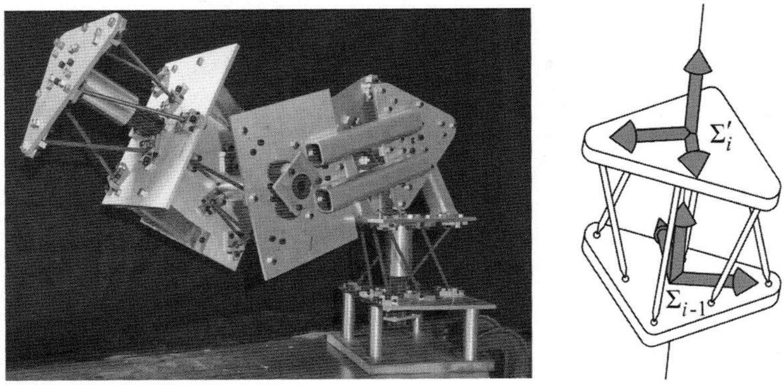

(a) (b)

Fig. 1 (a) Prototype 4A-chain. (b) Midline-to-vertex configuration GDP.

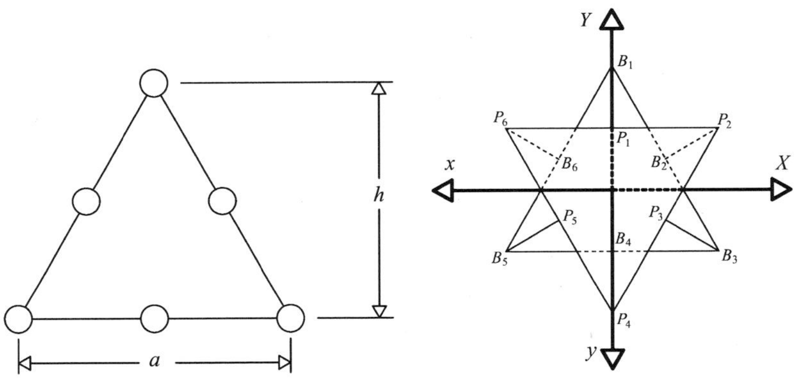

(a) (b)

Fig. 2 (a) Platform shape parameters. (b) Coordinate systems and leg anchor point.

The value of l is

$$l = h = \frac{a\sqrt{3}}{2}. \tag{1}$$

It turns out that the self-motions of this GDP couple rotation about an axis passing through the geometric centres of both the fixed base and moving platform triangles with translation along that axis. Using the coordinate systems illustrated in Figure 2b, it can be shown [2] that the separation of the fixed base and moving platform, d, is a function of the rotation angle, θ, about the axis common to both the fixed base and moving platform:

$$d = \rho \sin\left(\frac{\theta}{2}\right), \text{ where } \rho = \frac{a\sqrt{6}}{3}. \tag{2}$$

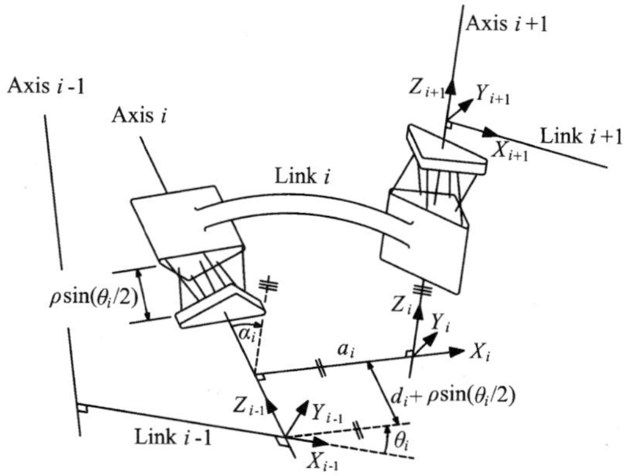

Fig. 3 DH-parameters of a link in an A-chain.

It is expected that A-chains will exhibit increased stiffness and positioning accuracy relative to R-chains. While we currently lack empirical proof, it appears to be true based on a visual comparison of the prototype manipulator with the first four R-pair joints in a Thermo CRS A465. The proposed actuation system consists of a central spline affixed to the moving platform that is constrained by three spur gears affixed to the base, all possessing identical pitch diameters. One of the spur gears is active, which rotates the spline. This arrangement allows the spline to translate along its axis of rotation.

2 The Jacobian of a Single A-Pair

The Jacobian matrix of a manipulator maps its joint rates to the linear and angular velocities of its end effector (EE). Standard methods from the literature, see [5] for example, can be adapted to account for the coupled translation and rotation of the A-pair. The Jacobian matrix of a 1A-chain can be determined by examining the rotation and translation components of the coupled motion separately. The orientation of the joint is directly expressed by the joint variable θ_1. The translation component of the joint motion is a function of θ_1, and computed with Equation (2).

In a 1A-chain, the linear velocity of the EE induced by $\dot{\theta}_1$ has two components: one due to the rotation of the joint, perpendicular to the axis of rotation as with a revolute joint; the other is due to the translation coupled to the rotation, and is expressed by $p_{e_z} = d_1 + \rho\sin(\theta_1/2)$, where p_{e_z} is the $\hat{\mathbf{z}}_0$-component of the EE position vector, d_1 is the offset of the EE from the base along $\hat{\mathbf{z}}_0$ when $\theta_1 = 0$, and $\hat{\mathbf{z}}_0$ is axis of rotation. There is only one joint rate $\dot{q}_1 = \dot{\theta}_1$. The influence of the rotation of the joint on the linear velocity is found as if it were a revolute joint:

$$\dot{q}_1 \mathbf{J}_{P_{1_r}} = z_0 \times (\mathbf{p}_e - \mathbf{p}_0)\dot{\theta}_1 = \begin{bmatrix} 0 \\ 0 \\ 1 \end{bmatrix} \times \begin{bmatrix} -\dot{\theta}_1 a_1 \cos\theta_1 - 0 \\ -\dot{\theta}_1 a_1 \sin\theta_1 - 0 \\ 0 \end{bmatrix} = \begin{bmatrix} \dot{\theta}_1 a_1 \sin\theta_1 \\ -\dot{\theta}_1 a_1 \cos\theta_1 \\ 0 \end{bmatrix}, \quad (3)$$

where, in general, $\mathbf{J}_{P_{1r}}$ is the is the 3×1 vector mapping the angular rate of joint i to its contribution to the linear velocity of the EE, $\mathbf{p}_e$ and $\mathbf{p}_0$ are the position vectors of the EE coordinate origin, and position vector of the joint coordinate system origin both expressed in the non moving frame, and a_1 is the DH-parameter for the link length of a link affixed to the moving platform of the single A-pair, illustrated in Figure 3. The time derivative of p_{e_z} yields the translation component of the Jacobian, $\mathbf{J}_{P_1t}$:

$$\dot{q}_1 \mathbf{J}_{P_1t} = \frac{d}{dt}\begin{bmatrix} 0 \\ 0 \\ d_1 + \rho \sin\left(\frac{\theta_1}{2}\right) \end{bmatrix} = \begin{bmatrix} 0 \\ 0 \\ \frac{\dot{\theta}_1}{2}\rho \cos\left(\frac{\theta_1}{2}\right) \end{bmatrix}, \tag{4}$$

therefore $\mathbf{J}_{P_1t} = [\,0\;0\;\frac{\rho}{2}\cos(\frac{\theta_1}{2})\,]^T$. Summing the two components yields the mapping from the joint rate $\dot{q}_1$ to the EE linear velocity:

$$\mathbf{J}_{P_1} = \mathbf{J}_{P_1r} + \mathbf{J}_{P_1t} = \begin{bmatrix} a_1 \sin\theta_1 \\ -a_1 \cos\theta_1 \\ \frac{\rho}{2}\cos\left(\frac{\theta_1}{2}\right) \end{bmatrix}, \tag{5}$$

and $\dot{\mathbf{p}}_e = \mathbf{J}_{P_1}(\mathbf{q})\dot{q}_1$.

The translation that is coupled with the rotation of the A-pair does not have an effect on the orientation of the EE, thus the contribution of the A-pair actuation rate to the angular velocity of the EE, $\boldsymbol{\omega}_e$, is the same as that of a revolute joint:

$$\mathbf{J}_{O_1} = \hat{\mathbf{z}}_0 = \begin{bmatrix} 0 \\ 0 \\ 1 \end{bmatrix}, \tag{6}$$

and $\boldsymbol{\omega}_e = \mathbf{J}_{O_1}(\mathbf{q})\dot{q}_1$. The full Jacobian is

$$\mathbf{J} = \begin{bmatrix} \mathbf{J}_{P_1} \\ \mathbf{J}_{O_1} \end{bmatrix}. \tag{7}$$

However, in this A-pair the EE coordinate system origin is located at the geometric centre of the moving platform (the EE frame is coincident with a base frame located at the geometric centre of the fixed base when the A-pair is in the theoretical home position). Hence, the origin of the EE lies on the joint axis rendering $a_1 = 0$ and the velocity relations simplify to

$$\mathbf{J}_{P_1} = \begin{bmatrix} 0 \\ 0 \\ \frac{\rho}{2}\cos\left(\frac{\theta_1}{2}\right) \end{bmatrix}, \quad \mathbf{J}_{O_1} = \begin{bmatrix} 0 \\ 0 \\ 1 \end{bmatrix}. \tag{8}$$

Table 1 DH-parameters.

Link i	a_i	α_i	d_i	θ_{fi}
1	0	90°	d_1	0°
2	a_2	180°	$-\rho$	−90°
3	0	−90°	$-\rho$	90°
4	0	0°	d_4	0°

The Jacobian matrix for the single A-pair is always rank deficient, which is expected because motion in an arbitrary direction will never be possible with just a single A-pair. If one considers only the two achievable degrees of freedom of the single A-pair with the EE origin on the joint axis (motion along the joint axis and rotation about the same axis) a more useful analysis can be performed.

The mapping to EE angular velocity from the joint rate is one-to-one and independent of the joint state. This implies that, if joint limits are ignored, the angular velocity of the EE can always be controlled one-to-one. However, the mapping of linear velocity is dependant on the joint state and cannot be continuously controlled. When $\cos(\frac{\theta_1}{2})$ goes to zero (*i.e.* when θ approaches 180°) the joint approaches a singular position. At the singularity the linear velocity of the EE is null and any rotation away from $\theta_1 = 180°$ in either direction will result in motion in the negative $\hat{\mathbf{z}}_0$-direction only. The singularity is also evident if the Jacobian is rearranged to solve for the joint rate required to achieve a certain velocity, v_1, along the $\hat{\mathbf{z}}_0$-axis:

$$\dot{\theta}_1 = \frac{2v_1}{\rho\cos\left(\frac{\theta_1}{2}\right)}. \tag{9}$$

As θ_1 approaches 180°, $\dot{\theta}_1$ approaches infinity.

3 The Jacobian of a 4A-Chain

The DH-parameters of the 4A-chain illustrated in Figure 1a, using the DH-parameter convention in [5], are listed in Table 1. Figure 3 shows the assignment of DH-parameters to a link in an A-chain. The position vector of the base frame origin is $\mathbf{p}_0 = \mathbf{0}$. The relative pose of the EE, $\mathbf{p}_e$, is found by transforming the homogeneous coordinates of its origin to the non moving base coordinate reference system. The homogeneous transformation, obtained using the methods of [4], has the form:

$${}^0\mathbf{T}_4 = \begin{bmatrix} 1 & 0 & 0 & 0 \\ p_{e_x} & -c_1c_{2-3}c_4 + s_1s_4 & c_1c_{2-3}s_4 + s_1c_4 & -c_1s_{2-3} \\ p_{e_y} & -s_1c_{2-3}c_4 - c_1s_4 & s_1c_{2-3}s_4 - c_1c_4 & -s_1s_{2-3} \\ p_{e_z} & -s_{2-3}c_4 & s_{2-3}s_4 & c_{2-3} \end{bmatrix}, \tag{10}$$

where

$$\mathbf{p}_e = \begin{bmatrix} p_{e_x} \\ p_{e_y} \\ p_{e_z} \end{bmatrix} = \begin{bmatrix} s_1 \rho\, s\frac{\theta_2}{2} + c_1 s_2 a_2 - s_1 \rho\, s\frac{\theta_3}{2} - c_1 s_{2-3} \rho\, s\frac{\theta_4}{2} - c_1 s_{2-3} d_4 \\ -c_1 \rho\, s\frac{\theta_2}{2} + s_1 s_2 a_2 + c_1 \rho\, s\frac{\theta_3}{2} - s_1 s_{2-3} \rho\, s\frac{\theta_4}{2} - s_1 s_{2-3} d_4 \\ \rho\, s\frac{\theta_1}{2} + d_1 - c_2 a_2 + \rho\, c_{2-3} s\frac{\theta_4}{2} + d_4 c_{2-3} \end{bmatrix}$$

is the position vector of the EE origin, and c_1, s_1, *etc.* are abbreviations for $\cos\theta_1$, $\sin\theta_1$, *etc.*, respectively. In addition to the EE pose the transformation matrices describing the pose of each intermediate reference frame (Σ_i, $i = 1, 2, 3$) are important. The pose of Σ_1 is given by

$$^0\mathbf{T}_1 = \begin{bmatrix} 1 & 0 & 0 & 0 \\ p_{1_x} & c_1 & 0 & s_1 \\ p_{1_y} & s_1 & 0 & -c_1 \\ p_{1_z} & 0 & 1 & 0 \end{bmatrix}, \quad \mathbf{p}_1 = \begin{bmatrix} p_{1_x} \\ p_{1_y} \\ p_{1_z} \end{bmatrix} = \begin{bmatrix} 0 \\ 0 \\ \rho\, s\frac{\theta_1}{2} + d_1 \end{bmatrix}. \tag{11}$$

The pose of Σ_2 is given by

$$^0\mathbf{T}_2 = \begin{bmatrix} 1 & 0 & 0 & 0 \\ p_{2_x} & c_1 s_2 & -c_1 c_2 & -s_1 \\ p_{2_y} & s_1 s_2 & -s_1 c_2 & c_1 \\ p_{2_z} & -c_2 & -s_2 & 0 \end{bmatrix}, \quad \mathbf{p}_2 = \begin{bmatrix} p_{2_x} \\ p_{2_y} \\ p_{2_z} \end{bmatrix} = \begin{bmatrix} \rho\, s_1 s\frac{\theta_2}{2} + a_2\, c_1 s_2 - \rho\, s_1 \\ -\rho\, c_1 s\frac{\theta_2}{2} + a_2\, s_1 s_2 + \rho\, c_1 \\ \rho\, s\frac{\theta_1}{2} + d_1 - a_2\, c_2 \end{bmatrix}. \tag{12}$$

Finally the pose of Σ_3 is

$$^0\mathbf{T}_3 = \begin{bmatrix} 1 & 0 & 0 & 0 \\ p_{3_x} & -c_1 c_{2-3} & s_1 & -c_1 s_{2-3} \\ p_{3_y} & -s_1 c_{2-3} & -c_1 & -s_1 s_{2-3} \\ p_{3_z} & -s_{2-3} & 0 & c_{2-3} \end{bmatrix}, \tag{13}$$

$$\mathbf{p}_3 = \begin{bmatrix} p_{3_x} \\ p_{3_y} \\ p_{3_z} \end{bmatrix} = \begin{bmatrix} \rho\, s_1 s\frac{\theta_2}{2} + a_2\, c_1 s_2 - \rho\, s_1 s\frac{\theta_3}{2} \\ -\rho\, c_1 s\frac{\theta_2}{2} + a_2\, s_1 s_2 + \rho\, c_1 s\frac{\theta_3}{2} \\ \rho\, s\frac{\theta_1}{2} + d_1 - a_2\, c_2 \end{bmatrix}. \tag{14}$$

The position vectors $\mathbf{p}_1$, $\mathbf{p}_2$ and $\mathbf{p}_3$ that describe the position of the origin of the corresponding intermediate reference frames are given in Equations (11), (12) and (14), respectively. The joint axes, taken from the respective transformation matrices, are

$$\hat{\mathbf{z}}_0 = \begin{bmatrix} 0 \\ 0 \\ 1 \end{bmatrix}, \quad \hat{\mathbf{z}}_1 = \begin{bmatrix} s_1 \\ -c_1 \\ 0 \end{bmatrix}, \quad \hat{\mathbf{z}}_2 = \begin{bmatrix} -s_1 \\ c_1 \\ 0 \end{bmatrix}, \quad \hat{\mathbf{z}}_3 = \begin{bmatrix} -c_1 s_{2-3} \\ -s_1 s_{2-3} \\ c_{2-3} \end{bmatrix}. \tag{15}$$

The vector mapping the rate of actuation of Joint 1 to the linear velocity of the EE due to the rotation of Joint 1 is

$$\mathbf{J}_{P_{1,r}} = z_0 \times (\mathbf{p}_e - \mathbf{p}_0) = \begin{bmatrix} \rho\, c_1 s\frac{\theta_2}{2} - a_2\, s_1 s_2 - \rho\, c_1 s\frac{\theta_3}{2} + \rho\, s_1 s_{2-3} s\frac{\theta_4}{2} + d_4\, s_1 s_{2-3} \\ \rho\, s_1 s\frac{\theta_2}{2} + a_2\, c_1 s_2 - \rho\, s_1 s\frac{\theta_3}{2} - \rho\, c_1 s_{2-3} s\frac{\theta_4}{2} - d_4\, c_1 s_{2-3} \\ 0 \end{bmatrix}, \quad (16)$$

and the vector mapping the rate of actuation of Joint 1 to the linear velocity of the EE due to the translation of Joint 1 is

$$\mathbf{J}_{P_{1,t}} = \begin{bmatrix} 0 \\ 0 \\ \frac{\rho}{2} c\frac{\theta_1}{2} \end{bmatrix}. \quad (17)$$

The total linear velocity Jacobian component for Joint 1 comes from the summation of Equations (16) and (17) for $i = 1$, giving

$$\mathbf{J}_{P_1} = \begin{bmatrix} \rho\, c_1 s\frac{\theta_2}{2} - a_2\, s_1 s_2 - \rho\, c_1 s\frac{\theta_3}{2} + \rho\, s_1 s_{2-3} s\frac{\theta_4}{2} + d_4\, s_1 s_{2-3} \\ \rho\, s_1 s\frac{\theta_2}{2} + a_2\, c_1 s_2 - \rho\, s_1 s\frac{\theta_3}{2} - \rho\, c_1 s_{2-3} s\frac{\theta_4}{2} - d_4\, c_1 s_{2-3} \\ \frac{\rho}{2} c\frac{\theta_1}{2} \end{bmatrix}, \quad (18)$$

and because only the rotational component of the joint motion impacts the orientation of the EE, the angular velocity component of the Jacobian is

$$\mathbf{J}_{O_1} = \begin{bmatrix} 0 \\ 0 \\ 1 \end{bmatrix}. \quad (19)$$

Similarly for Joints 2, 3, and 4:

$$\mathbf{J}_{P_2} = \begin{bmatrix} a_2\, s_1 c_2 - \rho\, s_1 c_{2-3} s\frac{\theta_4}{2} - d_4\, s_1 c_{2-3} - \frac{\rho}{2}\, c_1 c\frac{\theta_2}{2} \\ a_2\, s_1 c_2 - \rho\, s_1 c_{2-3} s\frac{\theta_4}{2} - d_4\, s_1 c_{2-3} - \frac{\rho}{2}\, c_1 c\frac{\theta_2}{2} \\ a_2 s_2 - \rho s_{2-3} s\frac{\theta_4}{2} - d_4 s_{2-3} \end{bmatrix}, \text{ and } \mathbf{J}_{O_2} = \begin{bmatrix} s_1 \\ -c_1 \\ 0 \end{bmatrix}.$$

$$\mathbf{J}_{P_3} = \begin{bmatrix} \rho\, c_1 c_{2-3} s\frac{\theta_4}{2} + d_4\, c_1 c_{2-3} - \frac{\rho}{2}\, c\frac{\theta_3}{2} s_1 \\ \rho\, s_1 c_{2-3} s\frac{\theta_4}{2} + d_4\, s_1 c_{2-3} + \frac{\rho}{2}\, c\frac{\theta_3}{2} c_1 \\ \rho s_{2-3} s\frac{\theta_4}{2} + d_4 s_{2-3} \end{bmatrix}, \text{ and } \mathbf{J}_{O_3} = \begin{bmatrix} -s_1 \\ c_1 \\ 0 \end{bmatrix}.$$

$$\mathbf{J}_{P_4} = \begin{bmatrix} -\frac{\rho}{2}\, c\frac{\theta_4}{2} c_1 s_{2-3} \\ -\frac{\rho}{2}\, c\frac{\theta_4}{2} s_1 s_{2-3} \\ \frac{\rho}{2}\, c\frac{\theta_4}{2} c_{2-3} \end{bmatrix}, \text{ and } \mathbf{J}_{O_4} = \begin{bmatrix} -c_1 s_{2-3} \\ -s_1 s_{2-3} \\ c_{2-3} \end{bmatrix}.$$

The full 6×4 Jacobian is assembled as

$$\mathbf{J} = \begin{bmatrix} \mathbf{J}_{P_1} & \mathbf{J}_{P_2} & \mathbf{J}_{P_3} & \mathbf{J}_{P_4} \\ \mathbf{J}_{O_1} & \mathbf{J}_{O_2} & \mathbf{J}_{O_3} & \mathbf{J}_{O_4} \end{bmatrix}. \quad (20)$$

A full examination of the singularities has yet to be conducted but a simple example of a singular configuration is easily found. With only four joint variables it is no surprise that there will be certain directions in which the EE cannot be moved at a given time, but in certain situations the capabilities are further diminished. When $\theta_{v_1} = \theta_{v_2} = \theta_{v_3} = \theta_{v_4} = 180°$ the Jacobian matrix becomes

$$\mathbf{J} = \begin{bmatrix} 0 & (a_2 + \rho + d_4) & -(\rho + d_4) & 0 \\ 0 & 0 & 0 & 0 \\ 0 & 0 & 0 & 0 \\ 0 & 0 & 0 & 0 \\ 0 & 1 & -1 & 0 \\ 1 & 0 & 0 & 1 \end{bmatrix}. \tag{21}$$

In this configuration instantaneous linear velocities along the y_0- and z_0-axes and angular velocity about the x_0-axis are not achievable.

4 Conclusions

In this paper the Jacobian for nA-chains in general, and in particular, a novel 4 DOF 4A-chain was derived. These chains are joined by A-pairs, which take advantage of the single DOF self motion of the architecturally singular midline-to-vertex configuration of the Griffis–Duffy platform. The self motion is a sinusoidally coupled rotation and translation. The coupling means that existing techniques for establishing the relationship between the joint rates and the resulting linear and angular velocity of the distal link in the chain have to be adapted. Linear and angular velocity relationships between links were considered distinctly and the results combined to reveal the manipulator Jacobian. With the Jacobian established, the manipulator singular configurations can now be investigated with the starting point based on the method reported in [3].

References

1. Griffis, M.W., Duffy, J.: Method and Apparatus for Controlling Geometrically Simple Parallel Mechanisms with Distinctive Connections. U.S. Patent Number: 5,179,525 (1993)
2. Husty, M.L., Karger, A.: Self-Motions of Griffis–Duffy Type Parallel Manipulators. In: Proceedings of ICRA '00 IEEE International Conference on Robotics and Automation, pp. 7–12. San Francisco, CA, USA (2000)
3. Husty, M.L., Karger, A.: Self Motions of Stewart–Gough Platforms, An Overview. In: Proceedings of the Workshop on Fundamental Issues and Future Research Directions for Parallel Mechanisms and Manipulators, pp. 131–141. Quebec City, Canada (2002)
4. Robinson, J.D., Hayes, M.J.D.: The Kinematics of A-Pair Jointed Serial Linkages. In: Proceedings of the ASME International Design and Technical Conferences, IDETC 2010. Montreal, Quebec, Canada (2010)
5. Siciliano, B., Sciavicco, L., Villani, L., Oriolo, G.: Robotics Modelling, Planning and Control. Springer-Verlag, London (1995)

Singular Manifold of the General Hexagonal Stewart Platform Manipulator

Viswanath Shanker and Sandipan Bandyopadhyay

Abstract The knowledge of the singular manifold of a manipulator is essential for its design, path-planning and control. However, due to the computational complexity, it is very difficult to obtain an analytical description of the manifold in terms of the architecture as well as configuration parameters of the manipulator. In this paper, the singularity of a general hexagonal Stewart platform manipulator (GHSPM) is studied and its singular manifold is obtained. The manifolds geometric structure is analysed by projecting it to the position and orientation subspaces, respectively. An explicit geometric characterisation is obtained in the former case, where is it found that the geometric nature of the singular surface is the same as in the case of the SRSPM. The other surface, however, defies such analysis at this point due to its high degree. The theoretical results are illustrated with numerical examples and plots.

Key words: Stewart platform, singular manifold, geometric characterisation

1 Introduction

The singular manifold of the Stewart platform manipulator (SPM) has been studied for decades. In 2000, St-Onge and Gosselin [3] pointed out that the singular locus of the Stewart platform manipulator (GSPM) with planar base and moving platforms to be a polynomial expression that is cubic in z and quadratic in x, y; x, y, z being the coordinates of the origin of the moving platform. However, the polynomial was not derived in closed-form and the study of singularities in the orientation space was not included. Li et al. [2] have studied the orientation-space singularities in terms of Euler angles, and found that the highest degree of the cosines of the Euler angles in

Viswanath Shanker · Sandipan Bandyopadhyay
Indian Institute of Technology Madras, Chennai, India, e-mail: viswanath.shanker@gmail.com, sandipan@iitm.ac.in

J. Lenarčič, M. Husty (eds.), *Latest Advances in Robot Kinematics*,
DOI 10.1007/978-94-007-4620-6_50,

the singularity condition was three. The state of the art in this regard can be found in greater detail in [1, 2].

In 2006, Bandyopadhyay and Ghosal [1] obtained the closed-form expression for the singular manifold of a class of SPMs, known as the semi-regular SPM (SRSPM), in which the alternate sides of the moving as well as the fixed platforms have equal lengths. For this class of SPMs, they were able to obtain the expressions for the singular manifold in closed-form, both in the position as well as the orientation space. Further, given the low degree of the singular surface in the position space, they were able to characterise it completely, and also obtained an explicit parametrisation for the same.

This paper attempts to extend the results of [1] to the more general case of the GHSPM. Though most of the Stewart platform-based manipulators fall in the SRSPM category, there are practical examples outside of this category, e.g., the spot-welding robot F-200*i*B from FANUC Robotics Corporation, USA. It is, therefore, of theoretical as well as practical interest to study the singular manifold of the GHSPM class of manipulators.

Studies reveal that the singular surface in the position coordinates of the GHSPM has a very similar geometry to that of the SRSPM. In fact, the characters may be identical; however, while this could be demonstrated exactly over hundreds of random test cases, conclusive proofs on some aspects could not be obtained for the most general case. The orientation-singular surface shows some change in its algebraic structure from the SRSPM case. These results have been illustrated through numerical examples and plots.

The paper is organised as follows: In Section 2, the mathematical formulation of singularity is presented, followed by a description of the symbolic simplification of the singularity condition in Section 3. The geometric characterisation of the singular surface in the position space is discussed in Section 4. The theoretical results are illustrated by numerical examples and plots in Section 5, followed by the conclusions in Section 6.

2 Mathematical Formulation

In this section, the kinematics of the GHSPM and the formulation of the singularity are explained in brief. The approach is based on the formulation given in [1], which presents a more comprehensive description.

Stewart, or Stewart–Gough platform manipulators are among the best-known six-degrees-of-freedom parallel manipulators. The most general form of SPM consists of a rigid moving platform connected to a fixed rigid base by six identical legs, of UPS or SPS architecture. However, in practice, the fixed as well as the moving platforms are of the form of rigid hexagons[1] in most SPMs. This class of manipulators,

[1] The case of triangular platforms is obviously included in this group.

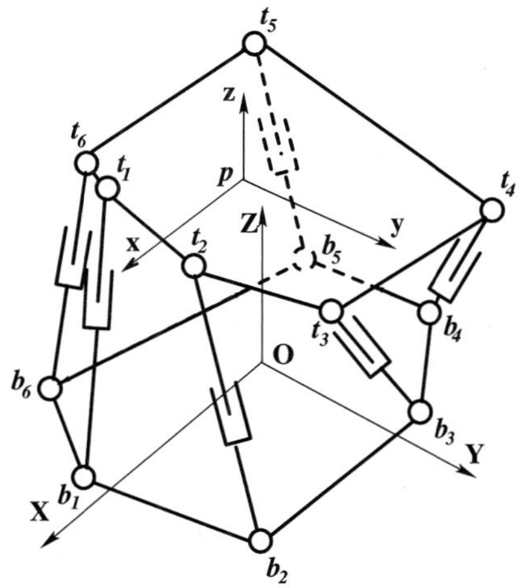

Fig. 1 General hexagonal Stewart platform manipulator (GHSPM).

known as the *general hexagonal Stewart platform manipulator* (abbreviated here as GHSPM), is studied in this paper.

The manipulator is shown in Fig. 1. The legs are assumed to have UPS architecture (which has the same kinematics as the SPS-legged SPM, except for the idle rotations of the legs about their respective axes). Let $\boldsymbol{t}_i = (xt_i, yt_i, 0)^T$ denote the position of the ith S-joint in the top platform with respect to the coordinate system attached to a point $\boldsymbol{p}(x, y, z)$ on the platform. Similarly, let $\boldsymbol{b}_i = (xb_i, yb_i, 0)^T$ denote the corresponding points in the base coordinate system. The rotation matrix $\boldsymbol{R}$, expressed in terms of the Rodrigue's parameters c_1, c_2, c_3, represent the orientation of the moving frame with respect to the fixed. The singularity condition is derived from the criterion of degeneracy of the statics of the manipulator. The wrench acting on the moving platform (w.r.t. $\boldsymbol{p}$) due to the force $\boldsymbol{F}_i$ in the ith leg is given in the base frame as $\boldsymbol{W}_i = (F_i\boldsymbol{s}_i; \boldsymbol{R}\boldsymbol{t}_i \times F_i\boldsymbol{s}_i)$ where $\boldsymbol{s}_i$ denotes the unit-vector along the ith leg. From the loop-closure condition applied to the loops $\boldsymbol{O} - \boldsymbol{b}_i - \boldsymbol{t}_i - \boldsymbol{p} - \boldsymbol{O}$ $(i = 1, \ldots, 6)$, one finds $\boldsymbol{s}_i = (\boldsymbol{p} + \boldsymbol{R}\boldsymbol{t}_i - \boldsymbol{b}_i)/l_i$ $(i = 1, \ldots, 6)$ where l_i denotes the length of the ith leg. The condition for static equilibrium is given by $\boldsymbol{W} = \sum_{i=1}^{6} \boldsymbol{W}_i$ $(i = 1, \ldots, 6)$, where $\boldsymbol{W} = (\boldsymbol{F}; \boldsymbol{M})$ denotes the net external wrench on the moving platform, in the same frame as $\boldsymbol{W}_i$. Breaking into the components, the equation of static equilibrium becomes $\boldsymbol{H}\boldsymbol{f} = \boldsymbol{W}$, $\boldsymbol{f} = (F_1, \ldots, F_6)^T$, where:

$$\boldsymbol{H} = \begin{pmatrix} \frac{1}{l_1}(\boldsymbol{p} + \boldsymbol{R}\boldsymbol{t}_1 - \boldsymbol{b}_1) & \ldots & \frac{1}{l_6}(\boldsymbol{p} + \boldsymbol{R}\boldsymbol{t}_6 - \boldsymbol{b}_6) \\ \frac{1}{l_2}((\boldsymbol{R}\boldsymbol{t}_1) \times (\boldsymbol{p} - \boldsymbol{b}_1)) & \ldots & \frac{1}{l_6}((\boldsymbol{R}\boldsymbol{t}_6) \times (\boldsymbol{p} - \boldsymbol{b}_6)) \end{pmatrix} \quad (1)$$

It is well-established in literature that the statics of the manipulator is degenerate (equivalently, the manipulator is in a singular configuration), when the *wrench transformation matrix* $\boldsymbol{H}$ is rank-deficient, i.e., $\det(\boldsymbol{H}) = 0$:

$$\frac{1}{l_1 l_2 l_3 l_4 l_5 l_6} \det \begin{pmatrix} (\boldsymbol{p} + \boldsymbol{R}\boldsymbol{t}_1 - \boldsymbol{b}_1)^T & \dots & (\boldsymbol{p} + \boldsymbol{R}\boldsymbol{t}_6 - \boldsymbol{b}_6)^T \\ ((\boldsymbol{R}\boldsymbol{t}_1) \times (\boldsymbol{p} - \boldsymbol{b}_1))^T & \dots & ((\boldsymbol{R}\boldsymbol{t}_6) \times (\boldsymbol{p} - \boldsymbol{b}_6))^T \end{pmatrix} = 0 \quad (2)$$

As the leg lengths may be assumed to be non-zero, one finds the condition for singularity as $S = 0$, where S is the determinant appearing in the L.H.S. of Eq. (2).

3 Symbolic Simplification of the Singularity Condition

While the derivation of the condition for singularity is fairly straight-forward, the analysis and understanding of it is extremely difficult. A number of previous attempts to simplify the expression S to a physically comprehensible, or mathematically amenable state have failed in obtaining S as a closed-form expression in terms of the architecture and pose parameters of the manipulator. To the best of the knowledge of authors, such expressions are available for only one class of SPMs, namely, the SRSPM [1], and a similar result on GHSPM is being reported for the very first time in this work.

The use of a computer algebra system (CAS) is imperative for the symbolic simplification of the singularity condition. However, as noted in [1], commercially available generic systems fail to deliver for problems of this magnitude, unless specifically designed simplification algorithms are used. The present work uses the algorithms developed in [1] to obtain the desired closed-form expressions. The singular manifold is obtained as surfaces in the position and the orientation spaces, respectively. The details of the steps involved are skipped for the sake of brevity, while the essential aspects are highlighted below. All computations were performed using the commercial CAS `Mathematica 8.0` [4].

3.1 *Singular Surface in the Position Space, $\mathcal{S}_p$*

- The original size of the expression for S is 2.69 MB (in `Mathematica`'s internal representation), or roughly 60 pages.[2]
- The expression is then transformed to the form $\mathcal{S}_p = \boldsymbol{C} \cdot \boldsymbol{P}$, where $\boldsymbol{C}$ is the 27-vector of the coefficients, and $\boldsymbol{P}$ the corresponding set of monomials in the variables x, y, z. In other words, S is cast in its *monomial-based canonical form* in the variables x, y, z, as described in [1]. The size of the expression at this stage is the largest (2.15 GB). However, as the expression is now broken into 27 smaller subexpressions in lesser number of variables (i.e., these do not contain any of x, y, z), the symbolic simplification works much better on these than on the original expression. In fact, in the experience of the authors, the CAS runs out of memory fairly soon while attempting to simplify expressions as big as the original form of S.

[2] This is only an estimate, as the output has never been printed for obvious reasons!

- The simplification process reveals that 11 of the elements of $\boldsymbol{C}$ vanish identically. The remaining 16 monomials with non-zero coefficients are:

$$\{1, z, z^2, z^3, y, yz, yz^2, y^2, y^2z, x, xz, xz^2, xy, xyz, x^2, x^2z\}$$

  It can be observed that the structure of $\mathcal{S}_p$, expressed as a polynomial in the coordinates of $\boldsymbol{p}$ is the same as in the case of the SRSPM. This confirms that *the generalisation of the SRSPM to the GHSPM does not change the structure of the singular manifold, when analysed as a surface in the position coordinates of the origin of the moving platform.*
- However, a key difference between the cases of SRSPM and GHSPM is that the later involves a large number of architecture parameters, i.e., xt_i, yt_i, xb_i, yb_i $(i = 1, \dots, 6)$ (out of which 17 are independent) as opposed to only 3 in the case of the SRSPM. Thus, in this case the final result is much larger (28.54 MB). It may appear that the simplification procedure is counter-productive, as it has increased the size of S by nearly 10 times. However, further analysis on geometric characterisation would justify the need for this particular form of the expression.
- The computation was carried out on a PC with a 3 GHz AMD Phenom II X6 Processor and 8 GB RAM. It took nearly 7 hours for the entire process.

3.2 Singular Surface in the Orientation Space, $\mathcal{S}_{\mathcal{O}}$

The steps involved are similar to the above, except that S is expressed as a polynomial in c_1, c_2, c_3 in this case. The end-result is somewhat different from that of the SRSPM in this case.

- There are 84 distinct monomials, as opposed to 77 in the case of the SRSPM.
- There is no identically vanishing coefficient; i.e., $\mathcal{S}_{\mathcal{O}}$ expressed as a polynomial in c_1, c_2, c_3 does have 84 non-zero terms in general.
- However, the total degree of any term does not exceed 6, the same as in the SRSPM.

4 Geometric Characterisation of $\mathcal{S}_p$

The low degree of $\mathcal{S}_p$ permits extensive analysis of its geometry. As the algebraic structure of $\mathcal{S}_p$ is the same as in the case of the SRSPM, there is not much of a difference in the geometry as well. However, due to the large size of the expressions in this case, not all of the results could be obtained conclusively. The key elements of this analysis are given below.

It can be seen from the expression of $\mathcal{S}_p$ that it may be expressed as a general conic section for a fixed value of z:

$$\mathcal{C} \triangleq ax^2 + 2hxy + by^2 + 2gx + 2fy + c = 0 \tag{3}$$

where the coefficients a, b, c, f, g, h are functions of z, architecture and orientation parameters. The geometric nature of $\mathcal{C}$ at various z-sections was studied as described below.

- The discriminant, $\delta = h^2 - ab$ was computed in closed-form, and was found to be of the form $\delta = e_0 z^2 + e_1 z + e_2$. The discriminant of this quadratic in z, $\Delta = e_1^2 - 4e_0e_2$, was also computed in closed form. However, since it has very high degree in the architecture parameters ($\simeq$ 400), it was not possible to establish the sign of Δ explicitly. Nevertheless, two sets of tests conducted suggest that Δ vanishes identically:
  - Random integer test: All the parameters appearing in e_0, e_1, e_2 above are assigned random integer values, such that these coefficients can be computed *exactly*. In each of the 700 tests conducted, the hypothesis $\Delta = 0$ held true.
  - Possible zero test of `Mathematica`: The CAS has a built-in routine `PossibleZeroQ` for testing if a function can vanish at some combination of its arguments. This test also returned a positive result.

  Thus, it can be *reasonably inferred* that $\Delta = 0$ and consequently $\delta = e_0(z - z_p)^2$, $z_p = -e_1/(2e_0)$, though it could not be proven conclusively.
- As per the above inference, the sign of δ is dependent on e_0 alone for all $z \neq z_p$. Once again, random integer tests are conducted for 1000 runs, and it is found that e_0 is a non-zero number and a perfect square. Thus $\delta \geq 0$ and consequently, $\mathcal{C}$ is a hyperbola at all $z \in \mathbb{R}$ except for $z = z_p$, where it is a parabola.
- A further possibility is that $\mathcal{C}$ degenerates to a pair of straight lines. The geometric condition for this case turns out to be a quintic equation in z, which upon simplification reveals itself as a quartic. Therefore, it is possible for $\mathcal{C}$ to degenerate into a pair of straight lines at 0, 2 or 4 values of z.

In summary, it may be stated that the geometric character of the surface $\mathcal{S}_p$ for GHSPM is (possibly) the same as that in the case of the SRSPM. However, some of its properties could not be established conclusively.

5 Visualisation and Illustrative Examples

The architecture parameters of the GHSPM considered are given below: $(xt_1, xt_2, xt_3, xt_4, xt_5, xt_6) = (0, 0, 0.4, 1.1, 1, 0.3)$, $(yt_1, yt_2, yt_3, yt_4, yt_5, yt_6) = (0, 1.2, 2, 0.6, -0.3, -0.9)$, $(xb_1, xb_2, xb_3, xb_4, xb_5, xb_6) = (0, 0, 0.3, 1, 0.8, 0.2)$, $(yb_1, yb_2, yb_3, yb_4, yb_5, yb_6) = (0, 1, 1.6, 0.3, -0.3, -0.7)$. For the orientation of the top platform given by $c_1 = 0,\ c_2 = 0.2,\ c_3 = 0.1$, $\mathcal{S}_p$ is found as:

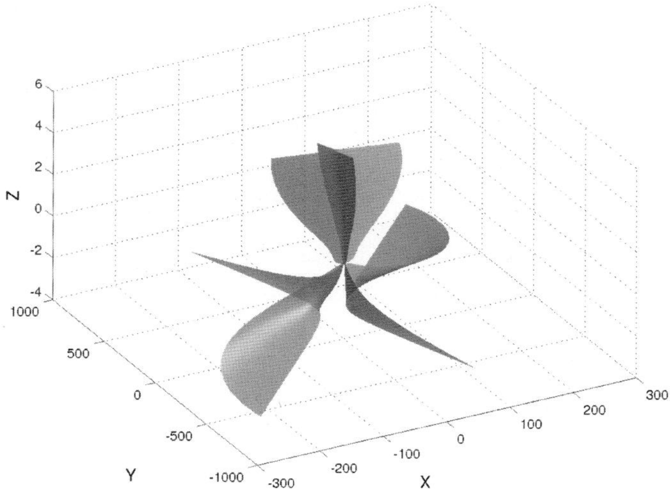

Fig. 2 $\mathcal{S}_p$ at $c_1 = c_2 = 0.2,\ c_3 = 0.1$.

$$2.921x^2z + 2.534x^2 + 3.426xyz + 9.509xy - 7.437xz^2 + 20.122xz + 9.871x + 0.313y^2z + 0.926y^2 + 6.147yz^2 + 10.584yz + 7.676y - 35.213z^3 + 12.602z^2 + 12.496z = 0$$

The resulting surface is shown in Fig. 2. Further, it is found that for this set of values, $\Delta = 6.3 \times 10^{-12} \simeq 0$, $z_p = -3.168$, and the sections at which the hyperbolas degenerate into pairs of straight lines are given by $z_s = -0.423, -2.683$. The first point can be seen clearly in Fig. 2. The singular surface in the orientation space, $\mathcal{S}_O$, is plotted in the coordinates (α, β, θ), where $(c_1, c_2, c_3) = (\cos\alpha\cos\beta, \cos\alpha\sin\beta, \sin\alpha)\tan(\theta/2)$, so that the plot is confined in a finite space. The surface $\mathcal{S}_O$ for $x = y = 0.1,\ z = 2$ is shown in Fig. 3. Further, for $c_2 = 0.1,\ c_3 = 0.1$, the singularity condition is given by

$$-2.266 + 8.476c_1 + 48.928c_1^2 - 0.595c_1^3 - 177.337c_1^4 - 353.120c_1^5 - 5.568c_1^6 = 0$$

This yields 4 real roots for c_1: $-62.913, -0.351, 0.151, 0.415$.

6 Conclusions

In this paper, closed-form expression for the singular manifold of a GHSPM has been derived. The geometric characterisation of the singular surface has been done in the position space, with results apparently identical with those of the SRSPM. However, the size and complexity of the expressions are much larger in this case, and

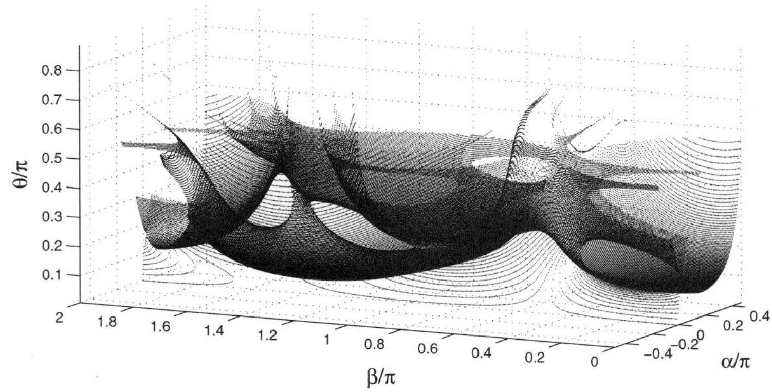

Fig. 3 $\mathcal{S}_{\mathcal{O}}$ for $x = y = 0.1,\ z = 2$

not all observations could be proven explicitly. Nonetheless, some of the analytical results presented here are new, and it is believed that the extension of the results of SRSPM to the more general case of GHSPM would help in further studies in, and applications of, this class of SPM.

References

1. Bandyopadhyay, S., Ghosal, A.: Geometric characterization and parametric representation of the singularity manifold of a 6–6 Stewart platform manipulator. Mech. Mach. Theory **41**, 1377–1400 (2006)
2. Li, H., Gosselin, C.M., Richard, M.J., St-Onge, B.M.: Analytic form of the 6-dimensional singularity locus of the general Stewart platform. J. Mech. Des. **128**, 279–287 (2006)
3. St-Onge, B.M., Gosselin, C.: Singularity analysis and representation of general Gough–Stewart platform. Int. J. Robot. Res. **19**(3), 271–288 (2000)
4. Wolfram Research Inc.: Mathematica, Version 8.0. Champaign, IL (2010)

Managing the Redundancy of $N-1$ Wire-Driven Parallel Robots

J-P. Merlet

Abstract We consider wire-driven parallel robot with $N \geq 4$ wires that are connected at the same point on the platform. Such robot has 3 d.o.f. but it is non-redundant (e.g. we cannot control the tension in the wires) as there will always be only at most 3 wires under tension simultaneously. We consider in this paper three approaches that make this robot really redundant: elasticity in the wires, using counterweights in the wires or attaching the redundant wires to a fixed point on the other wires. We show that these methods may be effective but still require further studies.

Key words: Cable robot, wire-driven parallel robot, redundancy

1 The $N-1$ Wire Driven Parallel Robot

A wire-driven parallel robot has the same mechanical structure as a parallel robot with rigid extensible legs but the linear actuators are substituted by wires that can be coiled and uncoiled. Such robot has the advantages of being mechanically simple and to allow for large workspace (the leg length variations being much larger than with rigid legs). Their main drawback is that wire can be pulled but cannot be pushed: hence kinematics cannot be decoupled from statics, especially for robot having less than 6 d.o.f., and this added complexity explains why the kinematics of such robot is still an open issue [5]. A large number of potential applications has led to a renewal of interest for wire-driven parallel robots for example for rescue crane [2, 9, 11], assistance robots and rehabilitation [3] or haptic devices [1, 6].

In this paper we are considering a special class of wire-driven parallel robot, called the $N-1$ robot, in which the N wires are all connected at the same point C on the platform. If $N \geq 3$, then the robot has 3 d.o.f., namely it allows to control the position of C but not the orientation of the platform. As soon as $N \geq 4$ such robot

J-P. Merlet
INRIA, Sophia-Antipolis, France, e-mail: jean-pierre.merlet@inria.fr

J. Lenarčič, M. Husty (eds.), *Latest Advances in Robot Kinematics*,
DOI 10.1007/978-94-007-4620-6_51,

is called *redundant*, whatever the definition of redundancy is [7, 10]. Redundant robot will be the topic of this paper, starting by an examination of the reality of this redundancy.

2 Is the $N-1$ Robot Redundant?

Using more than 3 non-elastic wires for an $N-1$ robot is a natural idea to improve the performances of this robot. Let us denote by A_i the fixed output points of the wires on the base and consider the workspace of this robot which is the volume V_ε spanned by moving the convex hull of the A_is along the downward vertical (with an upper limit which is the base plane and a lower limit determined by the maximal wire lengths). Hence adding a wire allows one to increase the workspace volume as soon as the added A_i is not located within the convex hull of the previous A_is.

It may also be sought that having redundant wire(s) allows one to control the tension distribution in the wires [4] and hence to improve the carrying capacity of the robot. Unfortunately we have shown both theoretically and experimentally in our ICRA paper from 2012 that this is not true if the wires are not elastic: *whatever the number of wires is, at a given pose there will always be at most 3 wires under tension while the other wires will be slack.* Without going into the proof let us explain intuitively this result. Consider a 3-1 robot in a pose C_3 that lies in its V_ε: the mechanical equilibrium at C_3 is satisfied and the tension in the 3 wires are uniquely determined. If we add a 4th wire its length is uniquely determined as the distance between A_4 and C_3. A wire system allows one to control either the wire length or the wire tension (but not both). Hence:

- if we impose the length, then the mechanical equilibrium will be satisfied with a 0 tension in the 4th wire
- if we impose a tension (i.e. the wire length is smaller than the distance between A_4, C_3), then C will move in a location different from C_3

Being unable to control the tensions in the wires is a disappointing result because this is typically one of the most obvious advantage of redundancy. We will propose in the next sections different ways to exploit the availability of additional wires for tension management while preserving the pose of the platform.

3 Tension Management

3.1 Elastic Wires

As mentioned previously a drawback of non-elastic wires is that tension control is difficult. This may be changed if we assume elasticity in the wires. Let τ_i be the tension in wire i, l_i its length at rest and ρ_i its length when under tension. If we assume that the wire is a perfect linear spring we have

$$\tau_i = k(\rho_i - l_i) \tag{1}$$

where k is the stiffness of the wire (assumed to be identical for all wires). Consider now a 3-1 robot submitted to a load of mass m and having all 3 wires connected at the center of mass of the load. The robot is submitted to a pure force $\mathscr{F} = (0,0,-mg)^T$ and the tension τ in the wires may be calculated as

$$\tau = \mathbf{J}^{\mathbf{T}}\mathscr{F} \tag{2}$$

where $\mathbf{J}^{\mathbf{T}}$ is the transpose of the Jacobian matrix of the robot. For the inverse kinematics (IK) the coordinates x,y,z of C are known, which allows one to calculate $\mathbf{J}^{\mathbf{T}}$ and then ρ_i as the distance between A_i, C. Using equation (2) we may then determine the τ. The control length of the wire may thus be calculated as

$$l_i = \rho_i - \frac{\tau_i}{k}$$

But there are sources of uncertainties in the modeling: on k, on l_i, on the location of the A_i and due to the fact that the wires are not exactly attached at the same point on the platform. We will focus on the influence of the uncertainties on k and l_i, assuming that the A_i have been calibrated while the influence of the colocation of the wire attachment points will be addressed in another paper. It is therefore necessary to investigate what is the influence of the stiffness on the pose of the robot for a given control input. We have thus solved the forward kinematics (FK) problem i.e. determine what are the possible coordinates of C for given l_is. Equation (2) allows one to calculate τ as functions of the coordinates of C while equation (1) has now as unknowns x,y,z,ρ_i. The geometrical IK of the robot provides an additional equation

$$\rho_i^2 = ||\mathbf{A_iC}||^2 \tag{3}$$

Equation (1) is linear in ρ_i and the result is reported in equation (3) to get a constraint equation in x,y,z. Repeating this process for all 3 wires leads to 3 constraint equations. Using resultant on these equations allows for successive elimination of x,y, leading to an univariate polynomial in z. This polynomial may be factored out as the product of 2 polynomials of degree 22, 34. Note that this approach is less efficient than the one proposed by Dietmaier [8] but has the advantage of providing directly the x,y,z. With this tool we may investigate the influence of uncertainties on k and l_i on the positioning.

As an example we consider the 3-1 wires robot with anchor points $A_1 = (0,0,0)$, $A_2 = (0,400,0)$, $A_3 = (400,0,0)$. The wire control values are given as $l_1 = 200$, $l_2 = 350$, $l_3 = 300$ which leads to the pose $x = 137.5$, $y = 96.875$, $z = -108.208$ for wires without elasticity. To take into account the uncertainty on the control l_i and on the stiffness k we have considered a possible ± 3 error on the l_i and a $\pm 0.1k$ error on the k. We have then solved the FK for a random sampling of 1000 sets of k,l within these ranges. For a nominal value of $k = 100$ we have found that the variations on x,y,z were in the ranges $[-3.86,2.7]$, $[-3.24,3.93]$, $[-5.16,3.74]$, while the τ variations were $[-15.49,14.23]$, $[26.9,28.96]$, $[-26.11,24.18]$. For a nominal value of $k=3000$ we found out that the variations on x,y,z were in the ranges $[-3.64,3.63]$, $[-4.3,3.47]$, $[-6.32,6.14]$, while the τ variations were $[-20.58,19.46]$, $[-40.1,36.2]$, $[-44.21,39.36]$. Hence even small uncertainties on the values of the l,k lead to significant positioning errors for the robot.

We consider now a 4-1 robot with the purpose of using the redundancy to adjust the wire tension e.g. to minimize the criteria $H=\sum_{j=1}^{j=4}\tau_j^2$. Using equation (2) one can obtain three wire tensions as a linear function of the remaining one. Without lack of exhaustivity we may calculate τ_2,τ_3,τ_4 as a function of τ_1. H is then a quadratic function in τ_1 and it is therefore trivial to determine τ_1 that leads to the minimum of H. For the IK, being given the pose of the load, the τ and equation (1) we may determine the four l_is. To determine the influence on the positioning of the uncertainties on k and on l_is we have to solve the FK problem.

In the FK problem the l_i's are given and we have to determine the pose of the load. For that purpose we note that the first equation of (1) allows to determine τ_1 as function of ρ_1, while equation (2) is used to determine τ_2,τ_3,τ_4. The three remaining equations of (1) are then linear in x,y,z. After solving this system we report the result in the IK equations (3) after subtracting the equation for wire 1 to the equations for wire 2, 3, 4. Together with (3) for wire 1 these equations constitutes a system of 4 equations in the unknowns $\rho_1,\rho_2,\rho_3,\rho_4$. One of this equation is linear in ρ_4 and is solved for this variable. The 3 remaining equations, denoted a_1,a_2,a_3, are of degree (6,6,2), (3,3,3), (9,9,3) in ρ_1,ρ_2,ρ_3. The four equations $a_1,\rho_3 a_1,a_2,a_3$ are linear in the monomials $1,\rho_3,\rho_3^2,\rho_3^3$ and hence the determinant of the matrix of this linear system should be 0, which leads to a polynomial $\mathscr{P}_1$ of degree 15 in ρ_1,ρ_2. Taking the resultant of a_1,a_2 in ρ_3 leads to a polynomial $\mathscr{P}_2$ of degree 18 in ρ_1,ρ_2. The resultant of $\mathscr{P}_1,\mathscr{P}_2$ factors out in 2 polynomials of degree 76 and 96 in ρ_1. Although this complete the theoretical solution, the degree of the involved polynomial is too high to be used in practice and consequently we have to resort to a numerical procedure. For that purpose we solve the linear system (2) to get τ_2,τ_3,τ_4 as function of τ_1. Then the first equation of (1) is used to determine τ_1 as a function of ρ_1. The three remaining equations of (1) together with the IK equations (3) constitute a system of 7 equations in the 7 unknowns $x,y,z,\rho_1,\rho_2,\rho_3,\rho_4$. As all unknowns may easily be bounded we have used an interval analysis approach to solve this system, all solutions being found in less than one second.

We have considered the 4-1 robot derived from the previous 3-1 by adding a 4th wire whose exit point on the base is $A_4=(400,400,0)$. We have then used the IK to determine what should be the l_i to reach the pose $x=100$, $y=200$, $z=-200$ while minimizing $\sum_{j=1}^{j=4}\tau_i^2$ for $k=1000$. The nominal values were determined as $l_1=l_2=299.558$, $l_3=l_4=412.1083$ which leads to $\tau_1=\tau_2=441.45$, $\tau_3=\tau_4=202.238$. Using the FK with these values of the ls leads also to solutions in which only 3 wires are under tension, namely wires (1, 2, 3) or (1, 2, 4), both cases leading to the same pose of the load with $x=99.6834$, $y=200.2192$, $z=-200.1581$. It should be noted that already elasticity does not allow for precise positioning as we are unable to determine the final pose of the platform for given control inputs.

We have then assumed similar errors on the ls and on k than in the previous example and have computed the FK solutions for 1000 sets of (l,k) chosen randomly. We found out that the variations on x, y, z were in the ranges $[-7.532, 4.579]$, $[-6.868, 5.75]$, $[-7.6, 2.875]$ for a nominal value of $k = 100$ and $[-5.488, 4.7948]$, $[-4.79, 4.262]$, $[-5.1429, 4.247]$ for $k = 1000$. Over the test set the mean values of the τ_i's for $k = 1000$ were 467.606, 463.06, 278.35, 256.35 with a variation of $[-192, 133.86]$, $[-187.45, 138.09]$, $[-274.185, 146.07]$, $[-241.75, 168.745]$. Here again we observe significant positioning errors and very large change in the tensions. As a conclusion adding elasticity in the wires for managing redundancy will require a very good wire length control together with a perfect stiffness calibration.

3.2 Using Counterweights

The principle here is to attach known weight(s) on some wire(s), close to the platform in order not to disturb the coiling of the wires. The purpose of the counterweight is to change the direction of the tension(s) applied on the platform in order to possibly control the value of the wire tensions (in this section a wire is under tension if its τ is negative).

Consider for example a 4-1 robot with a counterweight of mass m_4 on wire 4 located at point M at a distance d of C (Figure 2). For a given pose of C we are able to calculate the values of ρ_1, ρ_2, ρ_3. Wire 4 exerts on the platform a force τ_{14} that is directed along $\mathbf{MC}$ while it exerts a force τ_{24} on the counterweight that is directed along $\mathbf{A_4M}$. The mechanical equilibrium of the platform may be written as:

$$\sum_{j=1}^{j=3} \tau_j \mathbf{A_jC}/\rho_j + \tau_{14}\mathbf{MC}/d + (0,0,-mg)^T = \mathbf{0} \tag{4}$$

The mechanical equilibrium of the counterweight may be written as:

$$-\tau_{14}\mathbf{MC}/d + \tau_{24}\mathbf{A_4M}/(\rho_4 - d) + (0,0,-m_1 g)^T = \mathbf{0} \tag{5}$$

A direct consequence of equations (5) is that M must lie in the vertical plane that includes A_4, C. This constraint, together with the equations $||MC||^2 = d^2$, $||MA_4||^2 = (\rho_4 - d)^2$, allows one to determine the unique location of M as a function of ρ_4. Substituting the values of the coordinates of M into equations (4, 5) leads to a linear system of 5 equations in the unknowns $\tau_1, \tau_2, \tau_3, \tau_{14}, \tau_{24}$. Hence the wire tensions may be established as functions of ρ_4: their generic form is $\tau_i = P_i/W$, where P_i is a polynomial of degree 8 in ρ_4 while W is quadratic in this variable. Figure 1 shows the values of the tension τ_1, τ_2, τ_3 as a function of ρ_4 at the pose $(25, 125, -300)$ for a load of 80 kg and a counterweight of 5 kg located at a distance 50 from the platform together with the values of the tensions for the 3-1 robot with wires 1, 2, 3. It may be seen that even for a relatively low counterweight mass the tensions in the wire 2, 3 are substantially lower while the tension in wire 1 increases. We may

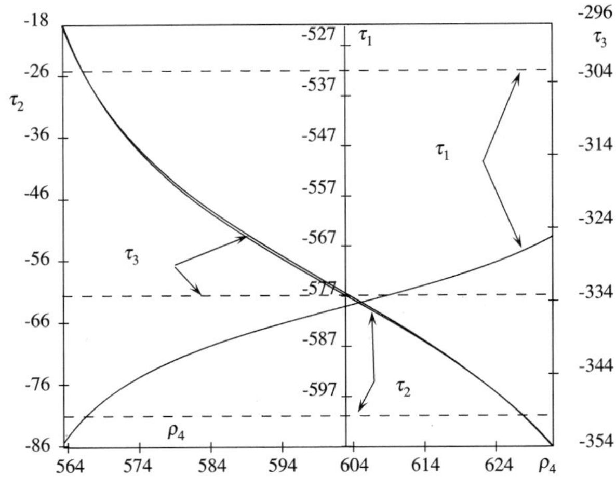

Fig. 1 The tensions in the wires of a 4-1 robot at the pose $(25, 125, -300)$ for a load of 80 kg and a counterweight of 5 kg on wire 4 located at a distance of 50 from the platform are shown as a function of ρ_4. The horizontal dashed lines show the value of the tensions in the wires for the 3-1 robot without wire 4.

consider the problem of determining the value of ρ_4 that maximizes $H = \sum_{j=1}^{j=3} \tau_j$, all tensions being negative or equal to 0. The derivative of H with respect to ρ_4 is a 14th order polynomial in ρ_4 and determining its positive roots allows one to find ρ_4 that maximizes H. We have then considered the cases where counterweights were attached to a single wire at different location or a counterweight was added to several wires. Extensive numerical tests confirmed that by adding counterweight(s) tension in some wires may be significantly reduced but at the cost of a large increase for the other wires and altogether no improvement for H. Note that, as mentioned by a reviewer, we may study the problem by assuming that the platform is a rigid line MC having 2R3T motion and this will lead to the same equations and results.

In conclusion adding counterweight is a possibility to deal with specific cases (e.g. decreasing the tension in one wire so that it can be disconnected) but is not a solution for overall improvement of the tensions in the wires.

3.3 Attaching Wires to Wires

The idea here is to have some wires that are not connected to C but to fixed location on other wires. As an example we will consider a 4-1 robot in which the 4th wire is attached at point M_1 on wire 1 so that the distance between M_1 and C when wire 1 is under tension is l_1 (Figure 2). The unknowns for the IK are the 3 coordinates of M_1 and the 5 tensions τ_1 to τ_5. First note that the mechanical equilibrium at M_1 imposes that M_1, A_1, A_4, C are coplanar and that for given M_1, τ_1 the tensions τ_4, τ_5 may be derived from the mechanical equilibrium. Hence we may consider only the

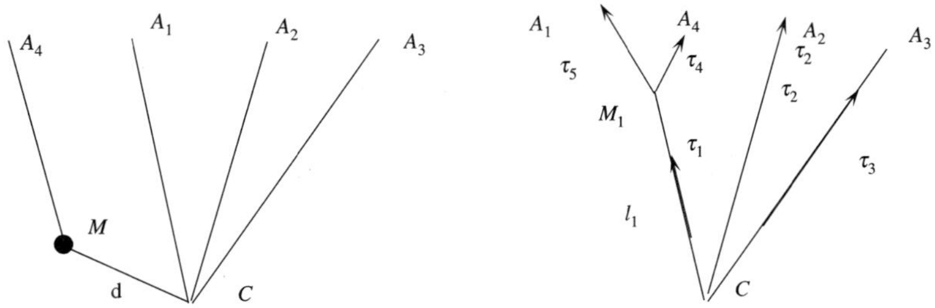

Fig. 2 A 4-1 robot with a counterweight attached on wire 4 and a 4-1 robot with wire 4 attached to wire 1. This robot reaches some pose under the influence of gravity.

following 6 unknowns: the coordinates x_1, y_1, z_1 of M_1 and the tensions τ_1, τ_2, τ_3. But for a given M_1 the equilibrium condition of the load (2) is a linear system in τ_1, τ_2, τ_3 that may be solved independently. Hence we may focus only on the constraints on M_1 i.e. $||\mathbf{M_1C}|| = l_1$ and the coplanarity condition between M_1, A_1, A_4, C. We end up with a system of 2 equations in 3 unknowns with one linear equation and one quadratic equation. Consequently we are able to express all tensions τ_1 to τ_5 as functions of any single variable in the set x_1, y_1, z_1. As we have a free variable we may choose it to optimize different criterion such as minimizing the sum H of the force exerted by the motors or minimizing the maximum $H_{\min}$ of the wire tensions. Note that finding the optimal choice for these criterion is easy: although the tensions are not algebraic expression of the free variable, the derivatives of H, $H_{\min}$ with respect to this variable are polynomials of degree 12 in the free variable.

As an example we have considered the pose $x = 110$, $y = 150$, $z = -200$ with $l_1 = 100$. If only 3 wires were used the criteria H was optimal when using wires 1, 2, 4 with a value of 1268.83 and wire tensions 669.85, 132.83, 466.14, while for $H_{\min}$ the best configuration is obtained when using wires 1, 2, 3 with a criteria value of 498.14 and wire tensions 375.116, 498.14, 413.17. We have then considered the 4-1 redundant case, examining all possible wire configurations. Surprisingly although we have tested numerous poses it appears that H is never improved when using redundancy, although we have not be able to figure out a theoretical explanation.

On the other hand $H_{\min}$ has been improved for the test pose with a value of 430.082 with as main wires 1, 2, 4 and wire 3 attached to wire 4. This corresponds to a gain of 13.66% compared to the non-redundant case. For this pose we have tested all values of l_1 between 10 and 130 with a step increment of 10 without observing any significant change in $H_{\min}$. Finally we have performed 100 test with random values for the coordinates of C within the ranges $[60, 340]$, $[60, 340]$, $[-300, -50]$ and random values for l_1 in the range $[10, 130]$. The mean value for the improvement on $H_{\min}$ was 13.27% with a minimum value of 0 and a maximal value of 37.714087%. Hence attaching wires to other wires seems to be a feasible solution to manage tension distribution in the wires.

4 Conclusion

Although apparently redundant an $N-1$ robot with $N \geq 4$ does not allow to manage tension distribution in the wires if they are not elastic. Tension management using the elasticity of the wires is quite difficult as the positioning of the platform is very sensitive to the stiffness of the wires and to wire lengths control. We have then investigated the use of adding counterweights in the wires, showing that the overall tension distribution is not improved, although this solution may lead to a decrease in some tensions. Then we have examined attaching redundant wires to fixed location on other wires: this simple solution is efficient to decrease the value of the maximal tension although the sum of the wire tensions is not improved. Management of redundancy opens numerous kinematics issues that are worth being investigated.

References

1. Arcara, P., et al.: Perception of depth information by means of a wire-actuated haptic interface. In: IEEE Int. Conf. on Robotics and Automation, pp. 3443–3348, San Francisco, 24–28 April 2000
2. Behzadipour, S., Khajepour, A.: Design of reduced dof parallel cable-based robots. Mech. Mach. Theory **39**(10), 1051–1065 (October 2004)
3. Bennour, S., Harshe, M., Romdhane, L., Merlet, J.P.: A robotic application for analysis and control of human motion. In: 4eme Congrès International Conception et Modélisation des Systèmes Mécaniques CMSM, Sousse, 30 May–1 June 2011
4. Bruckmann, T., Pott, A., Hiller, M.: Calculating force distributions for redundantly actuated tendon-based Stewart platforms. In: ARK, pp. 403–412, Ljubljana, 26–29 June 2006
5. Carricato, M., Merlet, J.P.: Direct geometrico-static problem of under-constrained cable-driven parallel robots with three cables. In: IEEE Int. Conf. on Robotics and Automation, pp. 3011–3017, Shangai, 9–13 May 2011
6. Castelli, G., Ottaviano, E.: Modeling and simulation of a cable-based parallel manipulator as an assisting device. In: Computational Kinematics, pp. 17–24, Duisburg, 6–8 May 2009
7. Corbel, D., et al.: Towards 100G with PKM. is actuation redundancy a good solution for pick-and-place? In: IEEE Int. Conf. on Robotics and Automation, pp. 4675–4682, Anchorage, 3–8 May 2010
8. Dietmaier, P.: An inverse force analysis of a spatial three-spring system. In: ARK, pp. 261–270, Ljubljana, 4–6 July 1994
9. Merlet, J.P., Daney, D.: A portable, modular parallel wire crane for rescue operations. In: IEEE Int. Conf. on Robotics and Automation, pp. 2834–2839, Anchorage, 3–8 May 2010
10. Pierrot, F.: Parallel mechanisms and redundancy. In: 1st Int. Colloquium, Collaborative Research Centre 562, pp. 261–277, Braunschweig, 29–30 May 2002
11. Tadokoro, S., et al.: A portable parallel manipulator for search and rescue at large-scale urban earthquakes and an identification algorithm for the installation in unstructured environments. In: IEEE Int. Conf. on Intelligent Robots and Systems (IROS), pp. 1222–1227, Kyongju, 17–21 October 1999

Dynamics of the Upper Limb with a Detailed Model for the Shoulder

Jorge Ambrósio, Carlos Quental, João Folgado and Jacinto Monteiro

Abstract The human muscle-skeletal system has a large number of redundant muscles, implying that the same motion may be obtained by different combination of muscle forces. As a consequence, the modeling of the kinematics of biomechanical models used for human motion task simulations has important implication on the distribution of the muscle forces and joint reaction forces. This work compares the performance and applicability of three biomechanical models, with different levels of complexity, in face of specific kinematic modeling assumptions for the anatomical joints and muscle geometry. The muscle contraction dynamics is simulated by the Hill-type muscle model, being the activation of each muscle a unknown in the redundant force sharing problem. An optimization technique is applied to minimize of an objective function related with muscle metabolic energy consumption. The input for the model analysis comprises the data for an abduction motion, kinematically consistent with the biomechanical models developed, acquired using video imaging at the Laboratory of Biomechanics of Lisbon.

Key words: Biomechanical model, force sharing, muscle models, optimization

1 Introduction

The upper limb is composed of an open chain mechanism that includes the shoulder, the elbow and the wrist, as presented in Figure 1. The shoulder includes four skeletal segments, the thorax, clavicle, scapula and humerus, connected through three anatomical articulations: sternoclavicular (SC), acromioclavicular (AC) and gleno-

Jorge Ambrósio · Carlos Quental · João Folgado
IDMEC-IST, Technical University of Lisbon, 1049-001 Lisbon, Portugal,
e-mail: jorge@dem.ist.utl.pt

Jacinto Monteiro
Medical School, University of Lisbon, Hospital de Santa Maria, 1649-028 Lisbon, Portugal

J. Lenarčič, M. Husty (eds.), *Latest Advances in Robot Kinematics*,
DOI 10.1007/978-94-007-4620-6_52,

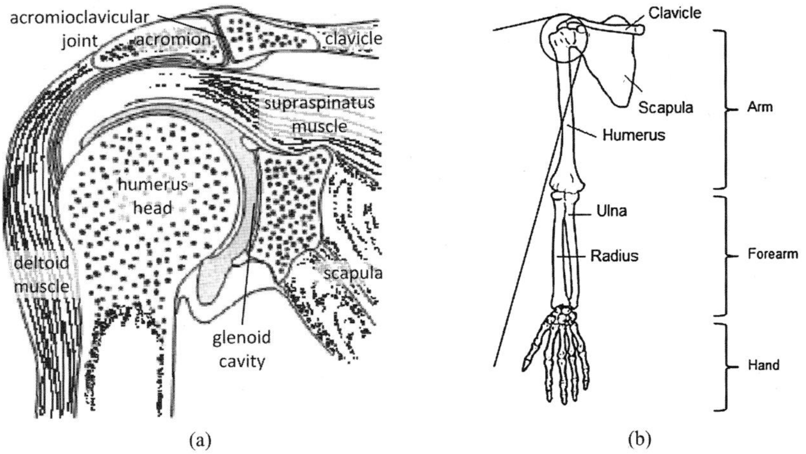

Fig. 1 Anatomy of the upper extremity (a) Glenohumeral (shoulder joint); (b) Bones of the upper extremity [4].

humeral (GH) joints. The scapula is constrained to slide over the thoracic cage. The coordinated actions in all articulations give the shoulder a large range of motion being several the mechanisms that provide additional stability to the glenoid fossa, such as the muscle coordination, elastic ligament tension, labrum deformation, joint suction, adhesion/cohesion, articular version, proprioception, or negative internal joint pressure [6]. Depending on the application, models with different complexity of the upper limb skeletal system can be envisaged. Typical application is crash-worthiness and ergonomy represent the shoulder joint as a spherical joint between the humerus and thorax [12]. In clinical applications the detailed anatomy of the shoulder is represented, being the anatomical joints between the different bony segments represented by mechanical joints [5]. Also the musculoskeletal system may be represented including more or less muscles represented by bundles eventually wrapping around complex obstacles [9].

In this work a simpler model is defined by 4 rigid bodies, thorax, humerus, ulna and radius, with the, glenohumeral (GH), humeroulnar (HU) and radioulnar (RU) articulations, and two other more complex models include 7 rigid bodies, thorax, rib cage, clavicle, scapula, humerus, ulna and radius, constrained by the sternoclavicular, acromioclavicular, scapulothoracic, GH, HU and RU articulations. The different muscular system in each of the biomechanical models, include 15 muscles, for the simpler model, and 21 muscles, for the most complex model. The objective of the models is the evaluation of the muscle and joint reaction forces. An optimization procedure is applied to solve the problem, in the framework of the inverse dynamic analysis of an experimentally acquired motion [11].

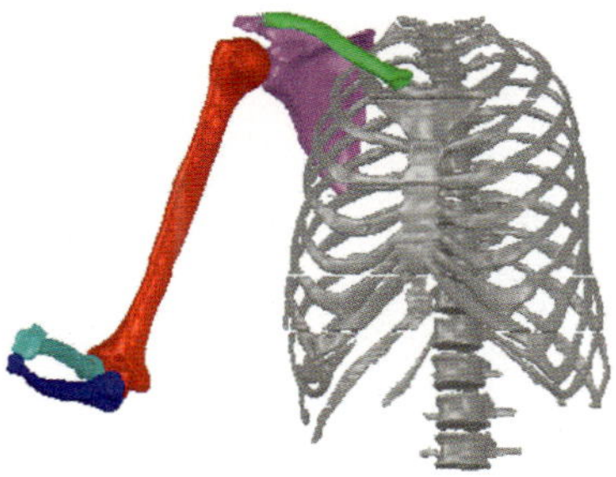

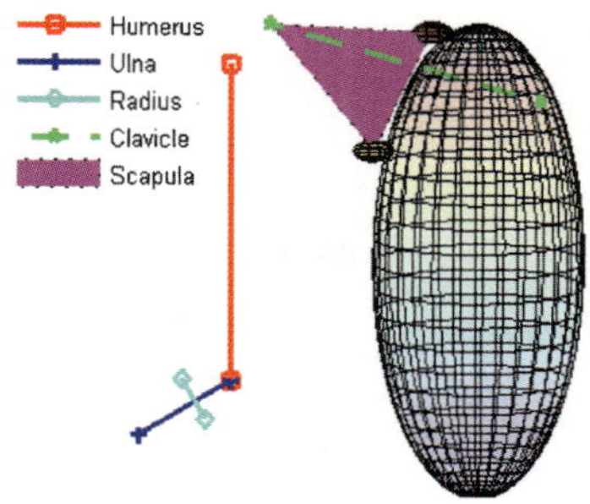

(a) (b)

Fig. 2 (a) geometrical description of the upper limb; (b) model of the upper limb.

2 Biomechanical Model

The skeletal model proposed here is based on data by Garner and Pandy [3] modified to be in accordance with the standardization of the upper limb data proposed by the International Society of Biomechanics (ISB) [4].The model includes the thorax, rib cage, clavicle, scapula, humerus, ulna and radius modeled as rigid bodies, as depicted by Figure 2.

Model 1 is defined by 4 rigid bodies, thorax, humerus, ulna and radius, constrained by 3 anatomical joints, glenohumeral joint (GH), humeroulnar joint (HU) and radioulnar joint (RU), being the clavicle and scapula stationary rigidly attached to the thorax. The remaining 2 models are similar as far as the skeletal system is concerned. They include 7 rigid bodies, thorax, rib cage, clavicle, scapula, humerus, ulna and radius, and 5 anatomical joints, sternoclavicular (SC), acromioclavicular (AC) and GH, represented by spherical joints, and HU and RU, represented by hinge joints. The scapulothoracic articulation is modeled by two holonomic constraints that ensure the scapula to glide over the rib cage. Cartesian coordinates with Euler parameters are used by the methods implemented in this work [7]. Being each body defined by 7 coordinates, Model 1 is described by 28 coordinates and 23 kinematic constraints being the two other models are defined by 49 coordinates and 40 kinematic constraints.

Figure 3 illustrates the biomechanical models developed. In Model 1 the clavicle and scapula are considered part of the thorax and, consequently, all the muscles between them are neglected. The model includes 15 muscles modeled by 24 bundles. The muscle path is defined by via points and uses the obstacle-set method to describe possible interactions of muscles with bones and soft tissues [2].

The musculoskeletal system of Model 2 includes all the muscles of the upper limb, i.e., 21 muscles defined by 37 bundles [9]. The musculoskeletal system of Model 3, characterized by 20 muscles modeled by 127 bundles, is based on the muscle data set published by the Delft Shoulder Group [5]. Although the muscle path is not as detailed as in Models 1 and 2, the number of bundles is significantly higher which may reproduce better muscles with broad origins or insertions. All

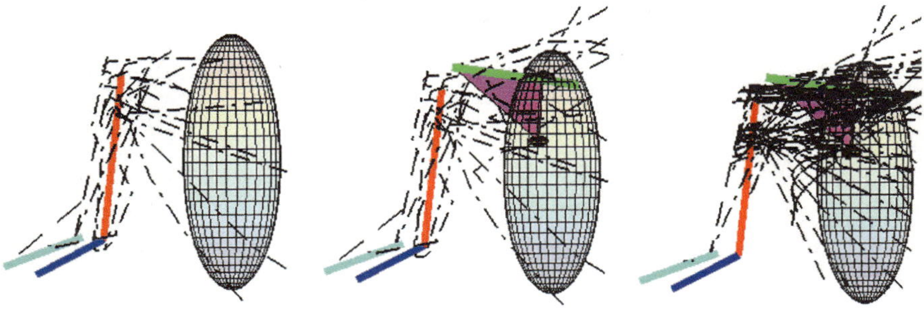

(a) Model 1 (b) Model 2 (c) Model 3

Fig. 3 Musculoskeletal models where black lines represent muscles: (a) Model 1; (b) Model 2; (c) Model 3.

muscle data, including origins and insertions coordinates as well as musculoskeletal parameters were scaled to the same anatomical dimensions of the skeletal model, being the data available in reference [9].

3 Methods

The arm abduction motion, analyzed here, is recorded for 7 seconds using 4 video cameras with sampling rates of 100 Hz while synchronous electromyography (EMG) signals of 9 superficial muscles are also acquired at the Laboratory of Biomechanics of Lisbon (LBL). Inverse dynamics is used in the solution of biomechanical problems when the motion is prescribed and only the forces responsible for such motion are unknown. The acquired kinematic data, positions in this case, is filtered and its kinematic consistency with the biomechanical models used is ensured [10]. The velocity and acceleration constraint equations, associated to the biomechanical models, are used afterwards to obtain the velocities $\dot{\mathbf{q}}$ and accelerations $\ddot{\mathbf{q}}$ of the anatomical segments.

The equations of motion of the biomechanical multibody system are given by

$$\mathbf{M}\ddot{\mathbf{q}} + \Phi_{\mathbf{q}}^{T}\boldsymbol{\lambda} = \mathbf{g} \tag{1}$$

where the masses, inertias and geometric characteristics are presented in reference [9]. Knowing the kinematics of the model and the external forces the problem can be solved for the unknown Lagrange multipliers $\boldsymbol{\lambda}$, associated to the joint reaction and muscle forces, i.e., can be solved using an inverse dynamics procedure. The major difference in the biomechanical model relatively to common mechanical models of other disciplines resides on the representation of the muscle forces.

The muscle contraction dynamics is represented using the Hill type muscle model that includes an active contractile element (CE) and a passive element (PE). The muscle force results from the contribution of both passive and active elements

$$\begin{cases} F^m = F^m_{CE}(L^m(t), \dot{L}^m(t), a^m(t)) + F^m_{PE}(L^m(t)) \\ F^m_{CE} = \dfrac{F^m_L(t) F^m_{\dot{L}}(t)}{F^m_0} a^m(t) \end{cases} \tag{2}$$

The passive element represents the non-controllable part of the muscle that only depends on the muscle length $L^m(t)$, if the muscle is stretched beyond its resting length. The contractile element depends upon the maximum isometric force of the muscle F^m_0, muscle length $L^m(t)$, muscle contractile velocity $\dot{L}^m(t)$ and activation $a^m(t)$. Functions $F^m_L(t)$ and $F^m_{\dot{L}}(t)$ describe the relationship force-length and force-velocity, respectively, and depend only in the kinematics of the anatomical segments and muscle physiology. The only unknown in the muscle force is the activation $a^m(t)$, which together with the joint reactions are part of vector $\boldsymbol{\lambda}$ in Equation (1).

Mathematically, the optimization problem associated to the solution of all muscle forces and joint reactions for all time frames, in which the kinematic data is acquired, is formulated as

$$\min J(t, \mathbf{a}) = \sum_{t=1}^{n} \sum_{i=1}^{m} c_1 \frac{V^i F^i_{CE}}{PCSA^i} + c_2 V^i a^i_t + c_3 a^{i2}_t$$

$$\text{subject to:} \begin{cases} \mathbf{f}_t(\mathbf{a}) = (\boldsymbol{\Phi}^T_{\mathbf{q}} \boldsymbol{\lambda} + (\mathbf{M}\ddot{\mathbf{q}} - \mathbf{g}))_t = \mathbf{0}, \quad t = 1, \ldots, n \\ \mathbf{0} \le \mathbf{a} \le \mathbf{1} \\ \left(\dfrac{\mathbf{f}^T_{GH} \boldsymbol{\tau}_{GH_d}}{\mathbf{f}^T_{GH} \mathbf{n}_{GH}} \right) \le th_d, \quad d = 1, 2, 3, 4 \quad \text{and} \quad t = 1, \ldots, n \\ (\mathbf{f}^T_{ST_p} \mathbf{n}_{ST_p})_t > 0, \quad p = 1, 2 \quad \text{and} \quad t = 1, \ldots, n \end{cases} \tag{3}$$

where the objective function represents the metabolic energy expenditure, V^i is the volume, F^i_{CE} the contractile element force, $PCSA^i$ the cross-sectional area and a^i the activation for muscle i, in a total of m muscles. Variables c_1, c_2 and c_3 are weighting factors defined according to [8]. The constraints of the optimization problem are the constraint equations, which must be fulfilled every time frame, the bounds of the muscle activations, which must be null or positive and the stability of the GH and ST joints that must be subjected to compressive forces all time. Regarding the GH joint stability constraint, $F^{GH}_{s,i}$ is the of the shear reaction force in the ith glenoid direction and F^{GH}_c the compressive reaction force at the GH joint; S_i represents the directional force ratio thresholds. In the ST joint stability constraint, $\mathbf{F}^{ST}$ stands for the ST reaction force and $\mathbf{n}^{ST}_c$ for the unitary vector that defines the direction of compression between the scapula and thorax.

The size of the optimization problem is proportional to the complexity of the biomechanical model. For each time frame: Model 1 is defined by 43 design variables (joint reaction and muscle forces), 24 equality constraints and 24 boundary constraints; Model 2 is defined by 70 design variables and 42 equality, 6 inequality and 37 boundary constraints; and Model 3 is defined by 161 design variables and

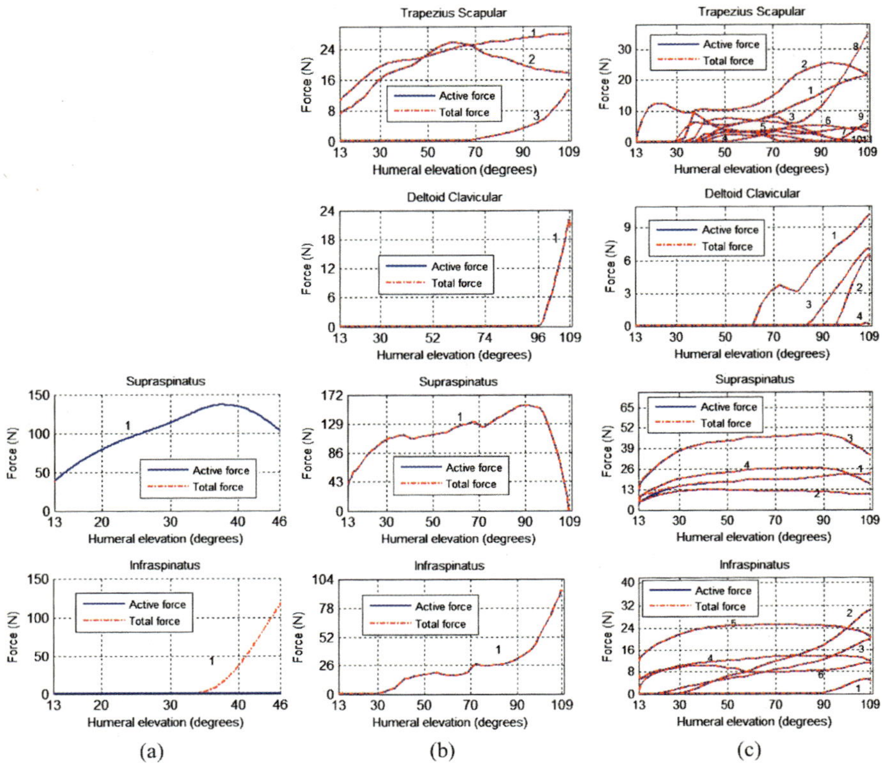

Fig. 4 Selected muscle forces predicted by the biomechanical models; (a) Model 1 (b) Model 2 (c) Model 3.

42 equality, 6 inequality and 128 boundary constraints. The optimization problem is suitably solved using the Sequential Quadratic Programming method.

4 Results

The application of the methods outlined in this work to the abduction motion acquired at the LBL lead to the estimation of the muscle forces for the complete musculoskeletal system of the upper limb. Forces for selected muscles are depicted in Figure 4 for the 3 models studied being the dashed line the contractile element contribution and the solid line the sum of the contractile and passive elements of the Hill muscle force model. Notice that the Trapezius Scapular and the Deltoid Clavicular are not present on Model 1 and, consequently, are not shown in Figure 4, for that model.

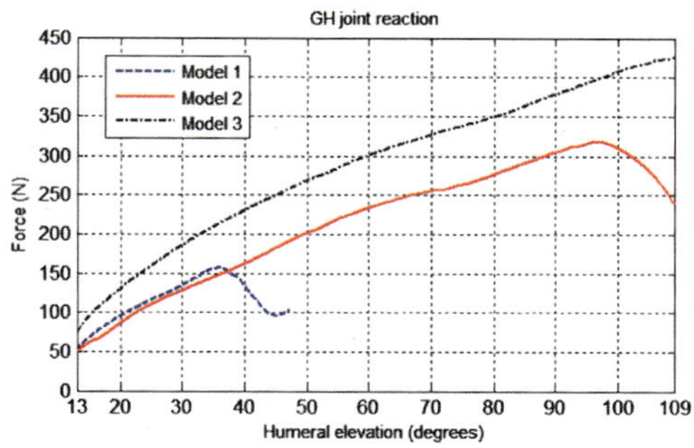

Fig. 5 Glenohumeral joint reaction force for the 3 biomechanical models.

Model 1 can only be applied for a range of the arm elevation between 13° and 46°, when the scapula does not contribute much to the shoulder motion, based on the translation of the GH joint. Despite its simplicity, the results highlight the supraspinatus muscle as the most active muscle. This muscle is responsible for starting the abduction motion and its activity is most important in the first 30° so the model, even being so simple, agrees with the literature [1]. After about 35° of arm amplitude, the teres minor and infraspinatus muscles start to produce a large amount of passive force which may mean that this model is not applicable in the full range of motion considered. Although Model 2 has a lower number of muscles bundles than Model 3, which is clearly a disadvantage, it has the advantage of including more via points and obstacles to define each muscle path and thus be more accurate representing muscles moment arms. In this motion, the results yielded by both models are very similar.

The GH joint reaction force is presented for the 3 models in Figure 5. Comparing Models 1 and 2 until 35° of arm amplitude shows very similar reaction forces. This result supports that a simpler model can indeed be applicable depending on the range of motion. The reaction forces from Models 2 and 3 are slightly shifted by 50–100 N but the pattern is similar. Both models are in close agreement with the results obtained by the Delft Shoulder group using the same optimization criterion [1]. The shift in the forces is due to the different moment arms considered in Models 2 and 3, which result from the application of the obstacle-set method.

5 Conclusions

Three biomechanical models of the upper limb, with different levels of complexity for both skeletal and musculoskeletal representation, are presented. Model 1 is applicable for a shorter range of motion of the Glenohumeral joint than Models 2 and 3. Model 2 defines the muscle path more accurately while Model 3 has a considerably higher number of muscle bundles that may simulate better the muscle function. For the type of motion considered both Models 2 and 3 perform well. However, in more complex motions the number of muscles used in Model 2 may present difficulties in the search for a feasible solution of the redundant force sharing problem. Also the mechanical definition of the anatomical joints has large implications on the results of the muscle force sharing problem. It is foreseen that the use of anatomically correct joints, modeled as contact joints, instead of the simple spherical and hinge mechanical joints used leads to a different muscle force distribution.

Acknowledgements This work was supported by the Foundation for Science and Technology through the project PTDC/SAU-BEB/103408/2008 and a PhD scholarship – SFRH/BD/46311/2008. The authors are grateful to Daniel Simões Lopes, Paulo Luzio de Melo and Nelson Ribeiro for their assistance at the Laboratory of Biomechanics of Lisbon.

References

1. De Groot, J.H.: The shoulder: a kinematic and dynamic analysis of motion and loading. Ph.D. thesis, Delft University of Technology (1998)
2. Garner, B.A., Pandy, M.G.: The obstacle-set method for representing muscle paths in musculoskeletal models. Comput. Methods Biomech. Biomed. Eng. **3**(1), 1–30 (2000)
3. Garner, B.A., Pandy, M.G.: Musculoskeletal model of the upper limb based on the visible human male dataset. Comput. Methods Biomech. Biomed. Eng. **4**(2), 93–126 (2001)
4. Van der Helm, F.C.T.: A standardized protocol for motion recordings of the shoulder. In: Proceedings of the First Conference of the ISG, pp. 7–12. Delft, The Netherlands (1997)
5. Van der Helm, F.C.T., Veeger, H.E.J., Pronk, G.M., Van der Woude, L.H.V., Rozendal, R.H.: Geometry parameters for musculoskeletal modelling of the shoulder system. J. Biomech. 25(2), 129–144 (1992)
6. Hong, Y., Bartlett, R.: Handbook of Biomechanics and Human Movement. Routledge, New York (2008)
7. Nikravesh, P.: Computer-Aided Analysis of Mechanical Systems. Prentice-Hall, Englewood Cliffs, New Jersey (1988)
8. Praagman, M., Chadwick, E.K., Van der Helm, F.C.T., Veeger, H.E.: The relationship between two different mechanical cost functions and muscle oxygen consumption. J. Biomech. **39**(4), 758–765 (2006)
9. Quental, C., Folgado, J., Ambrósio, J., Monteiro, J.: A multibody biomechanical model of the upper limb including the shoulder girdle. Multibody Syst. Dyn. (2012). DOI 10.1007/s11044-011-9297-0
10. Silva, M., Ambrósio, J.: Kinematic data consistency in the inverse dynamic analysis of biomechanical systems. Multibody Syst. Dyn. **8**(2), 219–239 (2002)
11. Silva, M., Ambrósio, J.: Solution of redundant muscle forces in human locomotion with multibody dynamics and optimization tools. Mech. Based Des. Struct. Mach. **31**(3), 381–411 (2003)
12. Silva, M., Ambrósio, J., Pereira, M.: Biomechanical model with penalized joint rotation for impact simulation. Multibody Syst. Dyn. **1**(1), 65–84 (1997)

Inverse Kinematics for the Control of Hyper-Redundant Binary Mechanisms with Application to Solar Concentrator Mirrors

Amy M. Bilton and Steven Dubowsky

Abstract Solar thermal systems require precision mirror concentrators. Their shape is expensive to manufacture and difficult to control. Here, a concept is proposed to control the shape of a solar concentrator using parallel binary actuators embedded in a mirror's structure. The actuators can only move a set distance between two bi-stable positions. Since the actuators have fixed displacements and the kinematic parameters of the hyper-redundant mechanism are not well known, determining the actuator inputs to correct a mirror shape error is a challenge. In this paper, a linearized kinematic model is developed that can be used for control. System models and laboratory experiments demonstrate the effectiveness of the approach.

Key words: Parallel binary kinematics, elastic averaging, solar concentrator

1 Introduction

Precision parabolic solar concentrator mirrors are an important component of many solar systems, see Figure 1. Shape errors from manufacturing, thermal warping and other effects can degrade their performance [2, 7]. It has been suggested that embedding bi-stable (binary) actuators in mirror structures could reduce solar concentrator manufacturing costs and improve system performance [5].

Using continuous actuators to correct mirror shape has been studied for telescopes such as the James Webb Space Telescope [3, 4]. In the Webb, complex and costly continuous actuators control the position and shape of mirror segments. Solar mirrors do not require high precision shape control and only the rigid body positions of segments need to be controlled. Here, this control is done with bi-stable actuators.

The precision of a binary system is determined by the number of actuators. As it increases, the precision approaches that of a continuous system. Generally, many

Amy M. Bilton · Steven Dubowsky
Massachusetts Institute of Technology, Cambridge, USA, e-mail: {bilton, dubowsky}@mit.edu

J. Lenarčič, M. Husty (eds.), *Latest Advances in Robot Kinematics*,
DOI 10.1007/978-94-007-4620-6_53,

Fig. 1 Faceted parabolic mirror system developed by SES and the US Department of Energy [7].

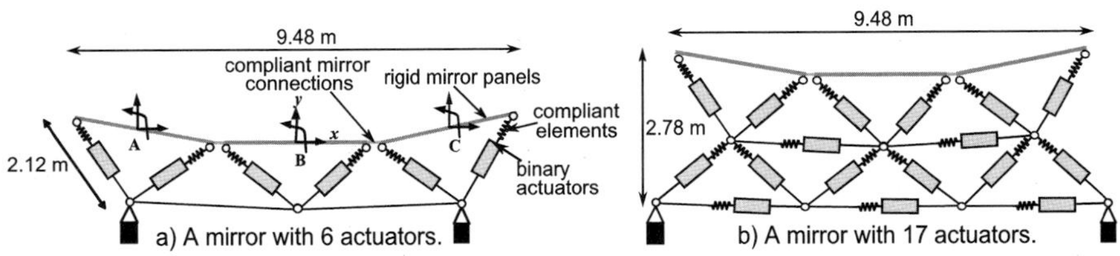

Fig. 2 Examples of a planar, segmented-mirror, parallel structures.

binary actuators are required and parameters of kinematic system are uncertain, making the system design and control challenging. Studies have shown the feasibility of binary solar mirrors, but control methods still need to be developed to operate them to correct manufacturing and other shape errors [5].

The earliest research on bi-stable actuator mechanism kinematics considered serial robotic chains actuated by solenoids, pistons and shape memory alloys [1, 6, 11]. More recently, studies demonstrated the potential of low-cost and simple bi-stable actuators using dielectric elastomer actuators (DEAs) [8, 9]. These actuators are highly efficient, as they only require power to change their current state. In addition, these actuators respond quickly, changing their output state in seconds.

A segmented mirror naturally has a parallel kinematic structure. Figure 2 shows simple examples with rigid mirror segments. Past examples of parallel binary devices include BRAID [11], and an MRI compatible manipulator [10]. The MRI concept improved its precision by using many more actuators than degrees-of-freedom and compliance to mediate the redundant actuation by elastic averaging [9]. Previous work on binary-actuated mirrors developed detailed system models without consideration of control requirements [5]. Control methods for parallel binary systems have been developed that rely on detailed system models [10]. The objective of this work is to develop methods to control binary structures without the need for detailed system models.

The workspace of binary mechanisms consists of a finite set of points and the inverse kinematics problem requires searching all possible discrete mechanism positions for the output closest to the desired shape [6, 11]. This paper presents a method to find the binary actuator input that best compensates for a mirror shape error. This method uses a linearized kinematic model with parameters determined

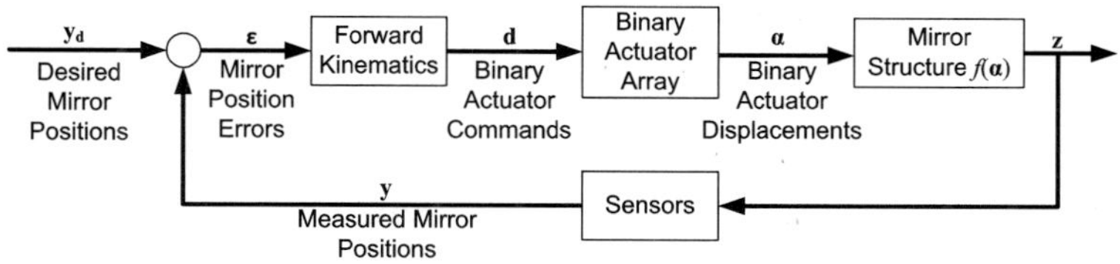

Fig. 3 Control system block diagram.

using system identification. The controller uses inverse kinematics to determine the actuator input for the mirror shape correction.

2 Analytical Development

Bi-stable actuators have fixed displacements and no actuator sensing is required. However, measurements of the mirror's shape are necessary for control. For manufacturing errors, such as accumulations of tolerances, this may be done off-line using laser industrial optical metrology. For real-time disturbances, such as to thermal distortions, indirect measurement may be required, such as monitoring the power collected. Figure 3 shows the control block diagram.

The problem is to find the binary-actuator commands required to correct an error in the mirror's shape. The challenge is the kinematic parameters of the binary-actuated parallel mechanisms are not well known. Since the workspace of a binary system consists of a set of 2^N (N is the number of actuators) points, the control problem reduces to determining the binary input that places the system closest to the desired position. Since the mechanism and actuators are not ideal, the parameters of the device also need to be identified. The approach developed exploits the fact that position errors are small and the system's behavior is nearly linear.

The representative examples, shown in Figure 2, have 6 and 17 embedded binary actuators. The mirror structure has three-segments where the mirrors are joined by flexible members. The supporting structure contains embedded actuators connected to the segments by pin joints. Table 1 gives the mirror properties.

To evaluate the performance of a solar thermal collector, a metric based the amount of light energy captured by the system's thermal receiver is formulated. In solar thermal systems, the receiver should ideally be as small as possible. Here, the receiver is assumed to lie in a plane and the mirror's control system should be

Table 1 Structural parameters of rigid panel model.

Parameter	Value	Parameter	Value
Structure Length	9.48 m	Binary Actuator Stroke Length (δ)	50 mm
Mirror Connector Stiffness (EI)	590 Nm^2	Elastic Element Stiffness (k)	1000–3000 N/m
Number of Binary Actuators (n)	6 and 17	Focal Plane Distance (L_p)	18.53 m

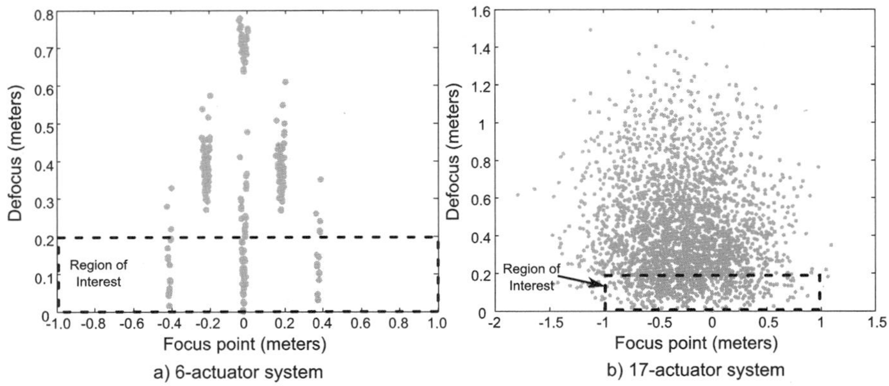

Fig. 4 Focal plane workspace for rigid-panel structures.

able to move each segment so that all of the sunlight is reflected on the smallest possible receiver. In other words, the mirror should be able to shift the location of segment's common focus and keep the defocus as small as possible. For example, for the mirrors shown in Figure 2, a receiver size (maximum defocus) of 20 cm and focal correction of 2 m are reasonable for solar collector applications.

The focal position and defocus can be evaluated by simple ray tracing. The position of the common focus is defined as the average position of the reflected rays for a given actuator input:

$$f_c(\mathbf{d}) = \frac{D_A(\mathbf{d}) + D_B(\mathbf{d}) + D_C(\mathbf{d})}{3} \tag{1}$$

where $D_A(\mathbf{d})$, $D_B(\mathbf{d})$ and $D_C(\mathbf{d})$ are the positions of the reflected rays in the focal plane for panels A, B, and C respectively. The defocus is defined as the maximum distance between two reflected rays:

$$F_d(\mathbf{d}) = \max(D_A(\mathbf{d}), D_B(\mathbf{d}), D_C(\mathbf{d})) - \min(D_A(\mathbf{d}), D_B(\mathbf{d}), D_C(\mathbf{d})) \tag{2}$$

The control system is intended to correct for manufacturing errors or slowly changing effects, such as thermal changes. Therefore, static or kinematic models are sufficient. These models can be developed analytically. However, real systems are complex, with dozens of actuators making closed-form analysis burdensome. For the simple example shown in Figure 2a, a system of 28 simultaneous equations must be solved. Hence, computer-based methods are more appropriate and ADINA Finite Element Analysis (FEA) is used. It will be shown below that linear models can be developed for control using the small displacement assumption.

ADINA was used to calculate the motions of reflected rays in the focal plane workspace. The analysis results for the 6-actuator mirror are shown in Figure 4a. It has been previously shown that a binary systems workspace can be more evenly distributed if the compliant elements are not symmetric [9]. Here, the stiffness of the elements is uniformly distributed between 1000 and 3000 N/m. Figure 4a shows that ability to shift the system focal point (the horizontal axis is limited to virtually five values). In addition, for most of these solutions, the system defocus exceeds the assumed limit of 20 cm, hence the light from the segments would miss the assumed receiver. Clearly, this 6-actuator design is ineffective.

Figure 4b shows the focal plane workspace of 17-actuator mirror with random compliant members. This workspace is more evenly distributed than the 6-actuator mirror. The 17-actuator mirror is able to correct for an almost continuous range of mirror distortions (2 m) with defocuses within the specification (20 cm).

The system control must determine how to switch the actuators to provide the required focal position and maintain an acceptable defocus range. This requires solving the forward kinematics problem with unknown mechanism parameters.

3 Mirror Control System

As shown in Figure 3, to control a binary mirror system, the forward kinematics must be computed. Analytical and FEA models of a spatial system with many actuators and embedded compliance are complex and depend on knowledge of system parameters. In addition, FEA models are too computationally intensive for online system control. Fortunately, solar mirrors should require only small corrections and the system is approximately linear for this range of movement [5]. In this case, linear approximation of the mirror displacements ($\hat{\mathbf{y}}$) can be written as $\hat{\mathbf{y}} = \mathbf{Ad}$ where $\mathbf{A}$ is a matrix representing the linear system kinematics, and $\mathbf{d}$ is a vector of zeros and ones indicating which actuators are extended.

Different methods can be used to identify the coefficients of $\mathbf{A}$. For example, $\mathbf{A}$ can be determined using system measurements when each actuator is individually deployed. For example, when actuator k is deployed, the kth column of $\mathbf{A}$ is determined directly as follows:

$$[a_{1,k}, \dots a_{m,k}]^T = [y_1, \dots, y_m]^T. \quad (3)$$

In many cases, it is more practical to determine $\mathbf{A}$ experimentally. Here, two methods are compared. In the first, the coefficients are directly measured, as shown in Equation (3). In the second, multiple actuators are deployed simultaneously and optimization methods are used to find $\mathbf{A}$ to minimize the mean error:

$$\varepsilon = \frac{1}{p}\sum_{i=1}^{p}\sqrt{(\hat{\mathbf{y}}_i - \mathbf{y}_i)^T(\hat{\mathbf{y}}_i - \mathbf{y}_i)} = \frac{1}{p}\sum_{i=1}^{p}\sqrt{(\mathbf{Ad}_i - \mathbf{y}_i)^T(\mathbf{Ad}_i - \mathbf{y}_i)} \quad (4)$$

where p is the number of calibration points, $\mathbf{y}_i$ is the ith vector of measured values, and $\mathbf{d}_i$ is the input vector for the ith measurement. Once the model is obtained, it is used to determine which actuator inputs minimize the distance to the desired system position. Methods to accomplish this include search routines such as genetic algorithms, which have been used in the area of binary robotics [6].

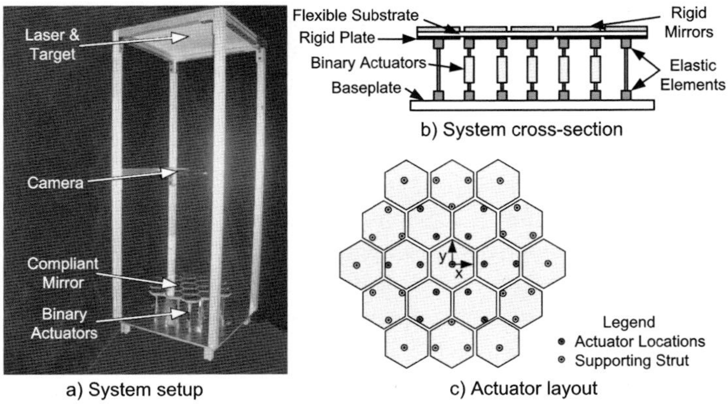

Fig. 5 Three-dimensional experimental setup.

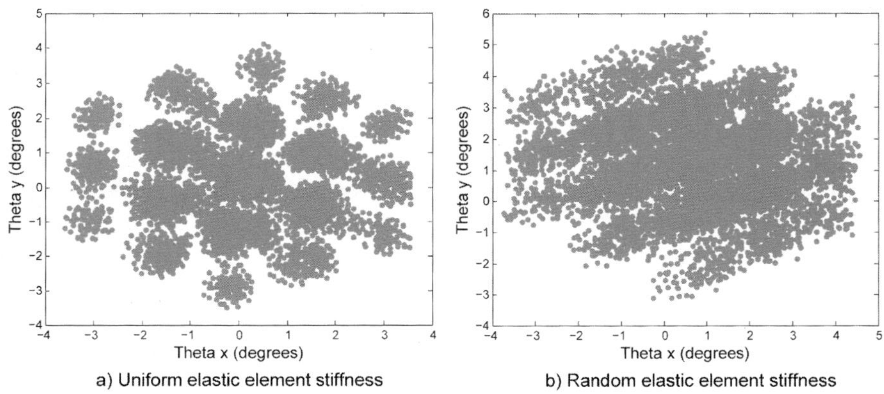

Fig. 6 Experimental system workspace.

4 Experimental Validation

The experimental system, shown in Figure 5, has 19 rigid hexagonal mirror segments mounted on a compliant plate. The plate is supported by 13 vertical binary actuators with rubber mounts that provide compliance. It is also supported by 30 vertical passive compliant struts. The binary actuators are assembled using micro-linear actuators manufactured by Firgelli Technologies.

The spatial rotation the center mirror is used as the system output. The rotations are measured using a laser beam which reflected off the center mirror onto a focal plane which is observed by a CCD camera. The camera data is transformed to yield the mirror's rotations.

The workspace of the center mirror segment with uniform elastic element stiffnesses is shown in Figure 6a. The mirror segment has 8192 (2^{13}) possible positions. Since the structure is symmetric, these positions are clustered and there are large un-

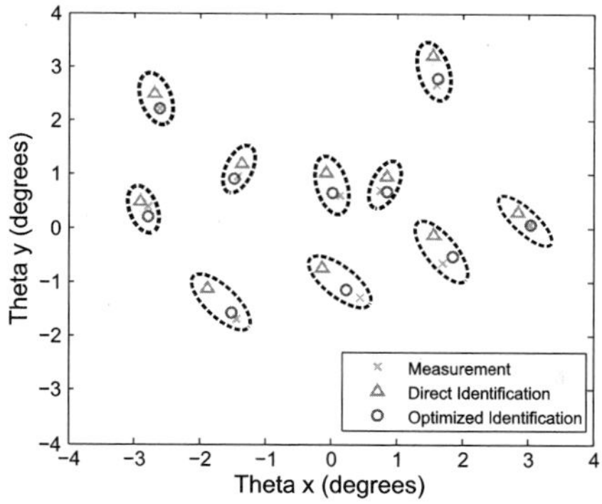

Fig. 7 Model results for 10 random points.

reachable areas. Breaking the symmetry by modifying the stiffnesses of the elastic elements will spread the system workspace, as shown in Figure 6b.

The experimental system is modeled using three methods: a FEA model and two linear models. The results of the FEA were compared to the experimental measurements. There was reasonably good agreement for the deflection magnitudes, but the FEA model had errors in individual measurements in excess of 1 degree. To improve the FEA accuracy, compliance of every element would need to be measured, which would be extremely cumbersome.

Two different linear models were developed. In the first, the coefficients of **A** were determined through direct measurement when the actuators were deployed individually, see Equation (3). In the second, the measurements were made for 52 randomly selected actuator combinations and **A** was determined using optimization methods, Equation (4). The results of these linear models are shown in Figure 7. The model outputs and measurements corresponding to the same actuator input are circled. The optimization approach resulted in more accurate results with an average error of less than 0.1 degrees. These small errors show the model is appropriate for system control.

5 Summary and Conclusions

A binary-actuation approach to correct shape deformations of solar concentrators due to manufacturing errors and thermal effects is presented. In this approach, binary actuators are embedded in a compliant mirror substructure and deployed in a specified pattern to correct the mirror shape. Analytical models and experimental studies demonstrate key features of binary-actuated solar concentrators, particularly their discrete workspaces. It is also shown that their workspaces can be more evenly distributed by varying elastic element stiffness to break the structure symmetry. A control approach for binary actuated mirror systems based on forward kinematics is also presented. This approach first uses system identification techniques to calibrate a linear model that can then be used to determine the binary inputs required to reach desired system positions. Experimental studies show the linear kinematic model has small errors and is appropriate for control.

Acknowledgements The support of the King Fahd University of Petroleum and Minerals in Dhahran, Saudi Arabia, through the Center for Clean Water and Clean Energy at MIT and KFUPM is acknowledged. The authors also thank Yuanyu Chen and Teresa Gomez for their assistance in this work and the Cyprus Institute for its partial financial support of Ms. Bilton.

References

1. Chirikjian, G.S.: Inverse kinematics of binary manipulators using a continuum model. J. Intell. Robot. Syst. **19**, 5–22 (1997)
2. Coventry, J.S.: Performance of a concentrating photovoltaic/thermal solar collector. Sol. Energy **78**(2), 211–222 (2005)
3. Gardner, J., Mather, J., Clampin, M., et al.: The James Webb Space Telescope. Space Sci. Rev. **123**(4), 485–606 (2006)
4. Irschik, H.: A review on static and dynamic shape control of structures by piezoelectric actuation. Eng. Struct. **24**(1), 5–11 (2002)
5. Lee, S.J., Bilton, A.M., Dubowsky, S.: On the kinematics of solar mirrors using massively parallel binary actuation. In: ASME Mechanisms and Robotics Conference (2010)
6. Lichter, M., Sujan, V., Dubowsky, S.: Computational issues in the planning and kinematics of binary robots. In: IEEE International Conference on Robotics and Automation, vol. 1, pp. 341–346 (2002)
7. Mancini, T., Heller, P., Butler, B., et al.: Dish-Stirling systems: An overview of development and status. J. Sol. Energy Eng. **125**(2), 135–151 (2003)
8. Plante, J.S.: Dielectric elastomer actuators for binary robotics and mechatronics. Ph.D. thesis, Massachusetts Institute of Technology (2006)
9. Plante, J.S., Dubowsky, S.: The calibration of a parallel manipulator with binary actuation. In: Lenarcic J., Wenger P. (eds.) Advances in Robot Kinematics: Analysis and Design, pp. 391–398. Springer, the Netherlands (2008)
10. Proulx, S., Miron, G., Girard, A., Plante, J.S.: Experimental validation of an elastically averaged binary manipulator for MRI-guided prostate cancer interventions. In: ASME Mechanisms and Robotics Conference (2010)
11. Wingert, A., Lichter, M., Dubowsky, S.: On the kinematics of parallel mechanisms with bistable polymer actuators. In: Lenarcic J., Thomas F. (eds.) 8th International Symposium on Advances in Robot Kinematics. Springer, the Netherlands (2002)

Mobile Robot Motion Primitives That Take into Account the Cost of Control

Sohee Lee and Frank Chongwoo Park

Abstract We propose a motion generation framework for nonholonomic wheeled mobile robots that takes into account the cost of control. Control implementation costs are formulated in terms of the minimum attention functional of [1]. Assuming a control law consisting of a general open-loop term added to a time-varying linear feedback term, we derive a set of minimum attention motion primitives for wheeled mobile robots of the Dubins type, and show how to utilize the primitives to generate minimum attention motions in real-time.

Key words: Minimum attention, motion primitives, wheeled mobile robot

1 Introduction

As the capabilities of mobile robots expand, the tasks being placed on the onboard computer of a mobile robot are becoming increasingly complex and computationally demanding. In particular, because real-time sensor data processing and object recognition are highly computation-intensive processes, one must take into account the cost of control implementation – loosely, the amount of "attention" required, which we make more precise below – when planning and controlling motions. Also, because motions need to be planned and generated in real-time, it is important to exploit as much as possible a basis of stored motions, or motion primitives, that can be combined in real-time to generate a rich class of motions.

In this paper we propose and derive a class of motion primitives for wheeled platforms (e.g., the moving base of typical mobile manipulators) that explicitly take into account the cost of control implementation. Control implementation costs are

Sohee Lee
Technische Universität München, München, Germany, e-mail: ssoheelee@gmail.com

Frank Chongwoo Park
Seoul National University, Seoul, Korea, e-mail: fcp@snu.ac.kr

J. Lenarčič, M. Husty (eds.), *Latest Advances in Robot Kinematics*,
DOI 10.1007/978-94-007-4620-6_54,

quantified in terms of the minimum attention functional of Brockett [1]. The basic premise behind the minimum attention functional is that the easiest control law to implement is a constant input; the more frequently the control changes, the more effort is required to implement it. Control laws typically depend on the state and time, so that the cost of implementation can be linked with the rate at which the control changes with respect to changes in both state and time ("attention").

To make things more precise, given a system with state equation $\dot{x} = f(x,u,t)$, where x and u are respectively the state and control vectors, and t denotes time, the attention functional is then formulated as

$$\int_0^{t_f} \int_{\Re^n} \alpha \left\| \frac{\partial u}{\partial x} \right\|^2 + (1-\alpha) \left\| \frac{\partial u}{\partial t} \right\|^2 dx\,dt. \quad (1)$$

Here, the first term reflects the dependence of the control on feedback, while the second term reflects the variation in the open-loop control term. It is interesting to point out the following connection between the weighting term $\alpha \in [0,1]$ and the human motor learning process: first attempts at a new motion rely almost entirely on feedback ($\alpha = 1$), but through practice (i.e., gradually decreasing α to zero) the motion eventually becomes an open-loop, refined motion.

While the attention functional is intuitively appealing, solutions are very difficult to come by – even the existence of solutions is not guaranteed in the general case – primarily because the integral is over both the state x and time t. In [1] analytic solutions are obtained for only the most basic second-order scalar system $\ddot{x} = u$ (and the solutions are far from straightforward). In subsequent work [2] it is shown that for linear systems, splitting of the control into a feedback term and an open-loop term leads to an algorithmic solution, but for more general nonlinear systems analytic characterizations appear to be difficult.

In this paper we consider a planar wheeled differential drive robot of the Dubins type and, assuming additive separability of the control term into a general feedforward and linear time-varying feedback term, we computationally derive a set of motion primitives – essentially, trajectories that minimize the attention functional for a representative set of boundary conditions – for a general class of boundary conditions. Our choice of control law is justified by observing that in most practical situations, it is necessary to restrict the set of possible feedback laws because of limitations on the gain or a desire for stability.

Moreover, for nonholonomic wheeled mobile robots it is well known that smooth stabilizing feedback laws do not exist, but that time-varying point-to-point stabilizing feedback laws can be constructed [3]. For our purposes we therefore adopt a control law consisting of a general open-loop term, and add to this a time-varying feedback term as proposed in [3]. While the focus of this work is not necessarily on stabilization, from both a physical and practical implementation perspective the additive feedforward-feedback control law structure is reasonable.

Finding, among this class of admissible controls, suboptimal solutions to the minimum attention functional can constitute a useful set of motion primitives. For example, control signal communications can be scheduled in such a way that less "attention" is given to updating the control, and more processing resources can be assigned to other tasks. Our overall motivation for determining the minimum attention primitives is in fact similar to that for the minimum-time motion primitives found in [4].

Before proceeding to our main results, we mention existing work on the related notion of limited communication control [5]; here the goal is to find an input control sequence to minimize the tracking error of the control system given a communication sequence. Whereas in minimum attention control the cost of control is defined as the objective criterion to be minimized, in the limited communication control framework this criterion is expressed in the form of a constraint.

2 Minimum Attention Motions for a Wheeled Mobile Platform

We consider a single-wheeled unicycle of the Dubins type (see Figure 2 (a)) for our differential drive mobile platform. The kinematic state equations are of the following form:

$$\dot{q} = \begin{pmatrix} \dot{x} \\ \dot{y} \\ \dot{\theta} \end{pmatrix} = \begin{pmatrix} r\cos\theta & 0 \\ r\sin\theta & 0 \\ 0 & 1 \end{pmatrix} \begin{pmatrix} v \\ \omega \end{pmatrix} = S(q)u, \tag{2}$$

where q represents the configuration of the unicycle, (x,y) denotes the Cartesian position of the reference frame attached to the mobile robot with respect to the global frame, θ denotes its orientation relative to the global frame, v is the velocity in the forward direction, and ω is the angular velocity with respect to the axis normal to the x-y plane. The minimum attention functional for our system assumes the following form:

$$\min_{u(q,t)} \int_0^{t_f} \int_{\Re^3} \alpha \left\| \frac{\partial u}{\partial q} \right\|^2 + (1-\alpha) \left\| \frac{\partial u}{\partial t} \right\|^2 dq\, dt \tag{3}$$

subject to

$$\begin{cases} \dot{q} = S(q)u \\ u_- \leq u \leq u^- \\ q(0), q(t_f), \dot{q}(0), \dot{q}(t_f) \text{ given,} \end{cases} \tag{4}$$

where u_- and u^- are lower and upper bounds on the control u, respectively.

To make the problem tractable, we assume that the control input, u, can be split into a closed-loop linear feedback term and an open-loop feedforward term. As noted earlier, this assumption on the form of the control is justifiable by observing that in most practical situations, it is necessary to restrict the set of possible feedback laws because of limitations on the gain or a desire for stability, and an additive feedforward-feedback control law makes sense from a practical implementation perspective.

We also restrict the feedback term to an admissible class of feedback controls that guarantee globally asymptotical stability of the system. As is well known, for nonholonomic systems there does not exist any smooth stabilizing feedback law, but various discontinuous or time-varying feedback controllers for point-to-point stabilization have been developed. For our purposes we consider a smooth time-varying feedback control term of the form proposed in [3]. To work with this feedback control law, we need to change the coordinates with respect to the body frame; we refer the reader to [3] for details of the coordinate transformation. The underlying state equation can be rewritten as

$$\dot{q}_b = \begin{pmatrix} \dot{x}_b \\ \dot{y}_b \\ \dot{\theta} \end{pmatrix} = \begin{pmatrix} 0 & \omega & 0 \\ -\omega & 0 & 0 \\ 0 & 0 & 0 \end{pmatrix} \begin{pmatrix} x_b \\ y_b \\ \theta \end{pmatrix} + \begin{pmatrix} 1 & 0 \\ 0 & 0 \\ 0 & 1 \end{pmatrix} \begin{pmatrix} v \\ \omega \end{pmatrix} = f(q_b, u, t). \tag{5}$$

With this model, the feedback control for stabilization to a desired configuration (without loss of generality, we may set $q_b(t_f) = (0,0,0)$) is of the form

$$\begin{pmatrix} v \\ \omega \end{pmatrix} = -\begin{pmatrix} k_1 & 0 & 0 \\ 0 & \sin t & k_3 \end{pmatrix} q_b = K(t) q_b, \tag{6}$$

where $k_1, k_3 > 0$. Therefore, the control becomes $u(q_b, t) = u_0(t) + K(t)q_b$ and the feedback term is related not only to the state but also to time.

The complete formulation of the minimum attention control problem for the nonholonomic mobile platform can be expressed as follows (here we replace q by the body coordinates q_b):

$$\min_{u_0, k_1, k_3} \int_0^{t_f} \alpha \|K(t)\|^2 + (1-\alpha) \|\dot{u} - K(t)\dot{q}_b\|^2 \, dt \tag{7}$$

subject to

$$\begin{cases} \dot{q}_b = f(q_b, u, t) \\ u = u_o(t) + K(t) q_b \\ u_- \leq u \leq u^-, k_1, k_3 > 0 \\ q_b(0), q_b(t_f), \dot{q}_b(0), \dot{q}_b(t_f) \text{ given.} \end{cases} \tag{8}$$

To convert the above into a finite-dimensional optimization problem, the control term $u(t)$ is parametrized with a relatively small set of control variables P. Here, we use B-spline curves depending on the choice of basis functions $B_i(t)$ and the control points $P = \{p_1, \ldots, p_m\}$, where $p_i \in \Re^n$ and n is equal to the dimension of the control. The trajectories then assume the form $u = u(t,P) = \sum_{i=1}^{m} B_i(t) p_i$.

By parameterizing $u(t)$ in terms of B-splines and solving the state equations (5) using $u(t,P)$ and with the initial conditions $q_b(0)$, we can derive the state and velocity trajectories. For the objective function, the derivative of the control trajectories can be derived straightforwardly from the B-spline basis functions. Then, the original optimal control problem reduces to a parameter optimization problem, in which the optimization variables are the control points.

Figure 1 (a) illustrates some sample minimum attention trajectories generated using our optimization procedure, for different start and goal configuration pairs.

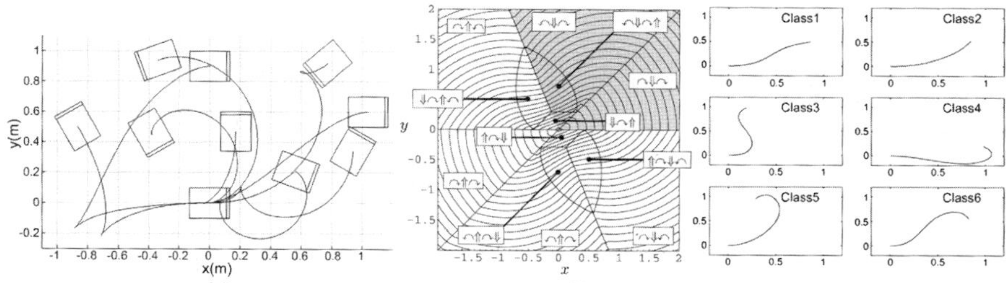

(a) (b) (c)

Fig. 1 For various start and goal configurations pairs, (a) some sample minimum attention trajectories and optimal trajectories of (b) minimum time motion primitives ($\theta = \frac{\pi}{4}$) [4] and (c) minimum torque motion primitives [6].

3 Motion Primitives for Mobile Robots

The first work to examine motion primitives for wheeled mobile robots is that of Balkcom and Mason [4], who find the minimum time trajectories between any two configurations in the plane. The optimal control problem is to minimize the integral functional

$$\min_{t_f,u} \int_0^{t_f} \mathbf{1}\, dt \quad \text{subject to} \quad \dot{q} = S(q)u, \;\; u \in \mathscr{U}, \tag{9}$$

where $\mathscr{U}$ denotes the set of admissible controls, and $q(0)$, $q(t_f)$, $\dot{q}(0)$, $\dot{q}(t_f)$ are given. The above optimal control problem is solved analytically, and it is proved that every nontrivial time-optimal trajectory is a finite sequence of linear paths and rotations about the robot center. Experiments are conducted for a planar differential drive nonholonomic wheeled robot of the Dubins type, in which the mobile robot starts from an arbitrary configuration (x, y, θ) and moves to the $(0,0,0)$ configuration; the configuration space is partitioned by optimal trajectories that consist of nine different symmetry classes. These are illustrated in Figure 1 (b).

Kim et al. [6] also propose a robot motion generation framework in which the training motion data consisting of minimum torque motions is constructed and clustered. Here the optimal control problem is to find the optimal torque profile for the following objective function:

$$\max_{t_f} \left\| \arg\min_{\tau} \int_0^{t_f} \|\tau(t)\|^2\, dt \right\|_{\infty} \quad \text{subject to } \tau(t) = M(q)\ddot{q} + C(q)\dot{q} + G(q), \tag{10}$$

where kinematic and torque limits and boundary conditions are given.

Using methods for dimension reduction, fast, torque-efficient motions for a nonholonomic mobile robot were obtained. The configuration space is divided into six regions even when the dynamics are taken into account.The average trajectories of each class are represented in Figure 1 (c).

Our aim is to develop a similar set of motion primitives for nonholonomic wheeled robots, but for the minimum attention functional. In our experiments, we

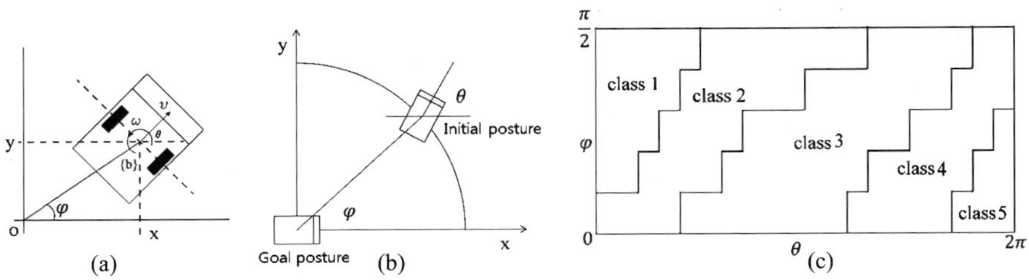

Fig. 2 The illustration of (a) a unicycle robot; (b) constructing the training motion data; (c) the result of partitioning the configuration space.

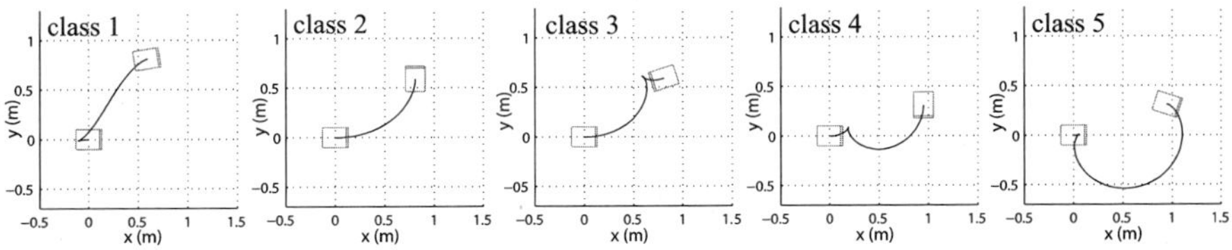

Fig. 3 The (averaged) minimum attention trajectories for each class.

assume that the robot starts at arbitrary points on the positive quadrant of an unit circle, with a user-specified initial heading angle, and moves to the origin with $\theta = 0$ as shown in Figure 2 (b). The position angle, ϕ, ranges from 0 to $\frac{\pi}{2}$ and the rotation angle, θ, ranges from zero to 2π. The test configurations are uniformly distributed in that region. For each initial configuration, the minimum attention path used as the training data for PCA is computed for (7). The paths with similar starting poses and similar trajectory shapes are grouped and classified so that five classes are determined as Figure 2 (c). The average trajectories are displayed as in Figure 3.

4 Real-Time Motion Generation Using Primitives

In this section we describe the general procedure for fast generation of minimum attention motions using the previously derived motion primitives. Elements of this approach were previously proposed in [6] and illustrated through the case of a wheeled mobile robot, and for space reasons we defer the technical details of this approach to [6]. Here, we describe how to generate training motion data for the minimum attention criterion, and based on results of numerical studies, discuss some practical aspects of the minimum attention principle as a paradigm for motion generation. The key steps are illustrated in the block diagram in [6].

By solving the optimization problem (7), the optimal trajectories used as training motion data can be obtained for user-defined initial and final poses. One difference with previous works is that it is the set of control input trajectories $u(t)$ that constitute the training motion data. Once a sufficiently large set of training motion data

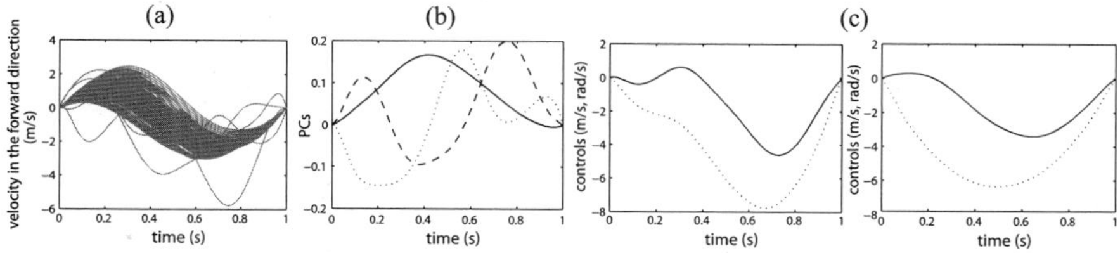

Fig. 4 (a) The training data profile and mean velocity; (b) Three dominant principal components; (c) The worst case: the optimal motion (left) and near optimal motion by PCA of class 3 (right)

is secured, these trajectories are grouped into several clusters: trajectories within a cluster will share several essential features, e.g., similar starting and ending poses, and similar trajectory shapes.

Assuming the training motion data is appropriately clustered, the next step is to derive a set of basis functions, or motion primitives, for each cluster that are representative of the trajectories contained. We achieve this via principal component analysis (PCA) of the trajectories in each cluster. A finite number of the most dominant principal components are then used as basis functions for generating the suboptimal motions.

Once the motion primitives are constructed, these can be linearly combined to form the control trajectories for the corresponding cluster and given initial configuration. Here, control trajectories can be interpolated with the three dominant principal components, e.g., the velocity in the forward direction $v(t) = v_{mean}(t) + x_1 \cdot pc1(t) + x_2 \cdot pc2(t) + x_3 \cdot pc3(t) + x_4$. The detailed formulation and notation are described in [6].

Figure 4 shows the principal components extracted from the training data of class 3 and the resultant minimum attention motion. For this case, the optimal value is increased only by 23% even for the worst case shown in Figure 4(c).

5 Discussion

As expected, the control profiles corresponding to minimum attention trajectories tend to be flat compared to, e.g. control profiles that minimize torque, energy, or wheel rotation (see Figure 5), while at the same time having smooth state trajectories.

This flatness of the control profile is closely related with an important application of the derived minimum attention motion primitives, communication scheduling. Once the motion is obtained, depending on the shape of the control trajectory, we can schedule updates to the control signal, thereby facilitating the use of limited computational resources and enabling mobile manipulators to perform multiple tasks as shown in Figure 5.

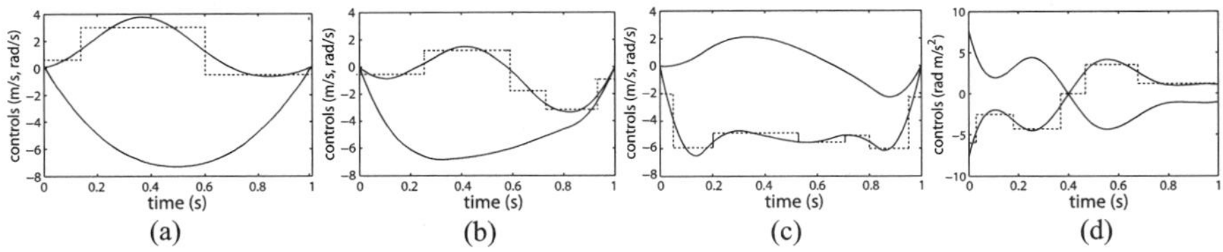

Fig. 5 The scheduling of communication of control in (a) the minimum attention, (b) minimum wheel rotation ($\int \frac{1}{2}|v+w|+|v-w|dt$), (c) the minimum energy ($\int \frac{1}{2}u^T u dt$), and (d) the minimum torque (10) motion.

6 Conclusion

In this paper, we have proposed a means of efficiently generating motions for nonholonomic wheeled mobile robots that explicitly takes into account the cost of control implementation. Control implementation costs are quantified in terms of the minimum attention functional [1], whose basic premise is that the easiest control law to implement is a constant input; the more frequently the control changes with respect to time and state, the more effort is required to implement it. Assuming additive separability of the control term into a general feedforward and linear time-varying feedback term, we first computationally derive a set of minimum attention motion primitives. A method for real-time motion generation based on linear combinations of these movement primitives is then described and application of the derived minimum attention motion primitives is discussed.

References

1. Brockett, R.W.: Minimum attention control. In: Proceedings of the 36th IEEE Conference on Decision and Control, Piscataway, NJ (1997)
2. Brockett, R.W.: Minimizing attention in a motion control context. In: Proceedings of the 42th IEEE Conference on Decision and Control. Maui, HI (2003)
3. Samson, C., Ait-Abderrahim, K.: Velocity and torque feedback control of a nonholonomic cart. In: Advanced Robot Control. Lecture Notes in Control and Information Sciences vol. 162, pp. 125–151. Springer (1991)
4. Balkcom, D.J., Mason, M.T.: Time optimal trajectories for bounded velocity differential drive vehicles. Int. J. Robot. Res. **21**, 199–217 (2002)
5. Hristu, D.: Optimal Control with Limited Communication, Ph.D. thesis, Division of Engineering and Applied Sciences, Harvard University (June 1999)
6. Kim, S., Park, F.C.: Fast robot motion generation using principal components: framework and algorithms. IEEE Trans. Ind. Electron. **55**(6), 2506–2516 (2008)
7. De Luca, A., Di Benedetto, M.D.: Control of nonholonomic systems via dynamic compensation. Kybernetica **29**(6), 593–608 (1993)
8. Brockett, R.W.: Asymptotic stability and feedback stabilization. In: Differential Geometric Control Theory, pp. 181–191. Birkhäuser, Boston (1983)

Synthesis of Spatial CC Dyads and 4C Mechanisms for Pick & Place Tasks with Guiding Locations

P. Larochelle

Abstract A novel dimensional synthesis technique for solving the mixed exact and approximate motion synthesis problem for spatial CC kinematic chains is presented. The methodology uses an analytic representation of the spatial CC dyad's rigid body constraint equation in combination with an algebraic geometry formulation of the perpendicular screw bisector to yield designs that exactly reach the prescribed pick & place locations while approximating an arbitrary number of guiding locations. The result is a dimensional synthesis technique for mixed exact and approximate motion generation that utilizes only algebraic geometry and does not require the use of any iterative optimization algorithms or a metric on spatial displacements. An example that demonstrates the synthesis technique is included.

Key words: Spatial mechanisms, 4C mechanisms, CC dyads

1 Introduction

As a product is assembled in an automated factory a common task that needs to be performed is the movement of parts or subassemblies from one location to another; this is commonly referred to as a *pick & place* task. For the assembly of a complex product the number of pick & place tasks that need to be performed could run into the thousands. Parts are picked out of bins and placed into subassemblies, subassemblies are picked up and placed into the final product, etc. One solution is to use devices with a high number of degrees of freedom such as industrial robots. Robots can perform these tasks but at penalties in costs, cycle time, and maintenance. A second solution is to use a cascading series of simple one degree of freedom devices;

P. Larochelle
Robotics & Spatial Systems Lab, Department of Mechanical and Aerospace Engineering, Florida Institute of Technology, Melbourne, USA, e-mail: pierrel@fit.edu

J. Lenarčič, M. Husty (eds.), *Latest Advances in Robot Kinematics*,
DOI 10.1007/978-94-007-4620-6_55,

e.g. a series of servo motors. Creating such a manipulation pipelines takes a longer design time and is often more art than science.

Spatial robotic mechanisms offer another alternative. The synthesis algorithm presented here is part of ongoing efforts directed at realizing the capability to design two degree of freedom robotic spatial mechanisms capable of performing spatial pick & place tasks. These low degree of freedom devices are capable of producing the necessary spatial motion for accomplishing pick & place tasks. Hence, spatial robotic mechanisms provide an alternative for solving spatial assembly tasks that might otherwise require a robot or multiple single degree of freedom devices.

A well known result from screw theory [1, 9] is that moving an object from one spatial location to another does not require six degrees of freedom. In fact, such motions can be accomplished with a single degree of freedom twist about a unique screw axis. However this solution is often impractical due to the location of the screw axis within the workspace and the collisions and interferences between objects that may result. Spatial robotic mechanisms are low degree of freedom machines that are a compromise between the 6 or more degree of freedom industrial robot and the series of single degree of freedom motion generators. Here, we focus on utilizing the spatial CC dyad as the motion generator for a class of spatial robotic mechanisms to achieve two desired locations exactly (i.e. pick & place) while approximating a set of guiding locations that take the workpiece from the pick location to the place location.

In a related work [12] presents the derivation of the constraint manifold for spherical RR dyads using the image space representation of displacements. This work was an extension of the ideas presented in [13]. In [11] the spatial generalization of the planar Burmester curves was presented from a geometric viewpoint. The focus of this work was the synthesis of CC and related dyads for exact motion generation through three and four locations. The synthesis of CC dyads for exact motion through 5 locations was presented in [9, 10]. In [3, 5, 6] the extension of Burmester theory, using Roth's line congruence approach [13], for the exact synthesis of 4C mechanisms for 4 locations is presented. The approximate motion synthesis of spatial 4C mechanisms for rigid body guidance was presented in [7]. Circuit and branch defects of the spatial 4C mechanism were investigated in [4] and the detection of self-collisions of the links was discussed in [2]. The methodology used here for performing the dimensional synthesis for mixed exact and approximate rigid body guidance is based upon the works of [14] and builds upon the spherical version presented in [8].

This paper proceeds as follows. First, the geometry and kinematics of the spatial CC dyad are reviewed. Next, the synthesis algorithm for solving the mixed exact and approximate motion generation problem for spatial CC dyads is presented. Finally, an example spatial robotic mechanism design is presented; the synthesis of a spatial 4C mechanism to accomplish a pick & place tasks exactly while approximating three guiding locations.

2 Synthesis Algorithm

A spatial 4C closed chain may be viewed as the combination of two CC dyads where each dyad consists of one link and two C joints; one fixed and the other moving, see Fig. 1. The approach taken here is to synthesize two dyads separately and then join their floating links to yield a kinematic closed chain. Let the fixed axis be specified by the dual vector $\hat{\mathbf{u}}$ measured in the fixed reference frame F and let the moving axis be specified by $\hat{\mathbf{v}}$ measured in the moving frame M. Moreover, let $\hat{\mathbf{l}}$ define the moving axis $\hat{\mathbf{v}}$ in the fixed frame F so that, $\hat{\mathbf{l}} = [\hat{A}]\hat{\mathbf{v}}$ where $[\hat{A}]$ is the dual orthogonal matrix that defines M with respect to F [9]. Because the link is rigid, the dual angle between the two axes of the dyad remains constant. This geometric constraint may be expressed analytically as,

$$\hat{\mathbf{u}} \cdot \hat{\mathbf{l}} = \hat{\mathbf{u}} \cdot [\hat{A}]\hat{\mathbf{v}} = \cos\hat{\alpha}. \tag{1}$$

This constraint equation is the foundation of the synthesis algorithm presented below. In order to solve the mixed exact and approximate synthesis problem we first solve the exact synthesis problem for 3 prescribed locations.

2.1 Exact Synthesis for Three Locations

Here we select a moving axis $\hat{\mathbf{v}}$ of a CC dyad and solve for the corresponding fixed axis $\hat{\mathbf{u}}$ such that the dyad guides the moving body exactly through 3 prescribed locations [6]. To solve this synthesis problem we first work with the real or direction part of the CC constraint equations and then subsequently address the moment part. We write the real part of Eq. (1) for each of the desired locations, $[\hat{A}]_i$, $i = 1,2,3$. Next, we subtract the first equation from the remaining two to arrive at a linear system of equations,

$$[P]\mathbf{u} = \mathbf{k} \tag{2}$$

where

$$[P] = \begin{bmatrix} (\mathbf{l}_2 - \mathbf{l}_1)^T \\ (\mathbf{l}_3 - \mathbf{l}_1)^T \\ 0\ 0\ \ 1 \end{bmatrix},$$

$\mathbf{l}_i$ is the direction of the moving axis in the ith location, $\mathbf{k} = [0\ 0\ 1]^T$, and $\mathbf{u}$ is the desired direction of the fixed axis. Note that we must solve Eq. (2) for each moving axis direction to find its corresponding fixed axis direction. Moreover, note that since we are using 3-vectors to define the axes when in fact they are directions that only require 2 independent coordinates, the last row of $[P]$ is chosen to yield the vector $\mathbf{u}$ that is the intersection of the fixed axis with the $z = 1$ plane. In the event that $[P]$ is rank deficient (i.e. when the fixed axis does not intersect the $z = 1$ plane) simply

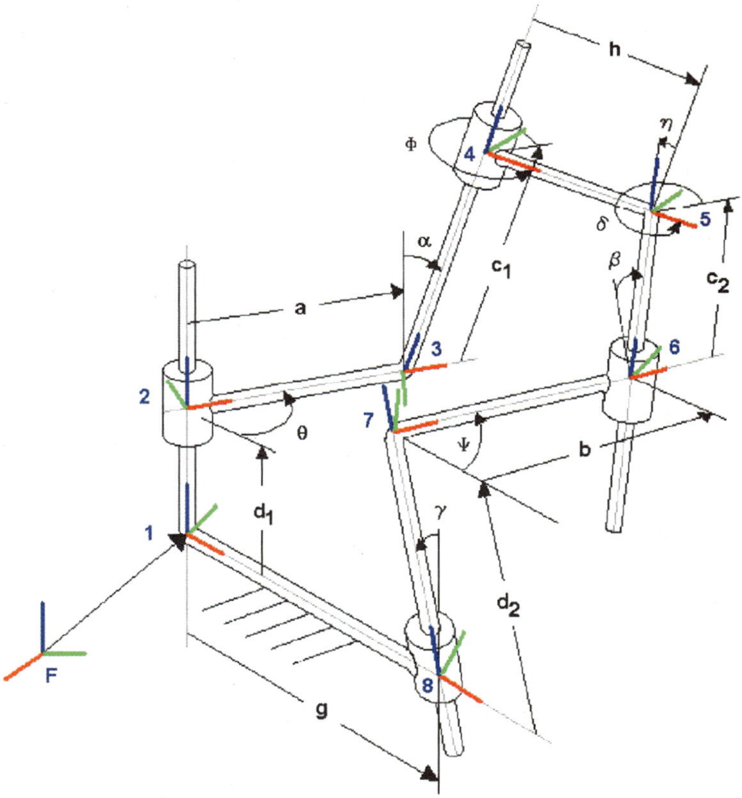

Fig. 1 Spatial 4C mechanism: geometry and nomenclature.

change the last row to any vector that does not lie in this plane (e.g. $[1\ 0\ 0]^T$). Next, we proceed to solve for the moment of the fixed axis.

We write the dual part of Eq. (1) for each of the desired locations, $[\hat{A}]_i$, $i = 1, 2, 3$ and then subtract the first equation from the remaining two to arrive at a linear system of equations,

$$[H]\mathbf{u}^0 = \mathbf{t} \tag{3}$$

where

$$[H] = \begin{bmatrix} (\mathbf{l}_2 - \mathbf{l}_1)^T \\ (\mathbf{l}_3 - \mathbf{l}_1)^T \\ \mathbf{u}^T \end{bmatrix}, \quad \mathbf{t} = \begin{bmatrix} -(\mathbf{l}_2^0 - \mathbf{l}_1^0)^T \mathbf{u} \\ -(\mathbf{l}_3^0 - \mathbf{l}_1^0)^T \mathbf{u} \\ 0 \end{bmatrix},$$

and $\mathbf{u}^0$ is the desired moment of the fixed axis. Solve Eqs. (2) and (3) for each desired moving axis of a CC dyad to find the unique corresponding fixed axis that guides the moving body exactly through the 3 prescribed locations.

2.2 Mixed Synthesis Algorithm

We now consider the synthesis of CC dyads that guide a moving body exactly through 2 pick & place locations and approximately through n guiding locations. First a desired moving axis $\hat{\mathbf{v}}$ is selected. Next, we seek a corresponding fixed axis $\hat{\mathbf{u}} = (\mathbf{u}, \mathbf{u}^0)$ for the dyad. We proceed by identifying the spherical image of the CC dyad. Duffy showed that associated with each spatial CC dyad there is a spherical image consisting of a spherical RR dyad whose link lengths are the angular twists of the CC dyad. Moreover, he proved that the spatial CC dyad and its associated spherical RR image have the exactly the same angular relationships and motions [1, 9]. Therefore the synthesis of the spatial CC dyad can be decomposed into two subproblems; (1) the angular synthesis or the synthesis of the link twist angles of the CC dyad and (2) the moment synthesis or the synthesis of the link length of the CC dyad. We address the former first.

The angular synthesis of the spatial CC dyad can be solved by performing the synthesis of its spherical RR image. The direction of the fixed axis is found by solving n 3 orientation problems to yield a set of fixed axis directions $\mathbf{u}_i$, $i = 1, 2, \ldots, n$. The 3 orientation problems are derived from the 2 pick & place locations along with 1 of the guiding locations. Hence, there are n unique 3 orientation problems (Eq. 2) that are solved to obtain n fixed axis directions $\mathbf{u}_i$, $i = 1, 2, \ldots, n$. It was shown in [8] that the direction of the fixed axis that will guide the moving body as desired is the normalized sum of these directions of $\mathbf{u}_i$, $i = 1, 2, \ldots, n$,

$$\mathbf{u} = \frac{\sum \mathbf{u}_i}{\|\sum \mathbf{u}_i\|}. \tag{4}$$

We now focus on the moment synthesis problem; finding the desired moment $\mathbf{u}^0$ of the fixed axis $\hat{\mathbf{u}}$.

The moment synthesis of the spatial CC dyad can be solved by utilizing the geometric interpretation of Eq. (1); that $\hat{\mathbf{u}}$ must lie on the screw perpendicular bisector associated with the pick & place locations of the desired moving axis $\hat{\mathbf{v}}$. For the CC dyad to reach exactly the pick & place locations Eq. (1) must hold true in both locations. Write Eq. (1) for the pick & place locations and take the difference to yield,

$$\hat{\mathbf{u}} \cdot (\hat{\mathbf{l}}_{\mathbf{place}} - \hat{\mathbf{l}}_{\mathbf{pick}}) = 0. \tag{5}$$

Eq. (5) is the equation of the screw perpendicular bisector of $\hat{\mathbf{l}}_{\mathbf{pick}}$ and $\hat{\mathbf{l}}_{\mathbf{place}}$ [9]. The set of screws $\hat{\mathbf{u}}$ that satisfy Eq. 5 is a two parameter set whose axes intersect and are orthogonal to $\hat{\mathbf{B}}$ as shown in Fig. 2. Note that $\hat{\mathbf{N}}$ is the common normal to $\hat{\mathbf{l}}_{\mathbf{pick}}$ and $\hat{\mathbf{l}}_{\mathbf{place}}$, $\hat{\mathbf{V}}$ is the midpoint screw, and $\hat{\mathbf{B}} = \hat{\mathbf{N}} \times \hat{\mathbf{V}}$. Recall that the direction of $\hat{\mathbf{u}}$ has been previously found from Eq. (4). Therefore finding a point on the fixed axis $\hat{\mathbf{u}}$ is sufficient for determining the unknown moment $\mathbf{u}^0$. From the properties of the screw perpendicular bisector it is known that $\hat{\mathbf{u}}$ must intersect and be orthogonal to

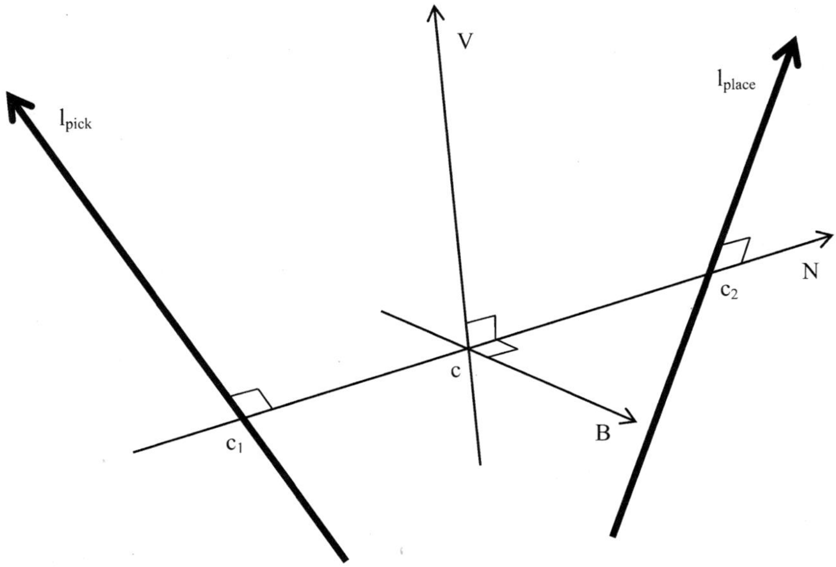

Fig. 2 Spatial CC dyad and the screw perpendicular bisector.

$\hat{\mathbf{B}}$; we now determine this point of intersection and use it to determine the unknown moment $\hat{\mathbf{u}}^{\mathbf{0}}$.

For the prescribed moving axis $\hat{\mathbf{v}}$ solve n 3 location problems to yield a set of fixed axes $\hat{\mathbf{u}}_i$, $i = 1, 2, \ldots, n$. The 3 location problems are derived from the 2 pick & place locations along with 1 of the guiding locations. Hence, there are n unique 3 location problems (Eqs. 2 and 3) that are solved to obtain n fixed axes $\hat{\mathbf{u}}_i$, $i = 1, 2, \ldots, n$. Because each of these CC dyads guide the body exactly through the pick & place locations their fixed axes also intersect $\hat{\mathbf{B}}$. Note that if each of these n CC dyads exactly reach all of the guiding locations then their n fixed axes $\hat{\mathbf{u}}_i$, $i = 1, 2, \ldots, n$ intersect $\hat{\mathbf{B}}$ in a unique point. In general the CC dyads will not be capable of exactly reaching the n guiding locations and the intersections of their fixed axes with $\hat{\mathbf{B}}$ will not be a unique point. Next, determine these n intersection points $\mathbf{p}_i$, $i = 1, 2, \ldots, n$. The desired point $\mathbf{p}$ on the fixed axis $\hat{\mathbf{u}}$ is the average of these intersection points,

$$\mathbf{p} = \frac{\Sigma \mathbf{p}_i}{n}. \tag{6}$$

Finally the unknown moment may be determined from $\mathbf{u}^{\mathbf{0}} = \mathbf{p} \times \mathbf{u}$. The CC dyad with prescribed moving axis $\hat{\mathbf{v}}$ and fixed axis $\hat{\mathbf{u}}$, as determined with the above algorithm, guides the moving body exactly through the pick & place locations and near the n guiding locations.

Table 1 Five prescribed locations and synthesis results.

Longitude	Latitude	Roll	X	Y	Z	Motion Type	Dyad #1 Constraint	Dyad #2 Constraint
0.00	0.00	0.00	3	5	-1	exact	$-0.5173+2.0058\varepsilon$	$-0.3854-4.3362\varepsilon$
0.00	25.00	10.00	4	4	-2	approximate	$-0.8074+1.9092\varepsilon$	$-0.1617-5.3040\varepsilon$
20.00	45.00	20.00	2	3	-3	approximate	$-0.8760+2.0811\varepsilon$	$-0.3371-4.6082\varepsilon$
65.00	65.00	10.00	5	2	-4	approximate	$-0.5801+2.2399\varepsilon$	$-0.5922-2.7565\varepsilon$
90.00	90.00	0.00	1	1	-5	exact	$-0.5173+2.0058\varepsilon$	$-0.3854-4.3362\varepsilon$

3 Example

We employ the preceding methodology and design a 4C spatial mechanism to guide a moving body exactly through two pick & place locations and near 3 guiding locations as defined in Table 1 where $[A] = [Rot_z(lng)][Rot_y(-lat)][Rot_x(rol)]$ and all angles are expressed in degrees. Two CC dyads are synthesized independently and then their floating links are joined to yield a 4C closed-chain mechanism.

For dyad #1 a moving axis was prescribed: $\hat{\mathbf{v}}_1 = [0.2673\ 0.5345\ -0.8018\ 0.0000\ 0.8018\ 0.5345]^T$. The mixed synthesis algorithm presented above yielded $\mathbf{u}_1 = [-0.1292\ 0.4342\ 0.8915]^T$, $\mathbf{p}_1 = [0.7935\ -0.0109\ -1.1956]^T$, and fixed axis $\hat{\mathbf{u}}_1 = [-0.1292\ 0.4342\ 0.8915\ 0.5094\ -0.5529\ 0.3431]^T$. The resulting CC dyad's link lengths are: $a = -2.34$ and $\alpha = 121.15$ (deg). For dyad #2 a different moving axis was chosen: $\hat{\mathbf{v}}_2 = [0.5774\ -0.5774\ 0.5774\ -0.5774\ -0.5774\ 0.0000]^T$. The mixed synthesis algorithm yielded $\mathbf{u}_2 = [-0.6675\ 0.5265\ 0.5265]^T$, $\mathbf{p}_2 = [2.5000\ -1.0968\ 0.0968]^T$, and fixed axis $\hat{\mathbf{u}}_2 = [-0.6675\ 0.5265\ 0.5265\ -0.6284\ -1.3809\ 0.5842]^T$. The resulting CC dyad's link lengths are: $b = 4.70$ and $\beta = 112.67$ (deg). When the two dyads are combined to form a spatial 4C mechanism the fixed link length is $g = 0.72$ and $\gamma = 38.34$(deg) and the length of the coupler link is $h = 0.78$ and $\eta = 128.11$(deg). This 4C mechanism has a non-Grashof $0-\pi$ double-rocker spherical four-bar image [1, 4].

To verify the motion of the moving body the CC dyad constraint equations were evaluated in each of the 5 locations; the left-hand side of Eq. (1), i.e $\hat{\mathbf{u}} \cdot [\hat{A}]\hat{\mathbf{v}}$, is reported in the right columns of Tb. 1. Note that the inner product between the fixed and moving lines of each CC dyad is identical in the pick & place locations thereby verifying that the moving body does in fact reach the pick & place locations exactly.

4 Conclusions

A novel dimensional synthesis technique for solving the mixed exact and approximate motion problem for spatial CC open and 4C closed kinematic chains has been presented. The methodology uses an analytic representation of the spatial CC dyad's rigid body constraint equation in combination with classical geometric motion synthesis techniques to yield designs that exactly reach two prescribed pick & place locations while approximating n guiding locations. Such tasks are common in automated assembly and production systems. An example was presented to demonstrate the synthesis procedure.

Acknowledgements Collaborations with Prof. Chintien Huang of National Cheng Kung University (NCKU) of Tainan, Taiwan are sincerely acknowledged. This material is based upon work supported by the National Science Foundation (NSF) under Grant #1132629. Any opinions, findings, and conclusions or recommendations expressed in this material are those of the author(s) and do not necessarily reflect the views of the NSF.

References

1. Duffy, J.: Analysis of Mechanism and Robotic Manipulators. Wiley and Sons (1980)
2. Ketchel, J., Larochelle, P.: Self-collision detection in spatial closed chains. ASME J. Mech. Des. **130**, CID #092305 (2008). doi:10.1115/1.2965363
3. Larochelle, P.: On the design of spatial 4C mechanisms for rigid-body guidance through 4 positions. In: Proc. of the ASME Design Engineering Technical Conferences, Boston, MA, DE-82, pp. 825–832 (1995)
4. Larochelle, P.: Circuit and branch rectification of the spatial 4C mechanism. In: Proc. of the ASME Design Engineering Technical Conferences. Baltimore, MD. Paper #MECH-14053 (2000)
5. Larochelle, P.: Spades: Software for synthesizing spatial 4c mechanisms. In: Proc. of the ASME Design Engineering Technical Conferences. Atlanta, GA. Paper #MECH-5889 (1998)
6. Larochelle, P.M.: Design of cooperating robots and spatial mechanisms. PhD Dissertation, University of California, Irvine (1994)
7. Larochelle, P., Dees, S.: Approximate motion synthesis using an SVD-based distance metric. In: Lenarčič, J., Thomas, F. (eds.) Advances in Robot Kinematics: Theory and Applications, pp. 483–490. Kluwer Academic Publishers (2002)
8. Larochelle, P.: Synthesis of part orienting devices for spatial assembly tasks. In: Lenarčič, J., Wenger, P. (eds.) Advances in Robot Kinematics: Analysis & Design, pp. 79–87. Springer (2008)
9. McCarthy, J.M., Soh, G.S.: Geometric Design of Linkages, 2nd edn. Springer-Verlag (2011)
10. Murray, A.P., McCarthy, J.M.: Five position synthesis of spatial CC dyads. In: Proc. of the ASME Design Engineering Technical Conferences, Minneapolis, MN. September, DE-70, pp. 143–152 (1994)
11. Nisbett, J.K.: A geometric approach to the spatial equivalent of Burmester curves. PhD Dissertation, The University of Texas at Arlington (1998)
12. Ravani, B., Roth, B.: Motion synthesis using kinematic mappings. ASME J. Mech. Transm. Autom. Des. **105**, 460–467 (1983)
13. Roth, B.: The design of binary cranks with revolute, cylindric, and prismatic joints. J. Mech. **3**(2), 61–72 (1968)
14. Tsai, L.W., Roth, B.: Design of dyads with helical, cylindrical, spherical, revolute, and prismatic joints. Mech. Mach. Theory **7**(1), 85–102 (1973)

On the Role of Passive Structures in the Knee Loaded Motion

Nicola Sancisi and Vincenzo Parenti-Castelli

Abstract The role of the passive structures of the human knee in loaded and unloaded motion has been deeply investigated in the literature. However, what makes its comprehension difficult is the inherent redundancy of the anatomical structures that constrain the motion itself. This paper, based on simulation and experimental data, provides some inferences on the role of the constraint redundancy of this complex anatomical joint. The results suggest that the knee behaves prevalently like a one-degree-of-freedom isostatic system when moderate external loads are applied, continuously constrained by 5 articular structures recruited according to the external loads and to the system configuration.

Key words: Knee, kinematics, statics, passive constraints, redundancy

1 Introduction

Human diarthrodial joints are complex systems that feature a number of involved anatomical structures such as ligaments, bones and muscles. The role of these structures in the joint kinematics and statics still rises a number of open questions and, in spite of a huge amount of papers devoted to this issue, it is still largely unknown.

A great attention has been devoted to the knee joint for its relevance in human locomotion. The kinematic and static behaviour of the knee, for instance, has been deeply investigated and many mathematical models together with both in vivo and in vitro experiments have been presented in the literature [3].

Since many years these authors pursued a systematic approach [1] which aims at modelling a joint by means of equivalent mechanisms. The approach involves three different steps starting from the definition of the kinematic model, then moving to

Nicola Sancisi · Vincenzo Parenti-Castelli
DIEM – Department of Mechanical Engineering, University of Bologna, 40136 Bologna, Italy, e-mail: {nicola.sancisi, vincenzo.parenti}@unibo.it

J. Lenarčič, M. Husty (eds.), *Latest Advances in Robot Kinematics*,
DOI 10.1007/978-94-007-4620-6_56,

a static model and finally to a dynamic model. The behaviour of the joint is determined by the constraints provided by the anatomical structures. When modelling the knee by equivalent mechanisms, redundant constraints appear which make the modelling a challenging task [8]. The behaviour of the knee has been modelled rather successfully in previous papers [6, 7], but the redundant nature of the passive/active anatomical structures makes a deeper comprehension still lacking of a definite assessment.

This paper presents a further advancement of the authors procedure that attempts to deepen the role of the anatomical structure redundancy in terms of kinematic and static behaviour of the knee. Namely, based on simulation and experimental data, the paper tries to shed light into this complex and challenging issue.

2 Model Definition

The study was performed by defining a static model of the knee. The model was recently proposed [9]: just the main information are reported here, while theoretical and mathematical details can be found in the cited literature.

The static model was obtained in two steps by means of a sequential approach [1, 9]. According to this procedure, a kinematic model of the joint is devised at the first step (M1). This model replicates the natural motion of the knee when no loads are applied (i.e. the knee passive motion) and thus it reproduces the mobility of the joint. The kinematic model features just a few anatomical structures, i.e. the structures that actually constrain the passive motion, and can replicate the knee behaviour in passive conditions only. This first model is then generalized at the second step (M2), by adding the remaining passive structures and by considering their viscoelastic properties. M2 represents the static model of the knee and can replicate the joint behaviour also when external static loads (muscular loads excluded) are applied: the new model reproduces the stability of the joint.

The most important feature of this sequential approach is that generalization from M1 to M2 model does not worsen or modify the results of the kinematic model. In other words, M2 reproduces the static behaviour of the joint, but it can also replicate the passive motion of the knee with the same accuracy as M1: as a consequence, M2 shows both the same mobility and the same stability as the original joint.

2.1 Kinematic Model M1

The kinematic model is based on the experimental evidence that the passive motion of the knee has 1 degree of freedom (dof) and it is mainly guided by a few articular structures: the two rigid articular contacts and the isometric fibres of three ligaments, namely the anterior cruciate (ACL), the posterior cruciate (PCL) and the medial collateral (MCL) ligaments [6, 7, 11]. The model features two rigid bodies, representing the tibia and femur, interconnected by five rigid links: two links represent the articular contacts respectively at the lateral (CL) and medial (CM) condyles; the remaining three links substitute the ACL, PCL, MCL isometric fibres [7] (Figure 1(a)). This model is basically a 1-dof mechanism and it replicates the knee passive motion with a high accuracy [6].

Five closure equations can be written by specifying that the length of each link remains constant. The passive motion can thus be obtained from the model by solving the closure equations of the mechanism. The model geometrical parameters are obtained from the anatomy by measuring the condyles and the isometric fibres of a specimen. This preliminary estimate is then refined by reducing the weighted mean squared differences between the model and the experimental passive motion to a minimum. A bounded optimization procedure is used to find the minimum, in order to keep the final geometry of the model close to the first estimate [1]. This procedure was applied to a specimen, and the relevant model M1 was defined.

2.2 Static Model M2

The kinematic model was generalized at the second step by the sequential approach, in order to obtain the static model of the knee [9]. In particular, all main ligaments were considered at this step, including the previously added ACL, PCL, MCL. Each ligament was modelled by a group of fibres that represent relevant fibre bundles of the original ligament. The ACL, PCL, MCL were also modelled as groups of fibres at this step, by adding new fibres beside the isometric ones obtained at the first step.

Thus, the knee ligaments were substituted by 24 fibres on the whole: 5 in the ACL (4 new, 1 isometric), 5 in the PCL (4 new, 1 isometric), 6 in the MCL (5 new, 1 isometric), 3 in the lateral collateral ligament (LCL), 1 in the arcuate ligament (AL), 2 in the popliteus tendon (PT), 2 in the oblique popliteus ligament (OPL). Moreover, elastic properties of ligaments were considered at this step: all fibres (isometric fibres included) can lengthen, when forces are applied. In particular, each fibre was modelled as a spring with a non-linear force-strain relation [5]:

$$\begin{aligned} F_j &= k_j\,\varepsilon_j^2 \quad && \varepsilon_j > 0 \\ F_j &= 0 && \varepsilon_j \leq 0 \end{aligned} \tag{1}$$

where k_j is a stiffness parameter and ε_j is the strain of the j-th fibre: $\varepsilon_j = \frac{L_j - L_{0j}}{L_{0j}}$ in which L_j and L_{0j} are respectively the length and the unloaded length of the fibre. Fibre-fibre and fibre-bone interactions were ignored.

The two M1 links at the condyles still were used to model the articular contacts. However, these two links are not rigid in M2 and, on the contrary, can change their length according to a force-strain relation similar to Eq. (1). This equation also makes it possible to reproduce bone separation at one or both condyles, as it may happen under certain loading conditions. The final static model is represented in

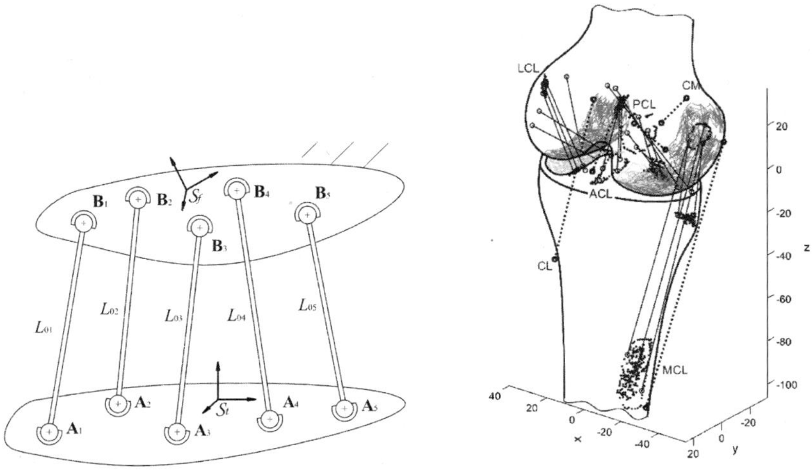

(a) Kinematic model M1 (b) Static model M2

Fig. 1 Kinematic and static model of the knee.

Figure 1(b): the dashed lines are the fibres identified at the first step (corresponding to the links of M1), the solid lines are the new fibres added at the second step.

The femur-tibia relative motion can be obtained by choosing a particular loading condition and by solving the equilibrium equations of the system. In this study, the loads related to the most common clinical tests (i.e. the anterior/posterior drawer, the in/external torsion and the ab/adduction tests) were applied to the joint. These loading conditions were chosen since they are clinically relevant. In particular, the loading conditions reported in a specific paper [2] were considered and the relevant experimental motions were chosen as a reference. For all clinical tests, the tibia was left free to move under the action of its weight and of the external forces corresponding to a particular test. The femur, on the contrary, was fixed at several flexion angles, from 0 to 90 degrees. A counterforce was also applied at the distal extremity of the tibia, in order to keep the tibia longitudinal axis approximately vertical.

A first estimate of the M2 geometrical (fibre insertions and unloaded lengths) and elastic (k_j) parameters was obtained from the literature and from the same specimen used to define M1 [4, 5, 9]. In particular, many posterior structures (AL, PT, OPL) were not recorded during the experiment and were thus reconstructed from photographic and published material [10]. The final value of the model parameters was obtained by minimizing the weighted mean squared differences between the model and the reference motions [2]. The procedure is similar to that used for M1, although the sequential approach introduces further constraints that have to be considered during parameter optimization to not modify the results obtained at the first step [9]. For instance, the unloaded lengths of the five springs that represent the ACL, PCL, MCL isometric fibres and the articular contacts were not optimized,

Table 1 Loaded passive structures at full extension for all considered clinical tests: • = major contribution, ○ = minor contribution.

Clinical test	ACL	PCL	sMCL	dMCL	LCL	AL	PT	OPL	CM	CL
Anterior drawer	•		•							•
Posterior drawer		•	•		○	•			•	•
External torsion	•		•		•		○		•	•
Internal torsion	•	•	•						•	
Abduction	•	•	•	○					•	•
Adduction	•				•	•			•	

since they were already defined at the first step. Moreover, the insertions of all fibres were not optimized to simplify the procedure.

The forces exerted by the passive structures during clinical tests were analysed by means of the optimized M2 model, in order to understand the role of the knee passive structures to guarantee both stability and mobility of the joint in loaded motion. For a more schematic and comprehensive representation of the results, the net force of all fibres of each ligament was computed. From a static point of view, this is the same as ignoring the moment exerted by each ligament on the bones. However, these moments are low in this case, and the simplification makes the analysis of the final results easier. Only the MCL was split into two different bundles, i.e. the deep (dMCL) and the superficial (sMCL) bundles, in order to account for the two distinct insertions that this ligament shows on the tibia (Figure 1(b)). Thus, a total amount of 10 forces were computed for all clinical tests and flexion angles; these forces correspond to the ligaments ACL, PCL, sMCL, dMCL, LCL, AL, PT, OPL and the articular contacts CM, CL.

3 Results

The static model M2 made it possible to replicate the specimen passive motion and the knee behaviour during clinical tests with a good accuracy. Detailed results on the model accuracy were already published [9] and are not reported here.

As for the analysis of the passive structure forces, the most loaded structures at the specimen reference pose (i.e. full extension, corresponding approximately to 5° of flexion for the considered specimen) are reported in Table 1. These results agree well in general with the experimental results from the reference paper [2]: both PCL and some posterior structures constrain the posterior drawer at full extension; some posterior structures (PCL excluded) also constrain the external rotation and the adduction of the knee. Moreover, Figures 2(a)–2(b) represent the magnitude of passive structure forces for the external torsion and abduction tests at all the considered flexion angles.

Overall, the results obtained from M2 show that in most cases only 5 structures exert the highest forces and support most of the loads. This seems almost a constant

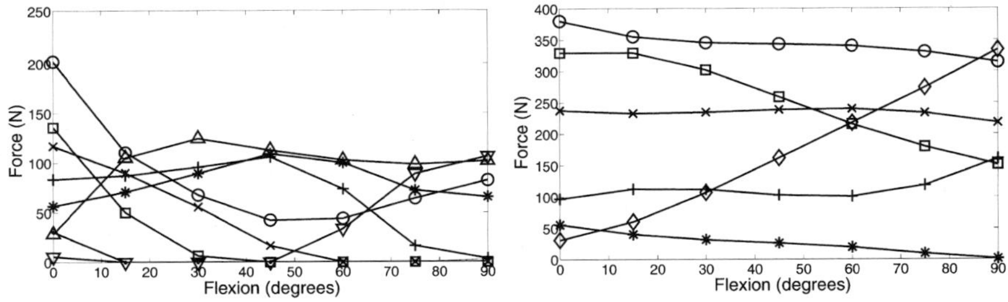

(a) External torsion test

(b) Abduction test

Fig. 2 Magnitude of the passive structure forces during clinical tests at all the considered flexion angles. For decreasing force magnitude at 0° of flexion: (a) CL, LCL, ACL, sMCL, CM, ALS, PT, dMCL; (b) CL, sMCL, ACL, CM, PCL, dMCL.

for the considered loading conditions and for the different flexion angles. In other words, despite the high number of passive structures in the knee, at any moment only 5 structures are prevalently tight during loaded motion (Figures 2(a)–2(b)). For instance, it can be noticed in Figures 2(a) that, approximately at about 50° of flexion, the ACL is substituted by the dMCL, overall keeping 5 structures tight during the external torsion test (apart from a brief overlapping). Only near full extension there are more than 5 structures tight: this can be justified by the fact that at full extension there is the need to improve the joint stability.

Table 1 confirms these considerations. Five anatomical structures prevail over the others for three tests; a sixth structure produces a minor contribution, to improve stability near full extension. On the contrary, for the other three tests, less than 5 structures are tight at full extension. However, it should be considered that the particular loading conditions analysed in the reference and in the present paper facilitate bone separation: no compressive preload is applied at the femur and tibia and, on the contrary, tibia weight tends to separate the bones. In this sense, it is probably no accident that the three tests that show less than 5 structures tight at full extension are also those that show bone separation at one articular surface (i.e. CM or CL do not exert forces). As a consequence, the ligaments are loaded more than in standard loading conditions, since they have to compensate for the unnatural bone separation.

These considerations have interesting outcomes both from a static and a kinematic point of view. Indeed, if flexion angle is fixed, the tibia has 5 dof with respect to the femur: at least 5 constraints are required to equilibrate this bone. Thus, from a static point of view, the forces exerted by the 5 tight passive structures constitute the minimum force complex that can guarantee the equilibrium of the joint in general. In other words, despite the high number of constraints that make the joint a potentially overconstrained hyperstatic system, the knee behaves as an isostatic structure when standard moderate loads are applied: the particular geometry of articular structures seems to guarantee that just the minimum number of structures is loaded at any moment, and no unnecessary constraints are active.

As for the kinematic point of view, if only 5 structures are prevalently tight, the knee behaves as a system that features two rigid bodies (i.e. the femur and tibia) interconnected by 5 links: this representation is quite similar to the one used for the definition of M1. Thus, the knee seems to behave almost as a 1-dof spatial mechanism similar to M1 also in loaded conditions, with the difference that, in this case, the links are not perfectly rigid. As a further consequence, the loads required to flex the knee are only those needed to equilibrate the external loads: apart from the strain in the mechanism links, no energy is spent (or received) to lengthen (or shorten) additional redundant elastic structures. This aspect could facilitate the flexion and the general mobility of the knee when loads are applied, since the joint remains intrinsically a 1-dof system: flexion forces can be controlled better, since redundant elastic forces (apart from minor secondary ones) do not interfere with the loaded motion.

In other words, alternate static M1 models could also be used to model the static behaviour of the knee. However, some differences arise with respect to the passive M1 model. Firstly, the articular structures featured in the static M1 mechanisms depend on the loading conditions and could change also during the flexion arc. On the contrary, as described in previous sections, the articular structures featured in the passive M1 may be retained for all the flexion arc. However, it should be noted that previous investigations showed that other articular structures can be considered also for M1, with no substantial change in the results [8]: this led to the conclusion that, in passive conditions, all articular structures are more or less kinematically equivalent and the knee moves as an overconstrained system with 1-dof. As a second difference, the strain of static M1 links should be considered during loaded motion. However, the static M1 models could still be considered rigid link mechanisms as a first approximation, for the computation of the joint loads and motion. The extent to which static M1 replicates the joint loads and motion has not been verified yet.

In conclusion, the results suggest that the knee behaves prevalently like a 1-dof isostatic system also when moderate external loads are applied. This aspect could improve the mobility of the joint. The knee behaves like a hyperstatic system at particular loading conditions or configurations (for instance, at full extension) when stability of the joint must prevail over mobility.

The present investigation shows some limitations. The static model M2 was indeed just partially based on data measured on a specimen: as explained, missing experimental data (posterior structures and loaded motion included) were taken from average measurements from the literature, and could be affected by large errors. Moreover, a stronger M2 validation is required: despite M2 replicated the reference motions with a good accuracy, the static model should be applied on more specimens and other reference motions, in order to be quantitatively reliable. However, it is believed that the qualitative analysis here provided could help to clarify the kinematic and static behaviour of the knee.

4 Conclusions

A static model of the knee was defined, to analyse the role of the joint passive structures to guide and restrain the knee motion under several loading conditions. Although preliminarily, the results show that when moderate external loads are applied, 5 articular structures provide most of the loads to guarantee the equilibrium: the knee prevalently behaves as a 1-dof isostatic system, in spite of the high number of passive structures that potentially could make it a hyperstatic system. This particular behaviour could improve joint mobility when loads are applied. On the contrary, more than 5 passive structures are recruited when the particular configuration or loading condition require a greater stability of the joint.

References

1. Franci, R., Parenti-Castelli, V., Sancisi, N.: A three-step procedure for the modelling of human diarthrodial joints. Int. J. Mech. Control **10**(1), 3–12 (2009)
2. Grood, E.S., Stowers, S.F., Noyes, F.R.: Limits of movement in the human knee. J. Bone Jt. Surg. **70-A**(1), 88–97 (1988)
3. Hefzy, M.S., Cooke, T.D.V.: Review of knee models: 1996 update. Appl. Mech. Rev. **49**(10-2), 187–193 (1996)
4. La Prade, R.F., Bollom, T.S., Wentorf, F.A., Wills, N.J., Meister, K.: Mechanical properties of the posterolateral structures of the knee. Am. J. Sports Med. **33**(9), 1386–1391 (2005)
5. Mommersteeg, T.J.A., Blankevoort, L., Huiskes, R., Kooloos, J.G.M., Kauer, J.M.G.: Characterization of the mechanical behavior of human knee ligaments: A numerical-experimental approach. J. Biomech. **29**(2), 151–160 (1996)
6. Ottoboni, A., Parenti-Castelli, V., Sancisi, N., Belvedere, C., Leardini, A.: Articular surface approximation in equivalent spatial parallel mechanism models of the human knee joint. Proc. Inst. Mech. Eng., H J. Eng. Med. **224**(9), 1121–1132 (2010)
7. Parenti-Castelli, V., Di Gregorio, R.: Parallel mechanisms applied to the human knee passive motion simulation. In: Lenarcic J., Stanisic M. (eds.) Advances in Robot Kinematics. Kluwer Academic Publishers (2000)
8. Sancisi, N., Parenti-Castelli, V.: On the role of ligaments in the guidance of the human knee passive motion. In: Proceedings of Euromech Colloquium 511, Ponta Delgada, Azores, Portugal, March 9–12, pp. 1–9 (2011)
9. Sancisi, N., Parenti-Castelli, V.: A sequentially-defined stiffness model of the knee. Mech. Mach. Theory **46**, 1920–1928 (2011)
10. Sudasna, S., Harnsiriwattanagit, K., May, C.: Ligamentous structures of the posterolateral aspect of the knee. Bull. Hosp. Joint Dis. Orthop. Inst. **50**(1), 35–40 (1990)
11. Wilson, D.R., O'Connor, J.J.: A three-dimensional geometric model of the knee for the study of joint forces in gait. Gait Posture **5**, 108–115 (1997)

Author Index

J. Lenarčič, M. Husty (eds.), *Latest Advances in Robot Kinematics*,
DOI 10.1007/978-94-007-4620-6,

Subject Index

J. Lenarčič, M. Husty (eds.), *Latest Advances in Robot Kinematics*,
DOI 10.1007/978-94-007-4620-6,